Methods in Enzymology

Volume 286
LIPASES
Part B
Enzyme Characterization and Utilization

METHODS IN ENZYMOLOGY

EDITORS-IN-CHIEF

John N. Abelson   Melvin I. Simon

DIVISION OF BIOLOGY
CALIFORNIA INSTITUTE OF TECHNOLOGY
PASADENA, CALIFORNIA

FOUNDING EDITORS

Sidney P. Colowick and Nathan O. Kaplan

Methods in Enzymology

Volume 286

Lipases

Part B
Enzyme Characterization and Utilization

EDITED BY

Byron Rubin

LIPOMED
SAN DIEGO, CALIFORNIA

Edward A. Dennis

DEPARTMENT OF CHEMISTRY AND BIOCHEMISTRY
UNIVERSITY OF CALIFORNIA, SAN DIEGO
SAN DIEGO, CALIFORNIA

ACADEMIC PRESS
San Diego London Boston New York Sydney Tokyo Toronto

This book is printed on acid-free paper.

Copyright © 1997 by ACADEMIC PRESS

All Rights Reserved.
No part of this publication may be reproduced or transmitted in any form or by any means, electronic or mechanical, including photocopy, recording, or any information storage and retrieval system, without permission in writing from the Publisher.
The appearance of the code at the bottom of the first page of a chapter in this book indicates the Publisher's consent that copies of the chapter may be made for personal or internal use, or for the personal or internal use of specific clients. This consent is given on the condition, however, that the copier pay the stated per copy fee through the Copyright Clearance Center, Inc. (222 Rosewood Drive, Danvers, Massachusetts 01923) for copying beyond that permitted by Sections 107 or 108 of the U.S. Copyright Law. This consent does not extend to other kinds of copying, such as copying for general distribution, for advertising or promotional purposes, for creating new collective works, or for resale. Copy fees for pre-1997 chapters are as shown on the chapter title pages. If no fee code appears on the chapter title page, the copy fee is the same as for current chapters.
0076-6879/97 $25.00

Academic Press
15 East 26th Street, 15th Floor, New York, New York 10010, USA
http://www.apnet.com

Academic Press Limited
24-28 Oval Road, London NW1 7DX, UK
http://www.hbuk.co.uk/ap/

International Standard Book Number: 0-12-182187-0

PRINTED IN THE UNITED STATES OF AMERICA
97 98 99 00 01 02 MM 9 8 7 6 5 4 3 2 1

Table of Contents

CONTRIBUTORS TO VOLUME 286.	ix
PREFACE.	xiii
VOLUMES IN SERIES	xv

Section I. Regulation and Characterization

1. Inhibition of Lipid Absorption as an Approach to the Treatment of Obesity	A. B. R. THOMSON, A. DE POVER, M. KEELAN, E. JAROCKA-CYRTA, AND M. T. CLANDININ	3
2. Regulation of Hormone-Sensitive Lipase Activity in Adipose Tissue	CECILIA HOLM, DOMINIQUE LANGIN, VINCENT MANGANIELLO, PER BELFRAGE, AND EVA DEGERMAN	45
3. Use of Gene Knockout Mice to Establish Lipase Function	DAVID Y. HUI	67
4. Properties of Pancreatic Carboxyl Ester Lipase in mRNA-injected *Xenopus* Oocytes and Transfected Mammalian Hepatic and Intestinal Cells	REZA ZOLFAGHARI, RAANAN SHAMIR, AND EDWARD A. FISHER	80
5. Noncatalytic Functions of Lipoprotein Lipase	GUNILLA OLIVECRONA AND AIVAR LOOKENE	102
6. Radiation Inactivation Studies of Hepatic Cholesteryl Ester Hydrolases	EARL H. HARRISON AND ELLIS S. KEMPNER	116
7. Immunological Techniques for the Characterization of Digestive Lipases	MUSTAPHA AOUBALA, ISABELLE DOUCHET, SOFIANE BEZZINE, MICHEL HIRN, ROBERT VERGER, AND ALAIN DE CARO	126

Section II. Substrate and Inhibitor Characterization

8. Physical Behavior of Lipase Substrates	DONALD M. SMALL	153
9. Phospholipase A_2 Activity and Physical Properties of Lipid-Bilayer Substrates	THOMAS HØNGER, KENT JØRGENSEN, DEBORAH STOKES, RODNEY L. BILTONEN, AND OLE G. MOURITSEN	168

10. Covalent Inactivation of Lipases	STÉPHANE RANSAC, YOUSSEF GARGOURI, FRANK MARGUET, GÉRARD BUONO, CHRISTOPH BEGLINGER, PIUS HILDEBRAND, HANS LENGSFELD, PAUL HADVÁRY, AND ROBERT VERGER	190
11. Mechanism-Based Inhibitors of Mammalian Cholesterol Esterase	SHAWN R. FEASTER AND DANIEL M. QUINN	231
12. On the Inhibition of Microbial Lipases by Tetrahydrolipstatin	LUTZ HAALCK AND FRITZ SPENER	252
13. Monolayer Techniques for Studying Lipase Kinetics	STÉPHANE RANSAC, MARGARITA IVANOVA, ROBERT VERGER, AND IVAN PANAIOTOV	263
14. Recovery of Monomolecular Films in Studies of Lipolysis	WILLIAM E. MOMSEN AND HOWARD L. BROCKMAN	292
15. Oil-Drop Tensiometer: Applications for Studying the Kinetics of Lipase Action	S. LABOURDENNE, A. CAGNA, B. DELORME, G. ESPOSITO, R. VERGER, AND C. RIVIÈRE	306
16. A Critical Reevaluation of the Phenomenon of Interfacial Activation	FRANCINE FERRATO, FRÉDÉRIC CARRIERE, LOUIS SARDA, AND ROBERT VERGER	327

Section III. Biocatalytic Utility

17. Screening Techniques for Lipase Catalyst Selection	U. ADER, P. ANDERSCH, M. BERGER, U. GOERGENS, B. HAASE, J. HERMANN, K. LAUMEN, R. SEEMAYER, C. WALDINGER, AND M. P. SCHNEIDER	351
18. Kinetics, Molecular Modeling, and Synthetic Applications with Microbial Lipases	KARL HULT AND MATS HOLMQUIST	386

19. Ester Synthesis via Acyl Transfer (Transesterification)	P. Andersch, M. Berger, J. Hermann, K. Laumen, M. Lobell, R. Seemayer, C. Waldinger, and M. P. Schneider	406
20. Cross-Linked Enzyme Crystals of Lipases as Catalysts for Kinetic Resolution of Acids and Alcohols	Jim J. Lalonde, Manuel A. Navia, and Alexey L. Margolin	443
21. Measurement and Control of Hydration in Nonaqueous Biocatalysis	Gillian A. Hutcheon, Peter J. Halling, and Barry D. Moore	465
22. Solvent Effect in Lipase-Catalyzed Racemate Resolution	Thorleif Anthonsen and Jaap A. Jongejan	473
23. Lipases in Supercritical Fluids	Enrico Cernia and Cleofe Palocci	495
24. Stabilization of Lipases against Deactivation of Acetaldehyde Formed in Acyl Transfer Reactions	H. K. Weber and K. Faber	509

Author Index 519
Subject Index 549

Contributors to Volume 286

Article numbers are in parentheses following the names of contributors.
Affiliations listed are current.

U. ADER (17), *Fb 9–Bergische Universität-GH-Wuppertal, D-42097 Wuppertal, Germany*

P. ANDERSCH (17, 19), *Fb 9–Bergische Universität-GH-Wuppertal, D-42097 Wuppertal, Germany*

THORLEIF ANTHONSEN (22), *Department of Chemistry, Norwegian University of Science and Technology, N-70345 Trondheim, Norway*

MUSTAPHA AOUBALA (7), *Laboratorie de Lipolyse Enzymatique, UPR 9025, IFRC1 du CNRS, 13402 Marseille Cedex 20, France*

CHRISTOPH BEGLINGER (10), *Division of Gastroenterology and Department of Research, University Hospital, CH-4031 Basel, Switzerland*

PER BELFRAGE (2), *Department of Cell and Molecular Biology, Section for Molecular Signalling, Lund University, S-221 00 Lund, Sweden*

M. BERGER (17, 19), *Fb 9–Bergische Universität-GH-Wuppertal, D-42097 Wuppertal, Germany*

SOFIANE BEZZINE (7), *Laboratorie de Lipolyse Enzymatique, UPR 9025, IFRC1 du CNRS, 13402 Marseille Cedex 20, France*

RODNEY L. BILTONEN (9), *Departments of Pharmacology and Biochemistry, University of Virginia, Charlottesville, Virginia 22908*

HOWARD L. BROCKMAN (14), *The Hormel Institute, University of Minnesota, Austin, Minnesota 55912*

GÉRARD BUONO (10), *Réactivité et Catalyse, ENSSPICAM URA, 13397 Marseille Cedex 20, France*

A. CAGNA (15), *Société IT Concept, Parc d'activités Chancolan, 69770 Longessaigne, France*

FRÉDÉRIC CARRIERE (16), *Laboratoire de Lipolyse Enzymatique, UPR 9025, IFRC1 du CNRS, 13402 Marseille Cedex 20, France*

ENRICO CERNIA (23), *Department of Chemistry, University of Rome "La Sapienza," 00185 Rome, Italy*

M. T. CLANDININ (1), *Nutrition and Metabolism Research Group, Division of Gastroenterology, Department of Medicine, University of Alberta, Edmonton, Canada T6G 2C2*

ALAIN DE CARO (7), *Laboratoire de Lipolyse Enzymatique, UPR 9025, IFRC1 du CNRS, 13402 Marseille Cedex 20, France*

EVA DEGERMAN (2), *Department of Cell and Molecular Biology, Section for Molecular Signalling, Lund University, S-221 00 Lund, Sweden*

B. DELORME (15), *Société IT Concept, Parc d'activités Chancolan 69770 Longessaigne, France*

A. DE POVER (1), *Ciba Pharma Division, Basel, Switzerland*

ISABELLE DOUCHET (7), *Laboratorie de Lipolyse Enzymatique, UPR 9025, IFRC1 du CNRS, 13402 Marseille Cedex 20, France*

G. ESPOSITO (15), *Société IT Concept, Parc d'activités Chancolan 69770 Longessaigne, France*

K. FABER (24), *Institute of Organic Chemistry, Graz University of Technology, A-8010 Graz, Austria*

SHAWN R. FEASTER (11), *Division of Biochemistry, Walter Reed Army Institute of Research, Washington, DC 20307*

FRANCINE FERRATO (16), *Laboratoire de Lipolyse Enzymatique, UPR 9025, IFRC1 du CNRS, 13402 Marseille Cedex 20, France*

ix

EDWARD A. FISHER (4), *Cardiovascular Institute, Mount Sinai School of Medicine, New York, New York 10029*

YOUSSEF GARGOURI (10), *Laboratoire de Biochimie, Ecole Nationale d'Ingénieurs de Sfax, 3038 Sfax, Tunisia*

U. GOERGENS (17), *Fb 9–Bergische Universität-GH-Wuppertal, D-42097 Wuppertal, Germany*

LUTZ HAALCK (12), *Institute of Chemical and Biochemical Sensor Research, 48149 Münster, Germany*

B. HAASE (17), *Fb 9–Bergische Universität-GH-Wuppertal, D-42097 Wuppertal, Germany*

PAUL HADVÁRY (10), *F. Hoffmann-La Roche & Co., Ltd., Pharmaceutical Research Department, CH-4002 Basel, Switzerland*

PETER J. HALLING (21), *Department of Pure and Applied Chemistry, University of Strathclyde, Glasglow G1 1XL, United Kingdom*

EARL H. HARRISON (6), *Department of Biochemistry, MCP-Hahnemann School of Medicine, Philadelphia, Pennsylvania 19129*

J. HERMANN (17, 19), *Fb 9–Bergische Universität-GH-Wuppertal, D-42097 Wuppertal, Germany*

PIUS HILDEBRAND (10), *Division of Gastroenterology and Department of Research, University Hospital, CH-4031 Basel, Switzerland*

MICHEL HIRN (7), *Immunotech, 13276 Marseille Cedex 9, France*

CECILIA HOLM (2), *Department of Cell and Molecular Biology, Section for Molecular Signalling, Lund University, S-221 00 Lund, Sweden*

MATS HOLMQUIST (18), *Department of Biochemistry and Biotechnology, Royal Institute of Technology, SE-100 44 Stockholm, Sweden*

THOMAS HØNGER (9), *Department of Chemistry, Technical University of Denmark, DK-2800 Lyngby, Denmark*

DAVID Y. HUI (3), *Department of Pathology and Laboratory Medicine, University of Cincinnati College of Medicine, Cincinnati, Ohio 45267-0529*

KARL HULT (18), *Department of Biochemistry and Biotechnology, Royal Institute of Technology, SE-100 44 Stockholm, Sweden*

GILLIAN A. HUTCHEON (21), *Department of Pure and Applied Chemistry, University of Strathclyde, Glasglow G1 1XL, United Kingdom*

MARGARITA IVANOVA (13), *Biophysical Chemistry Laboratory, University of Sofia, 1126 Sofia, Bulgaria*

E. JAROCKA-CYRTA (1), *Nutrition and Metabolism Research Group, Division of Gastroenterology, Department of Medicine, University of Alberta, Edmonton, Canada T6G 2C2*

JAAP A. JONGEJAN (22), *Kluyverlaboratory for Biotechnology, Delft University of Technology, NL-2628 Delft, The Netherlands*

KENT JØRGENSEN (9), *The Royal Danish School of Pharmacy, Department of Pharmaceutics, DK-2100 Copenhagen, Denmark*

M. KEELAN (1), *Nutrition and Metabolism Research Group, Division of Gastroenterology, Department of Medicine, University of Alberta, Edmonton, Canada T6G 2C2*

ELLIS S. KEMPNER (6), *National Institute of Arthritis and Musculoskeletal and Skin Disorders, National Institutes of Health, Bethesda, Maryland 20892*

S. LABOURDENNE (15), *CNRS Lipolyse Enzymatique, 13402 Marseille Cedex 20, France*

JIM J. LALONDE (20), *Altus Biologics, Cambridge, Massachusetts 02139*

DOMINIQUE LANGIN (2), *Institut Louis Bugnard, CHU–Rangueil, F-310 03 Toulouse Cedex 4, France*

K. LAUMEN (17, 19), *Fb 9–Bergische Universität-GH-Wuppertal, D-42097 Wuppertal, Germany*

HANS LENGSFELD (10), *F. Hoffman-La Roche & Co., Ltd., Pharmaceutical Research Department, CH-4002 Basel, Switzerland*

M. LOBELL (19), *Fb 9–Bergische Universität-GH-Wuppertal, D-42097 Wuppertal, Germany*

AIVAR LOOKENE (5), *Department of Medical Biochemistry and Biophysics, Umeå University, S-901 87 Umeå, Sweden*

VINCENT MANGANIELLO (2), *Pulmonary/Critical Care Medicine Branch, National Heart, Lung and Blood Institute, National Institutes of Health, Bethesda, Maryland 20892*

ALEXEY L. MARGOLIN (20), *Altus Biologics, Cambridge, Massachusetts 02139*

FRANK MARGUET (10), *Réactivité et Catalyse, ENSSPICAM URA, 13397 Marseille Cedex 20, France*

WILLIAM E. MOMSEN (14), *The Hormel Institute, University of Minnesota, Austin, Minnesota 55912*

BARRY D. MOORE (21), *Department of Pure and Applied Chemistry, University of Strathclyde, Glasglow G1 1XL, United Kingdom*

OLE G. MOURITSEN (9), *Department of Chemistry, Technical University of Denmark, DK-2800 Lyngby, Denmark*

MANUEL A. NAVIA (20), *Altus Biologics, Cambridge, Massachusetts 02139*

GUNILLA OLIVECRONA (5), *Department of Medical Biochemistry and Biophysics, Umeå University, S-901 87 Umeå, Sweden*

CLEOFE PALOCCI (23), *Department of Chemistry, University of Rome "La Sapienza," 00185 Rome, Italy*

IVAN PANAIOTOV (13), *Biophysical Chemistry Laboratory, University of Sofia, 1126 Sofia, Bulgaria*

DANIEL M. QUINN (11), *Department of Chemistry, The University of Iowa, Iowa City, Iowa 52242*

STÉPHANE RANSAC (10, 13), *Laboratoire de Lipolyse Enzymatique, UPR 9025, IFR 1 du CNRS, 13402 Marseille Cedex 20, France*

C. RIVIÈRE (15), *CNRS Lipolyse Enzymatique, 13402 Marseille Cedex 20, France*

LOUIS SARDA (16), *Laboratoire de Lipolyse Enzymatique, UPR 9025, IFRC1 du CNRS, 13402 Marseille Cedex 20, France*

M. P. SCHNEIDER (17, 19), *Fb 9–Bergische Universität-GH-Wuppertal, D-42097 Wuppertal, Germany*

R. SEEMAYER (17, 19), *Fb 9–Bergische Universität-GH-Wuppertal, D-42097 Wuppertal, Germany*

RAANAN SHAMIR (4), *Division of Gastroenterology and Nutrition, Schneider Children's Medical Center, Petach-Tiqua, Israel*

DONALD M. SMALL (8), *Department of Biophysics, Boston University Medical Center, Boston, Massachusetts 02118-2394*

FRITZ SPENER (12), *Department of Biochemistry, University of Münster, 48149 Münster, Germany*

DEBORAH STOKES (9), *Departments of Pharmacology and Biochemistry, University of Virginia, Charlottesville, Virginia 22908*

A. B. R. THOMSON (1), *Nutrition and Metabolism Research Group, Division of Gastroenterology, Department of Medicine, University of Alberta, Edmonton, Canada T6G 2C2*

ROBERT VERGER (7, 10, 13, 15, 16), *Laboratoire de Lipolyse Enzymatique, UPR 9025, IFRC1 du CNRS, 13402 Marseille Cedex 20, France*

C. WALDINGER (17, 19), *Fb 9–Bergische Universität-GH-Wuppertal, D-42097 Wuppertal, Germany*

H. K. WEBER (24), *Institute of Organic Chemistry, Graz University of Technology, A-8010 Graz, Austria*

REZA ZOLFAGHARI (4), *Department of Nutrition, Pennsylvania State University, University Park, Pennsylvania 16802*

Preface

The pace of lipase research has been accelerating. The powerful tools of molecular biology have been brought to bear, more new lipase amino acid sequences and three-dimensional structures are appearing, and new approaches for handling their complicated interfacial kinetics are being reported. In addition, more ways are being discovered and used to control lipase activity and for harnessing their catalytic prowess to pull greater efficiency into older chemical processes. Indeed, studies of heterogeneous lipase catalysis, long passed over by many academic researchers in favor of more experimentally tractable homogeneous, single-phase enzyme systems, are moving closer to the level of depth previously reserved for proteases, their hydrolytic cousins.

To the usual problems of abundance and purity that enzymologists and structural biologists generally face, lipases present the additional difficulty associated with multiphase systems. Unlike proteases, the substrates that are hydrolyzed by lipases are most efficiently presented to the enzyme in a separate, lipid phase. The presence of a suitable second phase appears to bring about an increase in lipase activity and, in some cases, effect a change in their three-dimensional structures. Part of the expanding interest in lipases derives from the increasing applications for these enzymes and from the success of new techniques for studying them.

Previous volumes of *Methods in Enzymology* have dealt specifically with phospholipids, their degradation (Volume 197, Phospholipases), and their biosynthesis (Volume 209, Phospholipid Biosynthesis). The recent explosion of interest in lipases led us to develop Volumes 284 and 286. The first, Biotechnology (Volume 284), includes sequencing, cloning, and structural studies of lipases, and the second, Enzyme Characterization and Utilization (Volume 286), includes the purification of novel lipases, kinetics and assay issues, aspects of lipid metabolism, and the use of lipases in organic synthesis.

Research in the lipase field has been dominated by European scientists and stimulated by the European Community Bridge Program. In addition, there has been a great deal of research emphasis on this field in industrial laboratories. Thus, the authorship of this volume is truly international and includes a diverse mixture of academic and industrial scientists.

Expert secretarial assistance from Mary Kincaid, Ophelia Chiu, and Vina Wong helped enormously with the development of this volume. Editorial assistance from Shirley Light is greatly appreciated.

BYRON RUBIN
EDWARD A. DENNIS

METHODS IN ENZYMOLOGY

VOLUME I. Preparation and Assay of Enzymes
Edited by SIDNEY P. COLOWICK AND NATHAN O. KAPLAN

VOLUME II. Preparation and Assay of Enzymes
Edited by SIDNEY P. COLOWICK AND NATHAN O. KAPLAN

VOLUME III. Preparation and Assay of Substrates
Edited by SIDNEY P. COLOWICK AND NATHAN O. KAPLAN

VOLUME IV. Special Techniques for the Enzymologist
Edited by SIDNEY P. COLOWICK AND NATHAN O. KAPLAN

VOLUME V. Preparation and Assay of Enzymes
Edited by SIDNEY P. COLOWICK AND NATHAN O. KAPLAN

VOLUME VI. Preparation and Assay of Enzymes (*Continued*)
Preparation and Assay of Substrates
Special Techniques
Edited by SIDNEY P. COLOWICK AND NATHAN O. KAPLAN

VOLUME VII. Cumulative Subject Index
Edited by SIDNEY P. COLOWICK AND NATHAN O. KAPLAN

VOLUME VIII. Complex Carbohydrates
Edited by ELIZABETH F. NEUFELD AND VICTOR GINSBURG

VOLUME IX. Carbohydrate Metabolism
Edited by WILLIS A. WOOD

VOLUME X. Oxidation and Phosphorylation
Edited by RONALD W. ESTABROOK AND MAYNARD E. PULLMAN

VOLUME XI. Enzyme Structure
Edited by C. H. W. HIRS

VOLUME XII. Nucleic Acids (Parts A and B)
Edited by LAWRENCE GROSSMAN AND KIVIE MOLDAVE

VOLUME XIII. Citric Acid Cycle
Edited by J. M. LOWENSTEIN

VOLUME XIV. Lipids
Edited by J. M. LOWENSTEIN

VOLUME XV. Steroids and Terpenoids
Edited by RAYMOND B. CLAYTON

VOLUME XVI. Fast Reactions
Edited by KENNETH KUSTIN

VOLUME XVII. Metabolism of Amino Acids and Amines (Parts A and B)
Edited by HERBERT TABOR AND CELIA WHITE TABOR

VOLUME XVIII. Vitamins and Coenzymes (Parts A, B, and C)
Edited by DONALD B. MCCORMICK AND LEMUEL D. WRIGHT

VOLUME XIX. Proteolytic Enzymes
Edited by GERTRUDE E. PERLMANN AND LASZLO LORAND

VOLUME XX. Nucleic Acids and Protein Synthesis (Part C)
Edited by KIVIE MOLDAVE AND LAWRENCE GROSSMAN

VOLUME XXI. Nucleic Acids (Part D)
Edited by LAWRENCE GROSSMAN AND KIVIE MOLDAVE

VOLUME XXII. Enzyme Purification and Related Techniques
Edited by WILLIAM B. JAKOBY

VOLUME XXIII. Photosynthesis (Part A)
Edited by ANTHONY SAN PIETRO

VOLUME XXIV. Photosynthesis and Nitrogen Fixation (Part B)
Edited by ANTHONY SAN PIETRO

VOLUME XXV. Enzyme Structure (Part B)
Edited by C. H. W. HIRS AND SERGE N. TIMASHEFF

VOLUME XXVI. Enzyme Structure (Part C)
Edited by C. H. W. HIRS AND SERGE N. TIMASHEFF

VOLUME XXVII. Enzyme Structure (Part D)
Edited by C. H. W. HIRS AND SERGE N. TIMASHEFF

VOLUME XXVIII. Complex Carbohydrates (Part B)
Edited by VICTOR GINSBURG

VOLUME XXIX. Nucleic Acids and Protein Synthesis (Part E)
Edited by LAWRENCE GROSSMAN AND KIVIE MOLDAVE

VOLUME XXX. Nucleic Acids and Protein Synthesis (Part F)
Edited by KIVIE MOLDAVE AND LAWRENCE GROSSMAN

VOLUME XXXI. Biomembranes (Part A)
Edited by SIDNEY FLEISCHER AND LESTER PACKER

VOLUME XXXII. Biomembranes (Part B)
Edited by SIDNEY FLEISCHER AND LESTER PACKER

VOLUME XXXIII. Cumulative Subject Index Volumes I–XXX
Edited by MARTHA G. DENNIS AND EDWARD A. DENNIS

VOLUME XXXIV. Affinity Techniques (Enzyme Purification: Part B)
Edited by WILLIAM B. JAKOBY AND MEIR WILCHEK

VOLUME XXXV. Lipids (Part B)
Edited by JOHN M. LOWENSTEIN

VOLUME XXXVI. Hormone Action (Part A: Steroid Hormones)
Edited by BERT W. O'MALLEY AND JOEL G. HARDMAN

VOLUME XXXVII. Hormone Action (Part B: Peptide Hormones)
Edited by BERT W. O'MALLEY AND JOEL G. HARDMAN

VOLUME XXXVIII. Hormone Action (Part C: Cyclic Nucleotides)
Edited by JOEL G. HARDMAN AND BERT W. O'MALLEY

VOLUME XXXIX. Hormone Action (Part D: Isolated Cells, Tissues, and Organ Systems)
Edited by JOEL G. HARDMAN AND BERT W. O'MALLEY

VOLUME XL. Hormone Action (Part E: Nuclear Structure and Function)
Edited by BERT W. O'MALLEY AND JOEL G. HARDMAN

VOLUME XLI. Carbohydrate Metabolism (Part B)
Edited by W. A. WOOD

VOLUME XLII. Carbohydrate Metabolism (Part C)
Edited by W. A. WOOD

VOLUME XLIII. Antibiotics
Edited by JOHN H. HASH

VOLUME XLIV. Immobilized Enzymes
Edited by KLAUS MOSBACH

VOLUME XLV. Proteolytic Enzymes (Part B)
Edited by LASZLO LORAND

VOLUME XLVI. Affinity Labeling
Edited by WILLIAM B. JAKOBY AND MEIR WILCHEK

VOLUME XLVII. Enzyme Structure (Part E)
Edited by C. H. W. HIRS AND SERGE N. TIMASHEFF

VOLUME XLVIII. Enzyme Structure (Part F)
Edited by C. H. W. HIRS AND SERGE N. TIMASHEFF

VOLUME XLIX. Enzyme Structure (Part G)
Edited by C. H. W. HIRS AND SERGE N. TIMASHEFF

VOLUME L. Complex Carbohydrates (Part C)
Edited by VICTOR GINSBURG

VOLUME LI. Purine and Pyrimidine Nucleotide Metabolism
Edited by PATRICIA A. HOFFEE AND MARY ELLEN JONES

VOLUME LII. Biomembranes (Part C: Biological Oxidations)
Edited by SIDNEY FLEISCHER AND LESTER PACKER

VOLUME LIII. Biomembranes (Part D: Biological Oxidations)
Edited by SIDNEY FLEISCHER AND LESTER PACKER

VOLUME LIV. Biomembranes (Part E: Biological Oxidations)
Edited by SIDNEY FLEISCHER AND LESTER PACKER

VOLUME LV. Biomembranes (Part F: Bioenergetics)
Edited by SIDNEY FLEISCHER AND LESTER PACKER

VOLUME LVI. Biomembranes (Part G: Bioenergetics)
Edited by SIDNEY FLEISCHER AND LESTER PACKER

VOLUME LVII. Bioluminescence and Chemiluminescence
Edited by MARLENE A. DELUCA

VOLUME LVIII. Cell Culture
Edited by WILLIAM B. JAKOBY AND IRA PASTAN

VOLUME LIX. Nucleic Acids and Protein Synthesis (Part G)
Edited by KIVIE MOLDAVE AND LAWRENCE GROSSMAN

VOLUME LX. Nucleic Acids and Protein Synthesis (Part H)
Edited by KIVIE MOLDAVE AND LAWRENCE GROSSMAN

VOLUME 61. Enzyme Structure (Part H)
Edited by C. H. W. HIRS AND SERGE N. TIMASHEFF

VOLUME 62. Vitamins and Coenzymes (Part D)
Edited by DONALD B. MCCORMICK AND LEMUEL D. WRIGHT

VOLUME 63. Enzyme Kinetics and Mechanism (Part A: Initial Rate and Inhibitor Methods)
Edited by DANIEL L. PURICH

VOLUME 64. Enzyme Kinetics and Mechanism (Part B: Isotopic Probes and Complex Enzyme Systems)
Edited by DANIEL L. PURICH

VOLUME 65. Nucleic Acids (Part I)
Edited by LAWRENCE GROSSMAN AND KIVIE MOLDAVE

VOLUME 66. Vitamins and Coenzymes (Part E)
Edited by DONALD B. MCCORMICK AND LEMUEL D. WRIGHT

VOLUME 67. Vitamins and Coenzymes (Part F)
Edited by DONALD B. MCCORMICK AND LEMUEL D. WRIGHT

VOLUME 68. Recombinant DNA
Edited by RAY WU

VOLUME 69. Photosynthesis and Nitrogen Fixation (Part C)
Edited by ANTHONY SAN PIETRO

VOLUME 70. Immunochemical Techniques (Part A)
Edited by HELEN VAN VUNAKIS AND JOHN J. LANGONE

VOLUME 71. Lipids (Part C)
Edited by JOHN M. LOWENSTEIN

VOLUME 72. Lipids (Part D)
Edited by JOHN M. LOWENSTEIN

VOLUME 73. Immunochemical Techniques (Part B)
Edited by JOHN J. LANGONE AND HELEN VAN VUNAKIS

VOLUME 74. Immunochemical Techniques (Part C)
Edited by JOHN J. LANGONE AND HELEN VAN VUNAKIS

VOLUME 75. Cumulative Subject Index Volumes XXXI, XXXII, XXXIV–LX
Edited by EDWARD A. DENNIS AND MARTHA G. DENNIS

VOLUME 76. Hemoglobins
Edited by ERALDO ANTONINI, LUIGI ROSSI-BERNARDI, AND EMILIA CHIANCONE

VOLUME 77. Detoxication and Drug Metabolism
Edited by WILLIAM B. JAKOBY

VOLUME 78. Interferons (Part A)
Edited by SIDNEY PESTKA

VOLUME 79. Interferons (Part B)
Edited by SIDNEY PESTKA

VOLUME 80. Proteolytic Enzymes (Part C)
Edited by LASZLO LORAND

VOLUME 81. Biomembranes (Part H: Visual Pigments and Purple Membranes, I)
Edited by LESTER PACKER

VOLUME 82. Structural and Contractile Proteins (Part A: Extracellular Matrix)
Edited by LEON W. CUNNINGHAM AND DIXIE W. FREDERIKSEN

VOLUME 83. Complex Carbohydrates (Part D)
Edited by VICTOR GINSBURG

VOLUME 84. Immunochemical Techniques (Part D: Selected Immunoassays)
Edited by JOHN J. LANGONE AND HELEN VAN VUNAKIS

VOLUME 85. Structural and Contractile Proteins (Part B: The Contractile Apparatus and the Cytoskeleton)
Edited by DIXIE W. FREDERIKSEN AND LEON W. CUNNINGHAM

VOLUME 86. Prostaglandins and Arachidonate Metabolites
Edited by WILLIAM E. M. LANDS AND WILLIAM L. SMITH

VOLUME 87. Enzyme Kinetics and Mechanism (Part C: Intermediates, Stereochemistry, and Rate Studies)
Edited by DANIEL L. PURICH

VOLUME 88. Biomembranes (Part I: Visual Pigments and Purple Membranes, II)
Edited by LESTER PACKER

VOLUME 89. Carbohydrate Metabolism (Part D)
Edited by WILLIS A. WOOD

VOLUME 90. Carbohydrate Metabolism (Part E)
Edited by WILLIS A. WOOD

VOLUME 91. Enzyme Structure (Part I)
Edited by C. H. W. HIRS AND SERGE N. TIMASHEFF

VOLUME 92. Immunochemical Techniques (Part E: Monoclonal Antibodies and General Immunoassay Methods)
Edited by JOHN J. LANGONE AND HELEN VAN VUNAKIS

VOLUME 93. Immunochemical Techniques (Part F: Conventional Antibodies, Fc Receptors, and Cytotoxicity)
Edited by JOHN J. LANGONE AND HELEN VAN VUNAKIS

VOLUME 94. Polyamines
Edited by HERBERT TABOR AND CELIA WHITE TABOR

VOLUME 95. Cumulative Subject Index Volumes 61–74, 76–80
Edited by EDWARD A. DENNIS AND MARTHA G. DENNIS

VOLUME 96. Biomembranes [Part J: Membrane Biogenesis: Assembly and Targeting (General Methods; Eukaryotes)]
Edited by SIDNEY FLEISCHER AND BECCA FLEISCHER

VOLUME 97. Biomembranes [Part K: Membrane Biogenesis: Assembly and Targeting (Prokaryotes, Mitochondria, and Chloroplasts)]
Edited by SIDNEY FLEISCHER AND BECCA FLEISCHER

VOLUME 98. Biomembranes (Part L: Membrane Biogenesis: Processing and Recycling)
Edited by SIDNEY FLEISCHER AND BECCA FLEISCHER

VOLUME 99. Hormone Action (Part F: Protein Kinases)
Edited by JACKIE D. CORBIN AND JOEL G. HARDMAN

VOLUME 100. Recombinant DNA (Part B)
Edited by RAY WU, LAWRENCE GROSSMAN, AND KIVIE MOLDAVE

VOLUME 101. Recombinant DNA (Part C)
Edited by RAY WU, LAWRENCE GROSSMAN, AND KIVIE MOLDAVE

VOLUME 102. Hormone Action (Part G: Calmodulin and Calcium-Binding Proteins)
Edited by ANTHONY R. MEANS AND BERT W. O'MALLEY

VOLUME 103. Hormone Action (Part H: Neuroendocrine Peptides)
Edited by P. MICHAEL CONN

VOLUME 104. Enzyme Purification and Related Techniques (Part C)
Edited by WILLIAM B. JAKOBY

VOLUME 105. Oxygen Radicals in Biological Systems
Edited by LESTER PACKER

VOLUME 106. Posttranslational Modifications (Part A)
Edited by FINN WOLD AND KIVIE MOLDAVE

VOLUME 107. Posttranslational Modifications (Part B)
Edited by FINN WOLD AND KIVIE MOLDAVE

VOLUME 108. Immunochemical Techniques (Part G: Separation and Characterization of Lymphoid Cells)
Edited by GIOVANNI DI SABATO, JOHN J. LANGONE, AND HELEN VAN VUNAKIS

VOLUME 109. Hormone Action (Part I: Peptide Hormones)
Edited by LUTZ BIRNBAUMER AND BERT W. O'MALLEY

VOLUME 110. Steroids and Isoprenoids (Part A)
Edited by JOHN H. LAW AND HANS C. RILLING

VOLUME 111. Steroids and Isoprenoids (Part B)
Edited by JOHN H. LAW AND HANS C. RILLING

VOLUME 112. Drug and Enzyme Targeting (Part A)
Edited by KENNETH J. WIDDER AND RALPH GREEN

VOLUME 113. Glutamate, Glutamine, Glutathione, and Related Compounds
Edited by ALTON MEISTER

VOLUME 114. Diffraction Methods for Biological Macromolecules (Part A)
Edited by HAROLD W. WYCKOFF, C. H. W. HIRS, AND SERGE N. TIMASHEFF

VOLUME 115. Diffraction Methods for Biological Macromolecules (Part B)
Edited by HAROLD W. WYCKOFF, C. H. W. HIRS, AND SERGE N. TIMASHEFF

VOLUME 116. Immunochemical Techniques (Part H: Effectors and Mediators of Lymphoid Cell Functions)
Edited by GIOVANNI DI SABATO, JOHN J. LANGONE, AND HELEN VAN VUNAKIS

VOLUME 117. Enzyme Structure (Part J)
Edited by C. H. W. HIRS AND SERGE N. TIMASHEFF

VOLUME 118. Plant Molecular Biology
Edited by ARTHUR WEISSBACH AND HERBERT WEISSBACH

VOLUME 119. Interferons (Part C)
Edited by SIDNEY PESTKA

VOLUME 120. Cumulative Subject Index Volumes 81–94, 96–101

VOLUME 121. Immunochemical Techniques (Part I: Hybridoma Technology and Monoclonal Antibodies)
Edited by JOHN J. LANGONE AND HELEN VAN VUNAKIS

VOLUME 122. Vitamins and Coenzymes (Part G)
Edited by FRANK CHYTIL AND DONALD B. MCCORMICK

VOLUME 123. Vitamins and Coenzymes (Part H)
Edited by FRANK CHYTIL AND DONALD B. MCCORMICK

VOLUME 124. Hormone Action (Part J: Neuroendocrine Peptides)
Edited by P. MICHAEL CONN

VOLUME 125. Biomembranes (Part M: Transport in Bacteria, Mitochondria, and Chloroplasts: General Approaches and Transport Systems)
Edited by SIDNEY FLEISCHER AND BECCA FLEISCHER

VOLUME 126. Biomembranes (Part N: Transport in Bacteria, Mitochondria, and Chloroplasts: Protonmotive Force)
Edited by SIDNEY FLEISCHER AND BECCA FLEISCHER

VOLUME 127. Biomembranes (Part O: Protons and Water: Structure and Translocation)
Edited by LESTER PACKER

VOLUME 128. Plasma Lipoproteins (Part A: Preparation, Structure, and Molecular Biology)
Edited by JERE P. SEGREST AND JOHN J. ALBERS

VOLUME 129. Plasma Lipoproteins (Part B: Characterization, Cell Biology, and Metabolism)
Edited by JOHN J. ALBERS AND JERE P. SEGREST

VOLUME 130. Enzyme Structure (Part K)
Edited by C. H. W. HIRS AND SERGE N. TIMASHEFF

VOLUME 131. Enzyme Structure (Part L)
Edited by C. H. W. HIRS AND SERGE N. TIMASHEFF

VOLUME 132. Immunochemical Techniques (Part J: Phagocytosis and Cell-Mediated Cytotoxicity)
Edited by GIOVANNI DI SABATO AND JOHANNES EVERSE

VOLUME 133. Bioluminescence and Chemiluminescence (Part B)
Edited by MARLENE DELUCA AND WILLIAM D. MCELROY

VOLUME 134. Structural and Contractile Proteins (Part C: The Contractile Apparatus and the Cytoskeleton)
Edited by RICHARD B. VALLEE

VOLUME 135. Immobilized Enzymes and Cells (Part B)
Edited by KLAUS MOSBACH

VOLUME 136. Immobilized Enzymes and Cells (Part C)
Edited by KLAUS MOSBACH

VOLUME 137. Immobilized Enzymes and Cells (Part D)
Edited by KLAUS MOSBACH

VOLUME 138. Complex Carbohydrates (Part E)
Edited by VICTOR GINSBURG

VOLUME 139. Cellular Regulators (Part A: Calcium- and Calmodulin-Binding Proteins)
Edited by ANTHONY R. MEANS AND P. MICHAEL CONN

VOLUME 140. Cumulative Subject Index Volumes 102–119, 121–134

VOLUME 141. Cellular Regulators (Part B: Calcium and Lipids)
Edited by P. MICHAEL CONN AND ANTHONY R. MEANS

VOLUME 142. Metabolism of Aromatic Amino Acids and Amines
Edited by SEYMOUR KAUFMAN

VOLUME 143. Sulfur and Sulfur Amino Acids
Edited by WILLIAM B. JAKOBY AND OWEN GRIFFITH

VOLUME 144. Structural and Contractile Proteins (Part D: Extracellular Matrix)
Edited by LEON W. CUNNINGHAM

VOLUME 145. Structural and Contractile Proteins (Part E: Extracellular Matrix)
Edited by LEON W. CUNNINGHAM

VOLUME 146. Peptide Growth Factors (Part A)
Edited by DAVID BARNES AND DAVID A. SIRBASKU

VOLUME 147. Peptide Growth Factors (Part B)
Edited by DAVID BARNES AND DAVID A. SIRBASKU

VOLUME 148. Plant Cell Membranes
Edited by LESTER PACKER AND ROLAND DOUCE

VOLUME 149. Drug and Enzyme Targeting (Part B)
Edited by RALPH GREEN AND KENNETH J. WIDDER

VOLUME 150. Immunochemical Techniques (Part K: *In Vitro* Models of B and T Cell Functions and Lymphoid Cell Receptors)
Edited by GIOVANNI DI SABATO

VOLUME 151. Molecular Genetics of Mammalian Cells
Edited by MICHAEL M. GOTTESMAN

VOLUME 152. Guide to Molecular Cloning Techniques
Edited by SHELBY L. BERGER AND ALAN R. KIMMEL

VOLUME 153. Recombinant DNA (Part D)
Edited by RAY WU AND LAWRENCE GROSSMAN

VOLUME 154. Recombinant DNA (Part E)
Edited by RAY WU AND LAWRENCE GROSSMAN

VOLUME 155. Recombinant DNA (Part F)
Edited by RAY WU

VOLUME 156. Biomembranes (Part P: ATP-Driven Pumps and Related Transport: The Na,K-Pump)
Edited by SIDNEY FLEISCHER AND BECCA FLEISCHER

VOLUME 157. Biomembranes (Part Q: ATP-Driven Pumps and Related Transport: Calcium, Proton, and Potassium Pumps)
Edited by SIDNEY FLEISCHER AND BECCA FLEISCHER

VOLUME 158. Metalloproteins (Part A)
Edited by JAMES F. RIORDAN AND BERT L. VALLEE

VOLUME 159. Initiation and Termination of Cyclic Nucleotide Action
Edited by JACKIE D. CORBIN AND ROGER A. JOHNSON

VOLUME 160. Biomass (Part A: Cellulose and Hemicellulose)
Edited by WILLIS A. WOOD AND SCOTT T. KELLOGG

VOLUME 161. Biomass (Part B: Lignin, Pectin, and Chitin)
Edited by WILLIS A. WOOD AND SCOTT T. KELLOGG

VOLUME 162. Immunochemical Techniques (Part L: Chemotaxis and Inflammation)
Edited by GIOVANNI DI SABATO

VOLUME 163. Immunochemical Techniques (Part M: Chemotaxis and Inflammation)
Edited by GIOVANNI DI SABATO

VOLUME 164. Ribosomes
Edited by HARRY F. NOLLER, JR., AND KIVIE MOLDAVE

VOLUME 165. Microbial Toxins: Tools for Enzymology
Edited by SIDNEY HARSHMAN

VOLUME 166. Branched-Chain Amino Acids
Edited by ROBERT HARRIS AND JOHN R. SOKATCH

VOLUME 167. Cyanobacteria
Edited by LESTER PACKER AND ALEXANDER N. GLAZER

VOLUME 168. Hormone Action (Part K: Neuroendocrine Peptides)
Edited by P. MICHAEL CONN

VOLUME 169. Platelets: Receptors, Adhesion, Secretion (Part A)
Edited by JACEK HAWIGER

VOLUME 170. Nucleosomes
Edited by PAUL M. WASSARMAN AND ROGER D. KORNBERG

VOLUME 171. Biomembranes (Part R: Transport Theory: Cells and Model Membranes)
Edited by SIDNEY FLEISCHER AND BECCA FLEISCHER

VOLUME 172. Biomembranes (Part S: Transport: Membrane Isolation and Characterization)
Edited by SIDNEY FLEISCHER AND BECCA FLEISCHER

VOLUME 173. Biomembranes [Part T: Cellular and Subcellular Transport: Eukaryotic (Nonepithelial) Cells]
Edited by SIDNEY FLEISCHER AND BECCA FLEISCHER

VOLUME 174. Biomembranes [Part U: Cellular and Subcellular Transport: Eukaryotic (Nonepithelial) Cells]
Edited by SIDNEY FLEISCHER AND BECCA FLEISCHER

VOLUME 175. Cumulative Subject Index Volumes 135–139, 141–167

VOLUME 176. Nuclear Magnetic Resonance (Part A: Spectral Techniques and Dynamics)
Edited by NORMAN J. OPPENHEIMER AND THOMAS L. JAMES

VOLUME 177. Nuclear Magnetic Resonance (Part B: Structure and Mechanism)
Edited by NORMAN J. OPPENHEIMER AND THOMAS L. JAMES

VOLUME 178. Antibodies, Antigens, and Molecular Mimicry
Edited by JOHN J. LANGONE

VOLUME 179. Complex Carbohydrates (Part F)
Edited by VICTOR GINSBURG

VOLUME 180. RNA Processing (Part A: General Methods)
Edited by JAMES E. DAHLBERG AND JOHN N. ABELSON

VOLUME 181. RNA Processing (Part B: Specific Methods)
Edited by JAMES E. DAHLBERG AND JOHN N. ABELSON

VOLUME 182. Guide to Protein Purification
Edited by MURRAY P. DEUTSCHER

VOLUME 183. Molecular Evolution: Computer Analysis of Protein and Nucleic Acid Sequences
Edited by RUSSELL F. DOOLITTLE

VOLUME 184. Avidin–Biotin Technology
Edited by MEIR WILCHEK AND EDWARD A. BAYER

VOLUME 185. Gene Expression Technology
Edited by DAVID V. GOEDDEL

VOLUME 186. Oxygen Radicals in Biological Systems (Part B: Oxygen Radicals and Antioxidants)
Edited by LESTER PACKER AND ALEXANDER N. GLAZER

VOLUME 187. Arachidonate Related Lipid Mediators
Edited by ROBERT C. MURPHY AND FRANK A. FITZPATRICK

VOLUME 188. Hydrocarbons and Methylotrophy
Edited by MARY E. LIDSTROM

VOLUME 189. Retinoids (Part A: Molecular and Metabolic Aspects)
Edited by LESTER PACKER

VOLUME 190. Retinoids (Part B: Cell Differentiation and Clinical Applications)
Edited by LESTER PACKER

VOLUME 191. Biomembranes (Part V: Cellular and Subcellular Transport: Epithelial Cells)
Edited by SIDNEY FLEISCHER AND BECCA FLEISCHER

VOLUME 192. Biomembranes (Part W: Cellular and Subcellular Transport: Epithelial Cells)
Edited by SIDNEY FLEISCHER AND BECCA FLEISCHER

VOLUME 193. Mass Spectrometry
Edited by JAMES A. MCCLOSKEY

VOLUME 194. Guide to Yeast Genetics and Molecular Biology
Edited by CHRISTINE GUTHRIE AND GERALD R. FINK

VOLUME 195. Adenylyl Cyclase, G Proteins, and Guanylyl Cyclase
Edited by ROGER A. JOHNSON AND JACKIE D. CORBIN

VOLUME 196. Molecular Motors and the Cytoskeleton
Edited by RICHARD B. VALLEE

VOLUME 197. Phospholipases
Edited by EDWARD A. DENNIS

VOLUME 198. Peptide Growth Factors (Part C)
Edited by DAVID BARNES, J. P. MATHER, AND GORDON H. SATO

VOLUME 199. Cumulative Subject Index Volumes 168–174, 176–194

VOLUME 200. Protein Phosphorylation (Part A: Protein Kinases: Assays, Purification, Antibodies, Functional Analysis, Cloning, and Expression)
Edited by TONY HUNTER AND BARTHOLOMEW M. SEFTON

VOLUME 201. Protein Phosphorylation (Part B: Analysis of Protein Phosphorylation, Protein Kinase Inhibitors, and Protein Phosphatases)
Edited by TONY HUNTER AND BARTHOLOMEW M. SEFTON

VOLUME 202. Molecular Design and Modeling: Concepts and Applications (Part A: Proteins, Peptides, and Enzymes)
Edited by JOHN J. LANGONE

VOLUME 203. Molecular Design and Modeling: Concepts and Applications (Part B: Antibodies and Antigens, Nucleic Acids, Polysaccharides, and Drugs)
Edited by JOHN J. LANGONE

VOLUME 204. Bacterial Genetic Systems
Edited by JEFFREY H. MILLER

VOLUME 205. Metallobiochemistry (Part B: Metallothionein and Related Molecules)
Edited by JAMES F. RIORDAN AND BERT L. VALLEE

VOLUME 206. Cytochrome P450
Edited by MICHAEL R. WATERMAN AND ERIC F. JOHNSON

VOLUME 207. Ion Channels
Edited by BERNARDO RUDY AND LINDA E. IVERSON

VOLUME 208. Protein–DNA Interactions
Edited by ROBERT T. SAUER

VOLUME 209. Phospholipid Biosynthesis
Edited by EDWARD A. DENNIS AND DENNIS E. VANCE

VOLUME 210. Numerical Computer Methods
Edited by LUDWIG BRAND AND MICHAEL L. JOHNSON

VOLUME 211. DNA Structures (Part A: Synthesis and Physical Analysis of DNA)
Edited by DAVID M. J. LILLEY AND JAMES E. DAHLBERG

VOLUME 212. DNA Structures (Part B: Chemical and Electrophoretic Analysis of DNA)
Edited by DAVID M. J. LILLEY AND JAMES E. DAHLBERG

VOLUME 213. Carotenoids (Part A: Chemistry, Separation, Quantitation, and Antioxidation)
Edited by LESTER PACKER

VOLUME 214. Carotenoids (Part B: Metabolism, Genetics, and Biosynthesis)
Edited by LESTER PACKER

VOLUME 215. Platelets: Receptors, Adhesion, Secretion (Part B)
Edited by JACEK J. HAWIGER

VOLUME 216. Recombinant DNA (Part G)
Edited by RAY WU

VOLUME 217. Recombinant DNA (Part H)
Edited by RAY WU

VOLUME 218. Recombinant DNA (Part I)
Edited by RAY WU

VOLUME 219. Reconstitution of Intracellular Transport
Edited by JAMES E. ROTHMAN

VOLUME 220. Membrane Fusion Techniques (Part A)
Edited by NEJAT DÜZGÜNEŞ

VOLUME 221. Membrane Fusion Techniques (Part B)
Edited by NEJAT DÜZGÜNEŞ

VOLUME 222. Proteolytic Enzymes in Coagulation, Fibrinolysis, and Complement Activation (Part A: Mammalian Blood Coagulation Factors and Inhibitors)
Edited by LASZLO LORAND AND KENNETH G. MANN

VOLUME 223. Proteolytic Enzymes in Coagulation, Fibrinolysis, and Complement Activation (Part B: Complement Activation, Fibrinolysis, and Nonmammalian Blood Coagulation Factors)
Edited by LASZLO LORAND AND KENNETH G. MANN

VOLUME 224. Molecular Evolution: Producing the Biochemical Data
Edited by ELIZABETH ANNE ZIMMER, THOMAS J. WHITE, REBECCA L. CANN, AND ALLAN C. WILSON

VOLUME 225. Guide to Techniques in Mouse Development
Edited by PAUL M. WASSARMAN AND MELVIN L. DEPAMPHILIS

VOLUME 226. Metallobiochemistry (Part C: Spectroscopic and Physical Methods for Probing Metal Ion Environments in Metalloenzymes and Metalloproteins)
Edited by JAMES F. RIORDAN AND BERT L. VALLEE

VOLUME 227. Metallobiochemistry (Part D: Physical and Spectroscopic Methods for Probing Metal Ion Environments in Metalloproteins)
Edited by JAMES F. RIORDAN AND BERT L. VALLEE

VOLUME 228. Aqueous Two-Phase Systems
Edited by HARRY WALTER AND GÖTE JOHANSSON

VOLUME 229. Cumulative Subject Index Volumes 195–198, 200–227

VOLUME 230. Guide to Techniques in Glycobiology
Edited by WILLIAM J. LENNARZ AND GERALD W. HART

VOLUME 231. Hemoglobins (Part B: Biochemical and Analytical Methods)
Edited by JOHANNES EVERSE, KIM D. VANDEGRIFF, AND ROBERT M. WINSLOW

VOLUME 232. Hemoglobins (Part C: Biophysical Methods)
Edited by JOHANNES EVERSE, KIM D. VANDEGRIFF, AND ROBERT M. WINSLOW

VOLUME 233. Oxygen Radicals in Biological Systems (Part C)
Edited by LESTER PACKER

VOLUME 234. Oxygen Radicals in Biological Systems (Part D)
Edited by LESTER PACKER

VOLUME 235. Bacterial Pathogenesis (Part A: Identification and Regulation of Virulence Factors)
Edited by VIRGINIA L. CLARK AND PATRIK M. BAVOIL

VOLUME 236. Bacterial Pathogenesis (Part B: Integration of Pathogenic Bacteria with Host Cells)
Edited by VIRGINIA L. CLARK AND PATRIK M. BAVOIL

VOLUME 237. Heterotrimeric G Proteins
Edited by RAVI IYENGAR

VOLUME 238. Heterotrimeric G-Protein Effectors
Edited by RAVI IYENGAR

VOLUME 239. Nuclear Magnetic Resonance (Part C)
Edited by THOMAS L. JAMES AND NORMAN J. OPPENHEIMER

VOLUME 240. Numerical Computer Methods (Part B)
Edited by MICHAEL L. JOHNSON AND LUDWIG BRAND

VOLUME 241. Retroviral Proteases
Edited by LAWRENCE C. KUO AND JULES A. SHAFER

VOLUME 242. Neoglycoconjugates (Part A)
Edited by Y. C. LEE AND REIKO T. LEE

VOLUME 243. Inorganic Microbial Sulfur Metabolism
Edited by HARRY D. PECK, JR., AND JEAN LEGALL

VOLUME 244. Proteolytic Enzymes: Serine and Cysteine Peptidases
Edited by ALAN J. BARRETT

VOLUME 245. Extracellular Matrix Components
Edited by E. RUOSLAHTI AND E. ENGVALL

VOLUME 246. Biochemical Spectroscopy
Edited by KENNETH SAUER

VOLUME 247. Neoglycoconjugates (Part B: Biomedical Applications)
Edited by Y. C. LEE AND REIKO T. LEE

VOLUME 248. Proteolytic Enzymes: Aspartic and Metallo Peptidases
Edited by ALAN J. BARRETT

VOLUME 249. Enzyme Kinetics and Mechanism (Part D: Developments in Enzyme Dynamics)
Edited by DANIEL L. PURICH

VOLUME 250. Lipid Modifications of Proteins
Edited by PATRICK J. CASEY AND JANICE E. BUSS

VOLUME 251. Biothiols (Part A: Monothiols and Dithiols, Protein Thiols, and Thiyl Radicals)
Edited by LESTER PACKER

VOLUME 252. Biothiols (Part B: Glutathione and Thioredoxin; Thiols in Signal Transduction and Gene Regulation)
Edited by LESTER PACKER

VOLUME 253. Adhesion of Microbial Pathogens
Edited by RON J. DOYLE AND ITZHAK OFEK

VOLUME 254. Oncogene Techniques
Edited by PETER K. VOGT AND INDER M. VERMA

VOLUME 255. Small GTPases and Their Regulators (Part A: Ras Family)
Edited by W. E. BALCH, CHANNING J. DER, AND ALAN HALL

VOLUME 256. Small GTPases and Their Regulators (Part B: Rho Family)
Edited by W. E. BALCH, CHANNING J. DER, AND ALAN HALL

VOLUME 257. Small GTPases and Their Regulators (Part C: Proteins Involved in Transport)
Edited by W. E. BALCH, CHANNING J. DER, AND ALAN HALL

VOLUME 258. Redox-Active Amino Acids in Biology
Edited by JUDITH P. KLINMAN

VOLUME 259. Energetics of Biological Macromolecules
Edited by MICHAEL L. JOHNSON AND GARY K. ACKERS

VOLUME 260. Mitochondrial Biogenesis and Genetics (Part A)
Edited by GIUSEPPE M. ATTARDI AND ANNE CHOMYN

VOLUME 261. Nuclear Magnetic Resonance and Nucleic Acids
Edited by THOMAS L. JAMES

VOLUME 262. DNA Replication
Edited by JUDITH L. CAMPBELL

VOLUME 263. Plasma Lipoproteins (Part C: Quantitation)
Edited by WILLIAM A. BRADLEY, SANDRA H. GIANTURCO, AND JERE P. SEGREST

VOLUME 264. Mitochondrial Biogenesis and Genetics (Part B)
Edited by GIUSEPPE M. ATTARDI AND ANNE CHOMYN

VOLUME 265. Cumulative Subject Index Volumes 228, 230–262

VOLUME 266. Computer Methods for Macromolecular Sequence Analysis
Edited by RUSSELL F. DOOLITTLE

VOLUME 267. Combinatorial Chemistry
Edited by JOHN N. ABELSON

VOLUME 268. Nitric Oxide (Part A: Sources and Detection of NO; NO Synthase)
Edited by LESTER PACKER

VOLUME 269. Nitric Oxide (Part B: Physiological and Pathological Processes)
Edited by LESTER PACKER

VOLUME 270. High Resolution Separation and Analysis of Biological Macromolecules (Part A: Fundamentals)
Edited by BARRY L. KARGER AND WILLIAM S. HANCOCK

VOLUME 271. High Resolution Separation and Analysis of Biological Macromolecules (Part B: Applications)
Edited by BARRY L. KARGER AND WILLIAM S. HANCOCK

VOLUME 272. Cytochrome P450 (Part B)
Edited by ERIC F. JOHNSON AND MICHAEL R. WATERMAN

VOLUME 273. RNA Polymerase and Associated Factors (Part A)
Edited by SANKAR ADHYA

VOLUME 274. RNA Polymerase and Associated Factors (Part B)
Edited by SANKAR ADHYA

VOLUME 275. Viral Polymerases and Related Proteins
Edited by LAWRENCE C. KUO, DAVID B. OLSEN, AND STEVEN S. CARROLL

VOLUME 276. Macromolecular Crystallography (Part A)
Edited by CHARLES W. CARTER, JR., AND ROBERT M. SWEET

VOLUME 277. Macromolecular Crystallography (Part B)
Edited by CHARLES W. CARTER, JR., AND ROBERT M. SWEET

VOLUME 278. Fluorescence Spectroscopy
Edited by LUDWIG BRAND AND MICHAEL L. JOHNSON

VOLUME 279. Vitamins and Coenzymes, Part I
Edited by DONALD B. MCCORMICK, JOHN W. SUTTIE, AND CONRAD WAGNER

VOLUME 280. Vitamins and Coenzymes, Part J
Edited by DONALD B. MCCORMICK, JOHN W. SUTTIE, AND CONRAD WAGNER

VOLUME 281. Vitamins and Coenzymes, Part K
Edited by DONALD B. MCCORMICK, JOHN W. SUTTIE, AND CONRAD WAGNER

VOLUME 282. Vitamins and Coenzymes, Part L
Edited by DONALD B. MCCORMICK, JOHN W. SUTTIE, AND CONRAD WAGNER

VOLUME 283. Cell Cycle Control
Edited by WILLIAM G. DUNPHY

VOLUME 284. Lipases (Part A: Biotechnology)
Edited by BYRON RUBIN AND EDWARD A. DENNIS

VOLUME 285. Cumulative Subject Index Volumes 263, 264, 266–289 (in preparation)

VOLUME 286. Lipases (Part B: Enzyme Characterization and Utilization)
Edited by BYRON RUBIN AND EDWARD A. DENNIS

VOLUME 287. Chemokines
Edited by RICHARD HORUK

VOLUME 288. Chemokine Receptors
Edited by RICHARD HORUK

VOLUME 289. Solid Phase Peptide Synthesis
Edited by GREGG B. FIELDS

VOLUME 290. Molecular Chaperones (in preparation)
Edited by GEORGE H. LORIMER AND THOMAS BALDWIN

Section I

Regulation and Characterization

[1] Inhibition of Lipid Absorption as an Approach to the Treatment of Obesity

By A. B. R. Thomson, A. De Pover, M. Keelan, E. Jarocka-Cyrta, and M. T. Clandinin

The treatment of obesity is an important therapeutic goal for reduction of secondary risk outcomes in patients with a variety of disorders such as hypertension, diabetes mellitus or atherogenic heart, and central nervous system or peripheral vascular disease. In some individuals, obesity may be associated also with psychological abnormalities due to a distortion of self-image. Dietary lipids represent a major source of often unwanted calories. The broad therapeutic approaches to reduce lipid absorption include reduction in the intake, digestion, or absorption of lipids. However, changes in fat intake or absorption modify the form and function of the gastrointestinal tract due to its physiological adaptive processes. For example, food substances may alter the intestinal response to bowel resection, to diabetes mellitus, to abdominal irradiation, or to chronic ethanol intake. Diet-induced changes in membrane composition modulate intestinal transport function as well as brush border membrane (BBM) digestive enzyme activities. The quantity and quality of lipid in the diet alter the activity of enterocyte microsomal lipid metabolic enzymes, the lipid composition of the intestinal BBM, the quantity of fatty acid-binding proteins in the cytosol of the enterocyte, and the transport capabilities of the intestine for lipids, sugars, and amino acids. Changes in dietary lipids also influence the function of other membranes from, for example, the liver, heart, adipocytes, and lymphocytes. The delivery of an altered type or amount of lipid into the colon may lead to a change in the bacterial flora, in bile acid concentrations, and in colonocyte turnover and, therefore, the potential for the tissue to become neoplastic. Thus, modification of the intake or absorption of lipids as a therapeutic end point for the management of patients with obesity may result in adaptive changes in the small intestine or colon, and this adaptation needs to be addressed before the long-term safety of such therapeutic programs can be accepted.

Introduction

The topic of lipid digestion and absorption has been reviewed[1-10] (Fig. 1). The types of dietary lipids may vary widely, with triglycerides enriched

[1] C. M. Bergholz, R. J. Jandacek, and A. B. R. Thomson *Can. J. Gastroenterol.* **5,** 137 (1991).

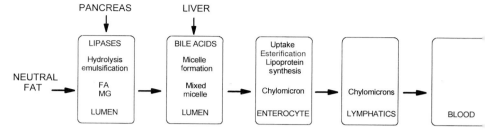

FIG. 1. Logical approach to understanding the steps involved in lipid digestion and absorption. (From Bergholz et al.[8])

with different ratios of saturated (S) to polyunsaturated (P) fatty acids varying widely depending on the diet choice. Because of the cholesterol-raising effect of the ingestion of saturated fatty acids, public awareness of the potential benefit of ingesting a greater proportion of dietary polyunsaturated fatty acids is growing. These may be derived from vegetable or from fish oil sources. The second perspective from which dietary fat content is highlighted in the public mind is in the area of weight control. Because of the energy density of lipids, weight loss programs often feature a reduction in total calories that is arrived at by decreasing the intake of dietary lipids. Apart from self-controlled restriction in total food intake, or the reduced intake of lipids resulting from the use of appetite modifiers, dieting and weight reduction may be achieved by substituting "fake fats" into the diet,[8] as well as by the inhibition of lipid absorption. Inhibition of lipid absorption can be achieved by reducing the digestion of dietary triglycerides by blocking pancreatic lipase activity, by blocking the activity of one of several enterocyte membrane or cytosol lipid binding proteins, or by changing the lipophilic properties of the BBM and thereby reducing the passive permeability properties of the intestine.

---

[2] B. J. Potter, D. Sorrentino, and P. D. Berk, *Annu. Rev. Nutr.* **9,** 253 (1989).
[3] P. Tso, *Physiol. Gastrointest. Tract* **3,** 1867 (1994).
[4] M. C. Carey and O. Hernell, *Semin. Gastrointest. Dis.* **2,** 189 (1992).
[5] E. Levy, *Can. J. Physiol. Pharmacol.* **70,** 413 (1992).
[6] A. B. R. Thomson, M. Keelan, M. L. Garg, and M. T. Clandinin, *Can. J. Physiol. Pharmacol.* **67,** 179 (1989).
[7] T. Ponich, R. N. Fedorak, and A. B. R. Thomson, *Curr. Gastroenterol.* 69 (1990).
[8] C. M. Bergholz, R. J. Jandacek, and A. B. R. Thomson *Can. J. Gastroenterol.* **5,** 137 (1991).
[9] A. B. R. Thomson, C. Schoeller, M. Keelan, L. Smith, and M. T. Clandinin, *Can. J. Physiol. Pharm.* **71,** 531 (1993).
[10] M. T. Clandinin, S. Cheema, C. J. Field, M. L. Garg, J. Venkatraman, and T. R. Clandinin, *FASEB J.* **5,** 2761 (1991).

The use of agents to alter the uptake, absorption, and utilization of lipids is underdeveloped as an approach to the management of the patient with obesity. Although dietary manipulation through education and behavior modification represent the current cornerstone of the management of the patient with obesity, these measures unfortunately are all too frequently unsuccessful in the medium or long term. The rising prevalence of obesity in the Western world is a concern because of the increasing evidence that obesity represents a risk factor for the development and complications of numerous disorders such as hypertension, non-insulin-dependent diabetes mellitus, artherosis of weight-bearing joints, or atheromatous disease of the cardiac, cerebral, and peripheral arteries. It is possible that the pathogenesis of obesity is directly, rather than indirectly, linked to these metabolic disorders. Even surgical procedures continue to be used in an attempt to reduce food intake, produce fat malabsorption, or remove excessive body fat.[11] With the recognition of new appetite regulatory systems [leptin agonists, neuropeptide Y (NPY) or NPY receptor antagonists, galanin antagonists, or serotonergic agents] and the need to achieve a reduction in metabolic risks to diminish the morbidity of these associated conditions, new efforts have focused on the medical treatment of the patient who is overweight or obese.

In this chapter we consider (1) the normal lipid composition of the diet, lipid digestion and absorption, and enterocyte lipoproteins and fatty acid-binding proteins; (2) new advances in the area of reduction of appetite using neurotensin, the ob gene product leptin, and apolipoprotein A-IV; (3) the effects of Olestra (a "fake fat") on the intestine; (4) the role of blocking lipid digestion by inhibiting the activity of pancreatic lipase; (5) methods to reduce the efficiency of the absorption of all nutrients; and (6) the potential risks of a reduced lipid intake or absorption on the small intestine, colon, or other tissues. We refer to data from animal and human studies that will alert us to possible important adverse effects of modifying lipid absorption, including intestinal morphology, the development of vitamin deficiencies, diarrhea, changes in the composition of bile, and possible alterations in colonocyte proliferation and, therefore, the malignant potential on the colon.

Dietary Lipids

The average so-called Western adult consumes about 100 g of triacylglycerol (TG) and 4–8 g of phospholipid [mostly phosphatidylcholine (PC), also known as lecithin] each day, of which about two-thirds are of animal

---

[11] G. Hellers, personal communication (1996).

origin.[10] This intake level represents about 40% of energy as fat, but in recent years fat intake has been reduced in the population and is better generalized as approximately 36% of energy as fat. Long-chain TGs are water-insoluble, nonswelling amphiphiles.[12] In water, TG forms crude unstable emulsion droplets, and PC disperses to form relatively stable concentric lamellar structures called liposomes. When these molecules of TG and PC interact physically in their dietary proportions of about 30:1, they form stable emulsions. The phospholipid acts as an emulsifier by forming a stable monolayer enveloping the TG droplets. Endogenous phospholipid (again, mostly lecithin) of hepatic origin (7–22 g/day) is also secreted into the intestinal lumen via the bile.[13,14] Together with cells sloughed into the intestinal lumen, this represents quantitatively important amounts of phospholipid destined for absorption. Oleate (18:1) and palmitate (16:0) are the major fatty acids present in dietary TG.

Dietary fats contain almost no free fatty acids; instead, these are present in the form of TG, which consists of one molecular of glycerol and three molecules of fatty acid. Dietary fats are mixed TG (i.e., the three positions on the glycerol molecule α, β, or α_1 are occupied by different fatty acids). For many fats, there is a preference for certain fatty acids to occupy positions α, β, or α_1 of the glycerol molecule. In other words, the source of dietary fat dictates the TG structure. For example, in many fats of animal origin, palmitic (16:0) and stearic acids (18:0) are found mainly in the α and α_1 positions, whereas in vegetable oils about 70% of the linoleic acid (18:2) occupied the β position. The position of the particular fatty acids in TG is of importance during the digestion and absorption of fats when monoacylglycerols and diacylglycerols are formed, and this position may also be of importance during the formation of phospholipids in the body.

Approximately 90% of the PC entering the intestinal lumen is derived from bile. Pancreatic phospholipase A_2 (in the presence of bile salts and calcium) hydrolyzes fatty acids from the β position to form lysophosphatidylcholine (LPC) and fatty acids. Chylomicron PC is formed primarily from absorbed luminal PC, and not from *de novo* synthesis.[15] Biliary phospholipid (again primarily PC) is absorbed from the lumen as LPC and is then reacylated in the enterocytes to PC for the formation of chylomicrons.[16–18]

---

[12] D. M. Small, *J. Am. Oil. Chem. Soc.* **45**, 108 (1968).
[13] T. C. Northfield and A. F. Hofmann, *Gut* **16**, 1 (1975).
[14] A. Noma, *J. Biochem.* **56**, 522 (1964).
[15] C. M. Mansbach, *J. Clin. Invest.* **60**, 411 (1977).
[16] O. Nilsson and G. Ronist, *Biochim. Biophys. Acta* **183**, 1 (1969).
[17] R. O. Scow, Y. Stein, and O. Stein, *J. Biol. Chem.* **242**, 4914 (1967).
[18] J. B. Rodgers, R. J. O'Brien, and J. A. Balint, *Am. J. Dig. Dis.* **20**, 208 (1975).

Dietary choline is used for the *de novo* synthesis of PC in the smooth endoplasmic reticulum of the enterocyte.[19]

Cholesterol in the intestinal lumen is both free and esterified. Esterified cholesterol is hydrolyzed by pancreatic cholesterol esterase to form free sterol, in which form the cholesterol is absorbed. Free cholesterol is absorbed and esterified at the endoplasmic reticulum of the enterocyte for formation of lipoproteins and chylomicrons; these then move toward the Golgi apparatus. The fatty acids used for esterification of cholesterol are influenced by the composition of dietary lipids. The Golgi vesicles fuse with the basolateral membrane of the enterocyte to release chylomicrons into the intracellular space, which then pass into lymph or portal blood.

Satiety Factors

Leptin

Two exciting new areas of appetite control relate to leptin and to NPY. The ob gene has been cloned and encodes for a 146-amino acid protein called leptin. Leptin is secreted into the blood exclusively by white fat adipocytes. A leptin receptor has also been cloned, and mRNA encoding for this receptor is present in choroid plexus, hypothalamus, lung, skeletal muscle, and kidney. Circulating leptin acts as an antiobesity agent by restraining appetite and by altering metabolic processes to burn fat. The hypothalamus appears to be a major target tissue for leptin, because binding sites appear to be present there and the intracerebroventricular injection of this hormone leads to a reduction of food intake. Part of the reduction in food intake produced by leptin may be due to a decrease in central NPY. Leptin may be the long sought after satiety factor released from the periphery (i.e., the adipocytes) to regulate long-term body weight. In this hypothesis, as body weight and fat mass increase, more leptin is secreted from adipocytes, appetite is inhibited, and metabolism is increased to bring the fat mass back to a desired set point. The incapacity to express a functional leptin appears to be the cause of obesity in the ob/ob mouse.

Neuropeptide Y

NPY is a 36-amino acid peptide that was originally isolated from porcine brain.[20] NPY is distributed throughout the peripheral and central nervous

[19] P. J. A. O'Doherty, G. Kakis, and A. Kuksis *Lipids* **8,** 249 (1973).
[20] K. Tatemoto, M. Carlquist, and V. Mutt *Nature (London)* **296,** 659 (1982).

system of mammals,[21] where it is costored and coreleased with noradrenaline. Four subtypes of NPY have been cloned (LY_1, LY_2, LY_4, LY_5). NPY is synthesized in the arcuate nucleus (ARC), which produces a dense projection of NPY-containing axons. These axons end mainly in the paraventricular nucleus (PVN) of the hypothalamus,[22] as well as in the dorsomedial nucleus and medial preoptic area.[23,24] These sites are known to be involved in the control of food intake and of energy expenditure.

NPY plays an important role as a physiological signal of feeding. Both food deprivation and restriction stimulate NPY mRNA levels within the ARC and PVN nuclei in rats.[25,26] Centrally (intracerebroventricular) injected NPY induces feeding, even in satiated animals, by increasing the size and duration of the meals.[27] This is accompanied by a reduction of energy expenditure and an increase in insulin release, which in turn facilitate fat deposition and formation of fat stores. NPY overrides peripheral satiety signals, and in the short term the peptide regulates the appetite for carbohydrate in particular: following central administration of NPY, rats prefer a high-carbohydrate diet over a high-fat or high-protein diet.[28]

Repeated central injections of NPY over several days cause hyperphagia, an increased intake of both carbohydrates and lipids, leading to fat deposition and body weight gain. NPY produces increased basal insulinemia and greater insulin response to meals.[29] NPY results in an increase in the *in vivo* insulin-stimulated glucose uptake by adipose tissue, and a marked decrease in glucose uptake by different muscle types. Selective centrally acting NPY antagonists would be expected to reduce food intake and body mass. NPY also stimulates the respiratory quotient, which reflects a diversion of metabolism toward carbohydrate utilization and fat synthesis,[30]

---

[21] T. S. Gray and J. E. Morley, *Life Sci.* **38,** 389 (1986).
[22] F. L. Bai, M. Yamano, Y. Shiotani, P. C. Emson, A. D. Smith, J. F. Powell, and M. Tohyama *Brain Res.* **331,** 172 (1985).
[23] B. J. Morris, *J. Comp. Neurol.* **290,** 358 (1989).
[24] B. M. Chronwall, D. A. Dimaggio, V. J. Massari, V. M. Pickei, D. Ruggiero, and T. L. O'Donohue, *Neuroscience* **15,** 1159 (1985).
[25] R. D. O'Shea and A. L. Gundlach, *J. Neuroendocrinol.* **3,** 11 (1991).
[26] L. S. Brady, M. A. Smith, P. W. Gold, and M. Herkenham, *Neuroendocrinology* **52,** 441 (1990).
[27] J. E. Morley, A. S. Levine, B. A. Gosnell, J. Kneip, and M. Grace *Am. J. Physiol.* **252,** R599 (1987).
[28] B. G. Stanley, D. R. Daniel, A. S. Chin, and S. F. Leibowitz, *Peptides* **6,** 1205 (1985).
[29] N. Zarjevski, I. Cusin, R. Vettor, F. Rohner-Jeanrenaud, and B. Jeanrenaud, *Diabetes* **43,** 764 (1994).
[30] J. A. Menendez, I. S. McGregor, P. A. Healey, D. M. Atrens, and S. F. Leibowitz, *Brain Res.* **516,** 8 (1990).

and it also decreases brown fat thermogenesis.[31] NPY reduces energy expenditure specifically by controlling sympathetic activity to intrascapular brown adipose tissue.[32]

The factors that regulate hypothalamic NPY are not known. Proposed modulatory agents include insulin, glucocorticoids, sex steroid, and corticotropin-releasing factor.[33] Insulin injected into the third ventricle prevents the rise in NPY mRNA in the ARC in food-deprived animals, suggesting that insulin may directly inhibit NPY gene expression.[34] Insulin's effect on hypothalamic NPY may be more complex, because hyperinsulinemia, although it maintains euglycemia, results in positive energy balance and does not inhibit NPY synthesis in the ARC.[35] Hypothalamic NPY mRNA and NPY protein content are increased in obese animals, in insulin-dependent diabetes, with intense exercise, and with lactation.[36,37] Under all of these conditions, increased hypothalamic concentration of NPY is followed by increased appetite.

NPY administration to normal rats produces a hormonal metabolic situation that is similar to that reported in the dynamic phase of the genetically obese fa/fa rat.[38] NPY is considered to be a potential link between the hyperinsulinemia, hypercorticosteronemia, and insulin resistance found in obese animals. NPY could be of primary importance in the establishment of obesity syndromes with incipient insulin resistance.[29] It has been postulated[39] that an increment in the hypothalamic levels of NPY in genetically obese animals may be caused by alterations in cerebral glucose utilization in the nuclei that regulate the autonomic nervous system, as well as the feeding and various endocrine systems. Thus, it is possible that hypothalamic NPY may contribute to the development of eating disorders and obesity.

---

[31] C. J. Billington, J. E. Briggs, M. Grace, A. S. Levine, *Am. J. Physiol.* **260,** R321 (1991).

[32] M. Egawa, H. Yoshimatsu, and G. A. Bray, *Am. J. Physiol.* **260,** R328 (1991).

[33] S. F. Leibowitz, *Ann. N.Y. Acad. Sci.* **739,** 12 (1994).

[34] M. W. Schwarz, A. J. Sipols, J. L. Marks, G. Sanacora, J. D. White, A. Scheurinck, S. E. Kahn, D. G. Baskin, S. C. Woods, D. P. Figlewica, and D. Porte Jr., *Endocrinology* **130,** 3608 (1992).

[35] I. Cusin, S. Dryden, Q. Wang, F. Rohner-Jeanrenaud, B. Jeanrenaud, and G. Williams, *J. Neuroendocrinol.* **7,** 193 (1995).

[36] J. D. White, D. Olchowvsky, M. Kershaw, and M. Berlowitz *Endocrinology* **126,** 765 (1990).

[37] S. P. Kalara, M. G. Dube, A. Sahu, C. P. Phelps, and P. S. Kalara, *Proc. Natl. Acad. Sci. U.S.A.* **38,** 10931 (1991).

[38] R. Vettor, N. Zarjevski, I. Cusin, F. Rohner-Jeanrenaud, and B. Jeanrenaud, *Diabetologia* **37,** 1202 (1994).

[39] B. Jeanrenaud, *Biobetologia* suppl **2,** S170 (1994).

Cholecystokinin

Lipids in the diet stimulate the release of several peptides, some of which influence appetite and therefore food intake.[40-46] Fat storage and metabolism are important for the long-term control of energy balance.[47] Dietary fats affect the short-term control of food intake during a test meal,[48] possibly as a result of the release of cholecystokinin (CCK). Fats in the small intestine release CCK,[49,50] and in humans unsaturated fats are particularly potent in releasing CCK.[51] Two types of vagal afferent fibers are sensitive to dietary lipids, and these have a preabsorptive satiating effect.[52-54] The abdominal vagus has CCK_A receptors,[55] so that intestinal fats activate vagal afferent fibers and also release CCK, thereby modifying food intake. Furthermore, pretreatment with potent, specific type A CCK antagonists decreases the satiating effects of fats in rats and pigs.[56-59] However, other workers have failed to show an effect of the type A CCK antagonist, loxiglumide, on food intake produced by infusion of corn oil into the jejunum of humans.[60] Clearly, more work needs to be performed before preparing a CCK designer drug as a means to control food intake.

[40] G. W. Aponte, I. L. Taylor, and A. H. Soll, *Am. J. Physiol.* **254,** G829 (1988).
[41] J. J. Holst, *Digestion* **17,** 168 (1978).
[42] Y. Koda, A. Wada, N. Yanagihara, Y. Uezono, and F. Izumi, *Neuroscience* **20,** 495 (1989).
[43] R. Liddle, I. Goldfine, M. Rosen, R. Taplitz, and J. Williams, *J. Clin. Invest.* **7,** 1144 (1985).
[44] A. Ohneda, A. Yanbe, Y. Maruhama, S. Ishi, Y. Kai, R. Abe, and S. Yamagata, *Gastroenterology* **68,** 715 (1975).
[45] R. Stubbs and B. Stabile, *Am. J. Physiol.* **248,** G347 (1985).
[46] I. Taylor, *Gastroenterology* **88,** 731 (1985).
[47] E. Scharrer and W. Langhans, in "The Control of Body Fat Content" (J. M. Forbes and G. R. Hervey, eds.), pp. 63–86, Smith-Gordon, London, 1990.
[48] I. Welch, K. Saunders, and N. W. Read, *Gastroenterology* **89,** 1293 (1985).
[49] I. McL. Welch, C. P. Sepple, and N. W. Read, *Gut* **29,** 306 (1988).
[50] C. G. Nicholl, M. J. Polak, and S. R. Bloom, *Annu. Rev. Nutr.* **5,** 213 (1985).
[51] K. Beardshall, G. Frost, Y. Morarji, S. Domin, S. R. Bloom, and J. Calam, *Lancet* **2,** 1008 (1989).
[52] J. Melone, *J. Auton. Nerv. Syst.* **17,** 331 (1986).
[53] D. P. Yox, H. Stokeberry, and R. C. Ritter, *Am. J. Physiol.* **260,** R503 (1991).
[54] D. P. Yox, H. Stokesberry, and R. C. Ritter, *Am. J. Physiol.* **260,** R681 (1991).
[55] T. H. Moran, G. P. Smith, and A. M. Hostetler, *Brain Res.* **415,** 149 (1987).
[56] R. D. Reidelberger, T. J. Kalogeris, P. M. B. Leung, and V. E. Mendel, *Am. J. Physiol.* **244,** R872 (1983).
[57] D. Greenberg, G. P. Smith, and J. Gibbs, *Am. J. Physiol.* **259,** R110 (1990).
[58] D. P. Yox, H. Stokesberry, and R. C. Ritter, in "The Neuropeptide Cholecystokinin (CCK), Anatomy and Biochemistry, Receptors, Pharmacology and Physiology," (J. Hughes and G. Dokray, eds.), pp. 218–222, Ellis Horwood, Chichester, 1989.
[59] P. C. Gregory and D. V. Rayner. The influence of gastrointestinal infusion of fats on regulation of food intake in pigs. *J. Physiol.* **385,** 47 (1987).
[60] J. Drewe, A. Gadient, L. C. Rovati, and C. Beglinger, *Gastroenterology* **102,** 1654 (1992).

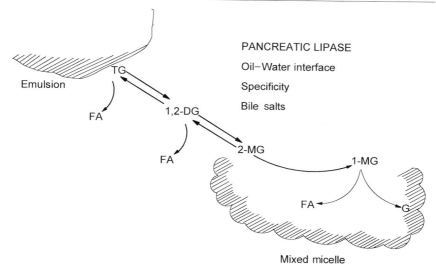

FIG. 2. The fatty acids (FA) in the α and α_1 position of triglyceride (TG) are removed by pancreatic lipase, leaving diglyceride (DG) and monoglyceride (MG). These are solubilized in the micelles, which serve as a means of moving the lipid digestive products across the intestinal unstirred water layer and into the enterocyte. (From Thomson.[340])

It is not known if dietary modifications (e.g., calorie restriction or low-fat diets) influence the secretion release or activity of leptin or NPY.

Fat Substitutes

A recently approved palatable "fake fat" is Olestra. Olestra is the generic name for the mixture of hexa-, hepta-, and octaesters of sucrose formed with long-chain fatty acids isolated from vegetable sources. Olestra has a sucrose core in place of the glycerol core of triglyceride, and is esterified with six to eight fatty acids rather than three (Fig. 2). These sucrose polyesters are not hydrolyzed by pancreatic lipase and, therefore, are not absorbed.

The discovery of Olestra resulted from research on lipid digestion and absorption. Mattson and Volpenhein[61] synthesized a series of polyols esterified with one to eight molecules of oleic acid, and investigated the rate of *in vitro* hydrolysis of each of these polyols by pancreatic lipolytic enzymes. When a mixture of rat pancreatic and bile fluid was added to esterified polyols, there was a sharp decrease in the rate of hydrolysis of polyols

---

[61] F. H. Mattson and R. A. Volpenhein, *J. Lipid Res.* **13**, 325 (1972).

having more than three fatty acid esters. Indeed, no measurable hydrolysis was observed for sorbitol esterified with six fatty acids, or for sucrose esterified with eight fatty acids. Steric hindrance due to the larger bulk of the Olestra molecule accounts for this lack of hydrolysis by lipases. Fat balance and absorption studies in thoracic duct-cannulated rats have shown that sucrose octaoleate is not absorbed.[62,63]

Olestra can be used interchangeably for fat in a wide variety of foods.[64] In contrast to the higher esters, the mono-, di-, and triesters of sucrose are fully digested and absorbed, and are used in foods as emulsifiers and to coat fruit. Because the physical properties of Olestra are similar to those of conventional triglycerides, Olestra can be blended with conventional fats and used in cooking, baking, or deep frying. Similarly, the heating and storage stabilities are comparable to those of conventional fat. Many foods made with Olestra have the same taste and texture as those made from fat from natural sources. Extensive results of research on absorption, chronic toxicity, carcinogenicity, reproduction, nutrition, and gastrointestinal physiology support the safety of Olestra in foods.[65–70] Because Olestra is not digested and is not absorbed, the gastrointestinal tract is the only organ system that is exposed.

The lack of absorption of Olestra is an important consideration in the assessment of the safety of Olestra and its effects on the gastrointestinal tract. Evidence that Olestra is not metabolized in the intestinal lumen or absorbed comes from animal studies with radiolabeled Olestra, from fat balance studies in animals and humans, and from the analysis of tissues from monkeys and rats in long-term feeding studies.[62,65,67] Olestra does not affect fecal or biliary profiles of bile acids, indicating that Olestra has no effect on microbial deconjugation or on the biotransformation of primary to secondary bile acids. Olestra is excreted unchanged in the feces, indicating that it is not metabolized by microorganisms in the colon, and the lack of an effect of Olestra on fecal water, electrolytes, and pH provides additional evidence that Olestra does not affect the colonic microflora.[65]

---

[62] F. H. Mattson and G. A. Nolen, *J. Nutr.* **102**, 1171 (1972).
[63] F. H. Mattson and R. A. Volpenhein, *J. Nutr.* **102**, 1177 (1972).
[64] C. A. Bernhardt, *Food Technol. Int.* **176**, 178 (1988).
[65] R. W. Fallat, C. J. Glueck, R. Lutmer, and F. H. Mattson, *Am. J. Clin. Nutr.* **29**, 1204 (1976).
[66] C. J. Glueck, F. F. Mattson, and R. J. Jandacek, *Am. J. Clin. Nutr.* **32**, 1636 (1979).
[67] F. E. Wood, E. J. Hollenbach, and R. M. Kaffenberger, *Food Chem. Toxicol.* **29**, 231 (1991).
[68] F. E. Wood, W. J. Tierney, A. L. Knezevich, H. F. Bolte, J. K. Maurer, and R. D. Bruce, *Food Chem. Toxicol.* **29**, 223 (1991).
[69] G. A. Nolen, F. E. Wood, and T. A. Dierckman, *Food Chem. Toxicol.* **25**, 1 (1987).
[70] M. R. Adams, M. R. McMahan, F. H. Mattson, and T. B. Clarkson, *Proc. Soc. Exp. Biol. Med.* **167**, 346 (1981).

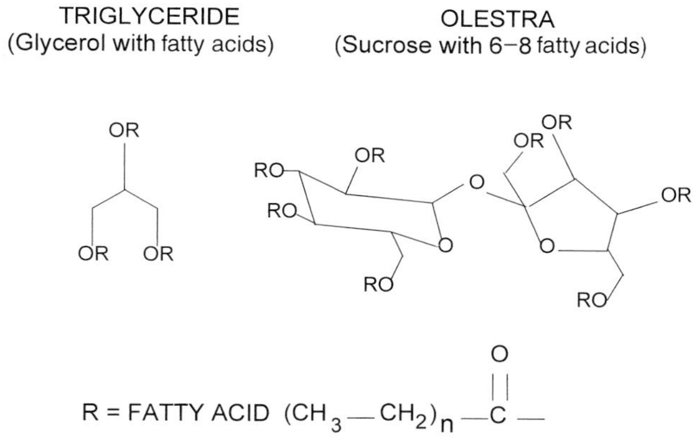

TRIGLYCERIDE
(Glycerol with fatty acids)

OLESTRA
(Sucrose with 6–8 fatty acids)

$$R = \text{FATTY ACID } (CH_3-CH_2)_n-\overset{O}{\underset{\|}{C}}-$$

FIG. 3. Comparison of the structure of a triglyceride molecule with that of Olestra. (From Berghy et al.[341])

Eating Olestra does not increase the load of unabsorbed lipid delivered to the small intestine and colon, but may result in occasional mild gastrointestinal adverse effects. At higher intakes of Olestra, there is a small decrease in the bioavailability of neutral sterols due to their partitioning into Olestra.[71] Thus, Olestra has the potential to decrease the absorption of lipophilic molecules.[72] Deficiencies of fat-soluble vitamins potentially may occur at very high levels of intake of Olestra, and supplementation of Olestra with some fat-soluble vitamins may be necessary, particularly in high-risk population groups. It remains to be established whether the consumption of Olestra-containing foods will lead to significant weight reduction, or will simply increase the dietary choices for a fatty food conscious public. The latter is likely to be the case, since the use of Olestra initially will be only for snack foods, so that the total reduction in fat intake will be limited by the availability of foods containing Olestra.

Digestion

Most lipid absorption occurs in the proximal small intestine, and normally about 95% of dietary fat is absorbed before the small intestinal contents enter the colon. The main lipid-digesting enzyme, pancreatic lipase, removes fatty acids from the α and α' position of dietary triglycerides, yielding the lipolytic product β-monoglyceride and long-chain saturated and polyunsaturated fatty acids (Fig. 3). These lipolytic products are solubi-

[71] R. J. Jandacek, F. H. Matatson, S. McNeely, L. Gallon, R. Yunker, and C. J. Glueck, *Am. J. Clin. Nutr.* **33**, 251 (1980).
[72] R. J. Jandacek, *Drug Metab. Rev.* **13**, 695 (1982).

lized in bile salt micelles, diffuse across the intestinal unstirred water layer, permeate the intestinal brush border membrane, and enter the cytosol of the enterocyte. In the enterocytes, the microsomal enzymes lead to the reformation of triglycerides, to the reesterification of cholesterol, and to the formation of lipoproteins from the apolipoproteins synthesized by the enterocyte (apo A-in, apo A-IV, and apo B). The lipoproteins then diffuse across the basolateral membrane of the enterocyte into the intestinal lymphatics or portal blood. Interestingly, in the rat, up to 39% of intestinally infused lipid is not transported via the lymph, but instead enters the portal vein.[73] The lipid in the portal vein is both absorbed fat as well as endogenous lipid, with the latter predominating.[73,74]

An immunologically distinct lingual lipase has been described that digests approximately one-third of ingested fat.[75,76] This lipase is derived from the serous gland located beneath the circumvallate papillae. Gastric lipase is secreted in response to mechanical stimulation, ingestion of food, the presence of a fatty meal, or by sympathetic agonists. Gastric lipolysis may be of physiological importance in the provision of fatty acids needed to trigger pancreatic lipase activity in the intestine, and preduodenal lipases are of importance for fat absorption in a variety of physiological and pathological conditions associated with pancreatic insufficiency.[77-80]

The activity of pancreatic lipase is inhibited by the presence of bile acids in concentrations above their critical micellar concentrations (CMC). This inhibition of lipase by bile acids is overcome by the presence of pancreatic colipase, which acts by attaching to the ester bond region of the TG molecule.[81] Pancreatic lipase strongly binds to colipase by electrostatic interactions, and allows hydrolysis of TG by lipase even in the presence of bile acids. Thus, intestinal fat digestion occurs as a result of the combined action of lipase and colipase.[82] Colipase is secreted by the exocrine pancreas

[73] C. M. Mansbach II and R. F. Dowell, *Am. J. Physiol.* **261,** G927 (1992).
[74] C. M. Mansbach II and S. Partharsarathy, *J. Lipid Res.* **23,** 1009 (1982).
[75] M. Hamosh, *Gastroenterology* **90,** 1290 (1986).
[76] M. Hamosh and R. O. Scow, *J. Clin. Invest.* **52,** 88 (1973).
[77] C. Figarella, A. De Caro, P. Deprez, M. Bouvry, and J. J. Bernier, *Gastroenterol. Clin. Biol.* **3,** 43 (1979).
[78] H. Hildebrand, B. Borgstrom, A. Bekassy, C. Erlanson-Albertsson and I. Helin, *Gut* **23,** 254 (1982).
[79] C. K. Abrams, M. Hamosh, V. S. Hubbard, S. K. Dutta, and P. Hamosh, *J. Clin. Invest.* **73,** 374 (1984).
[80] C. K. Abrams, M. Hamosh, S. K. Dutta, V. S. Hubbard, and P. Hamosh, *Gastroenterology* **92,** 125 (1987).
[81] B. Borgstrom, *FEBS Lett.* **71,** 201 (1976).
[82] B. Borgstrom, C. Erlanson-Albertsson, *in* "Lipases" (B. Borgstrom, and H. L. Brockman, eds.), p. 151, Elsevier, Amsterdam, 1984.

as the precursor, procolipase.[83] The procolipase propeptide (APGPR) may act as a satiety signal in humans.[84] The lipolytic products then distribute themselves between aqueous, oil, and intermediate phases. Within the aqueous phase, there are large micelles with hydrodynamic radii of 200 Å, which are saturated with mixed lipids and cholesterol; there is also a second phase containing unilamellar vesicles (liposomes, with radii of 400–600 Å), which are saturated with bile acids.[85]

Both pancreatic colipase and phospholipase A_2 are secreted in a precoenzyme and proenzyme form, respectively, and these enzymes require activation in the intestinal lumen by tryptic hydrolysis. Pancreatic lipase, colipase, phospholipase A_2, calcium, and bile salts act synergistically in the lipid hydrolytic reactions of the upper small intestine. In addition to the inhibiting effect of bile acids, the hydrolysis of TG by pancreatic lipase is inhibited by phospholipids and protein, which prevent the action of pancreatic lipase at the oil/water interface. Although pancreatic lipase is nearly completely inhibited by bile salts above their CMC, lingual lipase still exhibits some activity near this bile salt concentration. Gastric lipase accounts for the hydrolysis of 10–40% of dietary fat.[75] In fact, as much as 50–70% of dietary fat can be absorbed in the absence of pancreatic lipase, possibly because of the action of lingual lipase.[75] This, therefore, potentially limits the nutritional impact of the inhibition of lipid absorption that could result from the reduction in the activity of just pancreatic lipase.

THL (tetrahydrolipstatin orlistat, Hoffmann-LaRoche, Inc., New Jersey) has been developed to inhibit the action of gastrointestinal lipase.[86] THL is a chemically synthesized derivative of lipostatin, a natural product of *Streptomyces toxytricini*. THL is a potent inhibitor of intestinal lipase, which may be useful to limit fat absorption.[87] THL inhibits both gastric and pancreatic lipase, which reduces the hydrolysis of dietary triglycerides, and thereby limits the availability of free fatty acids available for absorption. Lipase is progressively inactivated through the formation of a long-lived covalent intermediary. THL inhibits liver esterase, but has no effect on trypsin, chymotrypsin, amylase, or pancreatic phospholipase A_2. Thus, normal hydrolysis and absorption of phospholipids occurs, and the hydrolysis and absorption of proteins and carbohydrates is also normal.

[83] J. S. Patton, P. A. Albertsson, C. Erlanson, and B. Borgstrom, *J. Biol. Chem.* **253,** 4195 (1978).
[84] R. C. Bowyer, W. M. Rowston, A. M. T. Jehanli, J. H. Lacey, J. Hermon-Taylor, *Gut* **34,** 1520 (1993).
[85] M. C. Carey, D. M. Small, and C. M. Bliss, *Annu. Rev. Physiol.* **45,** 651 (1983).
[86] J. B. Hauptman, F. S. Jeunet, and D. Hartmann, *Am. J. Clin. Nutr.* **55,** 3095 (1992).
[87] P. Hadvary, H. Lengsfeld, and H. Wolfer, *J. Biochem.* **256,** 357 (1988).

In subchronic experiments with lean and obese mice and with rats fed moderately high-fat diets, the magnitude of inhibition of fat absorption with THL remains relatively constant throughout the treatment period.[87–90] In diet-induced obese rats, THL increases the weight of the small and large intestines.[90] It is unknown whether this is associated with alterations in the percentage of intestinal wall comprised of mucosa, with changes in the intestinal surface area, or whether there were compensatory changes in the absorption of nutrients. With prolonged treatment of genetically obese mice and rats at a similar magnitude of inhibition of fat absorption, intestinal weights remained unaffected.[89]

With the moderate inhibition of pancreatic lipase activity with THL, modest body weight loss is achieved in a dose-dependent manner, above and beyond the weight loss achieved with diet alone; overall THL is well tolerated.[91] Dividing the dose or taking a single dose of orlistat gives a similar degree of increased food fat excretion; for example, 80 mg orlistat produces approximately 36% excretion of a fat intake of 76 g/day in healthy males.[92] Dietary fiber content and accessibility of fat (intracellular versus extracellular fat) do not influence the inhibitory effect of orlistat on lipid absorption.[93] Orlistat needs to be taken with meals to reduce the lipase activity in the intestinal lumen, which is stimulated by the intake of food. The issue of the possible need for supplementation of orlistat with fat-soluble vitamins needs to be established.

Absorption

Unstirred Water Layers, Membrane Permeation, and Diet Effects

The topic of lipid absorption has been reviewed.[9] While most lipid uptake is believed to be by a process of passive diffusion, a portion of cholesterol and phospholipid uptake may be protein mediated.[94] A new model has been proposed to explain the uptake of long-chain fatty acids

---

[88] P. Hadvary, W. Sidler, W. Meister, W. Vetter, and H. Wolfer, *J. Biol. Chem.* **266,** 2021 (1991).

[89] M. K. Meier, D. Blum-Kaelin, K. Bremer, D. Isler, R. Joly, P. Keller-Rupp, H. Lengsveld, and F. Hoffman, *Int. J. Obes.* **15**(suppl 1), 31 (1991).

[90] S. Hogan, P. Fleury, P. Hadvary, H. Lengsfeld, M. K. Meier, J. Triscari, and A. C. Sullivan, *Int. J. Obes.* **11,** 35 (1987).

[91] M. L. Drent, I. Larsson, T. William-Olsson, F. Quaade, F. Czubayko, K. von Bergmann, W. Strobel, L. Sjostrom, and E. A. van der Veen, *Int. J. Obesity* **19,** 221 (1995).

[92] Y. Hussain, C. Guzelhan, J. Odink, E. J. van der Beek, and D. Hartmann, *J. Clin. Pharmacol.* **34,** 1121 (1994).

[93] C. Guzelhan, J. Odink, J. J. N. Jansen-Zuidema, and D. Hartmann, *J. Int. Med. Res.* **22,** 255 (1994).

[94] H. Thurnhofer and H. Hauser, *Biochemistry* **29,** 2142 (1990).

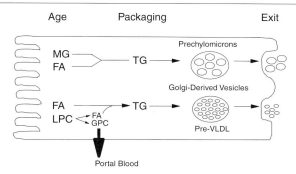

FIG. 4. Packaging and secretion of intestinal chylomicrons (CM) and very low density lipoproteins (VLDL) by enterocytes of small intestine. Absorbed monoglyceride (MG) and fatty acid (FA), major digestion products of triglyceride (TG), are reconstituted in the cell to form TG by the monoglyceride pathway, which is then subsequently packaged into predominantly pre-CM particles. This pathway is inhibited by the presence of Pluronic L-81. In contrast, absorbed FA and also FA derived from hydrolysis of absorbed lysolecithin are used to form TG via the α-glycerophosphate pathway, which is then packaged into pre-VLDL particles. Unlike formation of pre-CM, this pathway is not sensitive to Pluronic L-81. LPC, lysophosphatidylcholine. (Adapted from Tso and Simmonds.[220])

in the intestine (Fig. 4): the sodium/hydrogen exchangers (NHE) in the intestinal BBM are responsible for acidifying the unstirred water layer adjacent to the BBM.[95] This facilitates the partitioning of fatty acids out of the bile salt micelles, their protonation, and hence their greater permeation across the BBM. In addition, some fatty acid uptake is facilitated by the fatty acid-binding protein in the BBM (FABPpm). This protein is likely identical to mitochondrial aspartate aminotransferase.[96,97] Inhibition of NHE or FABPpm results in an approximate 30–40% decline in the uptake of fatty acids.[95,98] Another membrane protein has recently been described, the fatty acid transporter FAT, which has a molecular mass of 88 kDa, two putative transmembrane segments, and is extensively glycosylated.[99] FAT is highly expressed in the BBM at the tip of the jejunum villus and is

[95] C. Schoeller, M. Keelan, G. Mulvey, W. Stremmel, and A. B. R. Thomson, *Biochimica. Biophys. Acta* **1236,** 51 (1995).
[96] L. M. Isola, S. L. Zhou, C. L. Kiang, D. D. Stump, M. W. Bradbury, and P. D. Berk, *Proc. Natl. Acad. Sci. U.S.A.* **92,** 9866 (1995).
[97] S. L. Zhou, D. Stump, C. L. Kiang, L. M. Isola, and P. D. Berk, *Proc. Soc. Exp. Biol. Med.* **208,** 263 (1995).
[98] C. Schoeller, M. Keelan, G. Mulvey, W. Stremmel, and A. B. R. Thomson, *Clin. Invest. Med.* **18,** 380 (1995).
[99] N. A. Abumrad, M. R. Maghrabi, E. Z. Amri, E. Lopez, and P. A. Grimaldi, *J. Biol. Chem.* **268,** 17665 (1993).

strongly stimulated by a high-fat diet.[100] It is possible, therefore, that lipid uptake may be manipulated therapeutically by inhibiting NHE or FABPpm, or FAT.

Because lipid uptake has been viewed as being a passive process, attention has focused largely on ways to alter the lipid permeability properties of the intestine. This can be achieved by altering the lipophilic characteristics of the BBM. One promising means by which this may be accomplished is by a change in the fatty acids in the triglycerides in the diet: The intestinal transport of nutrients can be varied by isocaloric variations in the composition of individual dietary triglycerides.[101] Feeding rats a semisynthetic diet containing increased amounts of palmitic (16:0), stearic (18:0), and oleic (18:1) acids is associated with enhanced intestinal uptake of glucose, as well as increased uptake of fatty acids and cholesterol, as compared with the rates of uptake for animals fed an isocaloric diet containing increased amounts of linoleic (18:2) and linolenic (18:3) acids.[102]

Alterations in the uptake of lipids in response to changes in dietary fats may be mediated by variations in the physical properties of the BBM, which in turn might be influenced by changes in BBM total phospholipids, phospholipid fatty acids, phospholipid polar head groups, or in cholesterol content.[103] Such changes in BBM lipid composition have been obtained experimentally in homeotherms by feeding different dietary fats.[104,105] In models of intestinal adaptation (i.e., diabetes, aging, bowel resection, chronic ethanol intake, abdominal irradiation) where nutrient uptake changes, there are also modifications in the BBM phospholipids, enterocyte microsomal membrane (EMM) lipids, and EMM/BBM lipid metabolic enzyme activities.[106–112] Interestingly, the diet-induced changes in nutrient uptake are not associated with alterations in the BBM content of cholesterol

---

[100] H. Poirier, P. Degrace, I. Niot, A. Bernard, and P. Besnard, *Eur. J. Biochem.* **238**(2), 368 (1996).
[101] A. B. R. Thomson, M. Keelan, M. Garg, and M. T. Clandinin, *Can. J. Physiol. Pharmacol.* **65**, 2459 (1987).
[102] A. B. R. Thomson, M. Keelan, M. T. Clandinin, and K. Walker, *J. Clin. Invest.* **77**, 279 (1986).
[103] B. Fourcans and M. K. Jain, *Adv. Lipid Res.* **12**, 147 (1974).
[104] S. M. Innis and M. T. Clandinin, *Biochem. J.* **193**, 155 (1981).
[105] S. M. Innis and M. T. Clandinin, *Biochem. J.* **198**, 167 (1981).
[106] M. Keelan, K. Walker, and A. B. R. Thomson, *Compar. Biochem. Physiol.* **82**, 83 (1985).
[107] M. Keelan, K. Walker, and A. B. R. Thomson, *Mech. Aging Dev.* **31**, (1985).
[108] M. Keelan, K. Walker, and A. B. R. Thomson, *Can. J. Physiol. Pharmacol.* **63**, 1528 (1985).
[109] M. Keelan, K. Walker, and A. B. R. Thomson, *Can. J. Physiol. Pharmacol.* **63**, 1312 (1985).
[110] M. Keelan, C. I. Cheeseman, K. Walker, and A. B. R. Thomson, *Radiat. Res.* **105**, 84 (1986).
[111] M. Keelan, K. Doring, M. Tavernini, E. Wierzbicki, M. T. Clandinin, and A. B. R. Thomson, *Lipids* **29**, 851 (1994).
[112] M. Keelan, M. T. Clandinin, K. Walker, and A. B. R. Thomson, *Diabetes Res.* **14**, 165 (1990).

or phospholipids,[113] although there may be changes in the fatty acyl constituents in the BBM.[112,114]

Feeding isocaloric polyunsaturated fatty acid diets lowers lipid and sugar uptake into the intestine,[102] and these diets also have the same effects in rats with experimental diabetes mellitus, aging, or chronic ethanol ingestion, or following distal small bowel resection.[115] It is clear that the impact of any intervention aimed at reducing lipid absorption may be modified by the lipid composition of the diet, and that any such intervention may also change the intestinal absorption of other calorie-providing nutrients. Thus, the overall efficacy of a therapeutic intervention aimed to achieve weight reduction by the inhibition of calorie intake or of fat absorption may be limited by the associated adaptive processes that occur in the intestine in response to the alterations in nutrient intake or absorption.

Dietary levels of fat affect alkaline phosphatase activities in the intestine, lymph, and serum.[116,117] High-fat diets lead to greater concentrations of chymotrypsinogen and trypsinogen in the pancreas than do high-carbohydrate diets, with parallel changes in enzyme activity in pancreatic juice obtained from animals on the high-fat and high-carbohydrate diets.[118] Studies correlating increases in mucosal activities of lipid reesterifying enzymes with enhanced fat absorption suggest that the enzymatic adaptations that occur in the intestinal mucosa are biologically appropriate.[119,120] However, dietary alterations also lead to effects that are much less easy to explain,[121] such as the lower BBM disaccharidase activities with high-fat diets, the depression of peptide hydrolases and alkaline phosphatase activities with high-starch diets, and the induction of many disaccharidases by starch (which contains only glucose monosaccharidases). Some of these changes may be due to indirect nutritional effects on the metabolism, growth, and differentiation of intestinal epithelial cells, rather than to direct effects on their digestive activity. Other alterations may be due to changes in pancreatic function, or to variations in the proportion of carbohydrates in the diet. The multiplicity of enzyme changes and their regional nature (i.e.,

[113] T. A. Brasitus, N. O. Davidson, and D. Schachter, *Biochim. Biophys. Acta* **812,** 460 (1985).
[114] M. Keelan, A. A. Wierzbicki, M. T. Clandinin, K. Walker, R. V. Rojotte, and A. B. R. Thomson, *Diabetes Res.* **14,** 159 (1990).
[115] A. B. R. Thomson, M. Keelan, and G. E. Wild, *Clin. Invest. Med.* **19,** 331 (1996).
[116] M. Kaplan, *Gastroenterology* **62,** 452 (1972).
[117] N. Dickie, M. I. Robinson, and J. Tuba, *Can. J. Biochem. Physiol.* **33,** 83 (1955).
[118] J. T. Snook, *Am. J. Physiol.* **221,** 1381 (1971).
[119] C. A. Drevon, C.-C. Lilljeqvist, B. Schreiner, and K. Norum, *Atherosclerosis* **34,** 207 (1979).
[120] A. Singh, J. A. Balint, R. H. Edmonds, and J. B. Rodgers, *Biochim. Biophys. Acta* **260,** 708 (1972).
[121] D. M. McCarthy, J. A. Nicholson, and Y. S. Kim, *Am. J. Physiol.* **239,** G445 (1980).

jejunum vs ileum) argues against regulation by a single hormone, even though complex interactions of several hormones cannot be excluded as a possible mechanism to explain these diet-related alterations.

The intestinal uptake of hexoses and lipids is influenced by dietary lipid modification.[122,123] Feeding a low-fat diet increases the intestinal uptake of glucose and galactose.[124] Thus, quantitative as well as qualitative changes in dietary lipids (as would result from appetite suppression; dieting; dietary manipulation of macronutrients; use of "fake fats"; inhibition of pancreatic lipase; inhibition of NHE, FABPpm, or fatty acid-binding proteins in the cytosol; blocking enterocyte microsomal lipid metabolic enzymes; or interfering with lipoprotein metabolism) could all result in adaptive changes in the intestine, which might partially offset the expected alterations in lipid absorption, either by increasing the efficiency of lipid uptake, or the efficiency of uptake of carbohydrates or calorie-providing amino acids. After 6 weeks of feeding rabbits a high- or a low-carbohydrate diet, the uptake of lipid is altered, but the direction and extent of changes is different between the jejunum and the ileum.[125] The finding of lower jejunal cholesteral uptake in animals fed a high-carbohydrate as compared with a low-carbohydrate diet reflects the importance of dietary effects on intestinal permeation. Jejunal villous height, villous surface area, and mucosal surface area are higher in rabbits fed a low- as compared with a high-fat diet.[126] Feeding a low-fat diet reduces jejunal uptake of glucose. Modifying the dietary content of long-chain-length fatty acids is associated with changes in the DNA content of the intestine.[127] Thus, the long-term effect of changes in lipid uptake on intestinal morphology and function must be considered carefully.

Cholesterolemic and atherogenic implications are associated with modifying the dietary content of cholesterol, lipid, protein, carbohydrate, and calories.[128–130] In view of the sensitivity of the intestine to changes in the transport of nutrients in animals fed diets containing low or high amounts of fat, studies must now be performed to establish the levels of dietary fats,

[122] A. B. R. Thomson and R. V. Rajotte, *Am. J. Clin. Nutr.* **37**, 244 (1983).
[123] A. B. R. Thomson and R. V. Rajotte, *Am. J. Clin. Nutr.* **37**, 394 (1983).
[124] A. B. R. Thomson, M. Keelan, Y. McIntyre, and J. MacLeod, *Comp. Biochem. Physiol.* **84A**, 89 (1986).
[125] A. B. R. Thomson, *Dig. Dis. Sci.* **31**, 193 (1986).
[126] A. B. R. Thomson, Y. McIntyre, J. MacLeod, and M. Keelan, *Digestion* **35**, 78 (1986).
[127] C. L. Morin, V. Ling, and D. Bourassa, *Dig. Dis.* **25**, 123 (1980).
[128] R. M. G. Hamilton and K. K. Carroll, *Atherosclerosis* **24**, 47 (1976).
[129] D. Kritchevsky, S. A. Tepper, H. K. Kim, J. A. Story, D. Vesselinovitch, and R. W. Wissler, *Exp. Molec. Pathol.* **24**, 375 (1976).
[130] D. Vesselinovitch, G. S. Getz, R. H. Hughes, and R. W. Wissler, *Atherosclerosis* **20**, 303 (1974).

cholesterol, and calories that minimize these diet-related changes in plasma lipids. For example, can the reduced uptake of cholesterol observed in animals fed a high-fat diet be altered further if the animals are also fed a low-carbohydrate diet?[125] These questions are of particular value in the assessment of the importance of dietary manipulation in humans, and must be addressed during the establishment of the efficacy and safety of agents intended to be used therapeutically to modify lipid absorption.

When rats are fed a high-fat diet for 4 weeks, the *in vitro* uptake of oleic (18:1) acid is increased, and the activity of acyl-CoA-monoglyceride acyltransferase is enhanced.[131] The uptake of oleic acid in perfused rat ileum also rises after 4 weeks of feeding a high-fat diet, but the lipid content of the mucosa and the degree of reesterification of absorbed oleic acid are unaltered.[132] This implies that the intestine becomes more efficient in absorbing lipid. There is an increased ileal mucosal mass and protein content, with no change in DNA content or cell number. This suggests that the mucosa of the intestine responds to high-fat feeding by the development of hypertrophy, rather than hyperplasia. No increase is seen in the number of intestinal villi in animals given high-fat diets.[133]

Intestinal synthesis of cholesterol via HMG-CoA reductase is regulated by the complex interplay of dietary components, which may influence BBM lipid composition and therefore transport properties. The functional difference in the HMG-CoA reductase activity in the crypt and villus cell populations should be considered: crypt HMG-CoA reductase provides structural cholesterol for cell regeneration, whereas villus cholesterol is used primarily for lipoprotein synthesis. This potentially limits the usefulness of inhibitors of HMG-CoA reductase to modify intestinal cholesterol uptake, since structural cholesterol may be needed by the enterocyte. Interestingly, the administration of chenodeoxycholic acid to normal subjects (1 g/day for 7 days) does not reduce sterol synthesis in the jejunal mucosa.[134]

Cholesterol synthesis in the small intestine provides cholesterol for insertion into both lipoproteins and intestinal membranes. HMG-CoA reductase, the rate-limiting enzyme of cholesterol synthesis,[131] is induced by enterocyte regeneration and differentiation, probably through the supply of membrane cholesterol.[135] The cholesterol content of intestinal epithelial cells is responsible for a major rate-controlling effect on intestinal sterol

---

[131] S. Shefer, S. Hauser, V. Laper, and E. H. Mosbach, *J. Lipid Res.* **13,** 402 (1972).

[132] J. A. Balint, M. B. Fried, and C. Imai, *Am. Clin. Nutr.* **33,** 2276 (1980).

[133] J. M. Forrester, *J. Anat.* **111,** 283 (1972).

[134] D. J. Betteridge, W. Krone, C. Middleton, and D. J. Galton, *Eur. J. Clin. Invest.* **10,** 227 (1980).

[135] E. F. Stange, G. Preclik, A. Schneider, E. Seiffer, and H. Ditschuneit, *Biochim. Biophys. Acta* **678,** 202 (1981).

synthesis.[136,136a] There is a decrease in the villus cholesterol content from the jejunum toward the ileum in rats fed a nonpurified diet, whereas feeding a sucrose-enriched semipurified diet results in a reversed pattern of villus cholesterol distribution.[137–139] These disproportional patterns are not observed in the crypt cells, despite a change in their HMG-CoA reductase activity in response to dietary changes. This gradient for intestinal cholesterogenesis along the crypt–villous axis is also eliminated when rats are given a glucose-enriched diet.[137–139]

When animals are fed commercial nonpurified "chow" diets, the ileum appears to be the major site of intestinal sterol synthesis, and hence the activity of HMG-CoA reductase is higher here than in the proximal intestine.[131,140,141] The activity of this enzyme may also influence BBM fluidity,[142] and consequently integral membrane enzymes such as calcium- and magnesium-dependent ATPase, alkaline phosphatase, and the glucose transport system.[143]

Dietary cholesterol quickly inhibits intestinal and hepatic HMG-CoA reductase activity,[144,145] and this reduction in activity may be modified further by the fatty acid content of the diet. For example, when guinea pigs are fed a semisynthetic diet containing 10% cottonseed oil and 1% cholesterol for 2 days, there is an increase in both the free cholesterol and cholesterol esters in the intestinal mucosal cells, together with a fourfold increase in the activity of microsomal acyl-CoA : cholesterol acyltransferase. The addition of 5% corn oil to a 1% cholesterol diet has an additional suppressant effect on hepatic but not on intestinal reductase activity. In contrast, the addition of 5% coconut oil to a 1% cholesterol diet causes a further decrease of HMG-CoA reductase activity in the jejunum and ileum. When cholesterol is fed with polyunsaturated fats to rats, the activity of HMG-CoA reductase is suppressed in both the jejunum and ileum, whereas only the enzyme in the jejunum is suppressed when cholesterol is given

---

[136] E. F. Stange, M. Alavi, A. Schneider, H. Ditschuneit, and J. R. Poley, *J. Lipid Res.* **22**, 47 (1981).
[136a] H. Westergaard and J. M. Dietschy, *J. Nutr.* **104**, 1319 (1976).
[137] M. Sugano and Y. Fujisaki, *Experientia* **36**, 1399 (1980).
[138] M. Sugano, T. Ide, H. Okamatsu, Y. Fujisaki, and H. Takahara, *J. Nutr.* **110**, 360 (1980).
[139] P. R. Holt, A. A. Dominguez, and J. Kwartler, *Am. J. Clin. Nutr.* **32**, 1792 (1979).
[140] J. L. Merchant and R. A. Heller, *J. Lipid Res.* **18**, 722 (1977).
[141] H. Muroya, H. S. Sodhi, and R. G. Gould, *J. Lipid Res.* **18**, 301 (1977).
[142] T. A. Brasitus, A. R. Tall, and D. Schachter, *Biochemistry* **19**, 1256 (1980).
[143] T. A. Brasitus, D. Schachter, and T. G. Mamonneas, *Biochemistry* **18**, 4136 (1979).
[144] R. L. Gebhard and W. F. Prigge, *J. Lipid Res.* **22**, 1111 (1981).
[145] E. A. Stein, D. Mendelsohn, M. Fleming, G. D. Barnard, K. J. Carter, P. S. du Toit, J. D. L. Hansen, and I. Besohn, *Am. J. Clin. Nutr.* **28**, 1204 (1975).

with saturated fats.[146] Thus, dietary cholesterol has an inhibitory effect on intestinal cholesterol synthesis, and this process is modified further by changes in the content of saturated versus unsaturated fatty acids in the diet. This interaction between dietary cholesterol levels and the type and amount of dietary lipids needs to be carefully considered. Future studies of the effects of the reduction of lipid absorption on intestinal function, serum lipid concentrations, the levels of fat-soluble vitamins in the blood, and the properties of bile need to be reviewed against this background of the nature of the interactions between different dietary lipids, and between other dietary macronutrients.

The relationships between dietary cholesterol, the relative proportions of dietary polyunsaturated (P) to saturated (S) fatty acids (P/S ratio), and plasma cholesterol levels have been examined extensively.[145,147–152] Patients fed diets with high P/S ratios have lower cholesterol levels than those fed low P/S diets.[62,153] Increases in dietary cholesterol produce variable rises in plasma cholesterol,[154] whereas diets with high P/S ratios prevent the increase in plasma cholesterol due to added dietary cholesterol.[155]

Complex changes occur in the form and function of the digestive tract as animals develop.[156–158] It remains unclear by what mechanism(s) the intestine adapts with aging.[159] Changes in dietary content of lipids fed to nursing rats modifies the development of nutrient absorption in their suckling offspring, and dietary lipids may also be important in the expression of nutrient transport that occurs at weaning.[160] In young and in recently weaned rabbits, feeding a low-cholesterol diet alters the uptake of sugars

---

[146] W. J. Bochenek, J. B. Rodgers, *Biochim. Biophys. Acta* **575,** 57 (1979).

[147] D. Applebaum-Bowden, W. R. Hazzard, J. Cain, M. C. Cheung, R. S. Kushwala, and J. J. Alberts, *Atherosclerosis* **33,** 385 (1979).

[148] M. A. Flynn, G. V. Nolph, T. C. Flynn, R. Kahrs, and G. Krause, *Am. J. Clin. Nutr.* **32,** 1051 (1979).

[149] S. M. Grundy and E. H. Ahrens, Jr., *J. Clin. Invest.* **49,** 1135 (1970).

[150] M. W. Huff, R. M. G. Hamilton, and K. K. Carroll, *Atherosclerosis* **28,** 187 (1977).

[151] W. B. Kannel and T. Gordon, Section 24 *in* "The Framingham Diet Study: Diet and the Regulation of Serum Cholesterol," GDP, Washington, 1970.

[152] R. W. Mahley and K. W. Holcombe, *J. Lipid Res.* **18,** 314 (1977).

[153] D. S. Lin and W. E. Connor, *J. Lipid Res.* **21,** 1042 (1980).

[154] N. Spritz and M. A. Mishkel, *J. Clin. Invest.* **48,** 78 (1969).

[155] G. Schonfeld, W. Patsch, L. L. Rudel, C. Nelson, M. Epstein, and R. E. Olson, *J. Clin. Invest.* **69,** 1072 (1982).

[156] A. B. R. Thomson and M. Keelan, *Can. J. Physiol. Pharmacol.* **64,** 13 (1986).

[157] A. B. R. Thomson and M. Keelan, *Can. J. Physiol. Pharmacol.* **64,** 30 (1986).

[158] A. B. R. Thomson, *Am. J. Physiol.* **239,** G363 (1980).

[159] W. H. Karasov and J. M. Diamond, *Am. J. Physiol.* **8,** G443 (1983).

[160] N. Perin, E. Jarocka-Cyrta, M. Keelan, M. T. Clandinin, and A. B. R. Thomson, unpublished results (1997).

and lipids.[161] Furthermore, early nutritional experiences may influence later intestinal absorptive function, since some of these active and passive transport changes remain abnormal 2 weeks after the rabbits are returned to eating chow. Feeding a high-cholesterol diet for only a 2-week period shortly after weaning alters passive intestinal uptake.[162] Short-term feeding of young growing rabbits with a high-cholesterol diet reverses the normal developmental decline in jejunal uptake of glucose, and yet enhances the age-related uptake of leucine.[163] Thus, early feeding with a high-cholesterol diet alters the normal development of intestinal transport function. The same observation is seen with feeding of different types of dietary fatty acids.[164]

The reduced intestinal permeability that occurs with aging is not due to major alterations in mucosal surface area, although perturbations in lipid composition of the BBM may account for some of these changes in transport function.[107,165] Variations in the dietary saturated fat content alter lipid composition and fluidity of rat intestinal plasma membranes.[113] There is evidence that other dietary factors may influence intestinal absorption of nutrients.[166] For example, in mature rabbits feeding a diet low in cholesterol leads to an increase in the uptake of cholesterol and fatty acids.[122,167]

It is recognized that the effect of diet is specific to different tissues, membranes, and ages or stages of development. For example, the nature of the diet fed to young and developing animals may have a pronounced effect on the membrane composition and function of several organs including brain, heart, and liver.[168] The phospholipid composition of the intestinal mucosa changes with the age of the animal,[169] as well as with fasting and with diabetes[106] under these circumstances the intestinal uptake of hexoses and lipids is also modified. Future studies of the effect of dietary lipid modification of absorption will have to explore the duration of this effect on intestinal function, the effects of early nutrition, and the possibility that the normal adaptive response of the intestine that occurs in health and

[161] A. B. R. Thomson, *J. Lab. Clin. Med.* **107,** 365 (1986).
[162] A. B. R. Thomson and M. Keelan, *J. Pediatr. Gastroenterol. Nutr.* **9,** 98 (1989).
[163] A. B. R. Thomson, M. Keelan, and M. Tavernini, *J. Ped. Gastroenterol. Nutr.* **6,** 675 (1987).
[164] A. B. R. Thomson, M. Keelan, M. L. Garg, and M. T. Clandinin, *Biochim. Biophys. Acta* **1001,** 302 (1989).
[165] T. A. Brasitus, K.-Y. Yeh, P. R. Holt, and D. Schachter, *Biochim. Biophys. Acta* **778,** 341 (1984).
[166] A. B. R. Thomson and M. Keelan, *Surv. Dig. Dis.* **3,** 75 (1985).
[167] A. B. R. Thomson, *Am. J. Clin. Nutr.* **35,** 556 (1982).
[168] M. T. Clandinin, C. J. Field, K. Hargreaves, L. A. Morson, and E. Zsigmond, *Can. J. Physiol. Pharmacol.* **63,** 546 (1985).
[169] S. M. Schwarz, S. Ling, B. Hostetler, J. P. Draper, and J. B. Walkins, *Gastroenterology* **86,** 1544 (1984).

disease may be impaired, even irreversibly, depending on the late effects of early nutrition and the age of the animal or person being subjected to modification in the absorption of their dietary lipids.

Long-chain unsaturated fatty acids, but not saturated fatty acids and not glucose or amino acids, release neurotensin-like immunoreactivity (NTL_1) from primary cultures of canine enteric nerves. The release of NTL_1 is inhibited by somatostatin and is mediated at least in part by protein kinase C.[170] Ingestion of a fatty meal[171,172] and intraduodenal perfusion of lipids[173–175] elevate circulating levels of NTL_1.[175–177] This is surprising, because most NTL_1 is located in endocrine cells in the distal ileum.[178,179] Nonetheless, the NTL_1 released from the proximal intestine during ingestion of polyunsaturated fats increases the translocation of fatty acids from the lumen of the intestine into lymph.[180] The mechanism of this antiabsorptive effect is unknown. It remains to be determined whether blocking NTL_1 may be useful to reduce lipid absorption and, therefore, become a therapeutically effective treatment for obesity.

Cytosolic Fatty Acid-Binding Proteins

Once fatty acids are taken up across the BBM, they must be detached from the inner surface of the membrane and transported across the enterocyte cytosol, and then be formed into lipoproteins for exit across the basolateral membrane. Transport from the BBM to microsomal enzymes is facilitated by fatty acid-binding proteins (FABPcyt), which are present in the cytosol at very high concentrations exceeding the critical micellar concentrations of LCFAs.

[170] D. L. Barber, A. M. Cacace, D. T. Raucci, and M. B. Ganz, *Am. J. Physiol.* **261,** G497 (1991).
[171] R. A. Hammer, R. E. Carraway, and S. E. Leeman, *J. Clin. Invest.* **70,** 74 (1982).
[172] S. Rosell and A. Rokaeus, *Acta Physiol. Scand.* **107,** 263 (1979).
[173] B. Kihl, A. Rokaeus, S. Rosell, and L. Olbe, *Acta. Physiol. Scan.* **110,** 329 (1980).
[174] C. F. Ferris, R. A. Hammer, and S. E. Leeman, *Peptides* **2,** 263 (1981).
[175] M. Fujimura, T. Khalil, T. Sakamoto, G. H. Greeley, Jr., M. G. Salter, C. M. Townsend, Jr., and J. C. Thompson, *Gastroenterology* **96,** 1502 (1989).
[176] N. W. Read, A. McFarlane, R. I. Kinsman, T. E. Bates, N. W. Blackhall, G. B. Farrar, J. C. Hall, G. Moss, A. P. Morris, B. O'Neill, I. Welch, Y. Lee, and S. R. Bloom, *Gastroenterology* **86,** 274 (1984).
[177] J. P. Walker, M. Fujimura, T. Sakamoto, G. H. Greeley, Jr., C. M. Townsend, Jr., and J. C. Thompson, *Surgery* **98,** 224 (1985).
[178] R. A. Carraway and S. E. Leeman, Characterization of radioimmunoassayable neurotensin in the rat: its differential distribution in the central nervous system, small intestine and stomach. *J. Biol. Chem.* **25,** 7045 (1976).
[179] B. Frigerio, M. Ravazola, S. Ito, R. Buffa, C. Capella, E. Solcia, and L. Orci, *Histochemistry* **54,** 123 (1977).
[180] M. J. Armstrong, C. F. Ferris, and S. E. Leeman, in "Regulatory Peptides: Mode of Action on Digestive, Nervous and Endocrine Systems" (S. Bonfils, ed.), Elsevier, Amsterdam, 1985.

Two fatty acid-binding proteins exist in the enterocyte cytosol, a liver and an intestinal isoform (L- and I-FABPc, respectively). This topic has been reviewed.[181–183] Cytosolic fatty acid-binding proteins are a family of homologous hydrophobic ligand-binding proteins that are the product of an ancient gene family expressed in mammalian organs. Rat I-FABP and L-FABP have 25% amino acid homology, with molecular weights of 15,124 and 14,184, respectively. A crystal structure has been resolved at the 1–2 Å level for I-FABP.[184] I-FABP consists of single globular domains in the form of a flattened β barrel composed of 10 antiparallel sheets and two short α helices ("clamshell"). The bound fatty acid is buried in the interior of the β barrel, with the carboxylate group interacting with Arg-106, Gln-115, and two water molecules, and its hydrocarbon portion surrounded by hydrophobic residues. A relatively common substitution (~30%) of Ala54 by Thr54 in human I-FABP has been associated with a twofold higher apparent affinity for LCFAs.[185] A crystal structure has not yet been resolved for L-FABP. Binding studies indicate that L-FABP binds 2 mol of fatty acid, whereas I-FABP binds 1 mol.[186–188]

A key to understanding the physiological role of FABPs is the determination of their true affinities for fatty acids. The true affinity of FABP for fatty acids has been difficult to estimate because of the instability of the protein–ligand complex. Richieri and coworkers have recently developed a new method to measure FA binding: I-FABP was made fluorescent by labeling Lys27 with Acrylodan.[189] This fluorescent probe, called ADIFAB, is very sensitive to conformational changes induced by fatty acid binding. It has been used to titrate fatty acid binding to albumin[190] and to FABP[188] and to estimate plasma free fatty acid concentrations,[191] leading to binding constants and plasma-free concentrations in the low nanomole per liter range. A major characteristic of FA binding to FABP is the correlation between the variations of FA affinity and water solubility, which is very strong with I-FABP, and weaker with other FABPs, in particular L-

[181] N. M. Bass, *Int. Rev. Cytol.* **111,** 143 (1988).
[182] J. C. Sacchettini and J. I. Gordon, *J. Biol. Chem.* **268,** 18399 (1993).
[183] J. H. Veerkamp and R. G. Maatman, *Prog. Lipid Res.* **34,** 17 (1995).
[184] J. C. Sacchettini, G. Scapin, D. Gopaul, and J. I. Gordon, *J. Biol. Chem.* **267,** 23534 (1992).
[185] L. J. Baier, J. C. Sacchettini, W. C. Knowler, J. Eads, G. Paolisso, P. A. Tataranni, H. Mochizuki, P. H. Bennett, C. Bogardus, and M. Prochazka, *J. Clin. Invest.* **95,** 1281 (1995).
[186] J. B. Lowe, J. C. Sacchettini, M. Laposata, J. J. McQuillan, and J. I. Gordon, *J. Biol. Chem.* **262,** 5931 (1987).
[187] G. Nemecz, J. R. Jefferson, and F. Schroeder, *J. Biol. Chem.* **266,** 17112 (1991).
[188] G. V. Richieri, R. T. Ogata, and A. M. Kleinfeld, *J. Biol. Chem.* **269,** 23198 (1994).
[189] G. V. Richieri, R. T. Ogata, and A. M. Kleinfeld, *J. Biol. Chem.* **267,** 23495 (1992).
[190] G. V. Richieri, A. Anel, and A. M. Kleinfeld, *Biochemistry* **32,** 7574 (1993).
[191] G. V. Richieri and A. M. Kleinfeld, *J. Lipid Res.* **36,** 229 (1995).

FABP.[188] L-FABP has a higher affinity for polyunsaturated fatty acids (PUFAs) than I-FABP, and may thereby have a specific role regarding these essential FAs. This is of particular importance for the enterocytes after a lipid meal, where unbound fatty acid concentrations may reach high levels, close to their critical micellar concentrations. L-FABP has the unique property to bind in addition to FA, cholesterol, acyl-CoA, lysophospholipids, β-monoacylglycerol and other small lipophilic molecules, probably at the low-affinity site.[192–194] Because of the high affinity of FABP for fatty acids, the unbound FA concentrations in cytosol are probably in the nanomole per liter range, thus representing a small fraction of both their total and critical micellar concentrations. This suggests that FABP may function as an efficient buffer, protecting various cell functions against excessive concentrations and maintaining a pool of fatty acids necessary for lipid synthesis.[181]

Another key is the distribution of I- and L-FABP in the intestinal epithelium. The greatest quantities are in the proximal and middle two-thirds of the jejunum, where maximal lipid absorption occurs; it decreases distally.[195] High-fat diets have different effects on I- and L-FABP quantities—the former is increased in the ileum, but not in the jejunum, whereas the latter is increased similarly in both intestinal segments.[196] There is also a difference along the villus axis—in crypt cells only L-FABP is expressed, whereas I-FABP is maximally expressed at the tip of the villus.[197] It is also at the tip of the villus that the enzymes necessary for triacylglycerol synthesis from β-monoacylglycerol, the major postpandrial pathway, are maximally expressed. This suggests the I-FABP may have a specific role in postprandial lipid absorption.

The mechanism of FA transfer between membranes and FABP may also provide clues to understanding the function of FABP isoforms. Storch and colleagues have studied FA transfer from FABP to artificial membranes with anthroyloxy-labeled fatty acids.[198] The rate of transfer was much faster for adipocyte FABP than for L-FABP. To explain this difference, a "collisional" mechanism involving electrostatic interactions between membranes

---

[192] G. Nemecz, T. Hubbell, J. R. Jefferson, J. B. Lowe, and F. Schroeder, *Archiv. Biochem. Biophys.* **286,** 300 (1991).

[193] G. Nemecz and F. Schroeder, *J. Biol. Chem.* **266,** 17180 (1991).

[194] J. Storch, *Mol. Cell Biochem.* **123,** 45 (1993).

[195] N. M. Bass, J. A. Manning, R. K. Ockner, J. I. Gordon, S. Seetharam, D. H. Alpers, *J. Biol. Chem.* **260,** 1432 (1985).

[196] R. K. Ockner and J. A. Manning, *J. Clin. Invest.* **54,** 326 (1974).

[197] H. M. Shields, M. L. Bates, N. M. Bass, C. J. Best, D. H. Alpers, and R. K. Ockner, *J. Lipid Res.* **27,** 549 (1986).

[198] F. M. Herr, V. Matarese, D. A. Bernlohr, and J. Storch, *Biochemistry* **34,** 11840 (1995).

and adipocyte FABP was proposed. In principle, a lipid-phase diffusion or "collisional" mechanism should be more efficient to deliver the most insoluble fatty acids (i.e., long-chain saturated FAs) than simple water-phase diffusion. This may, at least in part, account for the fatty acid composition of tissues expressing different FABP isoforms. A "collisional" mechanism has also been suggested for heart FABP. It is intriguing to speculate that this mechanism may also apply to I-FABP.

Esterification, Lipoprotein Synthesis, and Secretion

During the process of normal lipid absorption, the developing chylomicrons appear in the rough and then in the smooth endoplasmic reticulum (ER), followed by the Golgi apparatus.[199–201] The nascent chylomicron particles leave the Golgi within vesicles, which traffic to the basolateral membrane where they fuse, cross by exocytosis into the lamina propria, and enter the intestinal lymph[201] or portal blood.[74] Brefeldin A (BFA), a macrocyclic lactone produced in fungi, causes the collapse of the Golgi into the ER.[202,203] Also, there may be changes in the trans-Golgi network.[204–206] Adding BFA to intraduodenally infused lipids in rats reduces lymphatic TG transport by more than 90%.[207] BFA blocks very low density lipoproteins (VLDL) as well as chylomicron (CM) transport out of the enterocyte.

The topic of the intestinal synthesis, secretion, and transport of lipoproteins has been reviewed.[208–210] During fasting, VLDL are the major TG-rich lipoproteins secreted by the intestine. Following fatty meals VLDL secretion continues but CM become the major intestinal lipoproteins. The small intestine also secretes high-density lipoprotein (HDL) lipoprotein into intestinal lymph.[211] The lipid droplets within the enterocyte have been

[199] R. A. Jersild, Jr., *Am. J. Anat.* **118,** 135 (1966).
[200] A. J. Ladam, A. Padykula, and E. W. Strauss, *Am. J. Anat.* **112,** 389 (1963).
[201] S. M. Sabesin and S. Frace, *J. Lipid Res.* **18,** 496 (1977).
[202] J. Lippincott-Schwartz, L. C. Yuan, J. S. Bonifacino, and R. D. Klausner, *Cell* **56,** 801 (1989).
[203] A. Tan, J. Bolscher, C. Feltkamp, and H. Ploegh, *J. Cell Biol.* **116,** 1357 (1992).
[204] J. Lippincott-Schwartz, L. Yuan, C. Tipper, M. Amherdt, L. Orci, and R. D. Klausner, *Cell* **67,** 601 (1991).
[205] K. Sandvig, K. Prydz, S. H. Hansen, and B. van Deurs, *J. Cell. Biol.* **115,** 971 (1991).
[206] S. A. Wood, J. E. Park, and W. J. Brown, *Cell* **67,** 591 (1991).
[207] C. M. Mansbach II and P. Nevin, *Am. J. Physiol.* **226,** G292 (1994).
[208] P. H. R. Green and R. M. Glickman, *J. Lipid Res.* **22,** 1153 (1981).
[209] C. L. Bisgaier and R. M. Glickman, *Annu. Rev. Physiol.* **45,** 625 (1983).
[210] D. Rachmilewitz, J. J. Albers, D. R. Saunders, and M. Fainaru, *Gastroenterology* **75,** 677 (1978).
[211] H. R. Bearnot, R. M. Glickman, L. Weinberg, and P. H. R. Green, *J. Clin. Invest.* **69,** 210 (1982).

called *pre-CM* to distinguish them from the CM in lymph and plasma. The pre-CM contain more free fatty acid, cholesterol, and protein and, therefore, have a different lipid composition from CM. CM are spherical TG-rich particles (750–6000 Å) formed within the intestine during TG absorption. CM consist of an oily core of TG and cholesterol esters, as well as some free cholesterol. Phospholipid is found in sufficient amounts to cover 80–100% of the surface of the CM, and protein covers the remaining surface.[212] Large CM are formed during the peak of lipid absorption. The composition of TG-rich lipoproteins largely reflects dietary fatty acids, whereas phospholipid and cholesteryl ester fatty acids in CM bear little relationship to dietary TG fatty acids.[213,214] Apolipoprotein (apo) A-I is the major apoprotein of HDL and is an important CM component, comprising about 30% of CM protein. Apo A-I is synthesized during TG absorption, along with other CM apoproteins such as apo B and apo A-IV.

The metabolism of intestinal TG-rich lipoproteins is a source of HDL.[215] Approximately 50% of rat plasma apo A-I, the principal apoprotein of HDL, is synthesized by the intestine. There are at least two intestinal forms of HDL particles.[216] Apo B is essential for the formation and secretion of CM by the enterocytes, and there may be a preformed pool of apo B in enterocytes.

Intestinal apo A-I secretion is maintained during biliary diversion, and the synthesis of this apoprotein occurs in the absence of CM formation.[211] Apo A-I, the principal apoprotein of plasma HDL, is the major apoprotein of mesenteric lymph lipoproteins (CM, VLDL, and HDL), and apo A-I is actively synthesized by the rat small intestine during TG absorption. Considerable amounts of apo A-I are present in bile,[217,218] and biliary components are likely to be the source of the lipid constituents of TG-rich lipoproteins formed by the intestine during periods of fasting. Despite reduction in CM and VLDL, which are known to transport up to 80% of fasting lymph apo A-I, there is no reduction in apo A-I output during biliary diversion. Thus, intestinal apo A-I secretion into lymph is maintained despite low levels of TG secretion. This suggests that factors other than TG resynthesis may regulate intestinal apo A-I secretion.

---

[212] D. B. Zilversmit, *J. Clin. Invest.* **44**, 1610 (1965).
[213] E. B. Feldman, B. S. Russell, C. B. Hawkins, and T. Forte, *J. Nutr.* **113**, 2323 (1983).
[214] E. B. Feldman, B. S. Russell, R. Chen, J. Johnston, T. Forte, and S. B. Clark, *J. Lipid Res.* **24**, 967 (1983).
[215] R. M. Glickman, *in* "Intestinal High Density Lipoprotein Formation. Recent Advances in Bile Acid Research" (L. Barbara, ed.), pp. 191–193, Raven Press, New York, 1985.
[216] A. M. Magun, T. A. Brasitus, and R. M. Glickman, *J. Clin. Invest.* **75**, 209 (1985).
[217] R. B. Sewell, S. J. Mao, T. Kawamoto, and N. F. LaRusso, *J. Lipid Res.* **24**, 391 (1983).
[218] T. Kawamoto, S. J. T. Mao, and . F. LaRusso, *Gastroenterology* **92**, 1236 (1987).

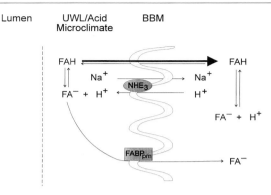

FIG. 5. A suggested model of fat absorption. (From Schoeller et al.[95,98])

The fatty acid composition of the TG in VLDL is different from that in CM, suggesting that there are different pathways of assembly for these particles. Rat enterocyte Golgi vesicles contain either CM or VLDL particles, with little mixing of particle sizes.[219] With increasing cholesterol absorption, more lipid is carried by VLDL, in contrast to the preferential increase in CM when more TG is absorbed.[213,214,220] The intracellular pathways for formation of CM and VLDL are different (Fig. 5): The pathway for CM is sensitive to the effect of Pluronic L-81, a hydrophobic surfactant that inhibits the intestinal formation and transport of chylomicrons but not VLDL. The latter pathway for VLDL is not sensitive to the inhibitory effect of this surfactant.[220] The pre-CM and pre-VLDL particles are then packaged in the Golgi complex, into either pre-CM- or pre-VLDL-containing vesicles. When TG and PC are infused separately, the fatty acids from the TG are used mainly for the production of CM, whereas those from PC predominantly stimulate VLDL formation. At a low rate of intestinal infusion of lipid, luminal fatty acids are incorporated into VLDL as well as into CM.[221–223] The potential for using agents to preferentially block the transport of CM formed from dietary lipids, preventing their exit across the BLM into lymph or portal blood, represents another potentially important therapeutic focus to inhibit lipid absorption.

[219] R. W. Mahley, B. D. Bennett, D. J. Morre, M. E. Gray, W. Thistlethwaite, and V. S. Lequire, *Lab. Invest.* **25,** 435 (1971).
[220] P. Tso and W. J. Simmonds, *Lab. Res. Meth. Biol. Med.* **10,** 191 (1984).
[221] Y.-F. Shiau, D. A. Popper, M. Reed, C. Unstetter, D. Capuzzi, and G. M. Levine, *Am. J. Physiol.* **248,** G164 (1985).
[222] Y.-F. Shiau, P. Fernandez, M. J. Jackson, and S. McMonagle, *Am. J. Physiol.* **248,** G608 (1985).
[223] R. K. Ockner, F. B. Hughes, and K. J. Isselbacher, *J. Clin. Invest.* **48,** 2079 (1969).

The pattern of lipoproteins in lymph and in the lamina propria does not mirror exactly the distribution of apo A-I and apo B among lipoproteins inside the enterocyte. Intracellular apoproteins may become disassociated from lipoproteins, or they may be associated with lipoproteins in various stages of assembly of protein with lipids.[224] Thus, not all apoproteins are secreted in association with identifiable lipoprotein particles from the enterocyte. For example, the distribution of apo A-I and apo B differ from each other in the enterocyte, and differ from those in the lamina propria and in the mesenteric lymph. Nearly all of the intracellular apo B appears to be associated with TG-rich lipoproteins, with smaller LDL-size lipoprotein or with membrane particles. Most apo A-I in the enterocyte is unassociated with any recognizable lipoprotein.

Factors that affect the time for CM to appear in the plasma include the rate of fat absorption from the intestinal lumen, the intracellular formation and secretion of CM, and the rate of CM movement from the intercellular spaces to the central lacteal (i.e., interstitial transport). This latter step is influenced by interstitial matrix hydration and by lymph flow, so that as lymph flow increases the CM appearance time is reduced.[225] The lower CM appearance time produced by expansion of the mucosal interstitium results from a diminished resistance of the interstitial matrix to CM movement.[226]

Apolipoprotein B is a large hydrophobic protein synthesized in mammalian liver and intestine. Apo B is a surface component of both TG-rich lipoproteins [chylomicrons and VLDL, and low-density lipoproteins (LDL)]. The intestinal rate of synthesis of apo B is unaffected by acute TG flux, but is influenced by external diversion of bile.[227] Key aspects of the regulation of apo B synthesis by cellular lipid flux may be mediated independently in jejunal and ileal enterocytes. During fasting, apo B is largely in the rough ER and Golgi cisternii of the enterocytes at the tip of the jejunal villi. In contrast, after fat feeding apo B is found in association with CM in the apical smooth ER.[228] This suggests that apo B is synthesized in the rough ER and is then transferred to the smooth ER to be added onto the pre-CM.

Apolipoprotein A-IV is a protein produced exclusively by the human small intestine.[229] The stimulation of apo A-IV synthesis and secretion by the small intestine after the ingestion of fat is probably mediated by events

---

[224] D. H. Alpers, D. R. Lock, N. Lancaster, K. Poksay, and G. Schonfeld, *J. Lipid Res.* **26**, 1 (1985).
[225] P. Tso, V. Pitts, and N. D. Granger, *Am. J. Physiol.* **249**, G21 (1985).
[226] P. Tso, J. A. Barrowman, and D. N. Granger, *Am. J. Physiol.* **250**, G497 (1986).
[227] N. O. Davidson, M. E. Kollmer, and R. B. Glickman, *J. Lipid Res.* **27**, 30 (1986).
[228] N. J. Christensen, C. E. Rubin, M. C. Cheung, and J. J. Albers, *J. Lipid Res.* **24**, 1229 (1983).
[229] J. R. Sherman and R. B. Weinberg, *Gastroenterology* **95**, 394 (1988).

involved in the packaging and secretion of CM, rather than by lipid uptake.[230] Apo A-IV may act as a physiological signal for satiety.[231–233] Designing drugs that mimic the effects of apo A-IV may represent an important new approach for the design of compounds to reduce food intake by decreasing meal size.

Potential Adverse Effects of Inhibiting Lipid Absorption

Effect of Oral Nutrition on Form and Function of Intestine

Enterocytes lining the small intestine are subjected continually to changes in the quality and quantity of nutrients in the diet. The process of intestinal adaptation has been examined by investigating alterations in nutrient transport, villous morphology, and membrane lipid composition. Intestinal adaptation has been studied in a variety of animal models such as diabetes, aging, short bowel syndrome, and chronic ethanol consumption and following external abdominal irradiation.[102,106–110,122–126,158,161,163,167,234–255]

---

[230] H. Hayashi, D. F. Nutting, K. Fujimoto, J. A. Cardelli, D. Black, and P. Tso, *J. Lipid Res.* **31,** 1613 (1990).
[231] K. Fujimoto, J. A. Cardelli, and P. Tso, *Am. J. Physiol.* **262,** G1002 (1992).
[232] K. Fujimoto, K. Fukagawa, T. Sakata, and P. Tso, *J. Clin. Invest.* **91,** 1830 (1993).
[233] K. Fujimoto, H. Machidori, R. Iwakiri, K. Yamamoto, J. Fujisaki, T. Sakata, and P. Tso, *Brain Res.* **608,** 233 (1993).
[234] A. B. R. Thomson, *Am. J. Physiol.* **236,** E685 (1979).
[235] A. B. R. Thomson, *Diabetes* **30,** 247 (1981).
[236] A. B. R. Thomson, *Dig. Dis. Sci.* **26,** 890 (1981).
[237] A. B. R. Thomson, in "Basic Mechanisms of Gastrointestinal Mucosal Cell Injury and Cytoprotection" (Grossman M, ed.), pp. 327–350, Williams and Wilkins, Baltimore, 1981.
[238] A. B. R. Thomson, *Am. J. Physiol.* **244,** G151 (1983).
[239] A. B. R. Thomson, *Diabetes* **32,** 900 (1983).
[240] A. B. R. Thomson, in "Mechanisms of Mucosal Protection in the Upper Gastrointestinal Tract" (A. Allen, G. Flemstrom, A. Gardner, W. Silen, L. A. Turnberg, eds.), pp. 233–239, Raven Press, New York, 1984.
[241] A. B. R. Thomson, *Am. J. Physiol.* **246,** G120 (1984).
[242] A. B. R. Thomson, *Clin. Invest. Med.* **8,** 296 (1985).
[243] A. B. R. Thomson, *Comp. Biochem. Physiol.* **82,** 819 (1985).
[244] A. B. R. Thomson, *Q. J. Exp. Physiol.* **71,** 29 (1986).
[245] A. B. R. Thomson, *Can. J. Physiol. Pharmacol.* **65,** 2281 (1987).
[246] A. B. R. Thomson, *Can. J. Physiol. Pharmacol.* **65,** 219 (1987).
[247] A. B. R. Thomson and R. V. Rajotte, *Am. J. Physiol.* **246,** (5Pt1), G627 (1984).
[248] A. B. R. Thomson, C. I. Cheeseman, and K. Walker, *Radiat. Res.* **107,** 344 (1986).
[249] A. B. R. Thomson, M. Keelan, C. I. Cheeseman, and K. Walker, *Int. J. Radiat. Oncol. Biol. Phys.* **12,** 917 (1986).

The changes in nutrient transport are often associated with alterations in BBM phospholipids. This underlines the importance of BBM lipids in intestinal function. We must stress that the effect of changes in dietary lipids is not limited to the intestine; dietary fat saturation influences the function of membranes isolated from other tissues such as the brain, heart, and liver.[10,104,256,257] The functional changes are related to alterations in their membrane phospholipids and/or phospholipid fatty acid composition. Variations in dietary fat saturation have specific effects on intestinal active and passive nutrient transport.[101,125,258–260] Changes in the dietary protein, carbohydrate, cholesterol, or essential fatty acid content also result in alterations in villous height, BBM enzyme activity, BBM lipid composition, and nutrient transport.[122,123,261] Thus, dietary constituents represent one of the major type of signals important for this adaptive process.

A model has been proposed for the mechanism of dietary lipid effect on nutrient uptake (Fig. 6). This adaptation may begin quickly, persist over time, and have lasting effects, even when the initial dietary stimulus has been removed. Thus, modifying lipid absorption by any means may alter the absorption of not only lipids, but also of carbohydrates and amino acids. The dietary changes that alter the form and function of the intestine include starvation and semistarvation, hyperalimentation, hyperphagia, or feeding diets enriched with or depleted of selected macronutrients such as lipids, carbohydrates, and proteins, or micronutrients such as iron or calcium. Whereas the direct effects imply trophic changes and possible chemical, physical, and mechanical actions of foodstuffs, the indirect effects of food are likely brought about via different mechanisms such as altered bacterial flora or modified release of local or systemic hormones and gastro-

---

[250] A. B. R. Thomson, Y. McIntyre, J. MacLeod, and M. Keelan, *Digestion* **35,** 89 (1986).

[251] A. B. R. Thomson, M. Keelan, and M. Tavernini, *Pediatr. Res.* **21,** 347 (1987).

[252] C. I. Cheeseman, A. B. R. Thomson, and K. Walker, *Radiat. Res.* **101,** 131 (1985).

[253] A. B. R. Thomson, C. I. Cheeseman, and K. Walker, *J. Lab. Clin. Med.* **102,** 813 (1983).

[254] A. B. R. Thomson, C. I. Cheeseman, and K. Walker, *Int. J. Radiat. Oncol. Biol. Phys.* **10,** 671 (1984).

[255] A. B. R. Thomson and R. V. Rajotte, *Comp. Biochem. Physiol.* **82,** 827 (1985).

[256] M. Foot, T. F. Cruz, and M. T. Clandinin, *Biochem. J.* **208,** 1 (1982).

[257] P. J. Neelands and M. T. Clandinin, *Biochem. J.* **212,** 572 (1983).

[258] A. B. R. Thomson, *Res. Exp. Med.* **186,** 413 (1986).

[259] A. B. R. Thomson, M. Keelan, M. L. Garg, and M. T. Clandinin, *Can. J. Physiol. Pharmacol.* **66,** 985 (1988).

[260] A. B. R. Thomson, M. Keelan, T. Cheng, and M. T. Clandinin, *Biochim. Biophys. Acta* **1170,** 80 (1993).

[261] M. Keelan, K. Walker, R. V. Rajotte, M. T. Clandinin, and A. B. R. Thomson, *Can. J. Physiol. Pharmacol.* **65,** 210 (1987).

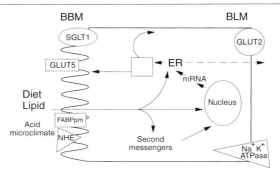

FIG. 6. A model is proposed for the control of intestinal adaptation in response to dietary lipid manipulations. This model is also applicable for the testing of other examples of intestinal adaptation, such as that occurrs in diabetes mellitus or following bowel resection. In this model the following steps are proposed: (1) Dietary lipids act as signals to release second messengers [such as activating ornithine decarboxylase (ODC) activity, intracellular calcium (Ca^{2+}), phospholipase C, phosphyatidyl inositol, or protein kinase C], which modify the genetic regulation of the expression of the mRNA for BBM and basolateral membrane (BLM) carrier proteins, as well as the enterocyte microsomal (EMM) lipid metabolic enzymes. (2) Dietary lipids also act as *substrate* for the diet-associated changes in EMM lipid metabolic enzymes, which thereby modify the EMM content of lipids available for trafficking to the BBM and the BLM. (3) The partial disassociation between diet-associated changes in EMM and BBM/BLM lipids is due to postmicrosomal modification of EMM lipids by lipid metabolic enzymes in the BBM, and by variations along the villus in the diet-associated changes in EMM and BBM/BLM lipids. (4) The partial disassociation between diet-associated changes in lipid uptake and BBM lipids is due to the contribution of the NHE and FABPpm to the total uptake of lipids, and to the variable expression of those proteins along the villus. (5) Posttranscriptional modulation of the sugar carriers occurs in the Golgi apparatus, as well as by the BBM and BLM composition of lipids. (6) The diet-associated changes in sugar uptake necessarily modify the activity of BLM Na^+-/K^+-ATPase, and the diet-associated alterations in lipid uptake modify the activity of BBM NHE activity, which further influences the required regulation of Na^+-K^+-ATPase. (Thomson et al.[115])

intestinal peptides. Thus, "luminal nutrition" comprises both an exogenous and an endogenous component, and there are many examples of conditions in which oral nutrition alters intestinal function (Table I).

Because alterations in lipid intake modify the uptake of other nutrients, and because weight-reducing diets may include a change in the amounts or ratios of carbohydrates as well as lipids, it is necessary to consider diet-associated variations in the uptake of other nutrients. What is the effect of variations in the type or amount of carbohydrate on intestinal function? Feeding glucose, galactose, or α-methyl glucoside has complex interactions on the kinetics of *in vivo* absorption of these sugars in rats fasted for 3 days.[262] Feeding isocaloric diets containing 65% glucose or 65% fructose

---

[262] E. S. Debnam and R. J. Levin, *Gut* **17,** 92 (1976).

TABLE I
CONDITIONS IN WHICH ORAL NUTRITION ALTERS
INTESTINAL FUNCTION

Fasting, starvation, and undernutrition
Protein deficiency and high-protein diets
Carbohydrate or sucrose-containing diets
Fiber
High-fat and high-cholesterol diets
Chemically defined "enteral" diets
Intestinal resection

---

increases the jejunal uptake of both glucose and fructose, as compared with feeding 30% glucose.[263] Fructose uptake is greater with feeding fructose than with feeding glucose, whereas the uptake of glucose is similar with feeding either fructose or glucose. The mechanism of this effect has not been established. Of note, the increased mucosal ornithine decarboxylase activity observed on refeeding following a fast occurs only when glucose, galactose, or 3-0-methyl glucose are perfused through the intestinal lumen. Thus, certain sugars may induce postfasting mucosal proliferation, which may further enhance carbohydrate absorption, and which may alter the uptake of other nutrients.

Karasov and Diamond[264] have also shown that intestinal sugar absorption in rats is influenced by the dietary carbohydrate content: In mice dietary carbohydrate does not affect villous structure, but mice fed a carbohydrate-free high-protein diet have a rate of glucose uptake that is relatively independent of position along the small intestine (i.e., jejunum versus ileum). A high-carbohydrate diet reversibly and within 1 day stimulates uptake in the jejunum but not into the ileum, restoring the normal proximal-to-distal gradient in glucose uptake. Passive permeability to glucose remains unchanged with carbohydrate feeding, but there is an increase in the value of the maximal transport rate (V_{max}) and of the Michaelis constant (K_m).

Feeding a sucrose-containing diet quickly increases the activity of BBM sucrase in enterocytes close to the crypts[265,266] or further up the villus.[267,268] The increase in sucrase activity is observed within 12 hours after the intro-

---

[263] C. Bode, J. M. Eisenhardt, F. J. Haberich, J. C. Bode, *Res. Exp. Med.* **179,** 163 (1981).
[264] W. H. Karasov and J. M. Diamond, *J. Physiol.* **394,** 419 (1984).
[265] F. Raul, P. M. Simon, M. Kedinger, J. F. Gremer, and K. Haffen, *Biochim. Biophys. Acta* **630,** 1 (1980).
[266] M. H. Ulshen and R. J. Grand, *J. Clin. Invest.* **64,** 1097 (1979).
[267] J. P. Cezard, J. P. Boyart, P. Cuisinier-Gleizes, and H. Mathieu, *Gastroenterology* **84,** 18 (1983).
[268] O. Koldovsky, S. Bustamante, and K. Yamada, *in* "Mechanisms of Intestinal Adaptation," MTP Press, London, 1982.

duction of the sucrose-containing diet, and the rate of enterocyte migration is unaffected when rats are switched from a low- to a high-carbohydrate diet.[268,269] Thus, it would appear that the "old" upper villus cells react in concert with the "new" lower villus cells, changing the activity of those BBM enzymes that would then help to facilitate the digestion of the higher levels of dietary carbohydrate.

With a change in the proportion of carbohydrate content of the diet, there is likely an inverse reciprocal alteration in the lipid content of the diet. It is likely, therefore, that variations in the quantity of carbohydrates in the diet also influence the uptake of lipids. If lipid absorption were reduced by some therapeutic maneuver, we do not know if the intestine would adapt to alter the efficiency of the uptake of lipids, or whether there might be adaptation of absorption of other nutrients, thereby modifying the expected weight loss resulting from the blocked lipid uptake.

Lessons from Chemically Defined Diets

One area in which the effect of low-fat diets on intestinal function has been explored is with the use of elemental diets. These defined formula diets are widely used in the treatment of a variety of nutritional disorders.[270] Most elemental diets contain no digestive residues (fiber), and these monomeric diets differ widely in their composition. Their fat content is variable: The percentage of calories available from vegetable fat ranges from 1 (Vivonex) to 30% (Flexical). Vivonex contains only amino acids, whereas Vital and Flexical contain protein hydrolysates (small peptides in addition to free amino acids). Other "nutritionally complete" polymeric preparations (i.e., Ensure, Isocal, Boost, Portagen) are liquid diets that contain all known required vitamins and minerals, in addition to calories and a nitrogen source from intact proteins.

These diets alter the form and function of the intestine.[271] For example, Isocal increases the mucosal mass of the jejunum[242] and is associated with an increased villous surface area in rabbits.[243] The low-fat content of these chemically defined diets points to the possible use of this literature to give suggestions as to how a low intraluminal content of fatty acid might influence intestinal function. When rats are fed Vivonex for 3 months, there is a 31% increase in villous height and a 15% decrease in crypt depth.[272] Alkaline phosphatase activity is increased about 25%, but there are no changes in lactase, sucrase, or maltase activities. Ensure, Flexical-HN, Isocal, and

[269] K. Yamada, S. Bustamante, and O. Koldovsky, *FEBS Lett.* **129,** 98 (1981).
[270] B. J. M. Jones, R. Lee, J. Andrew, P. Frost, and D. B. Silk, *Gut* **24,** 7 (1983).
[271] R. L. Koretz and J. H. Meyer, *Gastroenterology* **78,** 393 (1980).
[272] L. M. Nelson, H. A. Carmichael, R. I. Russell, and F. D. Lee, *Clin. Sci.* **55,** 509 (1978).

Portagen alter the *in vitro* uptake of hexoses, fatty acids, and cholesterol in both rabbits with an intact intestinal tract and animals previously subjected to a surgical resection of the ileum.

We do not know whether there is a specific nutrient present in these defined diets that contributes to the intestinal adaptive response or whether it is the depletion of a nutrient that results in the adaptive process.[273] Despite isocaloric feeding, body weight gain is lower in rats fed Vivonex and Vivonex-HN for 2 weeks, and higher in the Flexical- and Vital-fed groups, as compared with controls fed a casein-containing liquid diet.[274] The weights of the proximal intestinal segment, as well as the mucosal weight, protein, and DNA of Vivonex-fed animals, or of Vivonex-HN- and Vital-fed groups, were less than in control animals, whereas distal segments were similar in the various dietary groups. The Flexical-fed animals had a greater intestinal mass than did control rats, or those given the other defined formula diets. Intragastric feeding of these chemically defined diets leads to effects on intestinal mass that are distinct from those obtained with parenteral feeding or feeding chow.[127,269,275,276] Thus, differences in body weight gain, despite isocaloric feeding, may be due to variations in the nutritional composition of the formulas of these diets. One of these different components, as already noted, is the type and content of fats in these chemically defined diets.

Feeding a defined formula diet low in fat alters the jejunal uptake of LCFAs and cholesterol in rabbits.[258] For example, feeding Ensure reduces jejunal uptake of myristic (14:0) and palmitic acids, but increases the jejunal uptake of cholesterol; in contrast, feeding Flexical-HN increases jejunal uptake of 14:0 and cholesterol. The percentage of total calories derived from fat is approximately similar in Ensure and Flexical-HN, but corn oil is the dietary source of fat in Ensure, whereas the source of fat in Flexical-HN is soy oil, soy lecithin, and medium-chain triglyceride (MCT) oil. Thus, the dietary content of lipid in these chemically defined diets alters intestinal transport, so that even when animals are raised on a low-fat diet, the type of lipid in the diet influences absorption.

The addition of nonabsorbable bulk ("fiber") to the defined formula diets is associated with an increase in colonic mass equal to that observed in chow-fed animals.[272,276] After a 4-week period of feeding Vivonex to rats, there is a 75% decrease in the cell population of the colonic mucosa,

[273] L. R. Johnson, E. M. Copeland, S. J. Dudrick, L. M. Lichtenberger, and G. A. Costro, *Gastroenterology* **68**, 1177 (1975).
[274] E. A. Young, I. A. Cioletti, J. D. Traylor, and V. Balderas, *J. Parent. Enteral. Nutr.* **5**, 478 (1981).
[275] P. Janne, Y. Carpenter, and G. Willems, *Dig. Dis.* **22**, 808 (1977).
[276] B. Sircar, L. R. Johnson, and L. M. Lichtenberger, *Am. J. Physiol.* **238**, G376 (1980).

associated with decreases in the mitotic rates and DNA synthesis activity.[275] Intravenous or oral feeding of an elemental diet is associated with a 25% loss of colonic weight, as well as a decline in mucosal DNA synthesis, as compared with animals fed chow. Serum and antral gastrin levels fall. The addition of nonabsorbable bulk to the liquid diet stimulates colonic DNA synthesis and growth, without elevating serum or antral gastrin concentrations. Thus, dietary fiber may be necessary to maintain normal colonic growth and function. The nature of any possible interaction between dietary fiber and intraluminal concentrations of fatty acids is unknown, and awaits elucidation. In studies focusing on the effects of blocking lipid uptake and therefore changes in the intraluminal type or content of lipids on colonic growth and function, the fiber content of the diet is an important variable that will need to be taken into account.

Membrane Biochemistry

When the lipids in the intestinal lumen are modified, there are changes in the composition of the intestinal membranes as well as those in other tissues.[10] The fluid mosaic model of membranes has been extended and revised to include the concept of lipid–protein and lipid–lipid interactions. Evidence for a polymorphic organization of membrane structure, and therefore possibly of membrane function, has come from studies involving a wide variety of techniques including X-ray diffraction, electron microscopy, measurements of lateral diffusion, differential partitioning of lipid probes, differential scanning calorimetry, spin-label measurements, and differences in composition between inner and outer side of the membrane. It is presumed that the lipids and proteins have mobility within the membrane. Changes in the viscosity of the membrane may be achieved by alterations in its composition of either the lipids or the proteins.[277]

Membrane-associated proteins may be arranged peripherally, may be anchored at the membrane surface, or may be integral, penetrating into or through the structural matrix of the membrane lipid bilayer.[278] It is possible that the activity of membrane-associated enzymes may be dependent on an association with specific polar head groups at the BBM surface; may be modulated by the surrounding bilayer matrix, which is determined by its fatty–acyl content and/or phospholipid distribution; or may be dependent on an interaction with specific fatty–acyl chains at the protein–lipid inter-

---

[277] H. K. Kimelberg, "The Influence of Membrane Fluidity on the Activity of Membrane Bound Enzyme in Dynamic Aspects of Cell Surface Organization," Elsevier/North-Holland Biomedical Press, Amsterdam, 1977.

[278] S. J. Singer and G. L. Nicolson, *Science* **175**, 720 (1972).

face. Specific phospholipids are required before membrane-bound enzymes can express their full activity.[279] In addition, the asymmetrical distribution of phospholipids in membranes[280,281] further implies a specific purpose for phospholipid orientation.

Do changes in membrane macrostructure (i.e., morphology) or microstructure (i.e., membrane composition, including fluidity) result in variations in intestinal transport function? The relationship between BBM structure and function has been the subject of considerable interest.[279-283] BBM lipid fluidity may influence enzyme functions due to changes in bulk lipid fluidity, or to variations in the specific properties of the lipid in the microenvironment of the protein.[277] Such polymorphism of membrane structure gives rise to specific lipid domains that might enable lipids to influence independently specific protein functions. Thus, there may be an interrelationship between the membrane content of cholesterol, phospholipids, phospholipid fatty acids, and the physical properties of membranes, which in turn may influence membrane function.[284]

Demonstration of the importance of the fatty acid milieu surrounding the integral membrane proteins has been achieved through correlation of alterations in enzyme activity with changes in membrane fatty acid composition[285] alterations in membrane physical properties defined by the temperature of lipid phase transition; and differences in the value of the Arrhenius activation energy for enzymes from membranes with differing fatty acid distributions.[286] In artificial membranes, lipid phase transition decreases as the degree of membrane lipid unsaturation increases.[287] This indicates a fluidizing effect of unsaturated fatty–acyl chains. Thus, changing the dietary lipids modifies membrane phospholipids and their fatty–acyl chains, thereby altering membrane function.[104,105] It is likely that modifications in lipid absorption will have the same influence as a change in the ingested dietary lipids, and may thereby alter BBM lipid composition as well as digestive and transport functions.

There is considerable evidence that many of the protein-mediated functions of the intestine are influenced by the composition and physical state

[279] H. Sanderman, *Biochim. Biophys. Acta* **515**, 209 (1978).
[280] J. A. F. Op den Kamp, *Annu. Rev. Biochem.* **48**, 47 (1979).
[281] J. J. R. Krebs, H. Hauser, and E. Carafoli, *J. Biol. Chem.* **254**, 5308 (1979).
[282] D. Chapman, J. C. Gomez-Fernandez, and F. M. Goni, *FEBS Lett.* **98**, 211 (1979).
[283] J. E. Cronan and E. P. Gelmann, *Bact. Rev.* **39**, 232 (1975).
[284] C. D. Stubbs and A. D. Smith, *Biochim. Biophys. Acta* **779**, 89 (1984).
[285] E. W. Haeffner and O. S. Privett, *Lipids* **10**, 75 (1975).
[286] W. B. Im, J. T. Deutchler, and A. A. Spector, *Lipids* **14**, 1003 (1979).
[287] A. G. Lee, *Biochim. Biophys. Acta* **472**, 237 (1977).

of lipids in the BBM.[143,144,288-290] Variations in the diet produce extensive changes in intestinal digestive and transport functions.[166] In addition, modifications in the saturated-to-polyunsaturated fatty acid content in dietary triglycerides alter the fatty acid composition of the rat BBM acyl chains, cholesterol content, the cholesterol/phospholipid molar ratio, and BBM fluidity.[113]

For a variety of cell types, dietary fats influence membrane composition and lipid-dependent functions. For example, in a number of tissues the dietary fat level affects the plasma membrane phospholipid fatty acyl tail composition and membrane function,[10,168] including liver,[257] adipose tissue,[291] mitochondrial membrane,[104,105] brain,[292] and heart.[293] Thus, the inhibition of fat absorption may have unexpected effects on the function of tissues distant from the intestine. Furthermore, dietary fat modification influences T-cell-mediated immunity in insulin-dependent diabetes mellitus in rats.[294,295] T-cell functions are affected by the degree of saturation and the concentration of dietary fat, and linoleic acid appears to play a key role in modulating cellular immune responses.[296] An alteration in the dietary fat intake changes phospholipid fatty acid composition in a number of cellular and subcellular membranes and, consequently, affects membrane functions.[10,297,298] Thus, changing the type or amount of lipids in the diet alters the lipid composition of the intestinal BBM as well as the membrane composition and function of other tissues. Therefore, care must be taken to ensure that any maneuver targeted to modify lipid absorption does not adversely affect the membrane form and formation of intestinal as well as nonintestinal tissues.

Bile Acids and Cystic Fibrosis

Bile acids are essential for the absorption of lipids, and changes in the type or amount of bile acids in the intestinal lumen may alter lipid absorption, interrupt the enterohepatic circulation of bile acids, and lead to the

[288] T. A. Brasitus and D. Schachter, *Biochim. Biophys. Acta* **630**, 152 (1980).
[289] T. A. Brasitus and D. Schachter, *Biochemistry* **19**, 2763 (1980).
[290] T. A. Brasitus and D. Schachter, *Biochemistry* **21**, 2241 (1982).
[291] C. J. Field, A. Angel, and M. T. Clandinin, *Am. J. Clin. Nutr.* **42**, 1206 (1986).
[292] M. Foot, T. F. Cruz, and M. T. Clandinin, *Biochem. J.* **211**, 507 (1983).
[293] M. T. Clandinin, *J. Nutr.* **108**, 273 (1978).
[294] B. Singh, J. Lauzon, J. Venkatraman, A. B. R. Thomson, R. V. Rajotte, and M. T. Clandinin, *Diabetes Res.* **8**, 129 (1988).
[295] B. Cinader, M. T. Clandinin, T. Hosokawa, and N. M. Robblee, *Immunol. Lett.* **6**, 331 (1983).
[296] I. K. Thomas and K. L. Erickson, *J. Nutr.* **115**, 1528 (1985).
[297] P. V. Johnston, *Adv. Lipid Res.* **21**, 103 (1985).
[298] F. H. Faas and W. J. Carter, *Lipids* **18**, 339 (1983).

hepatic secretion of lithogenic bile and, therefore, the development of cholelithiasis. The active transport of bile acids in the ileum is responsive to alterations in membrane composition, and is sensitive to changes in dietary lipid constituents.[299] In conditions such as cystic fibrosis where the luminal concentration of TG is increased because of a deficiency of pancreatic lipase activity, sequestration of bile acids in the intestinal lumen may occur, with a partial interruption of the enterohepatic circulation. The liver is capable of compensating partially for this defective absorption. However, if the reduction in bile acid absorption is sufficient to overcome the compensatory secretory activity of the liver, the luminal concentration of bile acids may fall below their CMC and, therefore, steatorrhea develops. Furthermore, when the concentration of bile acids relative to cholesterol and phospholipid falls, lithogenic bile may result and gallstones may develop.

Feeding bile acids such as chenodeoxycholic (CDC) or ursodeoxycholic (UDC) acid may be used clinically in humans to dissolve cholesterol gallstones. UDC, when fed to normal human volunteers, has no effect on fecal neutral sterol excretion, whereas bile acid excretion is markedly increased.[300] Bran-rich diets increase fecal neutral sterol excretion and decrease cholesterol absorption. The bile acid sequestrant resin cholestyramine reduces the bile acid concentration within the aqueous phase of the intestinal contents, and binds phospholipids.[301] While cholestyramine may be effective when used to reduce elevated levels of plasma cholesterol and LDL, consideration must be given to a potential adverse effect on calcium, magnesium, iron, and zinc metabolism.[302] Changing the luminal content of bile acids by feeding CDC, UDC, or cholestyramine alters intestinal transport and morphology.[299] For example, the active ileal uptake of bile acids is enhanced by feeding UDC, CDC, or cholestyramine, whereas the jejunal passive permeability to bile acids is reduced by feeding cholestyramine.[299] Thus, changes in lipid absorption may alter bile acid absorption, and may be associated with an increased risk of development of gallstones.

Malabsorption of lipids occurs in patients with cystic fibrosis (CF) because of a deficiency of pancreatic lipase.[303,304] This serves as an example of more of the problems that may occur when fat is malabsorbed. There

---

[299] A. B. R. Thomson and M. Keelan, *Digestion* **38,** 160 (1987).

[300] G. Salvioli, R. Lugli, and J. M. Pradelli, *Dig. Dis. Sci.* **30,** 301 (1985).

[301] D. Gallagher and B. O. Schneeman. Effect of dietary cellulose on site of lipid absorption. *Am. J. Physiol.* **249,** G184 (1985).

[302] D. W. Watkins, R. Khalafi, M. M. Cassidy, *et al., Dig. Dis. Sci.* **30,** 477 (1985).

[303] K. Gaskin, P. Gurwitz, P. Durie, M. Corey, H. Levison, and G. Forstner, *J. Pediatr.* **100,** 857 (1982).

[304] G. Mastella, G. Barbato, C. Trabucchi, V. Mengoli, D. Olivieri, and F. A. Rossi, *Rendi. Gastroenterol.* **8,** 134 (1976).

is a large functional reserve capacity of the pancreas, because malabsorption of fat and protein in CF occurs only when lipase and trypsin outputs fall below 10% of normal.[305] In fact, some CF patients have normal fat absorption, even though lipase outputs are only 2% of normal.[306] CF patients have increased losses of bile acids, particularly the taurine conjugates,[307,308] and the bile acid loss is partly from sequestration of the bile acids in the stool.[309] The presence of unhydrolyzed triglycerides, phospholipid and protein, acidic precipitation, and an abnormal fecal flora have all been implicated in the explanation of the bile salt wastage in CF.[308,310] Essential fatty acid deficiency is common in CF,[311] and this may lead to alterations in the composition and metabolism of lipoproteins.[312] The deficiency of fat-soluble vitamins, such as vitamin E, and their correction play an important part in the therapy of CF.

Potential Effects on Colon

Several types of epidemiological studies have shown positive correlations with the incidence of colorectal cancer (CRC) and the intake of dietary fat. The strongest association is for saturated fat.[313,314] The phenotypic changes in the colonic mucosa leading to the development of CRC[315] probably reflect the genetic alterations associated with the multistage development of neoplasia.[316,317] The process of carcinogenesis involves initiation and promotion. Initiating substances interact with DNA and result in mutations. Stimulation of cell proliferation is an important mechanism of tumor

---

[305] E. P. DiMaggio, V. L. W. Go, and W. H. J. Summerskill, *N. Engl. J. Med.* **288,** 813 (1973).
[306] P. R. Durie, K. J. Gaskin, M. Corey, H. Kopelman, Z. Weizman, and G. G. Forstner, *J. Pediatr. Gastroenterol. Nutr.* **3,** 589 (1984).
[307] G. N. Thompson, *J. Pediatr. Gastroenterol. Nutr.* **7,** 214 (1988).
[308] A. M. Weber and C. C. Roy, *Acta Paediatr. Scand.* **317,** 59 (1985).
[309] S. K. Dutta, K. Armand, and T. R. Gadacz, *Gastroenterology* **91,** 1243 (1984).
[310] C. Leroy, G. Lepage, C. L. Morin, J. M. Bertrand, O. Dufour-Larue, and C. C. Roy, *Dig. Dis. Sci.* **31,** 911 (1986).
[311] G. Lepage, E. Levy, N. Ronco, L. Smith, N. Galeano, and C. C. Roy, *J. Lipid Res.* **30,** 1483 (1989).
[312] E. Levy, Y. L. Marcel, R. W. Milne, V. L. Grey, and C. C. Roy, *Gastroenterology* **93,** 1119 (1987).
[313] A. B. Miller, G. R. Howe, M. Jain, K. J. P. Craib, and L. Harrison, *Int. J. Cancer* **32,** 155 (1983).
[314] W. C. Willett, M. J. Stampfer, G. A. Colditz, B. A. Rosner, and F. E. Speizer, *N. Engl. J. Med.* **323,** 1664 (1990).
[315] M. Lipkin, *Cancer Res.* **34,** 878 (1974).
[316] E. R. Fearon, S. R. Hamilton, and B. Vogeistein, *Science* **238,** 193 (1987).
[317] B. Vogelstein, E. R. Fearon, and S. R. Hamilton, *N. Engl. J. Med.* **319,** 525 (1988).

development,[318] and subjects at higher risk for CRC have increased colonic epithelial proliferation.[319–322]

Bile acids and fatty acids potentially may act as cocarcinogens or as tumor promoters. In rats and mice, bile acids and fatty acids increase colonic proliferation rates.[323–325] Dietary fat given as a bolus to humans enhances colonic proliferation.[326] Calcium binds to bile acids and fatty acids,[327] and calcium inhibits the proliferation of human colonic epithelial cells[328,329] as well as in animal colon.[324,325,330–332]

Although controversy surrounds whether persons at increased risk of CRC have higher concentrations of bile acids in their stools,[333–335] a high-fat diet does increase the output and fecal concentrations of bile acids, as compared with a low-fat diet.[336,337] In animal models of CRC, increasing amounts of dietary fat enhance the incidence of tumors as well as the

---

[318] S. Preston-Martin, M. C. Pike, P. K. Ross, P. A. Jones, and B. E. Henderson, *Cancer Res.* **50**, 7415 (1990).

[319] M. Lipkin, W. E. Blattner, J. F. Fraumeni, Jr., H. T. Lynch, E. Deschner, and S. Winawer, *Cancer Res.* **43**, 1899 (1983).

[320] O. T. Terpstra, M. Van Blankenstein, J. Dees, and G. A. Eilers, *Gastroenterology* **92**, 704 (1987).

[321] M. Ponz de Leon, L. Roncucci, P. Di Donato, L. Tassi, O. Smerieri, M. G. Amorico, G. Malagoli, D. De Maria, A. Antonioli, N. J. Chahin, M. Perini, G. Rigo, G. Barberini, A. Manenti, G. Biasco, and L. Barbara, *Cancer Res.* **48**, 4121 (1988).

[322] G. Biasco, G. M. Paganelli, M. Miglioli, S. Brillanti, G. Di Febo, G. Gizzi, L. Ponz de, M. Campieri, and L. Barbara. Rectal cell proliferation and colon cancer risk in ulcerative colitis. *Cancer Res.* **50**, 1156 (1990).

[323] E. E. Deschner, B. I. Cohen, and R. F. Raicht, *Digestion* **21**, 290 (1981).

[324] A. W. Bull, L. J. Marnett, E. J. Dawe, and N. D. Nigro, *Carcinogenesis* **4**, 207 (1983).

[325] M. J. Wargovich, V. W. S. Eng, H. L. Newmark, and W. R. Bruce, *Carcinogenesis* **4**, 1205 (1983).

[326] J. Stadler, H. S. Stern, K. S. Yeung, V. Mcquire, R. Furrer, N. Marcon, and W. R. Bruce, *Gut* **29**, 1326 (1988).

[327] H. L. Newmark, M. J. Wargovich, and W. R. Bruce, *J. Natl. Cancer Inst.* **72**, 1323 (1984).

[328] M. Buset, M. Lipkin, S. Winawer, and E. Friedman, *Cancer Res.* **46**, 5426 (1986).

[329] M. Lipkin, E. Friedman, S. J. Winawer, and H. Newmark, *Cancer Res.* **49**, 248 (1989).

[330] F. R. DeRubertis and P. A. Craven, *J. Clin. Invest.* **74**, 1614 (1984).

[331] M. J. Wargovich, V. W. S. Eng, H. L. Newmark, and W. R. Bruce, *Cancer Lett.* **23**, 253 (1984).

[332] P. J. Hu, A. R. Baer, and M. J. Wargovich, *Nutr. Res.* **9**, 545 (1989).

[333] B. S. Reddy and E. L. Wynder, *Cancer* **39**, 2533 (1977).

[334] M. J. Hill, J. E. Lennard-Jones, D. M. Melville, K. Neale, and J. K. Ritchie, *Lancet* **ii**, 195 (1987).

[335] N. Breuer and H. Goebell, *Klin. Wochenschr.* **63**, 97 (1985).

[336] J. H. Cummings, H. S. Wiggins, D. J. A. Jenkins, H. Houston, T. Jivray, B. S. Drasar, and M. J. Hill, *J. Clin. Invest.* **57**, 953 (1978).

[337] B. S. Reddy, *Cancer Res.* **41**, 3766 (1981).

excretion of bile acids.[337,338,339] These studies therefore raise the possibility that an alteration in lipid absorption may change the type and amount of fatty acids as well as bile acids delivered to the colon. This may place the individual at an increased risk of enhanced cellular proliferation and, therefore, of the development of CRC.

Summary

A reduction in fat intake may be achieved by making educated choices to reduce total calorie intake, to consume a lower quantity of total fats, or to modify the ratio of saturated-to-polyunsaturated lipids. Leptin agonists or NPY or CCK antagonists may prove to be useful to diminish appetite and thereby reduce the total intake of food. But eating has such cultural, social, and hedonistic attributes that such a single-pronged approach is unlikely to be successful. The use of fat substitutes may prove to be popular to provide a wide range of snack food options, but these are likely to be of minimal use in weight reduction programs because of their distribution of additives in only a limited number of foods. The inhibitors of lipid digestion will be modestly successful in the short term; their long-term success will be influenced by gastrointestinal adverse effects and the need to consume fat-soluble vitamin supplements to prevent the development of fat-soluble vitamin deficiencies. The inhibition of lipid absorption is an attractive targeted approach for the treatment of obesity, since this would reduce the uptake of visible as well as invisible fats, which would potentially offer convenient dosing, and could also be a means to inhibit secondarily the uptake of carbohydrate calories.

Acknowledgments

The authors thank Chandra Messier for typing the manuscript. The original studies undertaken by the authors were supported in part by the Medical Research Council (Canada) and the National Science and Engineering Council of Canada.

[338] N. D. Nigro, D. V. Singh, R. L. Campbell, and M. S. Pak, *J. Natl. Cancer Inst.* **54**, 439 (1975).
[339] B. S. Reddy, *Cancer* **36**, 2401 (1975).

[2] Regulation of Hormone-Sensitive Lipase Activity in Adipose Tissue

By CECILIA HOLM, DOMINIQUE LANGIN, VINCENT MANGANIELLO, PER BELFRAGE, and EVA DEGERMAN

Short-Term Regulation of Hormone-Sensitive Lipase

Introduction

The quantitatively most important energy source in mammals is free fatty acids derived from adipose tissue triacylglycerols. Hormone-sensitive lipase (HSL)[1,2] plays a crucial role in the mobilization of free fatty acids from adipose tissue by controlling the rate of lipolysis. It catalyzes the first and rate-limiting step in the hydrolysis of the stored triacylglycerols, and also the subsequent hydrolysis of diacylglycerols and monoacylglycerols. However, to obtain complete hydrolysis of the monoacylglycerols, a second adipose tissue lipase is required, monoacylglycerol lipase.[3,4] HSL is under acute hormonal and neuronal control. Lipolytic hormones, such as catecholamines, cause an activation of cAMP-dependent protein kinase (cAMP-PK), which phosphorylates and activates HSL.[5,6] Insulin, in contrast, prevents this phosphorylation and activation by lowering cAMP levels, which are mediated mainly via phosphorylation and activation of cGMP-inhibited phosphodiesterase (cGI PDE), recently classified as PDE3.[7]

HSL is known to be phosphorylated at two distinct serine residues, termed the regulatory and basal sites, respectively.[5,6] cAMP-PK phosphorylates HSL at the regulatory site (serine 563 and serine 551 in rat and human HSL, respectively).[8-10] Studies in isolated rat adipocytes have shown

---

[1] M. Vaughan, J. E. Berger, and D. Steinberg, *J. Biol. Chem.* **239,** 401 (1964).
[2] P. Strålfors, H. Olsson, and P. Belfrage, *in* "The Enzymes" (P. D. Boyer and E. G. Krebs, eds.), Vol. 18, p. 147. Academic Press, New York, 1987.
[3] H. Tornqvist and P. Belfrage, *Methods Enzymol.* **71,** 646 (1981).
[4] G. Fredrikson, H. Tornqvist, and P. Belfrage, *Biochim. Biophys. Acta* **876,** 288 (1986).
[5] P. Strålfors and P. Belfrage, *J. Biol. Chem.* **258,** 15146 (1983).
[6] P. Strålfors, P. Björgell, and P. Belfrage, *Proc. Natl. Acad. Sci. U.S.A.* **81,** 3317 (1984).
[7] J. A. Beavo, M. Conti, and R. J. Heaslip, *Mol. Pharmacol.* **46,** 399 (1994).
[8] A. J. Garton, D. G. Campbell, P. Cohen, and S. J. Yeaman, *FEBS Lett.* **229,** 68 (1988).
[9] C. Holm, T. G. Kirchgessner, K. L. Svenson, G. Fredrikson, S. Nilsson, C. G. Miller, J. E. Shively, C. Heinzmann, R. S. Sparkes, T. Mohandas, A. J. Lusis, P. Belfrage, and M. C. Schotz, *Science* **241,** 1503 (1988).

that the extent of phosphorylation at the regulatory site correlates with the rate of lipolysis.[6] The mechanism underlying activation through phosphorylation of the regulatory site of HSL by cAMP-PK is poorly understood. The discrepancy between the increase in lipolytic rate in adipocytes on stimulation with lipolytic hormones (up to 50-fold in rat adipocytes) and the increase in triacylglycerol activity of HSL on phosphorylation by cAMP-PK *in vitro* (2- to 3-fold) has been an enigma for a long time.[2] A possible explanation for this discrepancy, which has been suggested by the authors, is that favorable conditions for substrate/enzyme interaction in the *in vitro* assay, that is, a large interfacial area (small particle size) and a high phospholipid: triglyceride ratio, result in relatively high enzyme activity of the dephosphorylated form and consequently only a small activation by phosphorylation. Support for this suggestion comes from calculations of basal and maximal lipolytic rates *in vivo* and *in vitro*, respectively, which show that the maximal lipolytic rates are the same *in vivo* and *in vitro* and that the difference is to be found in the basal lipolytic rate.[2]

An alternative, or additional, explanation could be that on purification, especially during the detergent solubilization step, HSL partly or fully undergoes the conformational change that is assumed to occur on phosphorylation by cAMP-PK *in vivo*. During recent years it has been demonstrated that HSL translocates from the cytosol to the lipid droplet in adipocytes after phosphorylation by cAMP-PK. The data suggest that redistribution of HSL is part of the mechanism behind activation via cAMP-PK phosphorylation.[11]

With regard to the molecular details behind the translocation process, it is not clear whether the phosphorylation-induced changes in the HSL molecule expose a hydrophobic lipid-binding region, which allows HSL to bind to the lipid droplet, or if conformational changes allow the exposition of a region that interacts with one or several other proteins, which aid HSL in translocating and binding to the lipid droplet. In this context we should point out that it has not been experimentally shown that HSL undergoes a conformational change on cAMP-PK phosphorylation. Finally we emphasize that when activation of HSL (rat, human, and bovine) is monitored using an *in vitro* assay (see discussion later) it is only the triacylglycerol lipase (and cholesteryl ester hydrolase) activity of HSL that increases by cAMP-PK phosphorylation, that is, the activity against diglycerides and

---

[10] D. Langin, H. Laurell, L. Stenson Holst, P. Belfrage, and C. Holm, *Proc. Natl. Acad. Sci. U.S.A.* **90**, 4897 (1993).

[11] J. J. Egan, A. S. Greenberg, M.-K. Chang, S. A. Wek, M. C. Moos, Jr., and C. Londos, *Proc. Natl. Acad. Sci. U.S.A.* **89**, 8537 (1992).

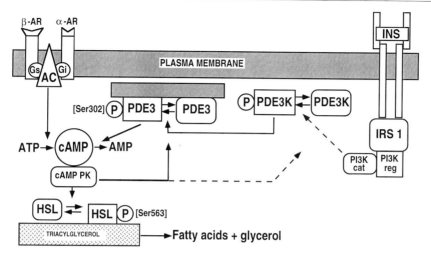

FIG. 1. Working hypothesis for short-term regulation of HSL activity by insulin and cAMP-increasing agents. For details, see text. HSL, Hormone-sensitive lipase; cAMP-PK, cAMP-dependent protein kinase; PDE3, cGMP-inhibited phosphodiesterase; PDE3K, cGMP-inhibited phosphodiesterase kinase; Gs, Gi, G proteins; AC, adenylate cyclase; AR, adrenoceptor; IRS1, insulin receptor substrate 1; PI3K, phosphatidylinositol-3-kinase [regulatory (reg) and catalytic (cat) subunit].

monoglycerides remains unchanged.[2] However, in some very recent monolayer experiments, we were able to show an increased activity against monomolecular films of dicaprin.[12] Whether the di- and monoglyceride lipase activity of HSL increases *in vivo* on phosphorylation by cAMP-PK is not known. In this context note that chicken adipose tissue HSL has shown to be activated 10-fold against tri- and diacylglycerols and also cholesteryl esters on phosphorylation by cAMP-PK,[13–15] suggesting that avian HSL differs from the HSL of mammals with regard to regulatory properties.

A working hypothesis for the hormonal regulation of HSL in adipocytes is presented in Fig. 1 and in the following text. The lipolytic response of the fat cell depends on the balance between stimulatory and inhibitory pathways. Stimulation of adipocytes with lipolytic agents, such as β-adrenergic agonists, results in receptor-mediated increases in cAMP. As described

[12] C. Holm, J. A. Contreras, R. Verger, and M. C. Schotz, *Methods Enzymol.* **284,** 272 (1997).
[13] J. C. Khoo, D. Steinberg, J. J. Huang, and P. R. Vagelos, *J. Biol. Chem.* **251,** 2882 (1976).
[14] J. C. Khoo and D. Steinberg, *J. Lipid Res.* **15,** 602 (1974).
[15] J. C. Khoo, D. Steinberg, and E. Y. C. Lee, *Biochem. Biophys. Res. Commun.* **80,** 418 (1978).

earlier, cAMP activates cAMP-PK and this kinase then phosphorylates HSL on serine 563, resulting in increased enzyme activity of HSL.[5,6,8,9] Insulin is the physiologically important antilipolytic hormone. The ability of insulin to antagonize hormone-induced lipolysis can to a large extent be explained by its ability to lower cAMP levels and thereby cAMP-PK activity,[16] leading to a decrease in the phosphorylation state of HSL.

The decrease in cAMP is mainly the result of an insulin-mediated activation of PDE3.[17,18] PDE3 is a member of one of at least seven different, but related, PDE gene families.[17] Two distinct PDE3 subfamilies, PDE3A and PDE3B, products of distinct but related genes, have been identified, of which PDE3B is expressed in adipocytes.[17,18] The importance of PDE3 as an upstream regulator of HSL activity in insulin's antilipolytic pathway has been shown with the use of specific and cell-permeable PDE3 inhibitors, such as OPC 3911.[19] Specific inhibition of PDE3 in intact cells completely blocks the antilipolytic action of insulin.[19] Activation of PDE3 is associated with phosphorylation of serine 302 in intact cells.[20] This site is also phosphorylated in response to cAMP-increasing hormones, leading to feedback regulation of the cAMP level.[20] In the presence of both hormones, that is, the physiological condition during which insulin exerts its antilipolytic action, there is a more than additive phosphorylation of serine 302 associated with a more than additive activation of PDE3, suggesting crosstalk between the two pathways upstream of PDE3.[21,22] Thus, the regulation of cAMP content, cAMP-PK activity, and thereby the activity of HSL is very complex and includes feedback loops as well as crosstalk between different pathways.

Upstream regulation of PDE3 can be summarized as follows: An insulin-stimulated PDE3 kinase (PDE3K) has been partially characterized but not yet molecularly identified.[23] Indirect evidence suggests that this kinase in

---

[16] C. Londos, R. S. Honnor, and G. S. Dhillon, *J. Biol. Chem.* **260**, 15139 (1985).

[17] V. C. Manganiello, T. Murata, M. Taira, P. Belfrage, and E. Degerman, *Arch. Biochem. Biophys.* **232**, 1 (1995).

[18] E. Degerman, M.-J. Leroy, M. Taira, P. Belfrage, and V. Manganiello, in "Diabetes Mellitus: A Fundamental and Clinical Text" (D. Le Roith, J. M. Olefsky, and S. Taylor, eds.), p. 197, Lippincott, Philadelphia, 1996.

[19] H. Eriksson, M. Ridderstråle, E. Degerman, D. Ekholm, C. J. Smith, V. Manganiello, P. Belfrage, and H. Tornqvist, *Biochim. Biophys. Acta* **1266**, 101 (1995).

[20] T. Rahn, L. Rönnstrand, C. Wernstedt, M.-J. Leroy, H. Tornqvist, V. Manganiello, P. Belfrage, and E. Degerman, *J. Biol. Chem.* **271**, 11575 (1996).

[21] E. Degerman, C. J. Smith, H. Tornqvist, V. Vasta, V. Manganiello, and P. Belfrage, *Proc. Natl. Acad. Sci. U.S.A.* **87**, 533 (1990).

[22] C. J. Smith, V. Vasta, E. Degerman, P. Belfrage, and V. Manganiello, *J. Biol. Chem.* **266**, 13385 (1991).

[23] T. Rahn, M. Ridderstråle, H. Tornqvist, V. Manganiello, G. Fredrikson, P. Belfrage, and E. Degerman, *FEBS Lett.* **350**, 314 (1994).

turn is activated via serine/threonine phosphorylation(s).[23] Another kinase, the dual specificity protein/lipid kinase, phosphatidylinositol-3-kinase (PI3K) is likely to be an upstream regulator of PDE3 and, consequently, HSL, since treatment of cells with the PI3K inhibitor wortmannin completely blocks the activation of PDE3K, phosphorylation/activation of PDE3, and finally the antilipolytic action of insulin.[23] Upon insulin stimulation the PI3K binds to specific tyrosine phosphorylated motifs on insulin receptor substrate 1 (IRS1), the major substrate for the insulin receptor tyrosine kinase. The signal transduction components between PI3K and PDE3K are not yet known. Preliminary data in our laboratories indicate that MAP kinases or p70 S6 kinases, well-known insulin-stimulated kinases, are not involved in the antilipolytic pathway of insulin.[24]

The basal site of HSL is located only two amino acids C terminal of the regulatory site (serine 565 and serine 553 in rat and human HSL, respectively).[8–10] Phosphorylation of this site occurs without any direct effect on enzyme activity. Studies in isolated rat adipocytes have indicated that the basal site is phosphorylated to a large extent in unstimulated cells.[6] This, together with the observation that phosphorylation of the two sites seems to be mutually exclusive, suggests that phosphorylation of the basal site has an antilipolytic role *in vivo*,[25,26] and that phosphorylation of the regulatory site by cAMP-PK has to be preceded by a dephosphorylation of the basal site. Among the kinases known to phosphorylate HSL at the basal site *in vitro,* that is, glycogen synthase kinase-4, Ca^{2+}/calmodulin-dependent protein kinase II, and AMP-activated protein kinase (AMP-PK),[25,26] the latter is the one most likely to be responsible for phosphorylation of the basal site *in vivo.* AMP-PK regulates two other enzymes involved in lipid metabolism, that is, acetyl-CoA carboxylase and 3-hydroxyl-3-methylglutaryl-CoA reductase, which both are inactivated through phosphorylation by AMP-PK resulting in an almost complete cessation of fatty acid and sterol biosynthesis, respectively.[27] AICAR (5-aminoimidazole-4-carboxamide ribonucleoside), a cell-permeable agent that within the cell is metabolized to ZMP (5-aminoimidazole-4-carboxamide ribonucleoside monophosphate), mimics the activating effects of AMP on AMP-PK. Recent studies on isolated rat adipocytes have shown that AICAR antagonizes isoprenaline-induced lipolysis, thus providing evidence that AMP-PK phosphorylation of HSL inhibits subsequent phosphorylation and activation by

---

[24] J. Wijkander and E. Degerman, unpublished observation (1996).
[25] A. J. Garton, D. G. Campbell, D. Carling, D. G. Hardie, R. J. Colbran, and S. J. Yeaman, *Eur. J. Biochem.* **129,** 249 (1989).
[26] A. J. Garton and S. J. Yeaman, *Eur. J. Biochem.* **191,** 245 (1990).
[27] D. G. Hardie, D. Carling, and A. T. R. Sim, *TIBS* **14,** 20 (1989).

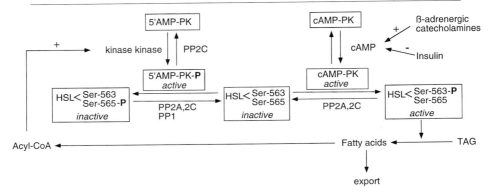

FIG. 2. Hypothesis regarding the interaction between the basal and regulatory site of HSL. The diagram illustrates the phosphorylation of the regulatory site of (rat) HSL (Ser-563) by cAMP-dependent protein kinase (cAMP-PK) and the hypothesis regarding the phosphorylation of the basal site (Ser-565) by 5' AMP-activated protein kinase (5' AMP-PK). For further details, see text. TAG, Triacylglycerol; PP, protein phosphatase.

cAMP-PK also *in vivo*.[28,29] The regulation of AMP-PK has not been described in detail, but seems to involve both allosteric activation by AMP and activation by phosphorylation.[30] The demonstration that the kinase responsible for phosphorylation of AMP-PK is activated by nanomolar concentrations of acyl-CoA is the basis for a proposed negative feedback regulation of HSL.[25,27] Fatty acids, either generated by lipolysis (i.e., by activation of HSL) or taken up from plasma, are converted to CoA esters, which at a certain concentration activate the kinase kinase and AMP-PK, leading to a phosphorylation of HSL at the basal site. This prevents phosphorylation at the regulatory site by cAMP-PK and thus inhibits further lipolysis (Fig. 2).

The data indicating that phosphorylation at the basal and regulatory sites is mutually exclusive[25] imply that activation of HSL by cAMP-PK has to be preceded by dephosphorylation of the basal site. Whether this dephosphorylation is an important regulatory step or merely a result of a high protein phosphatase activity in the cell is not clear. Data from studies using rat adipose tissue HSL and purified preparations of protein phosphatases (PP) or purified bovine adipose tissue HSL and phosphatases from

---

[28] J. E. Sullivan, K. J. Bricklehurst, A. E. Marley, F. Carey, D. Carling, and R. K. Beri, *FEBS Lett.* **353**, 33 (1994).

[29] J. M. Corton, J. G. Gillespie, S. A. Hawley, and D. G. Hardie, *Eur. J. Biochem.* **229**, 558 (1995).

[30] J. Weekes, S. A. Hawley, J. Corton, D. Shugar, and D. G. Hardie, *Eur. J. Biochem.* **219**, 751 (1994).

rat adipocyte extracts together indicate that (1) PP2A and PP2C are the major protein phosphatases acting on both sites, (2) the basal site can be selectively dephosphorylated by PP1, and (3) the basal site is dephosphorylated approximately threefold faster than the regulatory site.[31,32]

Methods

Phosphorylation and Activation of HSL by cAMP-PK in Vitro. The current protocol used in the authors' laboratories for phosphorylation by cAMP-PK is essentially as described previously[5] and was recently described by us in a modified version using recombinant rat HSL as substrate.[33] The optimal parameters for phosphorylation of HSL (i.e., salt concentration, ATP concentration, pH) have been described in detail before.[5] Described next is a protocol for phosphorylation of partially purified or homogenous preparations of HSL, followed by a protocol for monitoring the activation by assaying the triacylglycerol lipase activity. We should mention that recombinant HSL, purified as previously described,[12,13] can be phosphorylated directly after the phenyl-Sepharose step, without prior desalting, since it elutes from this column under low-salt conditions.

1. If necessary, desalt the HSL sample to be phosphorylated, either by dilution or using Centricon (PM30, Amicon, Danvers, MA) or a similar device. As a diluting/desalting buffer, we routinely use 5 mM imidazole, pH 7.2, 1 mM EDTA, 1 mM dithioerythritol, 5–50% glycerol, and 0.008–0.2% $C_{13}E_{12}$.[34] We have observed no adverse effects of the lower glycerol and detergent concentrations, and these are the preferred conditions for applications where high glycerol and detergent concentrations are undesirable, such as in monolayer experiments.[12]

2. Reconstitute the lyophilized powder of the catalytic subunit of cAMP-PK [bovine heart, Sigma P2645 (Sigma, St. Louis, MO), or another kinase source] to 10 U/μl in 60 mM dithioerythritol (10 min at room temperature). An alternative cAMP-PK source with equal performance in

---

[31] H. Olsson and P. Belfrage, *Eur. J. Biochem.* **168**, 399 (1987).
[32] S. L. Woods, N. Emmison, A. C. Borthwick, and S. J. Yeaman, *Biochem. J.* **295**, 531 (1993).
[33] T. Østerlund, B. Danielsson, J. A. Contreras, E. Degerman, G. Edgren, R. C. Davis, M. C. Schotz, and C. Holm, *Biochem. J.* **319**, 411 (1996).
[34] $C_{13}E_{12}$ is a polydisperse preparation of alkyl polyoxyethylenes with the indicated average composition; C, alkyl carbons; E, oxyethylene units. Detergents of this type used to be, but are no longer, available from Berolkemi AB, Stenungssund, Sweden. The authors still have a limited supply of $C_{13}E_{12}$ from this company. Other polyoxyethylene ethers, although not exactly $C_{13}E_{12}$, are available from Sigma and other companies. The authors have good experience with all tested variants of polyoxyethylene ethers with regard to their behavior with the purified enzyme.

the authors' hands is the catalytic subunit from bovine heart (Promega, Madison, WI) as a 100 U/μl solution.

3. Make up activation and control incubation buffers, which are 5× with regard to the final incubation conditions for the phosphorylation, that is:

5× *Activation buffer*
25 mM MgCl$_2$
1 mM ATP
1 U/μl catalytic subunit of cAMP-PK in 60 mM dithioerythritol
5× *Control buffer*
As above and, in addition, 5 μM protein kinase inhibitor [Sigma P0300 (Sigma)]

4. Add 0.2 vol of 5× activation and control buffer, respectively, to the desalted HSL sample. Incubate for 5–60 min at 37°.

5. Terminate the phosphorylation reaction by adding 3 vol of ice-cold stop solution. Keep the samples on ice until assayed.

Stop solution
25 mM EDTA
1 mM dithioerythritol
(Optional, 0.02% defatted bovine serum albumin)

6. Assay aliquots (in quintiplicates), corresponding to 5–20 mU of diacylglycerol lipase activity (i.e., approximately 25–100 ng of rat HSL) against a phospholipid-stabilized emulsion of triolein (0.5 mM) in a Tris-HCl buffer at pH 8.3, as described.[5,33]

Comments

1. The time course for the phosphorylation/activation performed as described is rapid. Maximal phosphorylation is obtained within minutes using partially purified or homogenous HSL as substrate.[5,33] Maximal phosphorylation/activation is maintained for at least 60 min.

2. To monitor the degree of phosphorylation, we include [^{32}P]ATP at a specific activity of $1-2 \times 10^6$ cpm/nmol, stop the reaction by adding 1 vol of SDS–PAGE sample buffer, and analyze the samples by SDS–PAGE using 8% polyacrylamide in the separation gel.

3. Using molar ratios between HSL : catalytic subunit of cAMP-PK of 15 : 1, or lower, we observe a stoichiometry of around 1.0 mol phosphate/mol HSL with no difference between the two kinase preparations mentioned. Under optimal conditions the activation obtained is around 200%.

4. The protocol given is, as mentioned, optimized for purified preparations of HSL. Phosphorylation and activation experiments in crude homogenates are much more difficult to perform and need to be optimized for each experimental system. As a general guideline, we recommend an increase in

the ATP concentration to up to 0.6 mM to compensate for the presence of ATPases in the homogenates and to include protein phosphatase inhibitors. As phosphatase inhibitors we prefer to use either 1 μM okadaic acid or 200 nM okadaic acid plus 0.2 μM inhibitor 2 (Promega), although 40 mM β-glycerophosphate, which is a cheaper alternative, can be used. The activation obtained in crude homogenates is in our experience optimally 50–80% (adipose tissue and insect cell homogenates).

5. As stated earlier, the only bulk substrates that can be used to monitor activation *in vitro* are triacylglycerol and cholesteryl ester substrates. Rather than the physiological pH, pH 8.3 is used in the activation assay because this was found to be the optimal pH for activation of rat adipose tissue HSL. We have not yet systematically checked to determine whether the optimal conditions for rat HSL are optimal for HSL from other species. Preliminary data from monolayer experiments indicate that dicaprin monomolecular films can be used to monitor the effect of phosphorylation in this system.[12]

6. Detergent present in the enzyme sample inhibits HSL activity and it is therefore important to dilute the enzyme to noninhibitory detergent concentrations, that is, to a final concentration of less than 0.0002% for $C_{13}E_{12}$.[35] Note that when phosphorylation/activation experiments are performed in crude homogenates, any endogenous triacylglycerols that remain in the fat-depleted homogenate will reduce the specific activity of labeled exogenous substrate.

Phosphorylation of HSL by AMP-PK. HSL can be phosphorylated on the basal site by AMP-PK, purified from rat liver, as described.[25,26,32]

Dephosphorylation of HSL by PP2A. As described earlier, PP2A dephosphorylates HSL on both the regulatory and the basal site *in vitro*. We routinely use a protocol that is a modification of one previously described for dephosphorylation of adipose tissue HSL. This protocol utilizes a commercially available partially purified preparation of PP2A. The sample to be dephosphorylated is either diluted in, or the buffer exchanged for, 5 mM Tris-HCl, pH 7.0, 0.1% 2-mercaptoethanol, 0.1 mM EDTA, 5% glycerol. PP2A (Upstate Biotechnology Incorporated, Lake Placid, NY (UBI 14-111), heterodimer) is added to obtain molar ratios of HSL : PP2A of around 30–40 : 1 and the mixture incubated for 1 hr at room temperature (22°). Okadaic acid (final concentration of 20 nM) can be used to stop the reaction.

Using recombinant rat HSL, phosphorylated on the regulatory site by cAMP-PK, as substrate, dephosphorylation of at least 80% is obtained using the preceding protocol.[33]

---

[35] G. Fredrikson, P. Strålfors, N. Ö. Nilsson, and P. Belfrage, *J. Biol. Chem.* **256**, 6311 (1981).

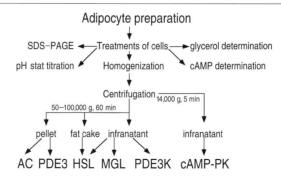

FIG. 3. Flow scheme for the use of isolated adipocytes for studying short-term regulation of HSL. PDE3, cGMP-Inhibited phosphodiesterase; AC, adenylate cyclase; HSL, hormone-sensitive lipase; MGL, monoacylglycerol lipase; PDE3K, cGMP-inhibited phosphodiesterase kinase; cAMP-PK, cAMP-dependent protein kinase.

Preparation of Adipocytes and Hormone Treatments. Adipocytes are prepared from epididymal fat pads of Sprague–Dawley rats by digestion with collagenase as originally described by Rodbell, but with modifications.[36,37] Small samples of the final cell suspension are aspirated into capillary hematocrit tubes and centrifuged for 3 min in a microhematocrit centrifuge in order to estimate the packed cell volume. Washed adipocytes (8–10% cell suspension) are incubated with hormones and other agents in Krebs–Ringer–Hepes buffer containing 2 mM glucose, 0.5–1 U/ml adenosine deaminase, 3–100 nM N^6-phenylisopropyladenosine and 1–3% bovine serum albumin. To study phosphorylation of proteins, the cells (8–10% cell suspension) are prelabeled for 90–120 min with 0.1–1 mCi/ml $^{32}P_i$ in low-phosphate Krebs–Ringer–Hepes buffer (300 μM KH_2PO_4).[38] Figure 3 represents an outline of how isolated adipocytes can be used to study the short-term regulation of HSL. Note that small differences with regard to buffer compositions, centrifugation speed, etc., in the following protocols are due to the fact that each protocol has been worked out separately and optimized for the specific component to be analyzed. With minor compromises it should be possible to adapt the protocols to be able to analyze several components from the same cell preparation. We should also emphasize that the amount of cells needed for each experiment, as well as the amount of $^{32}P_i$ for labeling experiments, varies according to the abundance of the component to be studied.

[36] M. Rodbell, *J. Biol. Chem.* **232,** 1065 (1964).
[37] R. C. Honnor, G. S. Dhillon, and C. Londos, *J. Biol. Chem.* **260,** 15122 (1985).
[38] T. J. Hopkirk and R. M. Denton, *Biochim. Biophys. Acta* **885,** 195 (1986).

SDS–PAGE Analysis of Adipocyte Phosphoproteins. Although HSL is found at low abundance in adipocytes, it is possible to visualize phosphorylated HSL directly by SDS–PAGE analysis of total extracts of ^{32}P-labeled adipocytes.[19] Described next is a protocol for preparation of total cell extracts from adipocytes for SDS–PAGE analysis. The advantage of this method is simplicity and that all subcellular compartments are taken into account; that is, the risk of trapping HSL in the fat cake is minimized. Obviously, this procedure cannot be used for studies of HSL translocation. Another disadvantage is that proteins are identified only by their molecular weights. This disadvantage can be at least partly overcome by performing parallel Western blots. Western blot analysis of HSL, using either enhanced chemiluminescence (ECL) or ^{125}I-labeled protein A or G as the detection system is described later. However, since ^{32}P proteins can interfere with both of these detection systems, we prefer to use a colorimetric detection system for Western blot analysis of ^{32}P-labeled proteins [for instance, the ProtoBlot Western Blot AP System (Promega) using alkaline phosphatase-conjugated antichicken IgG as secondary antibody. Sigma A1043 (Sigma)].

1. Transfer 250-μl aliquots of cell suspension (2%), labeled with 0.1 mCi/ml ^{32}P$_i$, to microcentrifuge tubes containing 100 μl dinonyl phthalate and centrifuge very rapidly (<10 sec).
2. Dissolve the cells floating on the dinonyl phthalate carefully in 50 μl of 8 mM Na-PP$_i$, 1.6 mM EDTA, 20% ethanol, 0.1% 2-mercaptoethanol, 4% SDS, 0.25% bromophenol blue, and 20 μg/ml each of leupeptin, antipain, and pepstatin A.
3. Freeze the tubes and slice through the dinonyl phthalate layer. Separate the top fraction, containing the solubilized cells, from remaining dinonyl phthalate by centrifugation, followed by boiling, centrifugation, and aspiration of the infranatant.
4. Reextract the dinonyl phthalate phase once more with 50 μl of fresh SDS solution (see step 2) to recover all cellular proteins. Pool the two extracts and analyze by SDS–PAGE with 8% polyacrylamide in the separation gel. Analyze the dried gel for ^{32}P by digital imaging.

Glycerol Release Determination. Glycerol is routinely measured according to the method described by Garland and Randle[39] using reagents from Boehringer Mannheim (Mannheim, Germany). This method is based on the conversion of NADH to NAD as glycerol is enzymatically converted to pyruvate, using glycerol kinase, pyruvate kinase, and lactate dehydrogenase, respectively, as catalysts.

[39] P. B. Garland and R. J. Randle, *Nature* **196,** 987 (1962).

For purposes requiring a very sensitive glycerol determination method, either a radiometric method[40] or a ultrasensitive bioluminescent technique[41] is used.

pH Stat Titration of Fatty Acid Release. The release of free fatty acids from isolated adipocytes can be measured continuously using a pH stat titration technique, as described previously.[42] An improvement of this technique, including a computerized on-line registration of the free fatty acid release from two incubations simultaneously, is under development and will be described elsewhere on completion.

Analysis of HSL in Infranatant, Pellet, and Fat Cake Fractions from Adipocytes. After treatment with various hormones and agents, adipocytes (1.2 ml of 2–10% cell suspensions) are washed twice with 5 ml of homogenization buffer [0.25 M sucrose, 50 mM Tris-HCl, pH 7.4 (optional), 1 mM EDTA, 1 mM dithioerythritol, protease inhibitors (10–20 μg/ml leupeptin, 2–20 μg/ml antipain, 1–10 μg/ml pepstatin A), and protein phosphatase inhibitors (either 1 μM okadaic acid or 10 mM sodium pyrophosphate and 50 mM NaF)][43] and homogenized at room temperature in 3–10 vol of homogenization buffer in a glass homogenizer using 20 strokes with a Teflon pestle. The homogenates are then centrifuged at 50,000–110,000g at +4° for 1 hr. The fat cakes are carefully transferred to Eppendorf tubes using a spatula and dissolved in 0.5–1 ml of SDS sample buffer (60 mM Tris-HCl, pH 6.8, 20% SDS, 10% glycerol, 5% 2-mercaptoethanol, and 0.025% bromophenol blue). The fat-free infranatant is carefully removed to a new Eppendorf tube. The pellet fraction is resuspended in the same homogenization buffer as before or dissolved in SDS sample buffer (as before but with 2% SDS). When adipocytes are prepared as described earlier, 65–80% of HSL distributes to the infranatant fraction, 10–25% to the fat cake fraction, and less than 10% to the pellet fraction from unstimulated cells.[35,44] As described in the introduction, HSL translocates to the fat cake on stimulation of adipocytes by lipolytic hormones and agents, and the distribution between the infranatant and fat cake is reversed.[11] As stated later, more than 90% of PDE3 is found in the pellet fraction, and more than 80% of monoacylglycerol lipase (MGL) is found in the infranatant fraction.

As discussed earlier, it is difficult to monitor the activation of HSL *in*

---

[40] D. C. Bradley and H. R. Kaslow, *Anal. Biochem.* **180,** 11 (1989).
[41] H. Kather, F. Schröder, and B. Simon, *Clin. Chim. Acta* **120,** 295 (1982).
[42] N. Ö. Nilsson and P. Belfrage, *Methods Enzymol.* **71,** 319 (1981).
[43] NaF should not be used as protein phosphatase inhibitor if HSL activity is to be measured since millimolar concentrations of NaF inhibit HSL activity.
[44] K. N. Frayn, D. Langin, C. Holm, and P. Belfrage, *Clin. Chim. Acta* **216,** 183 (1993).

vitro. The reason for this is not fully understood. Contributing to this difficulty are most likely both the marked difference between the *in vivo* substrate (i.e., the adipocyte droplet) and the *in vitro* substrate (i.e., phospholipid-stabilized trioleoylglycerol emulsions) and the redistribution of HSL to the fat cake on phosphorylation by cAMP-PK.[2,11] In practical terms, it is very difficult to study HSL activation in adipocyte preparations and other crude preparations. To some extent these difficulties are overcome by following HSL phosphorylation (using ^{32}P-labeled cells) instead of HSL activity. Using immunoprecipitation of HSL from fat-free infranatants of ^{32}P-labeled cells, combined with direct SDS–PAGE analysis of the fat cakes from the same cells, it is possible to quantify total phosphorylation of HSL. The basal site is typically labeled to about 50% in hormonally quiescent cells, whereas the regulatory site is labeled to close to 100% in cells that have been maximally stimulated.[6] Because phosphorylation of the two sites seems to be mutually exclusive, an approximate doubling of the HSL phosphorylation is observed in stimulated cells.

The proximity of the two phosphorylation sites makes it difficult to evaluate specific phosphorylation of one of the sites, since there is no possibility for generating fragments or peptides containing only one of the sites. The only possibility that remains for studying the phosphorylation state of each of the sites is to make phosphopeptide analysis of tryptic peptides followed by radiosequencing, which is very time consuming for routine purposes. Therefore, attempts to generate phosphorylation-state specific antibodies are ongoing in the authors' laboratories.

For quantification of the total amount of HSL in unlabeled cells, for instance in translocation experiments, we recommend Western blot analysis of the different fractions, as discussed later. For very dilute (infranatant) samples, HSL protein can be enriched by immunoprecipitation prior to Western blot analysis of the immunoprecipitates. The amount of enzymatically active HSL, with no discrimination between phosphorylated (activated) and nonphosphorylated HSL, can be measured in the infranatant fraction using an assay utilizing a diacylglycerol analogue as substrate. This assay provides high specificity and sensitivity and is described later in the section on long-term regulation.

Immunoprecipitation of HSL. For immunoprecipitation and Western blot analysis of HSL we use an affinity-purified antibody against recombinant rat HSL,[33] raised in hen. This antibody detects native as well as denatured enzyme, neutralizes activity, exhibits no discrimination between phosphorylated and unphosphorylated enzyme, and shows good cross-reactivity with human HSL. The following protocol describes immunoisolation of HSL from the infranatant fraction of rat adipocytes prepared as described earlier.

1. Aliquots of HSL corresponding to approximately 50–150 ng of HSL are incubated with affinity-purified antirat HSL (100–350 ng; 5 ng antibody/ng HSL) for 2 hr at room temperature.
2. Immobilized antichicken antibodies [antichicken IgY; Promega G1191 (Promega), prewashed according to the instructions from the manufacturer] are added. Immunocomplexes are allowed to form during 1 hr on ice, with the sample rotated every 10 min.
3. Immunoprecipitates are washed 5× with 1 ml of TBS (20 mM Tris-HCl, pH 7.4, 0.9% NaCl) and dissolved in SDS sample buffer.
4. Immunoprecipitates are analyzed by SDS–PAGE using 8% polyacrylamide in the separation gel followed by Western blot analysis as described later (unlabeled cells) or digital imaging of ^{32}P in dried gels (^{32}P-labeled cells).

Western Blot Analysis of HSL. We have systematically tested different protocols for Western blot analysis of HSL, and found the following protocol to yield the best results. All steps are performed at room temperature.

1. Block the nitrocellulose membrane for 2 hr in TBS with 5% defatted milk powder.
2. Wash the membrane 1 × 15 min, followed by 2 × 5 min in TBS, 0.25% defatted milk powder, 0.25% Tween 20 (wash buffer).
3. Incubate with affinity-purified anti-HSL, diluted to approx. 0.15 μg/ml in wash buffer, for 1 hr.
4. Wash 1 × 15 min, followed by 4 × 5 min.
5. Incubate with horseradish peroxidase-conjugated antichicken IgG [Promega G1351 (Promega)] diluted 5000× in wash buffer, for 1 hr.
6. Wash as in step 4.
7. Develop the membrane using the ECL system (Amersham RPN2109, Amersham, Buckinghamshire, UK) according to the instructions from the manufacturer.

Comment. Instead of the ECL system, ^{125}I-labeled protein A or G can be used. In that case, incubate with a rabbit antichicken antibody [Sigma C2288 (Sigma), 2000-fold dilution in wash buffer] as step 5, followed by washing (3 × 15 min), incubation for 1 hr with ^{125}I-labeled protein A protein G, diluted to 1 × 10^6 cpm/ml in wash buffer, and finally wash as before. Analyze the dried membrane for ^{125}I by digital imaging or autoradiography. Using both the ECL and ^{125}I-labeled protein A and G, according to the protocols described, we find the Western blot to be quantitative at least in the range from 0.5 to 100 ng (rat HSL).[45]

---

[45] C. Holm, unpublished observations (1996).

Measurement of Adenylate Cyclase Activity. Adenylate cyclase activity can be measured in the pellet fraction using $[\alpha^{32}P]ATP$ or $[^3H]ATP$ as substrate as described.[46,47]

Determination of cAMP Levels. To measure cAMP levels, aliquots of adipocyte suspension (0.5 ml) are treated with 50 μl 2 M HCl, heated to 100° for 1 min and neutralized. cAMP is measured using a protein kinase binding assay[48] or a radioimmunoassay.[49] The latter method is more sensitive, but also more tedious and expensive, Kits for determination of cAMP levels using radioimmunoassay, as well as enzyme immunoassay and scintillation proximity assay, are commercially available from Amersham and other manufacturers.

It is generally accepted that determination of the fraction of cAMP-PK activity that is activated by endogenous cAMP is a more relevant measurement than determination of cAMP levels, since it represents an indirect measurement of that cAMP relevant to metabolic events in the cell. Consequently, measurement of cAMP levels has to a large extent been replaced by determination of cAMP-PK activity ratios (see later discussion).

Determination of cAMP-PK Activity Ratios. Adipocyte incubations are terminated by homogenization of 800-μl cells (2.5–12% cell suspension) in 200-μl ice-cold 50 mM Tris-HCl (pH 7.4) containing 50 mM EDTA, 0.5 mM OPC3911 and 0.5 mM Ro 20-1724. The samples are immediately centrifuged (14,000g, 15 min, 4°) to obtain infranatant fractions in which the cAMP-PK activity ratios are assayed according to Eagan *et al.*[50] under conditions previously described.[37]

Analysis of Monoacylglycerol Lipase (MGL) in Infranatant Fractions of Adipocytes. MGL distributes ~80% to the infranatant fraction, prepared from rat adipocytes as described earlier.[3] To measure MGL activity in these crude infranatant fractions, 1(3)-mono-oleoyl[^{3}H]glycerol in mixed micelles with nonionic detergent is used.[3] This assay method is highly specific for MGL, since both HSL and lipoprotein lipase are strongly inhibited by the concentration of nonionic detergent used in the assay. Since cDNA probes and antibodies for MGL are not yet available, it is currently not possible to measure MGL mass and MGL mRNA levels.

Immunoisolation of PDE3 from ^{32}P-Labeled Adipocytes

1. The ^{32}P-labeled adipocytes (1.5–2 ml of a 10% cell suspension) are treated with various agents and then washed twice with 5 ml 50 mM

---

[46] R. A. Johnson and Y. Salomon, *Methods Enzymol.* **195**, 3 (1991).
[47] R. Alvarez and D. V. Daniels, *Anal. Biochem.* **187**, 98 (1990).
[48] A. G. Gilman, *Proc. Natl. Acad. Sci. U.S.A.* **67**, 305 (1970).
[49] E. K. Frandsen and G. Krishna, *Life Sci.* **18**, 529 (1976).
[50] J. J. Eagan, M.-K. Chank, and C. Londos, *Anal. Biochem.* **175**, 552 (1988).

TES, pH 7.4, 250 mM sucrose, 1 mM EDTA, 0.1 mM EGTA, 40 mM phenyl phosphate, 5 mM NaF, and 10 μg/ml each of antipain and leupeptin and 1 μg/ml of pepstatin A (buffer is kept at room temperature to avoid trapping of PDE3 in the fat cake).[20–22]

2. Incubations are terminated by homogenization in 1 ml of washing buffer at room temperature. Homogenates are transferred to centrifuge tubes on ice and centrifuged at 50,000–100,000g for 45–60 min.
3. Fat cakes and infranatants are removed (see earlier description). The crude membrane fractions, containing more than 90% of total cell PDE3 activity, are homogenized in solubilization buffer (50 mM Tris-HCl, pH 7.4, 5 mM MgCl$_2$, 1 mM EDTA, 100 mM NaBr, 50 mM NaF, 1% C$_{13}$E$_{12}$, 20% glycerol, and 10 μg/ml each of antipain and leupeptin and 1 μg/ml of pepstatin A) (1 ml/ml packed cell volume). After 1 hr on ice, supernatants (10,000g, 15 min, +4°) are mixed with specific PDE3 antibodies and incubated at 4°.[20–22]
4. Protein A Sepharose-CL4B (Pharmacia; 50% slurry) is added after 4–16 hr and the incubation continued for 15 min, after which the immunoprecipitates are washed 5× with 1 ml PBS, 0.1% N-lauryl sarcosine as described.[20] Washed immunoprecipitates are boiled in SDS sample buffer and subjected to SDS–PAGE analysis using 7% polyacrylamide in the separation gel. Dried gels are analyzed using digital imaging. For phosphorylation site analysis, the gels are subjected to Western blotting onto nitrocellulose membranes. ^{32}P-PDE3 is identified, cut out, and subjected to tryptic digestion. Tryptic phosphopeptides are then analyzed as detailed in Ref. 20.

PDE3 Assay. PDE3 activity is measured in the pellet fraction from unlabeled cells, as has been described in detail before.[51]

PDE3K Assay. Adipocyte incubations (8–10% cell suspension) are terminated by homogenization at room temperature in 50 mM TES-buffer, pH 7.5, containing 40 mM phenyl phosphate, 0.2 mM EGTA, 2 mM EDTA, 250 mM sucrose, 0.1 mM vanadate, 10 μg/ml each of antipain and leupeptin, and 1 μg/ml of pepstatin A (5–10 ml/ml packed cell volume). PDE3K activity is measured in fat-free infranatants using solubilized PDE3 from untreated adipocytes as substrate as detailed in Ref. 23.

Long-Term Regulation of Hormone-Sensitive Lipase

Introduction

Little is known of long-term regulation of HSL. The available data have recently been reviewed[52] so therefore we devote this section mainly to

[51] V. Manganiello, F. Murad, and M. Vaughan, *J. Biol. Chem.* **246**, 2195 (1971).
[52] D. Langin, C. Holm, and M. Lafontan, *Proc. Nutr. Soc.* **55**, 93 (1996).

methods used to study the long-term regulation of HSL, that is, methods to measure HSL mRNA levels, HSL mass, and HSL activity in adipose tissue biopsies or adipocytes. A brief summary of the current knowledge in the field will, however, be given.

Several descriptive studies are available of variation in HSL gene expression in different physiological situations. A reciprocal regulation of HSL and lipoprotein lipase has been demonstrated in pregnant rats, with increased HSL mRNA levels and activity during late pregnancy, concomitant with reduced LPL mRNA and activity levels.[53] A similar reciprocal regulation of HSL and lipoprotein lipase is seen in hibernating marmots, with increased HSL mRNA levels during the fasting period.[54] An increase in HSL mRNA levels, protein, and activity is also seen during long-term (>3 days) fasting in rats.[55] The levels of HSL mRNA, protein, and activity vary according to the anatomical location of the adipose tissue, with lower levels in subcutaneous rather than visceral fat.[56,57] A dramatic increase of HSL gene expression is observed during differentiation of preadipocytes to adipocytes, where HSL mRNA expression and activity is greatly enhanced concomitant with the appearance of other late markers (fatty acid-binding protein, glycerol-3-phosphate dehydrogenase) and the accumulation of triglycerides.[58] Furthermore, overexpression of HSL in a preadipocyte cell line prevents the appearance of late markers and the acquirement of a normal adipocyte phenotype.[59]

Several disorders in lipid metabolism are associated with impaired adipocyte lipolysis. It has been suggested that HSL could be involved in the pathogenesis of some of these disorders and studies supporting this notion are beginning to emerge.[60-63] The insulin-resistance syndrome and familial combined hyperlipidemia are two conditions that seem to involve postreceptor defects in lipolysis.[60-62] It has been shown that HSL activity is de-

---

[53] A. Martin-Hidalgo, C. Holm, P. Belfrage, M. C. Schotz, and E. Herrera, *Am. J. Physiol.* **266**, E930 (1994).
[54] B. E. Wilson, S. Deeb, and G. Florant, *Am. J. Physiol.* **262**, R177 (1992).
[55] C. Sztalryd and F. B. Kraemer, *Am. J. Physiol.* **266**, E179 (1994).
[56] C. Sztalryd and F. B. Kraemer, *Metabolism* **43**, 241 (1994).
[57] G. Tavernier, J. Galitzky, P. P. Valet, A. Remaury, A. Bouloumié, M. Lafontan and D. Langin, *Am. J. Physiol.* **268**, E1135 (1995).
[58] D. Langin, M. Dauzats, and M. Lafontan, unpublished observation (1994).
[59] C. Szatlryd, M. Komaromy, and F. B. Kraemer, *J. Clin. Invest.* **95**, 2652 (1995).
[60] S. Reynisdottir, K. Ellerfeldt, H. Wahrenberg, H. Lithell, and P. Arner, *J. Clin. Invest.* **93**, 2590 (1994).
[61] S. Reynisdottir, M. Eriksson, B. Angelin, and P. Arner, *J. Clin. Invest.* **95**, 2161 (1995).
[62] S. Reynisdottir, B. Angelin, D. Langin, H. Lithell, M. Eriksson, C. Holm, and P. Arner, *Arteriosc. Thromb. Vas.* in press.
[63] L. Hellström, D. Langin, S. Reynisdottir, M. Dauzats, and P. Arner, *Diabetologia* **39**, 221 (1996).

creased in patients with familial combined hyperlipidemia, but not in patients with the insulin-resistance syndrome, indicating that different mechanisms cause the impaired lipolysis in these conditions. In a very recent study, in which adipocyte lipolysis was investigated in nonobese subjects with and without a family trait for obesity, HSL activity was found to be reduced in the former group, suggesting that impaired function of HSL could be an early event in the development of obesity.[63]

Despite these descriptive studies of the long-term regulation of HSL very little is known about the factors controlling HSL gene expression. An extensive study regarding this was recently carried out in which the effect of cAMP and phorbol esters on HSL mRNA levels and HSL activity in two different adipocyte cell lines was measured.[64] Both cAMP and phorbol esters were shown to cause a substantial decrease in HSL mRNA levels and HSL activity, whereas insulin, growth hormone, and retinoic acid had no effect on HSL gene expression. The lack of effect of insulin is somewhat in contrast with a recent study showing that HSL mRNA, protein, and activity levels were increased twofold in streptozotocin-induced diabetic rats.[65] The effect on HSL activity was reversed on insulin treatment, although HSL mRNA and protein levels remained elevated, indicating that the effect observed was secondary to the hypoinsulinemia, rather than a direct effect of insulin. Alternatively, the apparent difference regarding the effect of insulin on HSL expression could be due to the large difference between the two experimental systems. Differences in experimental systems could also be the basis for the contrasting findings with regard to dexamethasone, which was shown to have no direct effect, although potentiating the cAMP effect, in the two adipocyte cell lines,[64] but an enhancing effect on HSL mRNA levels in isolated rat adipocytes.[66]

Methods

Methods used to analyze long-term regulation of HSL involve determination of HSL activity, HSL mass, and HSL mRNA levels. These are described next. To allow measurement of these parameters adipose tissue biopsy specimens are taken from animals or human subjects and divided in two parts, one of which is used for RNA extraction and measurement of mRNA levels, and the other for homogenization. The homogenate is separated into fat-free infranatant, fat cake, and pellet fractions as described earlier, and used for measurements of HSL activity and HSL mass. Cultured cells are prepared in the same way. We stress, however, that in cultured preadipocytes a redistribution of HSL from the infranatant fraction to the

[64] E. Plee-Gautier, J. Grober, E. Duplus, D. Langin, and C. Forest, *Biochem. J.* **318,** 1057 (1996).
[65] C. Szatlryd and F. B. Kraemer, *Metabolism* **44,** 1391 (1995).
[66] B. G. Slavin, J. M. Ong, and P. A. Kern, *J. Lipid Res.* **35,** 1535 (1994).

pellet fraction, rather than to the fat cake, has been described.[67] Adaptation of the methods for measuring HSL activity and HSL mRNA levels allows measurement of biopsies of approximately 100 mg wet weight.[44] Besides the methods described later, work is in progress in the authors' laboratories to develop phosphorylation state-specific HSL antibodies to allow estimation of the degree of phosphorylation of the basal and regulatory sites, respectively. Enzyme-linked immunosorbent assays for both human and rat HSL are also being developed. These new valuable tools should hopefully be available within the near future.

HSL Activity and HSL Mass

Fat-free infranatant pellet, and fat cake fractions are prepared from adipose tissue biopsies or adipocytes as described earlier. For homogenization of very small amounts of material, motor-driven pestles fitting microfuge tubes (Kontes, Vineland, NJ) are used (2 min). The high-speed centrifugation is performed in a table-top ultracentrifuge (Beckman TL-100, Beckman, Palo Alto, CA). Because we have frequently observed problems with proteolytic degradation of the HSL protein in fat-free infranatants prepared from human adipose tissue biopsies, we routinely supplement the homogenization buffer (0.25 M sucrose, 1 mM EDTA, 1 mM dithioerythritol) with excess protease inhibitors, that is, 40 μg/ml leupeptin, 20 μg/ml antipain, and 10 μg/ml pepstatin A. The protease inhibitors are always added fresh from concentrated stock solution, stored at $-20°$. Fat-free infranatants are used for measurement of HSL activity and for determination of HSL mass using Western blot analysis with or without prior immunoprecipitation. Fat cake fractions are used for determination of HSL mass by Western blot analysis.

The ideal substrate for measuring HSL activity in crude samples, such as fat-free infranatants prepared from isolated adipocytes or adipose tissue, is a monoether analogue of dioleoylglycerol: 1(3)oleoyl-2-O-oleylglycerol.[68] As described,[12,69] this substrate has several advantages over the dioleoylglycerol and trioleoylglycerol, and the use of this for measuring HSL activity in crude samples optimizes both the sensitivity and the specificity of the assay. The high sensitivity of this assay is explained by the fact that diacylglycerol is the preferred substrate for HSL and is hydrolyzed at least 10 times more rapidly than triacylglycerol.[70] The high specificity is due to the fact that no substrate for monoacylglycerol lipase is formed from monoacyl-

---

[67] A. H. Hirsch and O. M. Rosen, *J. Lipid Res.* **25**, 665 (1984).
[68] H. Tornqvist, P. Björgell, L. Krabisch, and P. Belfrage, *J. Lipid Res.* **19**, 654 (1978).
[69] G. Fredrikson, P. Strålfors, N. Ö. Nilsson, and P. Belfrage, *Methods Enzymol.* **71**, 636 (1981).
[70] G. Fredrikson and P. Belfrage, *J. Biol. Chem.* **258**, 14253 (1983).

monoalkylglycerol and that under the assay conditions (pH 7.0, without apoC-II) no lipoprotein lipase activity is measured. The synthesis of unlabeled and ^{3}H-labeled 1(3)oleoyl-2-O-oleylglycerol as well as the assay method has been described in detail.[12,33,68,69]

Other lipid substrates for HSL include dioleoylglycerol, trioloeylglycerol, cholesteryl esters, and monooleoylglycerol. Dioleoylglycerol is as sensitive as monoacylmonoalkylglycerol but is less specific since substrate for monoacylglycerol lipase is formed after hydrolysis of the first ester bond. Emulsified monooleoylglycerol can only be used for measuring HSL activity in preparations not containing monoacylglycerol lipase, that is, it is not useful for adipocyte/adipose tissue infranatant fractions. Cholesteryl esters and trioleoylglycerol are the only substrates that can be used to monitor the activation of HSL through cAMP-PK phosphorylation. However, as discussed earlier, it is generally difficult to measure the phosphorylation-induced activation of HSL in crude preparations. Because of the substrate specificity of HSL, with high preference for diacylglycerols,[69] assay methods based on trioloeylglycerol and cholesteryl esters are less sensitive than assay methods using diacylglycerol as substrate. Furthermore, small amounts of triacylglycerol remaining in the fat-free infranatant could interfere with the trioleoylglycerol assay. In addition to the lipid substrates described, two substrates can be used under conditions where they are fully water soluble; p-nitrophenyl butyrate (PNPB) and tributyrin.[33] The usefulness of these substrates in crude enzyme preparations is limited, however, by the fact that these substrates are hydrolyzed by a wide range of hydrolytic enzymes, making the assay method based on this substrates nonspecific.

We emphasize that the monoacylmonoalkylglycerol substrate does not discriminate between phosphorylated and unphosphorylated HSL. Therefore, measuring HSL activity with this substrate is to be considered as a mass determination assay for enzymatically active HSL. The problems with measuring the degree of activation of HSL in crude homogenates and samples were discussed earlier. Note that the redistribution of HSL on stimulation with lipolytic hormones and drugs presents a potential problem if measurement of HSL activity in fat-free infranatants is taken as a measure of total amount of enzymatically active HSL. First of all, it is possible that translocation effects could be added to any long-term effects on HSL expression, due, for instance, to catecholamine stress on killing of an animal or taking biopsies from patients. Secondly, nothing is known about any potential long-term regulation of the translocation process. To get an accurate determination of the total amount of HSL in adipocyte/adipose tissue preparations, we therefore recommend determination of HSL mass using Western blot analysis of the infranatant, fat cake, and pellet fractions. As stated, HSL distributes mainly to the infranatant and fat cake fractions in isolated adipocytes,[11] whereas it distributes mainly to the infranatant and pellet fraction in cultured preadipocyte cell lines.[67]

HSL mass is measured using Western blot analysis. For dilute infranatant fractions, the HSL antigen is enriched by immunoprecipitation prior to Western blot analysis. As mentioned earlier, the anti-HSL antibody, raised against recombinant rat HSL, cross-reacts well with human HSL in immunoprecipitation, Western blot, and immunoinhibition experiments. Analysis of HSL mass by Western blot method is linear from 0.5 ng to at least 100 ng.[45]

HSL mRNA: Northern Blot Analysis. We routinely extract RNA from adipose tissue and adipocytes using the acid guanidinium thiocyanate-phenol-chloroform method, described by Chomczynski and Sacchi.[71] We have had good experience with two extractions with chloroform[72] prior to extraction with phenol-chloroform and isopropanol precipitation.[71] For very small samples, the homogenization is performed using microfuge tubes and motor-driven pestles (Kontes, Vineland, NJ) (45 sec, 4°), as described earlier for preparing fat-free infranatants. All subsequent steps are performed in microfuge tubes.[44]

cDNA probes against rat, mouse, and human HSL are available.[9,10,73] Due to the high homology between the cDNA sequences from these three species (78–92%) the probes can be used across species (rodents, man, dog, guinea pig, and rabbit) with changes in the stringency for the hybridization and washings. For a 100% homologous probe we perform the Northern blot analysis as follows:

1. RNA samples are electrophoresed in 1% agarose, 2.2 M formaldehyde gels, transferred to nylon blots, and cross-linked by exposure to ultraviolet light.[74]
2. Blots are prehybridized in 0.5 M sodium phosphate, pH 7.0, 7% SDS, 1% bovine serum albumin, and 1 mM EDTA[74] at 65° for at least 1 hr and then hybridized in the same buffer using ^{32}P-labeled HSL cDNA (10^9 cpm/μg; 1–2 × 10^6 cpm/ml) overnight at 65°.
3. Blots are washed (2 × 20 min) with 2 × SSC (0.6 M NaCl, 0.06 M sodium citrate, pH 7.0), 0.1% SDS, 60° followed by washes (2 × 20 min) at 0.1 × SSC, 0.1% SDS, 60°. The membrane is exposed to autoradiography or subjected to digital image analysis.

[71] P. Chomczynski and N. Sacchi, *Anal. Biochem.* **162**, 156 (1987).
[72] I. Louveau, S. Chaudhuri, and T. D. Etherton, *Anal. Biochem.* **196**, 308 (1991).
[73] Z. Li, M. Sumida, A. Birchbauer, M. C. Schotz, and K. Reue, *Genomics* **24**, 259 (1994).
[74] G. M. Church and W. Gilbert, *Proc. Natl. Acad. Sci. U.S.A.* **81**, 1991 (1984).

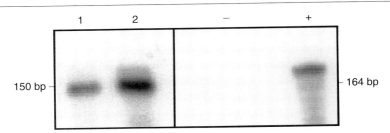

FIG. 4. Ribonuclease protection assay of human HSL mRNA. A human HSL cRNA probe was ^{32}P-labeled using T3 RNA polymerase. Protected products are shown for synthetic sense human HSL RNA (lane +), 10 μg yeast tRNA (lane −), 10 and 30 μg of human adipocyte total RNA (lanes 1 and 2).

HSL mRNA: Ribonuclease Protection Assay

A ribonuclease protection assay is a more sensitive method for measuring HSL mRNA levels from small human adipose tissue biopsies. This method is at least 10 times more sensitive than Northern blot analysis, and is more tolerant of limited degradation of the RNA, since the probe used is considerably shorter than the mRNA to be detected. This could be an important consideration for samples that are not collected under optimal conditions. The method described here in Fig. 4 is an improved version of our previously described assay.[44] The probe used in this modified assay was generated as follows: Human HSL cDNA was cut with *Xho*I and *Kpn*I. The resulting 150-bp fragment (nucleotides 694–844) was gel purified and subcloned into the pBluescript KS vector (Stratagene, La Jolla, CA). The plasmid containing the HSL cDNA was linearized using *Xho*I and used as template for making the labeled antisense probe.

1. Make the labeled antisense probe using T3 RNA polymerase and including [^{32}P]UTP in the reaction mixture. Eliminate the DNA template by RQ1DNase (Promega) digestion for 15 min at 37°. Check the integrity and quality of the probe on a 5% acrylamide–7 M urea gel.
2. Mix the labeled cDNA probe in 30 μl of hybridization buffer (80% deionized formamide, 0.4 M NaCl, 1 mM EDTA, 40 mM PIPES, pH 6.7) with the appropriate RNA sample.[75] Heat the mixture for 5 min at 85° and transfer to a 55° heating block.
3. After incubation at 55° for 14–18 hr, add 40 μg ribonuclease A (Boehringer Mannheim) and 700 U ribonuclease T1 (Boehringer

[75] J. J. Lee and N. A. Costlow, *Methods Enzymol.* **152**, 633 (1987).

Mannheim) in 0.3 ml of 700 mM NaCl, 5 mM EDTA, and 10 mM Tris-HCl, pH 7.5. Digest for 1 hr at 37°.
4. Stop the digestion by adding 50 μg of proteinase K, and incubate the samples for 30 min at 37°.
5. Add yeast tRNA (10 μg/tube) in 0.3 ml homogenization buffer (1.4 M guanidinium thiocyanate, 25 mM sodium citrate, pH 7.0, 0.5% N-lauryl sarcosine) and precipitate with isopropanol.
6. Wash the RNA pellets with 70% ethanol, air dry, and dissolve in 10 μl of sample buffer (97% deionized formamide, 0.1% SDS, 10 mM Tris-HCl, pH 7.0, 0.1% bromophenol blue, and 0.1% xylene cyanol FF).
7. Analyze the samples on a 5% acrylamide–7 M urea gel, dry the gel, and analyze using digital imaging.

Acknowledgments

We thank all coworkers and collaborators who have contributed to the work described, especially Marit Anthonsen, Juan Antonio Contreras, Birgitta Danielsson, Michéle Dauzats, Hans Eriksson, Keith Frayn, Jacques Grober, Henrik Laurell, Marie Karlsson, Tova Rahn, Ann-Helen Thorén, Jonny Wijkander, and Torben Østerlund.

The work in our laboratories referred to in this chapter was supported by grants from the Swedish Medical Research Council (No. 112 84 to C. H. and 3362 to P. B.), the Medical Faculty, Lund University, A. Påhlsson's Foundation, E. and W. Cornell's Foundation, the Crafoord Foundation, Groupe Danone, and the Ministère de L'Enseignement Supérieur et de la Recherche (grant to D. L.).

C. H., D. L., P. B., and E. D. are members of the BIOMED I Concerted Action EUROLIP supported by the European Union.

[3] Use of Gene Knockout Mice to Establish Lipase Function

By DAVID Y. HUI

Introduction

Dietary lipids exist as an emulsion of triglycerides, cholesterol, cholesteryl esters, and phospholipids. The emulsified lipid particle must be hydrolyzed before these nutrients can be absorbed through the gastrointestinal tract and transported to other tissues through the plasma. It is well recognized that the complete hydrolysis and digestion of dietary fat and lipids requires the concerted effort of many lipolytic enzymes located throughout the digestive tract. In humans, lipid digestion is initiated in the stomach through the action of gastric lipase.[1] Complete lipid digestion occurs in the proximal small intestine by lipolytic enzymes released from the pancreas. The pancreatic lipolytic enzymes include phospholipase A_2, pancreatic triglyceride lipase, and a nonspecific carboxyl ester lipase. Because the carboxyl ester lipase is the only enzyme in the intestinal lumen capable of hydrolyzing cholesteryl esters,[2] and because its activity is activated severalfold by bile salts present in the lumen,[3] this enzyme has also been referred to as cholesterol esterase or bile salt stimulated lipase. Readers interested in the structure–function relationship and other biochemical characteristics of these lipolytic enzymes can refer to other chapters in this volume and in Volume 284 for additional details. This chapter discusses the use of gene knockout technology to generate transgenic mice with defective expression of specific lipase genes and the use of these animals to address the physiologic function of these proteins.

Pancreatic Carboxyl Ester Lipase (Cholesterol Esterase)

The carboxyl ester lipase/cholesterol esterase (CEL) is synthesized in the acinar cells of the pancreas and is stored in zymogen granules. The CEL can be released into the intestinal lumen as a component of the pancreatic juice after food ingestion,[4] in response to elevated levels of gastric hormones such as cholecystokinin, secretin, and bombesin.[5,6] The role of the pancreatic CEL in dietary cholesterol absorption was first proposed based on observations that removal of the pancreas dramatically reduced cholesterol absorption,[7] and that the reinfusion of pancreatic juice containing CEL, but not CEL-depleted pancreatic juice, restored both the mucosal esterase activity and the intestinal cholesterol absorption process.[8] Furthermore, humans with exocrine pancreatic insufficiency usually display cholesterol malabsorption, which can be restored to nearly normal levels by exogenous enzyme substitution with preparations containing the pancreatic cholesterol esterase.[9]

A direct role for pancreatic CEL in mediating intestinal absorption of

---

[1] F. Carriere, J. A. Barrowman, R. Verger, and R. Laugier, *Gastroenterology* **105,** 876 (1993).
[2] C. Erlanson, *Scand. J. Gastroenterol.* **5,** 333 (1970).
[3] E. A. Rudd, N. K. Mizuno, and H. L. Brockman, *Biochim. Biophys. Acta* **918,** 106 (1987).
[4] C. R. Treadwell and G. V. Vahouny, *in* "Handbook of Physiology" (C. F. Code, W. Heidel, eds.), Vol. III, pp. 1407–1438, American Physiological Society, 1968.
[5] Y. Huang and D. Y. Hui, *J. Biol. Chem.* **266,** 6720 (1991).
[6] Y. Huang and D. Y. Hui, *Arch. Biochem. Biophys.* **310,** 54 (1994).
[7] L. L. Gallo, T. Newbill, J. Hyun, G. V. Vahouny, and C. R. Treadwell, *Proc. Soc. Exp. Biol. Med.* **156,** 277 (1977).
[8] L. L. Gallo, S. B. Clark, S. Myers, and G. V. Vahouny, *J. Lipid Res.* **25,** 604 (1984).
[9] M. Vuoristo, H. Vaananen, and T. A. Miettinen, *Gastroenterology* **102,** 647 (1992).

free and esterified cholesterol was also observed *in vitro* by incubating rat intestinal sacs with cholesterol containing micelles in the presence or absence of CEL. Experiments performed in the presence of CEL showed a three- to fivefold enhancement of intracellular cholesterol and cholesteryl ester accumulation compared to intestinal sacs incubated in the absence of the enzyme.[10] The ability of CEL to catalyze cholesterol uptake by enterocytes was reaffirmed in a study by Lange and his colleagues.[11] Using the Caco-2 cells as a model system for intestinal cells, these investigators showed that incubation of the cells with 1 nM unesterified cholesterol, provided in a mixed micelle with taurocholate and phosphatidylcholine, resulted in only minimal cellular uptake of the cholesterol in the medium.[11] The exogenous cholesterol associated with the Caco-2 cells was not transferred to a "physiologically important pool" that could be esterified and packaged into lipoproteins.[11] However, the level of cholesteryl ester formation was increased 39-fold when Caco-2 cells were incubated with cholesterol in the presence of exogenously added CEL.[11]

Although the literature data just summarized appeared to suggest a role for pancreatic CEL in mediating cholesterol absorption through the gastroinestinal tract, the physiologic importance of this enzyme in dietary cholesterol absorption remains controversial. Several independent studies have failed to confirm the earlier observations and yielded results that conflict with those previously described. For example, Watt and Simmonds also used pancreatic diverted rats and showed that cholesterol absorption efficiency was similar between the two groups, suggesting that pancreatic enzymes including the CEL are not involved in this process.[12] Research from the author's laboratory[13] and from Fisher and colleagues[14] also has failed to demonstrate an effect of exogenously added CEL on free cholesterol uptake by Caco-2 and enterocytic cells in culture. The only effect of CEL appeared to be its ability to facilitate cholesteryl ester transport.[13] However, other *in vitro* studies suggest that CEL may also act in concert with other lipolytic enzymes, including gastric lipase and pancreatic lipase, for complete digestion of dietary triglycerides before their absorption through the intestine. Whether CEL plays a physiological role in fat absorption remains to be determined.

In addition to its synthesis in the pancreas and its presence in the intestinal lumen, pancreatic-type CEL was also shown to be synthesized

---

[10] S. G. Bhat and H. L. Brockman, *Biochem. Biophys. Res. Commun.* **109**, 486 (1982).

[11] A. Lopez-Candales, M. S. Bosner, C. A. Spilburg, and L. G. Lange, *Biochemistry* **32**, 12085 (1993).

[12] S. M. Watt and W. J. Simmonds, *J. Lipid Res.* **22**, 157 (1981).

[13] Y. Huang and D. Y. Hui, *J. Lipid Res.* **31**, 2029 (1990).

[14] R. Shamir, W. J. Johnson, R. Zolfaghari, H. S. Lee, and E. A. Fisher, *Biochemistry* **34**, 6351 (1995).

in the liver and present in the serum.[15–19] In *in vitro* experiments, CEL was shown to mediate the conversion of large LDL particles to smaller LDL,[20] facilitate the selective uptake of HDL-associated cholesteryl esters by hepatic cells,[21] and hydrolyze lysophosphatidylcholine generated from LDL oxidation.[22] Taken together, these observations suggest that the systemic CEL may participate in the atherosclerosis process by modulating the composition and metabolism of plasma lipoproteins. The availability of animal models with different levels of CEL gene expression provides a useful tool to address the various functions of CEL, including its physiologic role in intestinal lipid transport, systemic lipoprotein metabolism, and atherosclerosis.

One approach to address the physiologic function of the pancreatic CEL is to produce animals defective in expression of the CEL gene. Production of CEL gene knockout mice was accomplished by homologous recombination of an exogenously added disrupted CEL gene with the endogenous CEL gene in mouse embryonic stem (ES) cells.[23] The CEL gene targeted ES cells were then used to produce a mouse model with defective CEL gene expression.[23] These animals are useful for determining the significance of the cholesterol esterase in lipid metabolism, including its function in dietary fat and cholesterol absorption.[23]

Principle of Gene Targeting by Homologous Recombination in Embryonic Stem Cells

Animals with specific genetic mutations and deletions can be produced by introducing site-specific mutations in the genome of embryonic stem cells. This approach takes advantage of observations that pluripotent ES cells obtained from mouse blastocysts can be cultured *in vitro* and remain viable for differentiation after their injection into a different embryo and reimplantation into a foster mother.[24] In a typical experiment, the ES cells

[15] E. A. Camulli, M. J. Linke, H. L. Brockman, and D. Y. Hui, *Biochim. Biophys. Acta* **1005**, 177 (1989).
[16] J. A. Kissel, R. N. Fontaine, C. Turck, H. L. Brockman, and D. Y. Hui, *Biochim. Biophys. Acta* **1006**, 227 (1989).
[17] E. H. Harrison, *Biochim. Biophys. Acta* **963**, 28 (1988).
[18] R. Zolfaghari, E. H. Harrison, A. C. Ross, and E. A. Fisher, *Proc. Natl. Acad. Sci. U.S.A.* **86**, 6913 (1989).
[19] K. E. Winkler, E. H. Harrison, J. B. Marsh, J. M. Glock, and A. C. Ross, *Biochim. Biophys. Acta* **1126**, 151 (1992).
[20] J. Brodt-Eppley, P. White, S. Jenkins, and D. Y. Hui, *Biochim. Biophys. Acta* **1272**, 69 (1995).
[21] F. Li, Y. Huang, and D. Y. Hui, *Biochemistry* **35**, 6657 (1996).
[22] R. Shamir, W. J. Johnson, K. Morlock-Fitzpatrick, R. Zolfaghari, L. Li, E. Mas, D. Lombardo, D. W. Morel, and E. A. Fisher, *J. Clin. Invest.* **97**, 1696 (1996).
[23] P. N. Howles, C. P. Carter, and D. Y. Hui, *J. Biol. Chem.* **271**, 7196 (1996).
[24] M. J. Evans, and M. H. Kaufman, *Nature* **292**, 154 (1981).

and the recipient embryo are obtained from animals carrying genes of different coat colors, such that the initial selection of chimeric mice can be based on the coat color of the offspring. The most commonly used ES cells to date are those derived from the mouse strain 129, which has an agouti coat color. These ES cells can then be microinjected into embryos obtained from C57BL/6J mice, which have a black coat color. Offspring with a high degree of agouti coat color, indicating the transmission of genes derived from the ES cell genome, can then be crossbred with each other to obtain animals with germline transmission of the ES cell genes.

Modification of specific genes in the ES cell genome depends on the ability of transfected DNA to recombine with the homologous gene in the chromosome.[25] Although such targeted recombination is rare compared with nonhomologous integration of transfected DNA, a methodology has been developed to optimize the chance of homologous recombination and to screen and select rapidly ES cells in which such an event has taken place. Most of the experiments to date utilized isogenic DNA for the targeting construct to maximize hybridization of the targeting DNA to the proper gene locus in the chromosome.[26] Furthermore, a minimum of 4-kB DNA homologous DNA sequence is usually used for optimal recombination to occur at the specific gene locus. A selectable gene marker, such as the neomycin-resistant gene, would then be inserted into an exon to disrupt the coding sequence of the gene of interest. The chimeric targeting gene construct can be used to transfect ES cells.

Homologous recombination of the transfected DNA with chromosomal DNA at the target locus will result in the replacement of a portion of the endogenous gene with the targeting construct, thus disrupting the coding sequence and inactivation of the endogenous gene. The use of a selectable gene marker allows for the selection of cells that have taken up and expressed the transfected DNA. Growth of the ES cells in the presence of antibiotic selection indicates the integration of the transfected DNA into the ES cell genome. In a successful experiment, approximately 0.01 to 0.001 of the antibiotic-resistant cells would have the transfected DNA targeted to the proper gene locus, whereas the remaining cells would have incorporated the DNA in a nonhomologous site. In some cases, investigators have included a negative selectable gene marker at the 5' or 3' end of the targeted construct to allow for selection against random insertion events. Homologous recombination at the targeted gene locus would result in deletion of the negative selectable gene marker, whereas integration at nonhomologous sites would have included this marker in the genome. The

---

[25] K. R. Thomas and M. R. Capecchi, *Cell* **51**, 503 (1987).
[26] C. Deng and M. R. Capecchi, *Mol. Cell. Biol.* **12**, 3365 (1992).

inclusion of a negative selection marker usually results in an additional 10-fold enrichment of homologous recombination clones.

The utilities for production of gene knockout mice are many-fold. First, gene knockout mice can be produced as an animal model for human diseases.[27] A second utility for the production of knockout mice is to explore the physiologic function and significance of specific genes.[28,29] This chapter summarizes research in the author's laboratory utilizing the gene knockout approach to study the physiologic significance of the CEL gene.

Description of Procedures

Materials and Reagents

1. Bacteriophage λ-DASH DNA cloning kit, including *Escherichia coli* LE392 cells (Stratagene, La Jolla, CA)
2. Restriction enzymes: *Bam*HI, *Bgl*II, *Eco*RI, *Hinc*II, *Hin*dIII, *Msc*I, *Pst*I, *Sac*I, *Sal*I, *Sph*I, *Ssp*I, and *Xba*I (New England BioLabs, Beverly, MA)
3. Klenow fragment of DNA polymerase I (Ambion, Inc., Austin, TX)
4. Nitrocellulose and Nytran Plus membrane filters (Schleicher & Schuell, Inc., Keene, NH)
5. PTZ18U plasmid DNA (United States Biochemicals, Cleveland, OH)
6. PMC1Neo plasmid DNA (Stratagene)
7. Agarose (FMC Biotechnology, Rockland, ME)
8. G418 neomycin analog (Sigma Chemicals, St. Louis, MO)
9. Taq Polymerase (Gibco BRL Life Technologies, Gaithersburg, MD)
10. ^{125}I-labeled anti-rabbit IgG (Amersham Life Science, Arlington Heights, IL)
11. [^{3}H]cholesterol, [^{3}H]cholesteryl oleate, and [^{14}C]sitosterol (Amersham Life Science)
12. C57BL/6 and the 129 strain of mice (Taconic Farms, Germantown, NY)

Buffers and Media

1. ES cell culture medium: Dulbecco's modified Eagle's medium (DMEM) with 4500 mg/liter glucose and 4 mM glutamine (HyClone Laboratories, Inc., Logan, UT)

---

[27] S. Ishibashi, J. L. Goldstein, M. S. Brown, J. Herz, and D. K. Burns, *J. Clin. Invest.* **93,** 1885 (1994).
[28] S. H. Zhang, R. L. Reddick, J. A. Piedrahita, and N. Maeda, *Science* **258,** 468 (1992).
[29] A. S. Plump, J. D. Smith, T. Hayek, K. Aalto-Setälä, A. Walsh, J. G. Verstuyft, E. M. Rubin, and J. L. Breslow, *Cell* **71,** 343 (1992).

2. ES cell qualified fetal bovine serum (Gibco BRL Life Technologies, Inc.)
3. SSC solution: 150 mM NaCl and 15 mM sodium citrate
4. 50× Denhardt's solution: 5 g Ficoll (type 400) 5 g polyvinylpyrrolidone, 5 g bovine serum albumin in 500 ml water
5. Hybridization buffer: 50% formamide, 5× SSC, 5× Denhardt's solution, 0.1% SDS, and 2 mg/ml denatured salmon sperm DNA
6. TAE buffer: 40 mM Tris-acetate, pH 7.5, 1 mM EDTA
7. Church and Gilbert hybridization buffer[30]: 500 mM sodium phosphate, pH 6.8, 1 mM EDTA, 7% SDS, 1% bovine serum albumin, 100 µg/ml salmon sperm DNA
8. Wash buffer A: 40 mM sodium phosphate, pH 6.8, 1 mM EDTA, 5% SDS, and 0.5% bovine serum albumin
9. Wash buffer B: 40 mM sodium phosphate, pH 6.8, 1 mM EDTA, and 1% SDS

Cloning of Mouse CEL Gene

The first step in generating a CEL gene knockout mouse is the isolation of the mouse CEL gene. Because efficiency of homologous recombination depends on the percentage of homologous sequence between the targeting DNA and the specific gene in the ES cell genome, a mouse genomic DNA library made with DNA from the mouse 129 strain was used for this purpose. Mouse DNA was partially digested with *Sau*3A and ligated to *Bam*HI-digested λ-DASH bacteriophage vector DNA (prepared by Dr. Thomas Doetschman, University of Cincinnati). The DNA was used to transfect *E. coli* LE392 cells. Approximately 1×10^6 recombinant bacteriophages were screened by means of the filter hybridization technique.[31] The full-length rat CEL cDNA was used as the probe.[16] The cDNA fragment was radiolabeled to a specific activity of approximately 1×10^8 cpm/µg with [^{32}P]dATP by the Klenow fragment of DNA polymerase I in the presence of a random oligonucleotide primer. Nitrocellulose filters containing the genomic library were hybridized at 40° for 60 hr with 1×10^6 cpm/ml of ^{32}P-labeled CEL cDNA probe in hybridization buffer.

At the end of the hybridization period, the filters were washed three times for 30 min each at 40° in $1 \times$ SSC buffer containing 0.1% SDS. The filters were air dried and then exposed to Kodak XAR-2 films for 18 hr to identify the positive clones. One positive clone that also hybridized with probes corresponding to both the 5'-flanking region and the 3'-flanking region of the rat CEL gene was selected for use. Restriction mapping,

---

[30] G. M. Church and W. Gilbert, *Proc. Natl. Acad. Sci. U.S.A.* **81**, 1991 (1984).
[31] W. D. Benton and R. W. Davis, *Science* **196**, 180 (1977).

Southern hybridization with probes corresponding to specific exons in the rat CEL gene,[32] and partial nucleotide sequencing were performed to identify the intron and exon locations of the mouse CEL gene.

Construction of CEL Gene Targeting Vector DNA

A 4.7-kB DNA fragment overlapping exons 1 through intron 7 of the mouse CEL gene was obtained by *Sac*I digestion of the genomic DNA clone. This DNA was ligated to similarly digested PTZ18U plasmid vector. Nucleotide sequence analysis of this DNA fragment revealed the presence of a unique *Msc*I restriction site within exon 4 of the CEL gene. Thus, the CEL gene targeting DNA could be constructed by insertion of the neomycin-resistant gene marker at this site to disrupt the coding sequence of the mouse CEL gene. In our laboratory, we obtained a neomycin-resistant gene cassette by digestion of the pMC1Neo (Stratagene) plasmid with *Ssp*I and *Hinc*II. The 1.3-kB DNA fragment was separated from vector DNA by agarose gel electrophoresis and used for ligation with an *Msc*I digested PTZ18U vector containing the 4.7-kB *Sac*I fragment of mouse CEL gene. A recombinant plasmid containing the neomycin-resistant gene in the same orientation as the CEL gene was selected for the gene targeting experiment. The DNA was purified by CsCl centrifugation, digested with *Sac*I, and the 6.5-kB DNA fragment containing the disrupted CEL gene sequence was isolated by agarose gel electrophoresis.

Targeted Disruption of CEL Gene in ES Cells

The R1 ES cell line[33] derived from the 129 mouse strain was used for targeted disruption of the CEL gene. The ES cells were grown on mitomycin-treated embryonic fibroblasts in 10% CO_2 atmosphere in ES cell culture medium supplemented with 0.14 mM monothioglycerol (Sigma Chemicals, St. Louis, MO) and 15% heat-inactivated fetal bovine serum.[34] The ES cells cultured at high density were supplemented with leukemia inhibitory factor at ~500 u/ml.[35-37] Approximately 5.5×10^7 actively growing ES cells were

---

[32] R. N. Fontaine, C. P. Carter, and D. Y. Hui, *Biochemistry* **30**, 7008 (1991).

[33] A. Nagy, J. Rossant, R. Nagy, W. Abramow-Newerly, and J. C. Roder, *Proc. Natl. Acad. Sci. U.S.A.* **90**, 8424 (1993).

[34] T. C. Doetschman, H. Eistetter, M. Katz, W. Schmidt, and R. Hemler, *J. Embryol. Exp. Morph.* **87**, 27 (1985).

[35] R. L. Williams, D. J. Hilton, S. Pease, T. A. Wilson, C. L. Stewart, D. P. Gearing, E. F. Wagner, D. Metcalf, N. A. Nicola, and N. M. Gough, *Nature* **336**, 684 (1988).

[36] A. G. Smith, J. K. Heath, D. D. Donaldson, G. G. Wong, J. Moreau, M. Stahl, and D. Rogers, *Nature* **336**, 688 (1988).

[37] J.-F. Moreau, D. D. Donaldson, F. Bennett, J. Witek-Giannotti, S. C. Clark, and G. G. Wong, *Nature* **336**, 690 (1988).

scraped into 0.5 ml of culture medium and then electroporated in the presence of 3 pmol of the CEL gene targeting DNA, using an IBI Gene Zapper 450 (VWR Scientific, Chicago, IL) set to deliver 200 μF at 800 V/cm.

The cells were transferred to fresh tissue culture flasks containing a feeder layer of G418-resistant embryonic fibroblasts and then cultured in ES cell medium for 24 hr. The cell culture was washed to remove dead cells and then a selection medium containing 300 μg/ml of the neomycin analog G418 was then added to the culturing flasks. The cells were allowed to grow in this selection medium and the G418 concentration was changed to 250 μg/ml after 3 days. Embryonic cell colonies resistant to G418 selection after 7 days of culture were isolated and expanded individually in 24-well dishes. When the cells reached near confluency, usually after 3 or 4 days of growth, approximately half of the cells in each colony were used to isolate DNA for characterization; the remaining cells were maintained for colony expansion and freezing.

Southern Blot Analysis

The initial screening for homologous recombination between the targeting DNA and the endogenous CEL gene was performed by Southern blot analysis. Genomic DNA from each colony was isolated and then digested with *Eco*RI according to the manufacturer's conditions. The digested genomic DNA was fractionated on 0.7–0.8% agarose gels in TAE buffer. After electrophoresis, DNA was denatured in 1.5 M NaCl, 0.5 N NaOH, and transferred overnight from agarose gels to Nytran Plus membranes. The membranes were marked for orientation, rinsed in 1.5 M NaCl, 0.5 M Tris-HCl, pH 7.0, dried, and baked for 1.5 hr at 80° under vacuum. The membrane filters containing the digested genomic DNA were prehybridized for 1 hr at 65° in Church and Gilbert's hybridization buffer[30] and then hybridized for 18 hr at 65° in the same buffer containing a ^{32}P-labeled 1.1-kB *Pst*I/*Sac*I DNA fragment that corresponds to the genomic sequence 5' from the targeting DNA. After hybridization, the filters were washed once at 65° with wash buffer A and three to four times with wash buffer B for 30 min each time. The Nytran membranes containing the genomic DNA were then exposed to X-ray films to identify reactive bands.

Using the hybridization procedure described, the 1.1-kB *Pst*I/*Sac*I DNA fragment would identified a fragment >30 kB in length with wild-type ES cell DNA, due to the absence of an *Eco*RI restriction site in the endogenous mouse CEL gene. Therefore, DNA isolated from any G418-resistant colonies that yielded a single *Eco*RI band at >30 kB with this radiolabeled probe would indicate a random insertion event, and homologous recombination did not take place in the CEL gene locus. However, if hybridization of the *Eco*RI-digested DNA resulted in a 6.7-kB positive signal in addition to the >30-kB band, then correct gene targeting has taken place in one CEL allele. This hybridization pattern would be the result of introducing two *Eco*RI sites present in the TK promoter of the Neo^R gene into exon 4 of the endogenous CEL gene. Further confirmation of correct targeting at the CEL gene locus was performed by digesting the ES colony DNA with *Xba*I and then hybridizing the DNA with an ~1100 bp *Bgl*II/*Sal*I DNA fragment corresponding to the 3' sequences from the targeting DNA. In the latter hybridization protocol, wild-type CEL allele yielded a 7.2-kB fragment as expected, while the correctly targeted allele resulted in an additional 9.0-kB band due to the insertion of 1.75 kB of DNA corresponding to the selectable marker cassette.

Additional confirmation for site-specific integration of the targeting DNA at the CEL locus could also be accomplished by Southern blot analysis of genomic DNA digested with various other restriction enzymes, using both the 5'- and 3'-flanking probes. The addition of the 1.75-kB neomycin-resistant gene cassette to the endogenous CEL gene resulted in an 8.3-kB *Hin*dIII digested fragment that hybridized with the 5' probe in addition to the 6.5-kB band observed for the endogenous CEL gene. Using the 3' probe, a 7.5-kB *Sph*I band, in addition to the endogenous 18-kB *Sph*I signal, was the expected result due to insertion of an *Sph*I site in the neomycin-resistant gene into the CEL gene locus.

Generation of Chimeric Mice

Mouse ES cells containing the disrupted CEL gene were microinjected into blastocysts obtained from female C57BL/6J mice. Approximately 100 of the injected blastocysts were then surgically implanted into a (C57BL/6 × C3H/HeN) F1 pseudopregnant female foster mother.[38] Because the ES cells used in this study were derived from the 129 mouse strain, chimeric mice were identified by the degree of agouti coat color on the black background. Male chimeric mice were mated with female outbred Black Swiss mice for germline transmission of the ES cell genome.

Test for Germline Transmission of Disrupted CEL Gene

Agouti coat color offspring were screened for transmission of the modified CEL gene by polymerase chain reaction (PCR) analysis of their DNA.

[38] A. Gossler, T. Doetschman, E. Serfling, and R. Kemler, *Proc. Natl. Acad. Sci. U.S.A.* **83**, 9065 (1986).

The mice were genotyped shortly before weaning. Ear punch tissue was incubated at 55° in 25 µl of digestion buffer containing 400 µg/ml proteinase K for 4 hr to overnight. After digestion, samples were vortexed, and the hair was allowed to settle. Two microliters of the DNA suspension was diluted with 48 µl of H_2O, heated at >90° for 5 min, and then cooled immediately on ice. One microliter of the boiled sample was used within 4 hr for PCR analysis. The PCR analysis was performed in 30 µl total volume with 2.5 µM of each oligonucleotide primer, 200 µM of all four deoxynucleotide triphosphates, and one unit of Taq polymerase. The reagents were added on ice and then preheated to 90° for 1 min before initiation of the PCR. The amplification was set for 33 cycles consisting of 30 sec at 94° and 2 min at 65°. Samples were then analyzed on 1.5% agarose gels.

Two sets of primers were used for genotyping. The first set of PCR primers was designed by Kim and Smithies[39] to detect the presence of a 555-bp fragment corresponding to the neomycin-resistant gene. The second set of primers was designed to amplify exon 4 of the mouse CEL gene. The upstream CEL primer was 5'-CCCTTTCAGTGTCCCACAACCT-3', and the downstream CEL primer was 5'-TCACTATTCCCGCTCTTA-CAGTC-3'. These primers amplify a 244-bp fragment from the wild-type exon 4 but, under the conditions used, do not amplify the targeted allele because of the insertion of the Neo gene between their cognate sequences. Identical conditions were used with both primer sets; however, the two amplifications could not be achieved in the same reaction. Thus, duplicate reactions were performed on each mouse DNA sample. A positive amplification signal with the exon 4 primers but not the Neo primers was scored as a wild type. Positive signals with both sets of primers indicated a CEL(+/−) genotype, and a positive signal with the Neo primers but not with the CEL exon 4 primers would indicate a homozygous CEL-null genotype.

In the author's laboratory, the modified CEL gene was transmitted by chimeric males at the expected frequency of approximately 50%. More important, crosses between CEL(+/−) male and CEL(+/−) female mice resulted in a 1:2:1 distribution of offspring with the CEL(+/+), CEL(+/−), and CEL(−/−) genotypes. Thus, CEL gene deletion did not confer any obvious signs of disadvantage to the embryos or any effect on reproduction.[23] The CEL(−/−) mice grew normally and appeared to be healthy by inspection.[23] Thus, these animals can be used for physiologic studies to determine the significance of CEL gene expression.

---

[39] H. S. Kim and O. Smithies, *Nucleic Acid Res.* **16,** 8887 (1988).

Cholesterol Ester Lipase Determination

The effect of CEL gene ablation on pancreatic CEL protein biosynthesis was confirmed by immunoblot assay. The mice were euthanized and the abdominal cavities opened. The pancreas was removed and homogenized on ice in a solution containing 10 mM sodium phosphate, pH 6.2, 0.1 M NaCl, 1 mM EDTA, 0.2% sodium azide, 1.5% glycerol, and 0.02% soybean trypsin inhibitor. The homogenate was centrifuged at 4° for 1 hr at 100,000g to precipitate particulate fractions. Twenty-five micrograms of the 100,000g supernatant fraction from each sample was analyzed by electrophoresis on a 10% SDS–polyacrylamide gel.[40] The electrophoresed samples were either stained with Coomassie blue or transferred onto nitrocellulose paper.[41] The nitrocellulose paper containing the samples was then used for immunoblotting using affinity purified rabbit antirat cholesterol esterase and ^{125}I-labeled antirabbit IgG. The results showed that CEL(+/−) mice displayed approximately half the level of CEL protein in their pancreas compared with the wild-type animals.[23] The pancreatic extract from CEL(−/−) mice did not show any anti-CEL cross-reactive material.[23]

Use of CEL Knockout Mice to Determine Physiologic Function of CEL

Cholesterol absorption efficiency in CEL-gene-deleted mice could be assessed using a standard absorption measurement protocol as described previously by several laboratories. In our laboratory, we used the single-dose dual-isotope feeding method, originally described by Zilversmit and Hughes[42] and validated for measuring cholesterol absorption in mouse by Dueland *et al.*,[43] to accomplish this task. Test meal was prepared either as a lipid emulsion or as sonicated vesicles. The emulsified substrate was prepared by mixing 35 μCi of [^{3}H]cholesterol (50 Ci/mmol) or [^{3}H]cholesteryl oleate (50 Ci/mmol) and 7 μCi of β-[^{14}C]sitosterol (56 mCi/mmol) with 2 mg of cholesterol, 8 mg of phosphatidylcholine (PC), and 50 mg of triolein in organic solvent. After evaporating the solvent under a nitrogen stream, 1 ml of deionized water was added and the sample was emulsified by sonication for 5 min. When cholesteryl ester absorption was being studied, the mix also included 0.2 mg of unlabeled cholesteryl oleate. For some of the cholesteryl ester studies, the emulsion consisted of 50 μCi of [^{3}H]cholesteryl oleate, 1 μCi of β-[^{14}C]sitosterol, 120 μg cholesterol, 12 μg cholesteryl oleate, 960 μg PC, and 5 μg triolein emulsified in 600 μl of water.

[40] U. K. Laemmli, *Nature* **227**, 680 (1970).
[41] W. N. Burnette, *Anal. Biochem.* **112**, 195 (1981).
[42] D. B. Zilversmit and L. B. Hughes, *J. Lipid Res.* **15**, 465 (1974).
[43] S. Dueland, J. Drisko, L. Graf, D. Machleder, A. J. Lusis, and R. A. Davis, *J. Lipid Res.* **34**, 923 (1993).

Sonicated vesicles were prepared by mixing 10 μCi of [^{3}H]cholesterol, 1 μCi of β-[^{14}C]sitosterol, 5 μg of cholesterol, and either 5 μg or 35 μg of PC, evaporating the solvent as above, adding 1 ml of water and sonicating ~1 min with a Heat Systems Ultrasonics W-380 probe sonicator (fine-tipped probe).

For absorption studies, mice were housed in metabolic cages where they had free access to food and water. Animals were allowed to adjust to the cages for at least 24 hr before beginning the test. For the experiment, each mouse received 50 μl of the test meal by stomach gavage at 3–4 hr before the beginning of their dark cycle. Feces were collected for a 24-hr period following administration of the test meal. The samples were homogenized in water and then extracted with an equal volume of chloroform/methanol (2/1 v/v). The aqueous phase was reextracted once with chloroform. The organic phases from each sample were combined, their volumes were measured, and an aliquot was used for scintillation counting. Counting efficiency was calculated using the external standard, channel ratio method. Cholesterol absorption efficiency, determined as percent of administered dose absorbed, was calculated based on the formula described by Grundy et al.[44] as follows: $\{1 - [(^3\text{H-dpm}/^{14}\text{C-dpm})$ in feces/$(^3\text{H-dpm}/^{14}\text{C-dpm})$ administered]$\} \times 100$. Results showed that the animals were able to absorb unesterified cholesterol from the diet with equal efficiency regardless of the CEL genotype and the type of diet used.[23] In contrast, the absorption of cholesteryl esters was dramatically reduced in CEL(−/−) mice in comparison with that observed in CEL(+/+) and CEL(+/−) mice.

Dietary cholesterol absorption efficiency could also be monitored by the appearance of radiolabeled cholesterol in serum after gastric infusion of the emulsified radiolabeled sterol. In these experiments, the amount of radiolabel sterol in 15 μl of serum was determined at various times after infusion of the substrate. The results of studies using this approach were similar to those obtained from fecal analysis of unabsorbed cholesterol. Thus, no radioactivity could be detected in the serum of CEL(−/−) mice if the radiolabel was supplied as [^{3}H]cholesteryl oleate. In contrast, if unesterified [^{3}H]cholesterol was used as the substrate, a similar level of radioactive sterol could be detected in the serum of all animals regardless of their CEL genotype.

The results of the studies summarized here indicate that CEL does not play a significant role in mediating intestinal absorption of dietary unesterified cholesterol. Although CEL appears to have a major impact in absorption of esterified cholesterol, this nutrient accounts for only a small

---

[44] S. M. Grundy, E. H. Ahrens, and G. Salen, *J. Lipid Res.* **9**, 374 (1968).

percentage of total cholesterol in the diet. The high level of CEL in the gastrointestinal tract suggests that CEL undoubtedly will have a significant impact on absorption of other nutrients in addition to cholesteryl esters. For example, fat-soluble vitamins exist mainly in esterified form and thus require hydrolysis by CEL before absorption into the enterocytes. Likewise, pancreatic lipase may not be sufficient for complete hydrolysis of dietary triglycerides, and thus may require CEL for complete digestion of dietary fat. The approach summarized here can be adopted to assess the role of CEL in absorption of dietary fat and fat-soluble vitamins. In addition, the availability of CEL gene knockout mice provides a unique opportunity to address the physiologic role of this enzyme in lipoprotein metabolism and atherosclerosis.

Acknowledgment

Research in the author's laboratory is supported by NIH grants RO1 DK40917 and RO1 DK46405.

[4] Properties of Pancreatic Carboxyl Ester Lipase in mRNA-injected *Xenopus* Oocytes and Transfected Mammalian Hepatic and Intestinal Cells

By REZA ZOLFAGHARI, RAANAN SHAMIR, and EDWARD A. FISHER

Introduction

The ability to express proteins *in vitro* has been valuable to the study of structural and functional aspects of a number of enzymes, including lipases. In our initial investigations of the pancreatic carboxyl ester lipase (CEL; E.C. 3.1.1.13), we were interested in translating CEL mRNA in an *in vitro* system likely to express an enzyme having appropriate posttranslational modifications, such as glycosylation.[1,2] In subsequent experiments, a particular requirement was to create specific cell lines stably expressing CEL cDNA to obtain convenient models for the study of the impact of CEL on lipid metabolism in tissues other than the pancreas.

[1] R. Zolfaghari, E. H. Harrison, A. C. Ross, and E. A. Fisher, *Proc. Natl. Acad. Sci. U.S.A.* **86**, 6913 (1989).
[2] R. Zolfaghari, E. H. Harrison, J. H. Han, W. J. Rutter, and E. A. Fisher, *Arterioscler. Thromb.* **12**, 295 (1992).

General Comments about in Vitro Systems to Express Proteins

mRNA as Substrate. The most common protein expression systems for mRNA are based on the lysates of either rabbit reticulocytes or wheat germ. These are widely available from commercial sources and typically include all constituents needed for translation except, of course, the mRNA of interest used to "program" the translational machinery. The mRNA can be provided directly to the reaction mixture or generated in an intermediate procedure by transcription *in vitro* of a plasmid containing a binding site for an RNA polymerase, such as SP6 or T7, that flanks the insert encoding the protein of interest. Only some posttranslational modifications can be achieved in these types of systems, however, and they depend on the presence of other components (i.e., microsomes for glycosylation) that have to be provided by the investigator.

An alternative *in vitro* translation system with a number of significant advantages for our purposes (see discussion later) is the oocyte of *Xenopus laevis*, the South African frog. The relatively large oocytes (~1.2 mm in diameter) are comparatively easy to microinject with either RNA (into the cytoplasm) or cDNA (into the nucleus) and typically express proteins (either from direct translation of the injected RNA sample or from translation of RNA transcribed from the injected cDNA sample) with many characteristics of the corresponding native forms, including the appropriate posttranslational modifications. In addition, besides similarities in structural features, other aspects of the expressed protein, such as vesicular transport of secretory proteins and insertion with the correct orientation into the plasma membrane for transmembrane proteins, also frequently resemble the properties of the native forms. For a more complete review of the virtues of this system and the wide spectrum of proteins that have been successfully studied with it, the reader is referred to Colman's work.[3,4]

DNA as Substrate. There are many systems in which a plasmid, containing both a promoter sequence active in the host cell and a DNA insert that encodes a protein of interest, is introduced into bacterial or eukaryotic cells with subsequent transcription to produce the mRNA of interest, which is then translated. Typically, this transfection approach is used with host cells either to produce large amounts of recombinant protein, such as in bacterial or insect cells, or to express a protein in specific types of eukaryotic cells to measure its metabolic effects. Frequently, it is also of interest to

---

[3] A. Colman, *Chap. 2 in* "Transcription and Translation" (B. D. Hames and S. J. Higgins, eds.), IRL Press, Oxford, UK, 1984.

[4] A. Colman, *Chap. 10 in* "Transcription and Translation" (B. D. Hames and S. J. Higgins, eds.), IRL Press, Oxford, UK, 1984.

observe how these effects may vary because of changes either in the level of expression or in the primary amino acid sequence of the protein.

DNA is introduced into cells in three major ways: electroporation, liposomes, and calcium salts. In electroporation, an electric current is passed through a chamber containing suspended cells and the DNA. A "pore" transiently forms that allows entry of the DNA. With liposomes, DNA is first incorporated into them. The liposomes are then added to the cell culture medium, and by interaction with the plasma membrane the DNA is delivered to the cell. In the third method, the DNA is mixed with calcium salts, resulting in the formation of microaggregates of DNA. When this material is added to the cell culture medium, the aggregations of DNA settle onto cell surfaces and enter by pinocytosis.

The optimal method to use depends on the plasmid, the protein to be expressed, and the cell type and, unfortunately, must be determined empirically. Because electroporation involves the expense of a special instrument and because the liposomes commercially provided for DNA transfection are relatively expensive, the calcium salt method is suggested as an initial approach that is cost effective and often scientifically successful.

The introduction of DNA into eukaryotic cells can lead to either the transient or long-term expression of the protein of interest. In the short term (less than 60–72 hr), the transfected plasmid remains as an episomal genetic element. If it is not integrated into the host cells' DNA, as cells continue to divide, the plasmid is not replicated along with the chromosomes and the effective "gene dosage" of the introduced DNA is progressively diluted. In a small percentage of cells (typically 1–2%), the plasmid DNA is chromosomally integrated, replicated with each cell cycle, and has the potential to be stably expressed in the descendant cells. To isolate cells in which plasmid DNA was successfully integrated, cotransfection of a plasmid that encodes a selectable marker is usually performed. The most common selection procedure is to cotransfect with a plasmid that encodes resistance to the neomycin class of anti-infective agents and to add the neomycin analogue G418 to the medium after transfection. Over the next few weeks, cells not expressing G418 resistance will die and those expressing it will survive and start to form small islands of clonal populations. The optimal concentration of G418 to use needs to be empirically determined, because the sensitivity of different cell types varies, not only because of endogenous factors preexisting the transfection procedure, but also because the copy number of the "resistance" plasmid or its site of integration into transcriptionally active or repressed regions of chromatin can vary among transfected cells. Typically, a G418 concentration of 0.4 mg/ml is recommended in many protocols. We have found that this concentration is appropriate for rat hepatoma Fu5AH cells, but is lethal in even successfully transfected Caco-2 cells, necessitating a concentration of 0.1 mg/ml. In general, a dose of G418 that causes slow attrition (1–2 weeks) of nontransfected cells will work well in the selection of transfected cells.

It is especially important to test cells selected for neomycin resistance for also expressing the DNA of interest, since the integration of one plasmid is independent of the other. The isolation of cells in which both plasmids were successfully integrated will be statistically favored, but not guaranteed, by using in the cotransfection a higher molar ratio of expression to resistance plasmid. Although expression plasmids do exist that contain a eukaryotic promoter, a multiple cloning site to insert the DNA to be expressed, and the neomycin resistance DNA with its own promoter, we have found it more convenient to use separate plasmids, one for the DNA to be expressed and one for neomycin resistance. In this way, DNAs of interest can be obtained from colleagues in a variety of expression plasmids and used directly, instead of having to reclone them into a new vector carrying the G418 resistance.

Commercially available expression vectors (either with or without a selectable marker) for use with mammalian host cells usually have promoters based on the viruses cytomegalovirus (CMV), simian virus 40 (SV40), or Rous sarcoma virus (RSV). These promoters usually lead to the robust expression of the DNA of interest in a wide variety of cell types. Also available from companies or colleagues are expression vectors that have special attributes, such as inductibility of expression or tissue-specific expression.

Finally, as mentioned earlier, cells surviving the selection procedure can be used to derive clonal populations. These clones are likely to vary in their level of expression of the transfected DNA encoding the protein of interest because of the same factors affecting the expression of G418 resistance, namely, plasmid copy number and integration site. A collection of well-characterized clones spanning a range of expression is a valuable resource with which to test "dosage" effects of the protein on the process under study.

Expression of Rat and Rabbit Hepatic and Intestinal RNA in *Xenopus* Oocytes

Initial Objective

When these studies first began, there was no molecular information on the nature of the hepatic bile salt-dependent cholesteryl ester hydrolase activity, although it had been shown in the rat to be remarkably similar in a number of properties to pancreatic CEL (e.g., see Harrison[5]). Further complicating matters, it had been hypothesized that this hepatic activity could represent uptake from the plasma of the pancreatic enzyme that had traversed the intestinal mucosa, since CEL was detected within enterocytes and its activity had been reported in both mesenteric lymph and plasma.[5-8] The initial question we addressed was whether the rat liver contained a species of mRNA that encoded CEL, thus indicating the capacity for endogenous synthesis of the enzyme. The strategy was to inject rat hepatic mRNA in *Xenopus* oocytes and then determine whether there was expression of an enzymatic activity with catalytic and immunologic features of pancreatic CEL.

Methods

Since translation *in vitro* is highly dependent on the integrity of the RNA, all precautions to limit RNase contamination should be taken in the preparation and handling of the RNA samples. For example, bake all glassware at 200° for at least 4 hr or use disposable sterile plasticware whenever possible, and use autoclaved distilled water to prepare all solutions. A useful summary of these procedures is given by Berger and Kimmel.[9] In addition, with the proliferation of convenient kits, many laboratories utilize prepackaged reagent sets to isolate RNA. We have found a convenient and low-cost alternative (especially if large amounts of tissue are involved) for the preparation of high-quality RNA from tissues and cells to be a guanidinium salt-precipitation technique we have published elsewhere.[10]

Solutions

Prepare the following stock solutions, filter through a 0.22-μm sterile filter, and autoclave for 20 min:

NaCl   5 M
KCl   3 M

---

[5] E. H. Harrison, *Biochim. Biophys. Acta* **963**, 28 (1988).
[6] L. L. Gallo, Y. Chiang, G. V. Vahouny, and C. R. Treadwell, *J. Lipid Res.* **21**, 537 (1980).
[7] P. Lechene de la Porte, N. Abouakil, H. Lafont, and D. Lombardo, *Biochim. Biophys. Acta* **920**, 237 (1987).
[8] D. Lombardo, G. Montalto, S. Roudani, E. Mas, R. Laugier, V. Sbarra and N. Abouakil, *Pancreas* **8**, 581 (1993).
[9] S. L. Berger and A. R. Kimmel, eds., "Guide to Molecular Cloning Techniques," Academic Press, Orlando, Florida, 1987.
[10] R. Zolfaghari, X. Chen, and E. A. Fisher, *Clin. Chem.* **39**, 1408 (1993).

CaCl$_2$ 1 M
MgCl$_2$ 1 M

Prepare ND96++ buffer as follows:

		Final Concentration
NaCl (5 M)	9.6 ml	96 mM
KCl (3 M)	0.335 ml	2 mM
CaCl$_2$ (1 M)	0.9 ml	1.8 mM
MgCl$_2$ (1 M)	0.5 ml	1.0 mM
Penicillin		100 units/ml
Streptomycin		100 μg/ml
HEPES	595 mg	5.0 mM
Sodium pyruvate	137.5 mg	2.5 mM

Add autoclaved water to about 500 ml and adjust pH to 7.6 with 1 N NaOH. Adjust final volume to 500 ml in a graduated cylinder and then filter sterilize through a 0.22-μm filter. Store at 4°. Make up this solution each week.

Prepare OR-2 buffer as follows:

		Final Concentration
NaCl (5 M)	8.2 ml	82 mM
KCl (3 M)	0.4 ml	4 mM
MgCl$_2$ (1 M)	0.5 ml	1 mM
HEPES	595 mg	5 mM

Add autoclaved water to about 500 ml and adjust the pH to 7.6 with 1 N NaOH. Adjust final volume to 500 ml in a graduated cylinder and filter sterilize through a 0.22-μm filter. Store at 4°. Make up this solution each month.

COLLAGENASE SOLUTION. Add 400 mg collagenase Type IA (Cat. #C-9891, Sigma, St. Louis, MO) into 200 ml OR-2 buffer and gently rock it into solution. Divide it into 10-ml aliquots and store it at −20°. This is stable for at least 1 year.

Animal

Oocyte positive/hormone-primed (injected by company 2 months prior to receipt) adult female *Xenopus laevis* frogs were obtained from Xenopus-I (Ann Arbor, MI). The frogs were individually placed in clear plastic containers (~5-liter capacity) that can be purchased from a pet or aquarium supply store. A container was filled with about 2 liters of dechlorinated water and covered with a screen on which a weight was placed so that the frog could not jump out. To dechlorinate the water, add 0.25 g sodium thiosulfate pentahydrate (Sigma) per liter of tap water and let stand for at least 24 hr at room temperature before use. For feeding, add two pieces of beef liver (about 3–4 g each piece) into the water every other day in the morning and change the water in the afternoon. The liver is purchased at any supermarket and kept frozen. The pieces are sliced from the frozen liver, and they thaw after being placed in the water.

Preparation and Microinjection of Xenopus laevis Oocytes

Collection of Oocytes. In a 2-liter beaker, add 1.7 g of the anesthetic TRICAINE (3-aminobenzoic acid ethyl ester, methanesulfonate salt; Sigma) and dissolve it in 1 liter of the dechlorinated water. Place the frog into the solution in the beaker and cover. The frog should be anesthetized in about 30 min. Check by pinching the upper limb with a pair of forceps.

Have sterile surgical scissors and fine-tipped forceps ready. Make a bed of ice on a diaper pad or other bench coating material and place the anesthetized frog on its back on the ice. With the scissors, cut a diagonal opening of abdominal skin just above the leg. Tent up the skin with forceps and then cut through the muscle. Take out the ovary lobes with a pair of forceps and cut off an appropriate amount; for a typical experiment, approximately 50 oocytes per experimental group (e.g., injection with no RNA control, with RNA samples, etc.) are used. Replace remaining ovarian tissue into the abdominal cavity. Put three to four stitches through the skin and muscle. You can use each frog three to four times to collect oocytes, alternating sides. Put the frog back in the water tank and she will be awake in about 2 hr.

Place the oocytes in a 10-cm-diameter petri dish containing ND96++ solution. Under a binocular low-power dissecting microscope, free the oocytes by teasing apart the lobes with two pairs of fine-tipped forceps. Put the oocytes in a 15-ml plastic tube and add ND96++ buffer and wash them with this buffer three times. After each wash, simply allow them to settle to the bottom of the tube and then aspirate the wash solution. Wash the oocytes similarly with OR-2 buffer three times. Aspirate as much liquid off without damaging the oocytes and then add collagenase solution (2 volumes of enzyme solution per volume of oocytes). Rock the oocytes gently at 19° for about 2 hr. Change the enzyme solution two times during this period. After enzyme treatment, wash the oocytes with OR-2 buffer and then with ND96++ buffer, three times with each buffer. Transfer the oocytes in ND96++ buffer to a 10-cm petri dish and store them at 4° overnight. This loosens the somatic cells covering the oocytes that may remain after the collagenase treatment removed most of the other adherent material.

Under a binocular microscope, peel the oocyte with two pairs of fine-tipped surgical forceps. This takes some practice to do without damaging the oocytes. Be patient! After a few sessions, 250–300 oocytes can be peeled in 2 to 3 hr. You will see some capillary blood vessels in the somatic layer surrounding the oocyte, which will help you visualize the layer to be peeled off. Also, the peeled oocytes are brighter in color than the unpeeled ones. After peeling, transfer the oocytes in the ND96++ buffer with a wide-bore Pasteur or transfer pipette into a 30-mm petri dish. The peeled oocytes are now ready for injection.

Injection of Oocytes. First prepare an injection pipette. This can be done in advance and the pipettes stored indefinitely for subsequent experiments. Alternatively, premade injection pipettes can be purchased from suppliers, such as Eppendorf. To make your own pipettes, first bake 10-μl Wiretrol calibrated micropipettes (Fisher Scientific, Pittsburgh, PA) at 200° for at least 4 hr and then pull it in a pipette puller (i.e., David Kopf Instruments, Pasadena, CA). Since this draws out the tip to a very fine fused tip, break open the tip with a pair of fine-tipped forceps under the binocular microscope so that the diameter of the opening is about 10 to 20 μm as estimated by the hairline calibrations in the eyepiece. The pipette can be stored in a covered petri dish by placing it across molding material ("mold mud") in the bottom of the dish so that the shaft is secured and the tip is free.

To hold the oocytes in place while they are being injected, place a piece of plastic filter mesh (mesh size 500 μm) at the bottom of a plastic 30-mm petri dish and fix it into position with chloroform. (Add approximately 1 ml, swish around, aspirate, and allow to air dry. Then wash the petri dish extensively with sterile water.) With this mesh size, each oocyte will settle into a square and be held in place relatively securely.

Put the injection pipette over the needle of a 10-μl Drummond digital microdispenser (Fisher Scientific) so that the tip of the needle goes close to the tip of the capillary pipette. Mark the position of the needle on the capillary pipette with a marking pen to allow for an approximate volume of 5 μl. Then with a thin needle syringe (e.g., 26-gauge) fill to the mark with mineral oil. Secure the microdispenser in a manipulatable holder placed close to the binocular microscope. Then under microscopic visualization, place 5 μl of RNA solution (2–3 mg/ml) or control solution (i.e., sterile water if that is what the RNA is dissolved in) as a droplet on a 30-mm petri dish and then take up the solution with the injection pipette. Thus, the mineral oil acts as a noncompressible interface between the displacement piston of the microdispenser and the aqueous solution to be injected.

Fill the petri dish containing the fixed plastic mesh with ND96++ buffer and place 10 to 15 peeled oocytes in it. Inject about 50 nl of RNA or control solution into the light-colored pole of an oocyte and then transfer it with a Pasteur pipette into a 30-mm petri dish containing ND96++ buffer. After moderate practice, it will take about 2 to 3 hr to inject 250 to 300 oocytes. Incubate the oocytes at 19° for a maximum of 3–5 days. Depending on the RNA of interest, 1 day may be sufficient to detect expression; this will have to be determined empirically. Change the buffer two times per day and discard any visibly damaged or discolored oocytes. Frequently, the discoloration takes the form of depigmented areas.

At the end of the incubation, the oocytes are assayed for the expression of the protein of interest and its relevant properties. For our studies, at the end of an experiment, oocytes were washed several times with 0.01 M Tris-maleate, pH 7.5, containing 0.25 M sucrose. A glass homogenizer was then used to homogenize a batch of 50 oocytes/ml of fresh wash buffer. The homogenate was centrifuged at low speed (15 min, 4°, 1000g) to pellet large granules of yolk protein. The supernatant was then collected and used for enzyme measurements, using a standard of 20 μl to represent one oocyte, since 50 oocytes are homogenized per milliliter of buffer and the yolk pellet was of small volume. The radiometric assay to measure the cholesteryl ester hydrolase activity of the expressed enzyme in both these and the transfection studies is described in detail elsewhere.[5]

In the case of secreted proteins, the buffer bathing the oocytes can be assayed. Finally, as controls, either water- or buffer-injected oocytes can be assessed. This will also give an indication of the expression of an activity endogenous to oocytes that may raise the "background" level, making less reliable the detection of the activity expressed from the injected RNA. In that case, if specific reagents (i.e., monoclonal antibodies) or specific assay conditions (i.e., a particular temperature for the assay, e.g., see Labarca and Paigen[11]) are available, the endogenous and exogenous activities may be differentiated from each other.

To ensure that a given batch of oocytes was translationally competent, if a mixed population of RNA were injected (e.g., total liver RNA), another protein that can be easily detected, such as lactate dehydrogenase (LDH; we used the Sigma assay kit), could be assayed. Alternatively, for single RNAs (such as that produced by *in vitro* transcription), another mRNA known to be readily expressed can be co-injected or injected separately in a parallel group of oocytes from the same batch used for the experimental RNA. Finally, for tissue or cell sources of RNA, although injection of total RNA is frequently sufficient in initial experiments, since the maximum concentration (2–3 mg/ml) and volume (50 nl) of the RNA sample are limited, using mRNA will increase the likelihood of expressing the protein in detectable amounts.

---

[11] C. Labarca and K. Paigen, *Proc. Natl. Acad. Sci. U.S.A.* **74,** 4462 (1977).

Results

In the initial set of studies,[1] we isolated a pool of total hepatic RNA from which we derived poly(A)⁺ mRNA using oligo-dT chromatography.[12] Injection of each oocyte with 50 nl of a 2 mg/ml solution of the mRNA led to the expression of a bile salt-dependent cholesteryl ester hydrolase activity, which was maximal in the neutral pH range. Expression increased with length of incubation after injection, up to 6 days, the longest period we tried. Since abundant activity was expressed by 3 days, we used this length of incubation to further characterize the expressed activity.

One of the remkarable (and unexplained) features of the hydrolysis of cholesteryl esters by CEL in homogenized tissue samples is the dependence for maximal activity on millimolar concentrations (sometimes as high as 100 mM) of trihydroxy bile salts, such as cholate and its derivatives. These concentrations far exceed the amount required to solubilize substrates and to induce conformational changes of the purified enzyme.[13] Because no other enzyme is known to demonstrate this behavior, the stimulation by trihydroxy bile salts has served as a unique distinguishing feature of CEL. As shown in Fig. 1A, the cholate maximum of the activity expressed in oocytes was 10 mM, similar to the optimal concentration found with rat liver homogenate.[5] Furthermore, as in rat liver homogenate, cholate was the most potent of the trihydroxy bile salts tested (Fig. 1B), and the dihydroxy bile salt, deoxycholate, was ineffective in stimulating CEL activity, although at the concentration used it would completely solubilize the substrate.

Another strong indication that bona fide CEl activity was expressed by the injected oocytes is shown in Fig. 2. Using an antibody specific for rat pancreatic CEL (kindly provided by Dr. Linda Gallo, George Washington University), the expressed activity was inhibited in a dose-dependent manner. Besides the catalytic and immunological similarities to CEL in mammalian tissue homogenate, another indication that the injected oocytes expressed CEL with fidelity was finding that CEL activity was also secreted, as it is by all mammalian systems so far reported. A final confirmation of the molecular nature of the expressed activity was our subsequent cloning from rat liver mRNA a cDNA species identical to the pancreatic cDNA (X. Chen, E. H. Harrison, E. A. Fisher, EMBL accession number Z22803).

The major aim of the studies discussed here was to detect in a pooled rat hepatic mRNA sample an expressed enzyme with the characteristics of

---

[12] A. Jacobson, *Methods Enzymol.* **152**, 254 (1987).
[13] P. W. Jacobson, P. W. Westerfield, L. L. Gallo, R. L. Tate, and J. C. Osborne, Jr., *J. Biol. Chem.* **265**, 515 (1990).

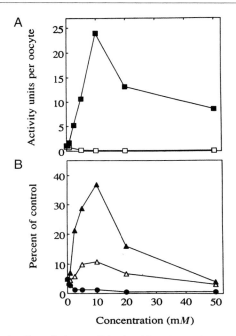

FIG. 1. (A) Effect of sodium cholate or (B) other bile salts on the cholesteryl ester hydrolase activity of rat CEL expressed in *Xenopus* oocytes. (A) Supernatant fractions of oocytes were prepared 3 days after injection with either rat liver mRNA (filled box) or water (open box) and were then incubated with 10 μM cholesteryl [^{14}C]oleate at 37° for 24 hr in the presence of 0–50 mM sodium cholate. (B) Supernatants of mRNA-injected oocytes were incubated in the same manner with 0–50 mM sodium taurocholate (filled triangle), sodium glycocholate (open triangle), or sodium deoxycholate (filled circle), and the cholesteryl ester hydrolase activity was then determined and related to the percent of activity in samples containing 10 mM sodium cholate. Points represent the means of four determinations from two separate microinjections. An activity unit is equal to 1 pmol of [^{14}C]oleate released per 24 hr. (Reprinted with permission from Zolfaghari et al.[1])

pancreatic CEL so that, in spite of the unavailability of a cDNA probe at the time for Northern blot analysis, we could determine whether the reported rat hepatic activity could have been synthesized from an endogenous source. Based on the initial studies, this was clearly the case. Our next focus was to enlarge the study to investigate whether (1) there were differences in hepatic CEL gene expression among individual rats that correlated with variations in the homogenate activity of the enzyme and (2) there was a species difference between the rat and rabbit in hepatic CEL expression, a possibility given that on a cholesterol-rich diet, rabbit, but not rat, liver accumulates a large store of cholesteryl ester.

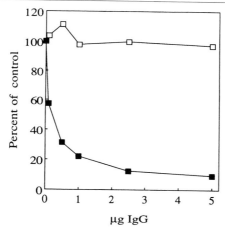

Fig. 2. Inhibition of rat liver bile salt-dependent cholesteryl ester hydrolase activity in injected *Xenopus* oocytes by immune IgG to rat pancreatic CEL. A fixed amount of oocyte supernatant fraction (about 18 cholesteryl ester hydrolase units) was incubated with various amounts of rabbit antirat pancreatic CEL IgG (filled box) or normal rabbit IgG (open box). Points represent the mean of two determinations. The sodium cholate concentration was 10 mM. (Reprinted with permission from Zolfaghari et al.[1])

In contrast to the earlier study, we now had available a CEL cDNA probe through a collaboration with Drs. Jan Han and William Rutter (UCSF) and we measured the enzyme activity and CEL mRNA abundance in rat and rabbit liver.[2] As shown in Table I, the enzyme activity was quite variable among the rats (consistent with a previous report[5]). In contrast, among the rabbits, hepatic activity was uniformly low. We were not surprised, then, that no rabbit liver had detectable CEL mRNA on blotting analysis (data not shown, but published in Zolfaghari et al.[1]). Quite surpris-

TABLE I
BILE SALT-DEPENDENT CHOLESTERYL ESTER HYDROLASE
ACTIVITY OF LIVER HOMOGENATES FROM RATS AND RABBITS[a,b]

Animal	N	Activity	Range
Rat	10	1345 ± 1324	(5–3350)
Rabbit	15	3 ± 1	(2–5)

[a] Reprinted with permission from Zolfaghari et al.[2]
[b] Data are mean ± SD, with range in parentheses. One activity unit equals 1 nmol oleic acid released from cholesteryl oleate per hour by 1 gram of tissue. The sodium cholate concentration was 20 mM.

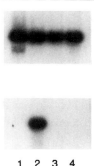

1 2 3 4

Fig. 3. Northern blot analysis of rat liver RNA. Samples of poly(A)$^+$ RNA (3 μg/lane) were separated on denaturing agarose gels and transferred to nylon embranes. Hybridization to β-actin (top panel) or CEL (bottom panel) DNA ^{32}P-labeled probes was followed by very stringent washes and autoradiography. (Reprinted with permission from Zolfaghari et al.[2])

ing, in contrast, was the Northern blot analysis of rat hepatic RNA samples (Fig. 3). Of the samples shown, only one has CEL mRNA. As summarized in Table II, when the liver homogenates were tested for CEL activity, that sample had no enzyme activity, but the other three samples did. That these three samples contained bona fide CEL activity was confirmed not only by its dependence on millimolar concentrations of cholate, but also by the ability to inhibit >95% of the activity in each homogenate with the anti-CEL antiserum used earlier (data not shown).

The *Xenopus* expression system proved valuable to investigate these unexpected findings. As shown in Table II, injection of only the sample that was Northern blot positive led to expression of CEL activity. Another,

TABLE II
BILE SALT-DEPENDENT CHOLESTERYL ESTER HYDROLASE ACTIVITY IN HOMOGENATES OF RAT LIVER AND INJECTED *Xenopus* OOCYTES[a,b]

Rat	Liver homogenate activity	Oocyte expression
1	2328	<5
2	5	555
3	3188	<5
4	580	<5

[a] Reprinted with permission from Zolfaghari et al.[2]
[b] Animal numbers correspond to lanes in Fig. 3. The activity units for the liver homogenates and oocytes are those of Table I and Fig. 1, respectively. The sodium cholate concentrations were 10 mM (oocytes) or 20 mM (liver homogenate).

later, confirmation that the Northern blot signal from sample 2 represents CEL mRNA was the cloning of the CEL cDNA noted earlier from the hepatic RNA of that animal. Therefore, we concluded that the signal in sample 2 represented authentic CEL mRNA and also that no other hepatic message expressing a CEL-like activity in the tissue homogenate was present. Some of the possibilities the former finding suggested include that (1) the translation of CEL mRNA can be regulated, (2) the protein can be quickly degraded (as has been suggested[14] based on the PEST-sequence hypothesis[15]), (3) the enzyme can be quantitatively secreted, or (4) that the intracellular enzyme can exist in an inactive state. The latter finding suggested that the homogenate CEL activity in the absence of CEL mRNA was not the result of the expression of another gene encoding a CEL-like activity. In addition, the results were compatible with either the CEL mRNA being recently degraded [from a relatively specific process since ribosomal RNA and mRNA of another gene (actin) were intact] or that the activity was derived from outside the liver. Overall, although there were species differences in hepatic CEL gene expression and enzyme activity, the complicated results precluded the formulation of a functional significance for these differences. A clearer result, though still surprising, was that the level of CEL mRNA and enzyme activity in rat liver appeared to be independently regulated.

Expression of Rat CEL cDNA in Transfected Hepatic and Intestinal Cells

Objective

At the time these studies were initiated, the majority of the CEL literature was concerned with CEL's actions in the intestine, particularly its effects on the processing of dietary cholesterol in the free and esterified forms. With the availability of the cDNA for rat CEL, we were interested in establishing an *in vitro* model to investigate mechanisms for the putative intestinal functions of the enzyme. In addition, the reports (see earlier discussion) that rat liver contained an enzymatic activity highly similar, if not identical, to pancreatic CEL raised the issue of its function in that organ. Thus, an *in vitro* model for studying CEL in the liver was also sought. The cell lines chosen were Caco-2, a widely used model of human intestine, and rat hepatoma Fu5AH, which is well characterized in terms of cholesteryl

---

[14] J. A. Kissel, R. N. Fontaine, C. W. Turck, H. L. Brockman, and D. Y. Hui, *Biochim. Biophys. Acta* **1006,** 227 (1989).

[15] S. Rogers, R. Wells, and M. Rechsteiner, *Science* **234,** 364 (1986).

ester metabolism and which does not express either CEL mRNA or a bile salt-dependent cholesteryl ester hydrolase activity.

With both model systems we wanted either to conduct experiments of fairly long duration or to examine the metabolic effects of a range of CEL activity, so the isolation of stably expressed clonal populations of cells was desired. If only transient expression is necessary for the aim of the study, then a previous approach we took[16] should be consulted.

Methods

Solutions and Reagents

Prepare 100 ml of 2× HEPES-buffered saline (HBS):

		Final Concentration
NaCl (5 M)	5.6 ml	280 mM
KCl (1 M)	1.0 ml	10 mM
Na_2HPO_4 (0.1 M)	1.5 ml	1.5 mM
Dextrose	0.22 g	12 mM
HEPES	1.2 g	50 mM

Add autoclaved water to about 90 ml and mix. Adjust the pH to 7.05 with 1 N NaOH and then adjust final volume to 100 ml with water. Sterilize by passage through a 0.22-μm filter and store in 5-ml aliquots at $-20°$.

CACL$_2$. Dissolve 10.8 g of $CaCl_2:6H_2O$ in 20 ml of autoclaved water. Sterilize by passage through a 0.22-μm filter. Store in 1-ml aliquots at $-20°$.

CELL CULTURE MEDIUM. The medium was minimal essential medium (MEM) supplemented with basal medium Eagle vitamins (GIBCO-BRL, Gaithersburg, MD). Stock cultures were maintained in growth medium [MEM containing 50 mg/ml gentamicin, 1% L-glutamine (stock solution = 100%), and 5% heat-inactivated fetal bovine serum (FBS) for Fu5AH and 10% for Caco-2 cells] at 37°, 5% CO_2. The heat-inactivated FBS was prepared by incubation of the FBS at 55° for 2 hr, after which time it contained no detectable CEL activity.

G418. Mix 250 mg G418 (geneticin; GIBCO-BRL) in 5 ml water for a 50 mg/ml stock solution and filter sterilize through a 0.22-μm filter. Store at $-20°$.

Transfection Procedure

PLASMIDS. For transfection of rat CEL cDNA, we used an expression plasmid containing a SV40-promoter and the full-length cDNA, originally cloned as the rat lysophospholipase[17] but later shown to be identical to

---

[16] K. Morlock-Fitzpatrick and E. A. Fisher, *Proc. Soc. Exp. Biol. Med.* **208**, 186 (1995).
[17] J. H. Han, C. Stratowa, and W. J. Rutter, *Biochemistry* **26**, 1617 (1987).

pancreatic CEL.[14] This plasmid is described in detail elsewhere.[18] The plasmid encoding resistance to G418 was obtained from Clontech (Palo Alto, CA) and is listed in their catalog as pMAMneo, but for brevity is referred to here as pNeo. As explained in the introductory remarks, we used cotransfection of pCEL and pNeo and ultimately isolated cells expressing both plasmids. To reduce the selection of cells expressing only pNeo, an approximately 10:1 molar ratio of pCEL:pNeo plasmids was used, thereby increasing the likelihood that cells resistant to G418 also expressed pCEL. Some laboratories use as much as a 100:1 ratio.

TRANSFECTION. All incubations of the cells are performed in a standard tissue culture incubator (37°, 5% CO_2). Twenty-four hours before transfection, plate the cells at about 75,000 cells/well in a 12-well plate in MEM with 5 or 10% FBS for Fu5AH or Caco-2 cells, respectively. In the tissue culture hood, to a disposable sterile 5-ml plastic tube add 266 μl supercoiled DNA plasmid solution (40 μg/ml; total concentration whether it contains pCEL, pNeo, or both in a 10:1 molar ratio, pCEL:pNeo) and 300 μl of 2× HBS solution. Slowly add 37 μl of 2 M $CaCl_2$ with gentle swirling for about 30 sec. Incubate the mixture for 30 min at room temperature and then pipette it up and down to resuspend the DNA–salt precipitate that formed. Transfer 0.1 ml (which contains approximately 2 μg of DNA) of this mixture into the medium in each well and then swirl the plate gently, which will disperse the precipitated DNA evenly over the cell monolayer. The precipitates eventually sink onto the cell surfaces and are taken up.

After 16–24 hr, rinse the cells three times with growth medium and then incubate the cells in fresh growth medium for 24 hr. Then change the medium to selection medium, which is growth medium containing 0.4 mg/ml G418 for Fu5AH or 0.1 mg/ml for Caco-2 cells. (We found these two concentrations were optimal for these two cell lines. As noted in the introductory material, the optimal dose is determined empirically for each cell line. In general, a useful dose is one that causes gradual death of nontransfected cells so that few viable cells remain after 7–10 days.) Incubate the cells in the selection medium containing G418 for 3 weeks, changing the medium every 2–3 days. By this point, there should be small islands of proliferating cells, each one representing a clonal population.

To isolate clonal populations representing different levels of CEL expression, we used an end-limiting dilution method: trypsinize cells and count them. Serially dilute them to about 5 to 10 cells/ml in selection medium. Then transfer 0.2 ml of the diluted cell suspension (i.e., an average of one cell/well) to each well of a 96-well plate and add more selection medium to fill each well approximately two-thirds. After 24 hr of incubation,

[18] L. Ellis, E. Clauser, D. O. Morgan, M. Edery, R. A. Roth, and W. J. Rutter, *Cell* **45**, 721 (1986).

carefully examine the wells and mark those containing one cell. After cells in these marked wells become confluent, aspirate medium samples (50 μl) for assay of CEL activity (see Harrison[5] or Zolfaghari et al.[19] for assay). Trypsinize cells from each positive well to a separate T-25 flask and grow the cells to provide stock cultures for other confirmatory assays (such a Southern, Northern, and Western blotting) and for further experimental study. As is summarized in the results section, this procedure led to the identification of Fu5AH and Caco-2 clones that spanned a wide range of CEL expression.

At the end of the 3-week selection period, alternative procedures for deriving clones of stably expressing cells include (1) the use of cloning cylinders to isolate physically one island of cells on a plate from the others; each "cylinder compartment" can be screened, trypsinized, and replated into a separate well or flask; or (2) the use of a sterile cotton-tipped applicator to "pick off" an individual cell or a small island of cells. This can then be used to inoculate medium in a separate well or flask.

Results

Hepatic Cell Studies. We identified more than 20 transfected Fu5AH clones stably secreting CEL and divided the activity results into terciles to classify clones as low, medium, or high expressors. Representatives of each group were selected for further study (reported in more detail in Zolfaghari et al.[19]). As shown in Table III, the clones spanned a range of approximately six- to eightfold for both the cell lysate and medium CEL activities. These cells were loaded in cell culture with cholesterol using methods described in detail elsewhere (Zolfaghari et al.[19] and references therein), thereby stimulating the formation of cholesteryl esters (CE). If CEL were active as an intracellular cholesteryl ester hydrolase, either the level of accumulated CE or its rate of hydrolysis would be expected to be lower or higher, respectively, and to vary with the level of expression. The results summarized in Tables IV and V, however, showed that the level of expression from none (either nontransfected Fu5AH cells or cells designated as NEO, which were transfected with only the G418 resistance vector, pNeo) to relatively high (clone 1-G1) did not affect either cellular levels of CE or its fractional hydrolysis. In contrast, HDL added to the medium of cells secreting high amounts of enzyme activity had a slightly increased capacity to accept cellular cholesterol compared to results with HDL incubated with control or low expressing cells (data not shown).

---

[19] R. Zolfaghari, J. M. Glick, and E. A. Fisher, *J. Biol. Chem.* **268,** 13532 (1993).

TABLE III
BILE SALT-DEPENDENT CHOLESTERYL ESTER
HYDROLASE ACTIVITY OF Fu5AH RAT HEPATOMA
CELL LINES STABLY TRANSFECTED WITH
CEL cDNA[a,b]

Cell lines	Hydrolysis of cholesteryl ester (nmol/hr/mg cell protein)	
	Medium	Cell lysate
2-H3	113 ± 12	1.53 ± 0.27
1-C6	129 ± 04	2.14 ± 0.02
2-A10	524 ± 87	6.77 ± 0.16
1-C1	528 ± 22	6.12 ± 2.81
1-G1	845 ± 43	5.53 ± 0.47
2-C6	921 ± 23	8.40 ± 1.67

[a] Reprinted with permission from Zolfaghari et al.[19]

[b] Fu5AH rat hepatoma cells stably cotransfected with rat pancreatic CEL cDNA and pNeo plasmid were grown in MEM medium containing 5% FBS. After 3 days, the media were collected and the cells washed with PBS, harvested, and then homogenized in 0.25 M sucrose solution. The bile salt-dependent (at 20 mM sodium cholate) cholesteryl ester hydrolase activity of CEL was determined in both medium and cell lysate using cholesteryl [^{14}C]oleate as the substrate. Each value is the average of three separate determinations (± SD) for a 24-hr incubation period. The nontransfected cells and the cells transfected with only pNeo plasmid had no detectable CEL activity.

The lack of an effect on cellular CE metabolism may be explained, at least in part, by the finding of relatively low activity in the cell lysate of any clone (Table III). This low activity could not be explained by rapid and quantitative secretion of the enzyme, resulting in a small intracellular pool, since on Western blot analysis (data not shown, but given in Zolfaghari et al.[19]) there was abundant intracellular protein; a rough estimate was that on a mass basis the intracellular enzyme had 500 times lower activity than the secreted form. Overall, the results supported a model in which the intracellular enzyme does not play a significant role in hepatic CE metabolism, despite years of speculation

TABLE IV
CHOLESTEROL CONTENT OF Fu5AH RAT HEPATOMA CELLS AFTER LOADING WITH CHOLESTEROL-RICH PHOSPHOLIPID DISPERSIONS[a,b]

Cell line	μg/mg Cell protein	
	Free cholesterol	Esterified cholesterol
Fu5AH	20.5 ± 0.8	87.8 ± 4.3
NEO	19.7 ± 1.0	82.1 ± 1.6
2-H3	20.6 ± 0.7	82.1 ± 6.1
2-A10	23.7 ± 1.2	90.2 ± 2.2
1-G1	20.2 ± 0.7	80.3 ± 5.6

[a] Reprinted with permission from Zolfaghari et al.[19]
[b] Cells were loaded by exposure to cholesterol-rich phospholipid dispersions for 48 hr. After an 18-hr incubation with MEM containing 0.1% BSA, cells were harvested, lipids extracted, and cholesterol quantitated by gas liquid chromatography. Values are the means ± SD, $n = 4$.

to the contrary, but that the secreted enzyme may affect the properties of lipoproteins. This would be consistent with finding CEL in the plasma of many mammals, including human[8,20,21] and the modification of human lipoproteins *in vitro* by CEL.[20,21]

[20] J. Brodt-Eppley, P. White, S. Jenkins, and D. Y. Hui, *Biochim. Biophys. Acta* **1272,** 69 (1995).
[21] R. Shamir, W. J. Johnson, K. Morlock-Fitzpatrick, R. Zolfaghari, L. Li, E. Mas, D. Lombardo, D. W. Morel, and E. A. Fisher, *J. Clin. Invest.* **97,** 1696 (1996).

TABLE V
HYDROLYSIS OF CHOLESTERYL ESTERS IN Fu5AH RAT HEPATOMA CELLS[a,b]

Cell line	Fraction of esterified cholesterol hydrolyzed per 12 hr
Fu5AH	0.184 ± 0.024
NEO	0.201 ± 0.020
2-H3	0.212 ± 0.044
2-A10	0.212 ± 0.036
1-G1	0.216 ± 0.046

[a] Reprinted with permission from Zolfaghari et al.[19]
[b] Cells were loaded with cholesterol and ester hydrolysis measured in the presence of a cholesterol esterification inhibitor (Sandoz compound 58-035) to prevent reesterification. Values are the means ± SD, $n = 3$.

TABLE VI
PERCENT UPTAKE AND DISTRIBUTION OF MICELLAR-DERIVED CHOLESTEROL IN
Tr AND Neo CELLS INCUBATED WITH MIXED MICELLES PROVIDING DIFFERENT
CONCENTRATIONS OF FC[a]

	FC concentration		
	50 μM	5 μM	1 nM
Tr			
Percent uptake	22.4 ± 1.6	21.1 ± 2.3	16.2 ± 1.7
Free cholesterol	20.7 ± 0.13	2.6 ± 0.03	0.16 ± 0.005
Cholesteryl ester	1.6 ± 0.04	0.028 ± 0.003	0.003 ± 0.001
Neo			
Percent uptake	21.4 ± 1.5	21.6 ± 2.0	17.4 ± 1.2
Free cholesterol	19.8 ± 0.12	2.8 ± 0.02	0.17 ± 0.002
Cholesteryl ester	1.6 ± 0.04	0.14 ± 0.02	0.005 ± 0.001

[a] Reprinted with permission from Shamir et al.[24]
[b] Tr and Neo cells were incubated for 12 hr to allow enzyme accumulation in Tr-cell medium. Mixed micelles containing labeled FC were then added to the conditioned media of Tr and Neo cells to provide a final cholesterol concentration of 1 nM, 5 μM, or 50 μM. After a 4-hr incubation, uptake of labeled cholesterol from the micelles was determined by scintillation counting of extracted cellular lipids. The distribution of cellular cholesterol between FC and CE in the extracts was assayed by TLC, and the molar quantity of each species calculated. The percentage of micellar cholesterol taken up by the cells is given as mean ± SD, normalized to cell protein ($n \geq 6$). The calculated masses for FC and CE for the 50 and 5 μM conditions are given in nmoles per milligram cell protein. The calculated masses for the 1 nM condition are given in pmoles per milligram cell protein. The mass results for each cell type are displayed as mean ± SD, $n \geq 3$. The mixed micelles provided a final concentration of sodium taurocholate of 6 mM.

Intestinal Cell Studies. A controversial hypothesis from animal[22] and cell culture[23] studies is that in addition to its ability to hydrolyze dietary CE, and thereby liberate the free cholesterol moiety preparative to its aborption by enterocytes, CEL may directly promote the absorption of free cholesterol (FC) by acting, for example, as a carrier protein or as a factor maintaining a favorable concentration gradient across the enterocyte plasma membrane (see introduction of Shamir et al.[24] for review and references). We developed a model system[24] to test this *in vitro*. CEL cDNA transfected into Caco-2 cells led to the isolation of approximately 16 clones

---

[22] L. L. Gallo, S. B. Clark, S. Myers, and G. V. Vahouny, *J. Lipid Res.* **25**, 604 (1984).
[23] A. Lopez-Candales, M. S. Bosner, C. A. Spilburg, and L. G. Lange, *Biochemistry* **32**, 12085 (1993).
[24] R. Shamir, W. J. Johnson, R. Zolfaghari, H. S. Lee, and E. A. Fisher, *Biochemistry* **34**, 6351 (1995).

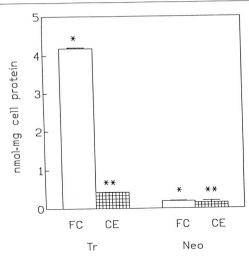

FIG. 4. Distribution of micellar-derived cholesterol in Tr and Neo cells after incubation with mixed micelles providing 10 μM CE. Tr and Neo cells were incubated for 12 hr to allow enzyme accumulation in Tr-cell medium. Mixed micelles containing labeled CE (as cholesteryl oleate) were then added to the conditioned media of Tr and Neo cells to provide a final CE concentration of 10 μM. After a 4-hr incubation, uptake of labeled cholesterol from the micelles was determined by scintillation counting of extracted cellular lipids. The distribution of cellular cholesterol between FC and CE in the extracts was assayed by TLC, and the mass of each species calculated. The results for each cell type were normalized to cell protein and are displayed as mean ($\pm$ SD), $n = 6$. There was significantly more FC and CE in Tr cells (*$P < 0.001$, **$P < 0.0001$, respectively). The mixed micelles provided a final concentration of sodium taurocholate of 6 mM. (Reprinted with permission from Shamir et al.[24])

stably secreting a wide range of CEL activity. Because the concentration of CEL in intestinal juice is estimated to be at least 100 nM,[23] we chose a clone expressing a sufficiently high level so that in a typical experiment the concentration in the conditioned medium would be at least that; in fact, we estimated that with the selected clone (designated as Tr in the tables and graphs), at the start of cholesterol uptake experiments, the concentration of CEL in the conditioned medium was approximately 700 nM.

Mixed micelles were formulated to mimic the human intestinal environment and were based on the comprehensive studies of intestinal juice published elsewhere[25] by Dr. Martin Carey and his colleagues. As shown in Table VI, neither the fraction of micellar free cholesterol taken up by the cells nor the intracellular distribution (free versus ester) of the free

[25] O. Hernell, J. E. Staggers, and M. C. Carey, *Biochemistry* **29,** 2041 (1990).

cholesterol taken up depended on CEL activity. The major conclusion from these data was that CEL did not promote the uptake of micellar free cholesterol.

Two important experiments to support this conclusion were (1) to add human CEL to the medium, in case there was a species difference in a cholesterol absorption-facilitating property, since the cDNA we used encoded the rat CEL; the pattern of the results in Table VI were unchanged (data not shown here, but summarized in Shamir et al.[24]); and (2) to verify that the established role of CEL in the intestinal lumen, that of cholesteryl ester hydrolysis, was observable in our system. As shown in Fig. 4, when mixed micelles containing cholesterol in the form of CE were presented to Tr and control cells, only the cells expressing CEL had significant cholesterol uptake. That this was a result of the hydrolysis of micellar CE was shown by finding on TLC analysis that at the end of the incubation, 99.6% of the CE in the conditioned medium had been converted to free cholesterol and fatty acids. Overall, based on these studies, the major role for CEL in cholesterol absorption appeared to be the hydrolysis of dietary CE. This result *in vitro* was recently confirmed *in vivo* in a mouse model in which the CEL gene was inactivated by homologous recombination (reported in Howles *et al.*[26] and elsewhere [3] in this volume).

Concluding Remarks

In this chapter, we described *in vitro* expression systems that can be programmed with either RNA or DNA molecules encoding proteins of interest. Although we had goals relevant to our particular research program on CEL, it should be clear that the same approaches can be used to study a plethora of features associated with many other lipases and proteins.

Acknowledgments

We thank our coauthors of the published works we discussed in this chapter and Ms. Jill F. Fisher for editorial assistance. The financial support of the National Institutes of Health, the American Heart Association, SEPA Affiliate, the Howard Heinz Endowment, and the W. W. Smith Charitable Trust is gratefully acknowledged.

---

[26] P. N. Howles, C. P. Carter, and D. Y. Hui, *J. Biol. Chem.* **271**, 7196 (1996).

[5] Noncatalytic Functions of Lipoprotein Lipase

By GUNILLA OLIVECRONA and AIVAR LOOKENE

Introduction

Lipoprotein lipase (LPL) is a member of the mammalian triglyceride lipase family.[1] LPL is closely related to hepatic (heparin-releasable) lipase and to pancreatic lipase and more distantly related to some yolk proteins from *Drosophila*. The physiologic function of LPL with regard to hydrolysis of lipids in plasma lipoproteins has been extensively reviewed elsewhere. LPL acts extracellularly, attached to the vascular endothelium, through interaction with heparan sulfate proteoglycans on cell surfaces.[2] The role of LPL in interconversion of plasma lipoproteins has been revealed by experiments in bulk phase *in vitro*[3,4] and *in vivo* from studies of individuals with genetic LPL deficiency.[5] Early studies indicated, however, that LPL may also have important noncatalytic functions as a transfer protein of lipoprotein core constituents (cholesteryl esters, cholesteryl ethers, retinyl esters) to cells without hydrolysis of the lipids.[6] Beisiegel *et al.*[7] showed that LPL was able to stimulate binding of different classes of lipoproteins to cultured cells. In addition, LPL stimulated binding of lipoproteins to a cell surface receptor called the low-density receptor-related protein (LRP). Chemical cross-linking showed that LPL itself bound to LRP and that this binding could be prevented by heparin.[7] In later studies, the site in LPL for interaction with LRP was localized to the carboxy-terminal end, somewhere between residues 382–425.[8]

[1] T. G. Kirchgessner, J.-C. Chuat, C. Heinzmann, J. Etienne, S. Guilhot, K. Svenson, D. Ameis, C. Pilon, L. d'Auriol, A. Andalibi, M. C. Schotz, F. Galibert, and A. J. Lusis, *Proc. Natl. Acad. Sci. U.S.A.* **86,** 9647 (1989).
[2] T. Olivecrona and G. Bengtsson-Olivecrona, *in* "Heparin" (D. Lane *et al.*, eds.), p. 335. Edward Arnold Publishers Ltd., London, 1989.
[3] R. J. Deckelbaum, S. Eisenberg, M. Fainaru, Y. Barenholz, and T. Olivecrona, *J. Biol. Chem.* **254,** 6079 (1979).
[4] R. J. Deckelbaum, R. Ramakrishnan, S. Eisenberg, T. Olivecrona, and G. Bengtsson-Olivecrona, *Biochemistry* **31,** 8544 (1992).
[5] J. D. Brunzell, *in* "Metabolic Basis of Inherited Disease" (C. R. Scriver *et al.*, eds.), p. 1913. McGraw-Hill, New York. 1995.
[6] O. Stein, G. Friedman, T. Chajek-Shaul, G. Halperin, T. Olivecrona, and Y. Stein, *Biochim. Biophys. Acta* **750,** 306 (1983).
[7] U. Beisiegel, W. Weber, and G. Bengtsson-Olivecrona, *Proc. Natl. Acad. Sci. U.S.A.* **88,** 8342 (1991).
[8] A. Nykjær, M. Nielsen, A. Lookene, N. Meyer, H. Roigaard, M. Etzerodt, U. Beisiegel, G. Olivecrona, and J. Gliemann, *J. Biol. Chem.* **269,** 31747 (1994).

The mediated binding of lipoproteins by LPL appears to be dependent on the noncovalent, homodimeric structure of the native molecule, but independent of catalytic function.[9] A lipid-binding region important for interaction with lipoproteins and lipid emulsions has been localized close to or partly overlapping the region engaged in binding of the receptor LRP (around residue 390).[10] Another lipid-binding region is present around the active site and/or on the lid structure, but this alone is not sufficient for high-affinity interaction with lipoproteins.[11]

With regard to binding to heparin or to heparin-like proteoglycans, there are at least four clusters of positively charged amino acid residues per subunit of LPL that could be involved in the interaction.[12] Site-directed mutations of several of these residues disturbs the interaction with heparin, but no residue alone has been shown to be of decisive importance.[13,14] The interaction with heparin probably occurs simultaneously with several positively charged residues on both subunits in a cooperative fashion. This explains the high affinity of LPL for heparin. LPL monomers, created by mild dissociation of the noncovalent dimer, has about a 6000-fold lower affinity for proteoglycans compared to the affinity of the active dimer.[15] Inactive LPL monomers with decreased heparin affinity are found in tissues,[16] in cell culture media, and as a result of site-directed mutagenesis of a number of residues in LPL, which are not directly involved in the interaction with heparin but which do disturb the conformation of the dimer.[17,18] The minimal heparin fragment that can completely satisfy the binding region in dimeric LPL must contain 10 saccharide units.[15]

Theoretically LPL monomers should be able to mediate binding between lipoproteins and heparin/proteoglycans because the binding sites do

---

[9] A. Nykjær, G. Bengtsson-Olivecrona, A. Lookene, S. K. Moestrup, C. M. Petersen, W. Weber, U. Beisiegel, and J. Gliemann, *J. Biol. Chem.* **268**, 15048 (1993).

[10] A. Lookene and G. Bengtsson-Olivecrona, *Eur. J. Biochem.* **213**, 185 (1993).

[11] K. A. Dugi, H. L. Dichek, G. D. Talley, H. B. Brewer, Jr., and S. Santamarina-Fojo, *J. Biol. Chem.* **267**, 25086 (1992).

[12] H. van Tilbeurgh, A. Roussel, J.-M. Lalouel, and C. Cambillau, *J. Biol. Chem.* **269**, 4626 (1994).

[13] Y. Ma, H. E. Henderson, M.-S. Liu, H. Zhang, I. J. Forsythe, I. Clarke-Lewis, M. R. Hayden, and J. D. Brunzell, *J. Lipid Res.* **35**, 2049 (1994).

[14] A. Hata, D. N. Ridinger, S. Sutherland, M. Emi, Z. Shuhua, R. L. Myers, K. Ren, T. Cheng, I. Inoue, D. E. Wilson, P.-H. Iverius, and J.-M. Lalouel, *J. Biol. Chem.* **268**, 8447 (1993).

[15] A. Lookene, O. Chevreuil, P. Østergaard, and G. Olivecrona, *Biochemistry* **35**, 12155 (1996).

[16] M. Bergö, G. Olivecrona, and T. Olivecrona, *Biochem. J.* **313**, 893 (1996).

[17] A. Hata, D. N. Ridinger, S. D. Sutherland, M. Emi, L. K. Kwong, J. Shuhua, A. Lubbers, B. Guy-Grand, A. Basdevant, P.-H. Iverius, D. E. Wilson, and J.-M. Lalouel, *J. Biol. Chem.* **267**, 20132 (1992).

[18] A. Krapp, H. F. Zhang, D. Ginzinger, M. S. Liu, A. Lindberg, G. Olivecrona, M. R. Hayden, and U. Beisiegel, *J. Lipid Res.* **36**, 2362 (1995).

not interfere with each other as in the case of lipoproteins and LRP. However, the affinity for proteoglycans of monomeric LPL is probably too low. Only low amounts of LPL monomers appear to be bound to the vascular endothelium in a heparin-releasable manner.[19] The ability of dimeric LPL to mediate binding of lipoproteins to cell surfaces is due to its high-affinity binding sites for lipids, for proteoglycans, and for LRP. A similar ability has recently been found for hepatic lipase, which also binds both to LRP and to lipoproteins.[8,20,21] Furthermore, it binds to heparin and heparin-like proteoglycans but with much lower affinity than LPL.[2]

A prerequisite for studies of the noncatalytic function of these lipases is to generate inactive, but conformationally intact or native, forms. Our results suggest that blocking of the active site of LPL with specific inhibitors does not induce major conformational changes in the enzyme.[22] Experiments have thus far mostly been done with LPL and we therefore concentrate on LPL in the following presentation.

Active Site Inhibitors

Principle

LPL is, like the other members of the triglyceride lipase gene family, a serine hydrolase with a catalytic triad consisting of Ser-132, Asp-156, and His-241.[23] LPL is slowly inhibited by common inhibitors for serine hydrolases like diisopropyl fluorophosphate (DFP) and phenylmethanesulfonyl fluoride (PMSF), but high concentrations (mM) are needed and the inhibition is usually incomplete.[24] Low concentrations of LPL can be inhibited by diethyl-p-nitrophenyl phosphate (paraoxan, E600).[25] However, the phosphorylated enzyme hydrolyzes slowly to regenerate active enzyme.[25] LPL can be almost completely inhibited by equimolar concentrations of the inhibitor tetrahydrolipstatin or Orlistat® (Hoffman La Roche, Basel,

---

[19] G. Olivecrona, M. Hultin, R. Savonen, N. Skottova, A. Lookene, Y. Tugrul, and T. Olivecrona, in "Atherosclerosis X" (F. P. Woodford et al., eds.), p. 250. Elsevier, New York, 1995.

[20] M. Z. Kounnas, D. A. Chappell, H. Wong, W. S. Argraves, and D. K. Strickland, *J. Biol. Chem.* **270**, 9307 (1995).

[21] A. Krapp, S. Ahle, S. Kersting, Y. Hua, K. Kneser, M. Nielsen, J. Gliemann, and U. Beisiegel, *J. Lipid Res.* **37**, 926 (1996).

[22] A. Lookene, N. Skottova, and G. Olivecrona, *Eur. J. Biochem.* **222**, 395 (1994).

[23] J. Emmerich, O. U. Beg, J. Peterson, L. Previato, J. D. Brunzell, H. B. Brewer, Jr., and S. Santamarina Fojo, *J. Biol. Chem.* **267**, 4161 (1992).

[24] D. Quinn, K. Shirai, and R. L. Jackson, *Prog. Lipid Res.* **22**, 35 (1983).

[25] D. M. Quinn, *Biochim. Biophys. Acta* **834**, 267 (1985).

Switzerland),[22] which was originally developed as a drug to inhibit digestive lipases for weight reduction.

A main drawback of this inhibitor is that it is slowly turned over by LPL. Reversibility with THL was earlier shown for inhibition of pancreatic cholesteryl hydrolase.[26] Regeneration of active LPL from THL-inhibited LPL appeared to be stimulated by the presence of a lipid substrate.[22] Reactivation is therefore, unfortunately, more rapid under most experimental conditions under which one would be likely to use the inhibited lipase. The regeneration of activity can be prevented by the presence of a high excess of inhibitor during the experiments.[18] Another drawback is that inhibition of LPL by low concentrations of THL requires that the enzyme be dissolved in detergent-containing buffer, preferably in deoxycholate. A third drawback is that the inhibited LPL appears to form tetramers or even higher oligomers.[22] Thus, the solubility of LPL is reduced by the inhibitor from being low for the native enzyme to being even lower for the THL-inhibited form. This can create problems under experimental conditions when concentrations of THL–LPL higher than 0.1 mg/ml are needed. Practical limits for the solubility of THL–LPL in 5 mM deoxycholate appeared to be 0.5 mg/ml. In buffer (Bristris or Tris, pH 6.5–7.5) with a high concentration of salt (1 M NaCl) the limit is around 0.1 mg/ml.

For more recent studies of the noncatalytic function of LPL, we have used the sulfonylfluoride-based inhibitor hexadecylsulfonylfluoride (HDS).[27] This can be synthesized according to Horrevoets et al.[28] Like THL, HDS almost completely inhibits LPL when added in equimolar concentrations over the monomer concentration of LPL (Fig. 1). In practice we add a 10-fold molar excess of inhibitor over the lipase. As with THL, inhibition is more efficient in buffer containing detergent, probably because of the limited solubility of the inhibitors. Inhibition with HDS is, as expected, irreversible. HDS has negative effects on the solubility of LPL similar to those of THL, probably because two long hydrocarbon chains from the inhibitor are added to the LPL dimer. In practice, we have found that inhibited LPL has highest solubility in bile salt-containing buffers. Detergents are in some cases incompatible with the experimental conditions in which the inhibited lipase is used. For that reason, and also for removal of unreacted inhibitor, the inhibited lipase may have to be repurified. We usually use adsorption to heparin-agarose followed by gradient elution of

---

[26] B. Borgström, *Biochim. Biophys Acta* **962**, 308 (1988).
[27] N. Skottova, R. Savonen, A. Lookene, M. Hultin, and G. Olivecrona, *J. Lipid Res.* **36**, 1334 (1995).
[28] A. J. G. Horrevoets, H. M. Verheij, and G. H. De Haas, *Eur. J. Biochem.* **198**, 247 (1991).

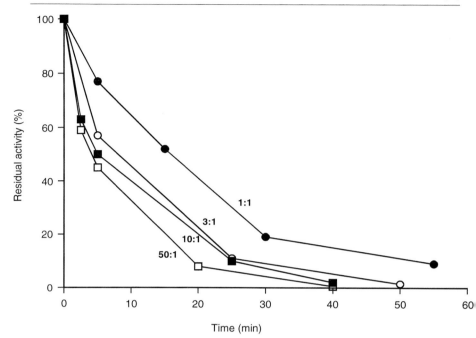

Fig. 1. Time courses for inhibition of LPL with different concentrations of HDS. HDS and LPL were incubated at molar ratios (1:1, 3:1, and 50:1, calculated on LPL monomers) in 10 mM Tris, 4 mM deoxycholate, 0.1 mM linoleic acid, pH 8.5 at 25°. LPL activity was determined using tributyrylglycerol in gum arabic.[10]

the inhibited LPL with NaCl to ensure that only dimeric LPL with native structure is used for experiments.

For studies of the time course and the completeness of the inhibition, an assay for LPL activity is recommended. This can be any assay for LPL with long-chain triacylglycerols or phospholipids as substrate.[29] For practical purposes, and to be able to check rapidly the amount of remaining activity in the preparations, we often use tributyrylglycerol emulsified in gum arabic as substrate in a pH stat.[10] The released butyric acid is directly determined by titration. For studies of the amount of LPL activity under experimental conditions, a more sensitive assay with radioactively labeled lipids is often required.[29]

[29] G. Bengtsson-Olivecrona and T. Olivecrona, in "Lipoprotein Analysis: A Practical Approach" (C. A. Converse et al., eds.), p. 169. Oxford University Press, New York, 1992.

Reagents

Buffer for dialysis: 10 mM Tris–HCl containing 5 mM deoxycholate, 0.1 mM linoleate, pH 8.5

Stock solution of HDS or THL (10–20 mg/ml) in dimethyl sulfoxide

Procedure

Bovine LPL isolated from milk is the most convenient source of pure enzyme.[30] LPL from other sources can also be used. The enzyme is dialyzed against the bile salt-containing buffer at 4° for 48 hr using dialysis tubing with an appropriate cutoff range. During the first 24 hr of dialysis, the enzyme precipitates in a flocculate precipitate that sediments in the dialysis bag. After extended dialysis, the precipitate dissolves and both LPL activity and mass is usually almost completely recovered. LPL is then stable in the bile salt-containing buffer, even at room temperature for several days. It is also stable during freezing at $-20°$ or at lower temperatures. The maximal solubility of LPL in 5 mM deoxycholate is around 1 mg/ml. This limit can be increased about 10-fold by using taurodeoxycholate instead of deoxycholate.[31] Reaction with either THL or HDS is performed by adding the inhibitor from the stock solution in dimethyl sulfoxide. As shown in Fig. 1 the reaction with HDS is rather rapid at room temperature. A 1:1 molar ratio of HDS (or THL) to LPL monomer is sufficient for almost complete inhibition. In practice we use a 10-fold molar excess of inhibitor over LPL monomers (active sites) and carry out the incubation with the inhibitor for 6 hr at 25° with HDS and for 30 min at 25° with THL. By titration of LPL with the inhibitors it can be demonstrated that both active sites on LPL are catalytically active[22] (Fig. 1). The inhibited HDS–LPL is stable in the deoxycholate-containing buffer even in the frozen state. Reactivation does not occur with LPL inhibited by HDS but does occur with LPL inhibited by THL, as discussed earlier.

Determination of LPL Mass: Labeling of LPL with Na-^{125}I

Principle

For determination of concentration of inactive LPL, a regular protein determination can be used or absorbancy measurements having an extinction coefficient at 280 nm ($E_{1\%}$) for bovine milk LPL of 16.8.[32] For determi-

---

[30] G. Bengtsson-Olivecrona and T. Olivecrona, *Methods Enzymol.* **197**, 345 (1991).
[31] C. Martinez, C. Cambillau, A. Lookene, and G. Olivecrona, unpublished results.
[32] T. Olivecrona, G. Bengtsson-Olivecrona, J. C. Osborne, Jr., and E. S. Kempner, *J. Biol. Chem.* **260**, 6888 (1985).

nation of low concentrations, an immunoassay might be needed that can detect a few nanograms per milliliter. Enzyme-linked immunosorbent assays on microtiter plates have been developed in several laboratories.[33–35] For experiments with inactive LPL, a label on the lipase protein might be necessary, for example, for binding studies to cells in culture.[7] We describe here iodination of LPL with Na-^{125}I by the mild lactoperoxidase/glucose oxidase method, which retains LPL in an active form.[36] The iodinated lipase has to be repurified by chromatography on heparin-agarose to remove not only unreacted iodine but also damaged LPL in forms with lower heparin affinity than the catalytically active dimers.

Reagents

LPL (bovine, 0.3–0.5 mg/ml) in 10 mM Bistris, 1.3 M NaCl, pH 6.5
1 M Tris–HCl, pH 7.4
Glucose
Na-^{125}I (1 mCi) and unlabeled NaI (1 mM)
Lactoperoxidase (ammonium sulfate precipitate, Boehringer Mannheim, Germany)
Glucose oxidase (1 mg/ml 10 mM Tris–HCl, pH 7.4, Sigma, St. Louis, MO)
Sodium metabisulfite (10 mg/ml 10 mM Tris–HCl, pH 7.4)
10 mM Tris–HCl buffer, pH 7.4, containing 20% (w/v) glycerol and 0.1% (w/v) Triton X-100
Heparin-Sepharose (Pharmacia Fine Chemicals, Uppsala, Sweden)

Procedure

Add 50 μl of cold 1 M Tris–HCl, pH 7.4, to 0.5 ml LPL to increase buffer capacity and pH. Add 5 mg glucose. Keep sample on ice. Add (in a well-ventilated hood) in the following order: 1 mCi Na-^{125}I, 5 μl unlabeled NaI, 5 μl suspension of lactoperoxidase, and finally 5 μl glucose oxidase to generate small amounts of H_2O_2. Cap the tube and mix gently every 5 min for 20 min. Add 10 μl sodium metabisulfite to stop the reaction. Dilute the sample after 2 min with cold 10 mM Tris–HCl, pH 7.4, containing 20% glycerol and 0.1% Triton X-100 (2 ml) to reduce the salt concentration

---

[33] J. Peterson, W. Y. Fujimoto, and J. D. Brunzell, *J. Lipid Res.* **33**, 1165 (1992).
[34] E. Vilella, J. Joven, M. Fernández, S. Vilaró, J. D. Brunzell, T. Olivecrona, and G. Bengtsson-Olivecrona, *J. Lipid Res.* **34**, 1555 (1993).
[35] M. Kawamura, T. Gotoda, N. Mori, H. Shimano, K. Kozaki, K. Harada, M. Shimada, T. Inaba, Y. Watanabe, Y. Yazaki, and N. Yamada, *J. Lipid Res.* **35**, 1688 (1994).
[36] L. Wallinder, J. Peterson, T. Olivecrona, and G. Bengtsson-Olivecrona, *Biochim. Biophys. Acta* **795**, 3524 (1984).

enough to allow binding of LPL to heparin. Immediately apply the sample on a small column of heparin-Sepharose (1 ml) equilibrated in the same buffer, preferentially at 4°. Wash the column extensively with buffer before the gradient is started. The gradient should run from 0.1 to 2.0 M NaCl in the Tris–HCl buffer with glycerol and Triton X-100 (50 ml + 50 ml) at a flow rate of at least 1 ml/min. Measure radioactivity in aliquots of the collected fractions in a gamma counter. The peak of radioactivity eluting at the highest concentration of NaCl (0.8–1.2 M NaCl) corresponds to catalytically active, dimeric LPL.

To improve stability during storage of the radiolabeled LPL, 2 mg bovine serum albumin is usually added per milliliter and the preparation is then frozen at −70°. The specific radioactivities of the preparations are usually between 2000–10,000 cpm/ng (as determined by immunoassay). This corresponds to incorporation of 1–2 mol iodine/mol LPL dimer. The catalytic activity can be checked in any type of lipase assay and should be as high as in unlabeled preparations. The labeled LPL is intact on SDS–PAGE for at least 2 months. With time material with lower affinity for heparin (inactive LPL monomers) appears in the preparations (Fig. 2). This amount usually corresponds to less than 5% after storage for 1 month. We usually discard the preparation at this time. For preparation of active site-inhibited LPL, ^{125}I-labeled LPL, without addition of bovine serum albumin, is directly dialyzed against the bile salt-containing buffer as described earlier.

Mild Dissociation of LPL to Inactive Monomers

Principle

Active LPL is composed of noncovalent homodimers. It was found that loss of catalytic activity is associated with dissociation of the dimers to monomers.[37] The rate of inactivation is dependent on the concentration of LPL, indicating that active monomers are in rapid equilibrium with dimers.[37] It was concluded that the monomer is in a metastable state and readily and irreversibly decomposes to inactive monomer after a minor conformational change.[37] The inactive monomers have a tendency to form oligomers at higher concentrations. The dissociation of active LPL dimers occurs easily, for example, on exposure to elevated temperature. Dissociation can also be initiated and controlled by addition of dissociating agents like guanidinium

---

[37] J. C. Osborne, G. Bengtsson-Olivecrona, N. S. Lee, and T. Olivecrona, *Biochemistry* **24**, 5606 (1985).

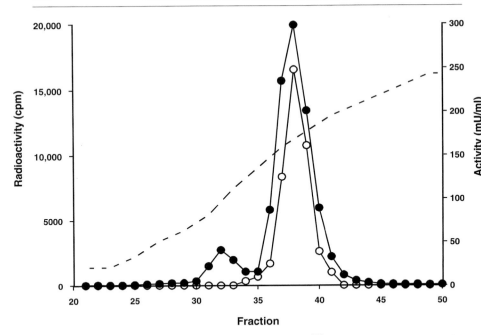

FIG. 2. Distribution of radioactivity in a preparation of ^{125}I-labeled LPL as separated by chromatography on heparin-Sepharose. A trace amount of radiolabeled LPL was applied on a heparin-Sepharose column together with unlabeled active LPL. The protein was eluted by a linear NaCl gradient from 0.1 to 2.0 M. The gradient is shown by the dashed line. Distribution of radioactivity (filled circles) represents elution profile for LPL mass. The activity profile (open circles) shows elution of active, dimeric LPL. The first peak of LPL (eluted at 0.6 M NaCl) represents inactive LPL monomers. The second larger peak of radioactivity elutes together with the peak of active LPL dimers at 1.0 M NaCl.

chloride or urea. In the following sections, we describe dissociation of LPL to inactive monomers by treatment with 1 M guanidinium chloride.

Reagents

10 mM Tris–HCl buffer, pH 7.4, containing 0.5 M $(NH_4)_2SO_4$, 0.2 M NaCl, 1 mM EDTA

Stock solution of guanidinium chloride (analytical grade), 6 M in 10 mM Tris–HCl, pH 7.4

Procedure

For preparation of monomeric LPL in high concentration it was previously found that the most stable conditions were obtained using buffer containing ammonium sulfate in addition to NaCl,[37] since the sulfate ion has a stabilizing effect on LPL. The lipase is dialyzed overnight at 4° against this buffer. Then guanidinium chloride is added to a final concentration of 1 M. In most cases, inactivation is accomplished on incubation at 10°. The inactivation rate is dependent on the concentration of LPL and on the temperature. At a concentration of 0.1 mg/ml, less than 10% of the activity remains after 2 hr at 10°. Incubation temperatures above 20° should be avoided since the stability and solubility of LPL is poor in salt-containing buffers at elevated temperature. The inactivation process has to be followed by measurements of remaining catalytic activity, for example, using tributyryl glycerol as substrate in a pH stat.[10]

Purification of Inactivated LPL

Principle

Repurification of active site-inhibited LPL (by HDS or by THL) or of monomeric LPL should be avoided due to the low solubility of the modified variants. We recommend that the inhibited LPL be stored in bile salt-containing buffer and that the purification be done just before the lipase is needed for experiments. Monomeric LPL in salt-containing buffer cannot be frozen at high concentration because of low solubility and stability. Therefore, it has to be freshly prepared for each experiment. If repurification is necessary for experimental reasons, we use adsorption to heparin-Sepharose (see earlier discussion). To increase recovery of the inactive LPL variants (based on experiments with ^{125}I-labeled LPL[38]) the gradient should also contain 1 mg bovine serum albumin per milliliter in addition to glycerol and Triton X-100. In cases when only the excess of inhibitor or the dissociating agent (guanidinium chloride) has to be removed, the sample can either be dialyzed or it can be passed over a small gel permeation column equilibrated with the appropriate buffer.

Reagents

    Gel permeation columns, for example, NAP-5 (Pharmacia Fine Chemicals, Uppsala, Sweden)
    Heparin-Sepharose or small columns of heparin-Sepharose for HPLC (Pharmacia Fine Chemicals)
    10 mM Tris–HCl buffer, pH 7.4, with 20% glycerol, 0.1% Triton X-100, 1 mg bovine serum albumin (fraction V, Sigma) per milliliter and also containing 0.1 M NaCl or 1.6 M NaCl for the gradient

---

[38] G. Olivecrona and A. Lookene, unpublished results.

Procedure

For exchange of buffer by gel permeation chromatography or dialysis the desired buffer is selected. For higher concentrations of LPL, it should always contain a relatively high salt concentration ($>0.3\ M$ and preferably $1\ M$ NaCl) and be cold (4 to 10°) since the solubility and stability of LPL is low at low salt concentration (in buffer without detergents) and at elevated temperature.

The inactive LPL is bound to a small column of heparin-Sepharose from the deoxycholate-containing buffer or from the salt-containing buffer with guanidinium chloride. The latter solution has to be diluted 10-fold to allow adsorption. The inactive LPL is then eluted by a salt gradient as described earlier under purification of radiolabeled LPL. Monomeric LPL elutes in the position of inactive LPL (around $0.6\ M$ NaCl; see Fig. 2) and active-site inhibited LPL variants in the position of dimeric LPL (around $0.9\ M$ NaCl; see Fig. 2). If a more concentrated solution of the repurified material is needed, batch elution with buffer containing $1.6\ M$ NaCl in a single pass can be used instead of a gradient.

Separation of LPL Monomers and Dimers by Density Gradient Ultracentrifugation

Principle

Separation of LPL dimers from inactive monomers is not possible with the conventional technique (gel permeation chromatography) probably because of the extended shape of the LPL subunit.[37] Therefore, separation by chromatography on heparin-Sepharose as described earlier or by sucrose density gradient centrifugation has to be used to allow studies of the monomer/dimer distribution of LPL.[22,37]

Reagents

Sucrose solutions 5 and 10% (w/v) in Bistris, pH 6.5, containing 1.5 M NaCl and 1 mg bovine serum albumin (fraction V, Sigma) per milliliter

Beckman ultracentrifuge (L5-65B), SW50.1Ti rotor, Ultra Clear tubes

Procedure

Linear sucrose gradients from 10 to 5% (w/v) are made in the ultracentrifuge tube (total volume 4.8 ml) at 10°. Samples of 0.1 ml containing LPL (preferably in the same buffer as the gradient) are carefully layered on top of the gradients. Centrifugation (at 40,000 rpm) is performed for 18 hr at 10°. After the run, the tubes are emptied from the bottom through puncture with a syringe needle. Fractions of 0.25 ml are collected for analysis. Standard proteins can be used together wtih LPL or in separate tubes.[22] Active LPL dimers are detected by activity assay. Inactive LPL monomers are seen as a peak of LPL mass sedimenting more slowly than the peak of active LPL.

Studies of Interactions of LPL Molecule Using Surface Plasmon Resonance Technique

Principle

The noncatalytical interactions of LPL can be studied in a variety of systems like cell cultures,[21] organ perfusions,[27] ligand blots,[8] and microtiter plates.[9] Also interaction with emulsified substrates or lipoproteins can be studied after separation by centrifugation or gel permeation chromatography of the lipid from the water phase.[10,39] In the following sections, we describe the use of the biosensor technique on a BIAcore (Pharmacia) to study noncatalytic interactions of LPL. Its detection is based on surface plasmon resonance (SPR), a quantum mechanical phenomenon that detects changes in the refractive index close to the surface of thin gold film on a glass support (i.e., sensor chip). The chip surface is coated by carboxymethylated dextran polymer to which one of the reactants can be covalently linked, while the other is injected in flow over the surface. Binding of a soluble component to the immobilized one leads to an increase of the refractive index. This refractive index change alters the SPR, which can be detected optically. Binding is measured in arbitrary response units.

We concentrate here on the interaction with heparin-like molecules, but the system can be used for studies of the interaction of LPL with different classes of lipoproteins[40] and presumably also for interactions of LPL with other molecules like cell surface receptors. The system is well suited for studies of the interaction of LPL with different variants of glycosaminoglycans as well as for the interaction of a specific glycosaminoglycan with different variants of LPL, for example, for investigation of mutations affecting the heparin-binding regions in LPL. It is also possible in this system to study competition with free ligands, for example, fragments of heparin, and thereby measure their relative affinity for LPL. There are

[39] P. Carrero, D. Gómez-Coronado, G. Olivecrona, and M. A. Lasunción, *Biochim. Biophys. Acta* **1299,** 198 (1996).

[40] A. Lookene, R. Savonen, and G. Olivecrona, *Biochemistry* **36,** 5267 (1997).

at least two reasons for studies of the interaction of LPL with heparan sulfate. One is for detailed studies of the interaction of LPL with heparin and heparin-like substances, since LPL has a relatively high affinity for these substances and because interaction with heparan sulfate is of great importance for the normal physiologic function of LPL.[2] The other reason is that LPL attached to a heparan sulfate-covered dextran chip can be considered an *in vitro* model for LPL attached to the vascular side of blood vessels. In this position LPL interacts with several ligands, among them lipoproteins. The nondestructive way of attaching LPL to heparan sulfate or directly to streptavidin after biotinylation of LPL offers a great possibility for interaction studies with other ligands. In the BIAcore system, association and dissociation rate constants can be directly determined in real time.

We have mostly used the biotin/streptavidin interaction to immobilize heparin/heparan sulfate or LPL to the dextran-covered sensor chips. Heparin or heparan sulfate can be biotinylated via their amino groups according to the method of Lee and Conrad.[41]

Reagents

Heparin or Heparan Sulfate. For biotinylation the heparin has to be a relatively crude preparation containing peptides from the core proteins. Thus, heparin preparations manufactured for clinical use are not suitable for this purpose. Heparan sulfate can be prepared from endothelial cells as described.[42]

  Sodium carbonate buffer, 0.2 M, pH 8.2
  Sulfo-NHS-biotin (Pierce, Rockford, IL)
  Sensor chips (CM5) (Pharmacia Biosensor, Uppsala, Sweden)
  EDC/NHS (0.2 M/0.05 M) (freshly prepared) (Pharmacia Biosensor)
  Streptavidin (Sigma), 0.1 mg/ml in 10 mM sodium acetate buffer, pH 4.5
  Ethanolamine, 1 M, pH 8.5 (Pharmacia Biosensor)
  Gel permeation column for buffer exchange, for example, NAP-5 (Pharmacia Fine Chemicals, Uppsala, Sweden)
  BIAcore (Pharmacia Biosensor)

Procedure

For biotinylation a solution of heparin or heparan sulfate (5–10 mg/ml in 0.2 M NaCO$_3$, pH 8.6) is incubated with a 10-fold molar excess of sulfo-

---

[41] W. T. Lee and D. H. Conrad, *J. Exp. Med.* **159**, 1790 (1984).
[42] A. Lindblom and J. Fransson, *Glycoconjugate J.* **7**, 545 (1990).

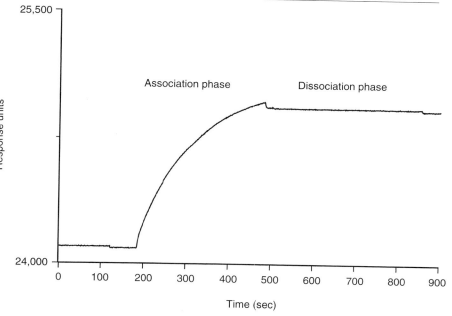

Fig. 3. Sensogram for the binding of LPL to heparin. Biotinylated heparin was immobilized to matrix-bound streptavidin. In the association phase LPL was injected over the layer of the sensor chip with immobilized heparin. The increase in response units is proportional to the amount of bound LPL. In the dissociation phase the sensor chip was washed with the running buffer. The apparent dissociation rate is very low due to rapid rebinding of LPL to the surface.[15]

NHS-biotin for 3 hr at 25°. Nonreacted biotin is removed by gel permeation chromatography on a NAP-5 column. For immobilization of streptavidin on the sensor chip, the carboxylated dextran matrix has to be activated. We use the methodology described by Johnsson et al.[43] For this 30 μl of the EDC/NHS mixture is injected over the surface of the sensor chip at a flow rate of 5 μl/min. Then the solution of streptavidin is injected at the same flow rate (total volume 30 μl). Nonreacted activated groups are blocked by injection of 1 M ethanolamine (30 μl). For immobilization of heparin or heparan sulfate to the sensor chip, the solution of purified, biotinylated heparin/heparan sulfate is injected. The increase in response units (RU) is then about 100–200, corresponding to binding of the glucosaminoglycan. Sensor chips derivatized by heparin/heparan sulfate can be stored at 4° for several months. Binding of LPL to the immobilized glycosaminoglycan is performed in 20 mM HEPES buffer, pH 7.4, 0.15 M NaCl, 3.4 mM EDTA,

[43] B. Johnsson, S. Löfås, and G. Lindquist, *Anal. Biochem.* **198**, 268 (1991).

using enzyme concentrations between 5 and 50 nM. Because of rapid inactivation of LPL at low salt concentration, dilutions of stock solutions of LPL should be made immediately before injection.

Figure 3 shows the result of a typical experiment where LPL is bound to a sensor chip covered by heparin. The association is rapid but dissociation is very slow indicating very high apparent affinity. Since the dissociation is rapid in the presence of free ligand (heparin or fragments of heparin) it was previously concluded that slow dissociation is only apparent because of rapid rebinding of LPL.[15] The surface of the sensor chips can be regenerated by injection of either of 1.5 M NaCl, 0.1% SDS, or 0.1 M NaOH, which releases the bound LPL. Thus, the glycosaminoglycan-covered chips can be used for several experiments.

[6] Radiation Inactivation Studies of Hepatic Cholesteryl Ester Hydrolases

By EARL H. HARRISON *and* ELLIS S. KEMPNER

Introduction

Radiation inactivation is a technique capable of determining the mass of protein molecules that are involved in particular biochemical reactions, independent of the presence of all other proteins. This method of radiation target analysis has been used to determine the molecular size and oligomeric state of several lipases. In particular, we present here the use of this technique to measure the mass of the catalytically active species of cholesteryl ester hydrolases in rat liver.

The liver plays a central role in cholesterol metabolism and contains enzymes that catalyze the formation and hydrolysis of cholesteryl esters. The latter process is catalyzed by cholesteryl ester hydrolases (CEHs) and a number of such activities have been reported to be present in the liver.

Four hepatic enzymes that catalyze hydrolysis of cholesteryl esters have been purified to homogeneity: the acid lipase localized in lysosomes,[1] hepatic lipase,[2] a soluble, bile salt-independent CEH,[3] and a neutral bile salt-

---

[1] S. D. Fowler and W. J. Brown, *in* "Lipases" (B. Borgstrom and H. L. Brockman, eds.), pp. 329–364. Elsevier Science Publishers B. V., Amsterdam, 1984.
[2] P. K. J. Kinnunen, *in* "Lipases" (B. Borgstrom and H. L. Brockman, eds.), pp. 307–328. Elsevier Science Publishers B. V., Amsterdam, 1984.
[3] S. Ghosh and W. M. Grogan, *Lipids* **26,** 793 (1991).

dependent CEH isolated from rat liver cytosol.[4] The latter enzyme is now appreciated to be the enzyme carboxyl ester lipase.[5]

Carboxyl ester lipase (also called bile salt-activated lipase) is secreted from the pancreas, lactating breast, and liver of several mammalian species. The enzyme catalyzes the hydrolysis of a wide variety of lipid esters, including cholesteryl esters, triacylglycerols, fat-soluble vitamin esters, and lysophospholipids, as well as acting on water-soluble esters (see Wang and Hartsuck[5] for a recent review). The human, bovine, and rat enzymes have been cloned and sequenced,[6-9] and these and the porcine enzyme have been the object of enzymological studies.

In contrast to its broad substrate specificity, carboxyl ester lipase (CEL) is markedly and specifically activated by bile salts, particularly trihydroxylated bile salts such as cholate and its conjugates. The mechanism of bile salt activation of catalytic activity is complex and poorly understood. Although the presence of bile salts can clearly effect the physical state of hydrophobic substrates, it is also clear that they have direct effects on the enzyme itself. For example, when present in the nanomolar to micromolar range (i.e., below the critical micellar concentration or CMC) they have been reported to bind to enzyme monomers and cause changes in thermal stability[10] and secondary structure.[11] At concentrations above the CMC (i.e., in the millimolar range), activating bile salts induce the formation of dimers and higher oligomers of the enzyme.[12-14] However, the relationship of the bile salt-induced oligomerization of the enzyme and the activation of catalytic activity was still unclear. The molecular size and oligomeric state of active CEL in rat liver cytosol has now been ascertained by radiation target analysis.

[4] E. D. Camulli, M. J. Linke, H. L. Brockman, and D. Y. Hui, *Biochim. Biophys. Acta* **1005**, 177 (1989).
[5] C.-S. Wang and J. A. Hartsuck, *Biochim. Biophys. Acta* **1166**, 1 (1993).
[6] J. H. Han, C. Stratowa, and W. J. Rutter, *Biochemistry* **26**, 1617 (1987).
[7] J. A. Kissel, R. N. Fontaine, C. W. Turck, H. L. Brockman, and D. Y. Hui, *Biochim. Biophys. Acta* **1006**, 227 (1989).
[8] E. M. Kyger, R. C. Wiegand, and L. G. Lange, *Biochem. Biophys. Res. Comm.* **164**, 1302 (1989).
[9] T. Baba, D. Downs, K. W. Jackson, J. Tang, and C.-S. Wang, *Biochemistry* **30**, 500 (1991).
[10] T. Tsujita, N. K. Mizuno, and H. L. Brockman, *J. Lipid Res.* **28**, 1434 (1987).
[11] P. W. Jacobson, P. W. Wiesenfeld, L. L. Gallo, R. L. Tate, and J. C. Osborne, *J. Biol. Chem.* **265**, 515 (1990).
[12] K. B. Calme, L. Gallo, E. Cheriathundam, G. V. Vahouny, and C. R. Treadwell, *Arch. Biochem. Biophys.* **168**, 57 (1975).
[13] D. Lombardo and O. Guy, *Biochim. Biophys. Acta* **611**, 147 (1980).
[14] C. J. Rojas and E. H. Harrison, *Proc. Soc. Exp. Biol. Med.* **206**, 60 (1994).

In addition to the soluble enzymes described earlier, there are many reports of both acid and neutral CEH activities localized in microsomal fractions of rat liver homogenates.[15–18] We reported the presence of distinct acid and neutral CEH activities in rat liver microsomes and showed that these activities were associated with endosomes and/or plasma membrane vesicles present in the microsomal fraction.[19] None of the membrane-bound CEHs have been purified and hence little is known of their biochemical properties and structure.

We have used radiation inactivation to study the molecular sizes of these acid and neutral CEHs in rat liver microsomes. The results lend further support to the suggestion that distinct acid and neutral CEHs are present, and they allow us to compare the molecular sizes of these enzymes with those of other lipases in the liver.

Radiation Target Analysis: Theory and Applications

High-energy electrons interact randomly with all molecules in their path. In each event considerable energy is transferred to the molecule, resulting in extensive structural damage and loss of biological activity. The rate of loss of the activity is directly proportional to the mass of the functional unit.

Ionizing radiation in the form of gamma rays or high-energy electrons collides principally with orbital electrons in molecules lying along the trajectory of the radiation. The interactions occur randomly, depending only on the number of electrons per molecule. Thus, the greater the mass of the molecule, the greater the chance of a molecule being hit. In each event, energy is transferred to the struck molecule and is absorbed. On the average, about 60 electron volts are deposited in this primary ionization, half of which ultimately cause breaks in covalent bonds elsewhere in the molecule (the other half is absorbed by reversible processes that leave the molecule unaltered). In proteins, covalent bonds are broken randomly throughout the molecule regardless of where the original damage occurred. Many covalent bonds are broken due to a single primary ionization and molecules suffering such harsh damage lose all biological function. If one of the

---

[15] D. Deykin and D. S. Goodman, *J. Biol. Chem.* **237,** 3649 (1962).
[16] M. C. Riddle, E. A. Smuckler, and J. A. Glomset, *Biochim. Biophys. Acta* **338,** 339 (1975).
[17] A. Nilsson, *Biochim. Biophys. Acta* **450,** 379 (1976).
[18] R. A. Coleman and E. B. Hayes, *Biochim. Biophys. Acta* **751,** 230 (1983).
[19] M. Z. Gad and E. H. Harrison, *J. Lipid Res.* **32,** 685 (1991).

covalent bonds in the polymer backbone is broken, the protein will be fragmented.

There also will be radiation damage to solvent and other molecules in the sample. If the radiation exposure is conducted at very low temperatures, all radiolytic products are frozen and unable to diffuse. During subsequent storage at low temperature, these products react slowly with only those chemical species in the immediate vicinity. When later thawed, these samples reveal that the only damaged protein molecules are those in which the radiation caused a primary ionization. Molecules that do not suffer a primary ionization under these conditions are unaltered.

It follows directly from these principles that the surviving biological activity in irradiated samples will decrease exponentially with radiation exposure, and the rate of this decrease is directly related to the mass of those proteins responsible for the measured biological activity. More complex systems have been observed in which a single exponential did not fit the surviving activity in irradiated samples[20] (See later discussion for microsomal, neutral CEH). Such results can be analyzed by an extension of target theory to more complicated situations involving two or more different size proteins that catalyze the reaction being measured.[21] To establish these cases it is first necessary to confirm that the inactivation curve is not a single exponential. This is best determined by using a wide range of radiation doses.

Lipases as well as all other molecules exposed to ionizing radiation are severely damaged with consequent loss of function. Under appropriate conditions this inactivation can be analyzed by means of target theory, yielding the mass of only those structures associated with the measured function. If samples are irradiated at sufficiently low temperatures, if the measurement of radiation dose is accurate, and if the enzymatic assay used gives an accurate measurement of the V_{max} of the reaction then the analysis can yield the mass of only those structures associated with the measured function. Application of this technique to various lipases has led to the determination of the mass of those proteins required for the expression of enzymatic activity, whether *in vivo* or *in vitro*.

A variety of lipases have been examined in irradiated tissues. The calculated radiation target sizes were compared with the enzyme subunit size (where known). These data indicated the number of subunits required for the lipolytic reaction. Earlier work has been discussed in a previous

---

[20] E. S. Kempner, *Trends Biochem. Sci.* **18**, 236 (1993).
[21] E. S. Kempner, *Bull. Math. Biol.* **57**, 883 (1995).

volume in this series.[22] More recent studies have focused on phospholipase A_2[23–25] and hepatic lipase.[26]

Experimental Procedures

In the experiments described here, individual rat liver homogenates were prepared and fractionated to yield the microsomal and cytosolic fractions as cited in Harrison *et al.*[27] In some experiments, a portion of the cytosol fraction had sodium cholate added to a final concentration of 20 mM prior to irradiation as described next.

All samples (0.5 ml) were placed in glass ampules, frozen on dry ice, and sealed under air. Samples were kept at $-70°$ until irradiation at $-135°$ with 13-MeV electrons as described in Harmon *et al.*[28] Typical dose rates were 20–30 Mrad/hr as determined by thermoluminescent dosimetry. Following irradiation samples were kept at $-70°$ until they were thawed once, immediately before the enzyme assays described later. All activities are expressed relative to the unirradiated controls. Target size determinations in each experiment were from a least-squares fit of the data as described by Harmon *et al.*[28]

Ester hydrolase activity was determined using the radiometric assay described by Harrison.[29] Carboxyl ester lipase activity assays were conducted in reaction mixtures containing 50 mM Tris-maleate, pH 7, and final concentrations of 20 mM sodium cholate. Concentrations of the substrate (cholesteryl oleate, retinyl palmitate, or triolein) were 40–80 μM, and these concentrations gave maximal apparent initial velocities. It is important to note that *all* enzyme assays were conducted in the presence of final concentrations of 20 mM cholate. For samples that were also irradiated in the presence of cholate, the concentration was adjusted by taking into account the amount of cholate added along with the enzyme source. For the microsomal, bile salt-independent CEH activities, the same conditions were used with the omission of bile salts. Neutral CEH was assayed in 50

[22] E. S. Kempner, J. C. Osborne, Jr., L. J. Reynolds, R. A. Deems, and E. A. Dennis, *Methods Enzymol.* **197,** 280 (1991).
[23] N. M. Tremblay, D. Nicholson, and M. Potier, *Biochem. Biophys. Res. Commun.* **183,** 121 (1992).
[24] E. J. Ackerman, E. S. Kempner, and E. A. Dennis, *J. Biol. Chem.* **269,** 9227 (1994).
[25] L. J. Reynolds, E. S. Kempner, L. L. Hughes, and E. A. Dennis, *Biophys. J.* **68,** 2108 (1995).
[26] J. S. Hill, R. C. Davis, D. Yang, J. Wen, J. S. Philo, P. H. Poon, M. L. Phillips, E. S. Kempner, and H. Wong, *J. Biol. Chem.* **271,** 22931 (1996).
[27] E. H. Harrison, C. J. Rojas, M. Z. Gad, and E. S. Kempner, *J. Biol. Chem.* **268,** 17867 (1993).
[28] J. T. Harmon, T. B. Nielsen, and E. S. Kempner, *Methods Enzymol.* **117,** 65 (1985).
[29] E. H. Harrison, *Biochim. Biophys. Acta* **963,** 28 (1988).

mM Tris-maleate, pH 7, and acid CEH activity was assayed using 50 mM sodium acetate, pH 5.

Studies on Carboxyl Ester Lipase

The soluble fraction of pooled rat livers was divided into aliquots, frozen, and exposed at −135° to various doses of radiation. One to 2 weeks later, the samples were thawed and assayed for lipase activity using three different substrates. Figure 1 shows the results from a single experiment. All activities showed a simple exponential decrease with radiation exposure. Cholesteryl ester hydrolase and retinyl ester hydrolase (REH) activities showed very similar responses with radiation dose to less than 1% of the initial activities. These data indicate that only a single-size structure is involved in greater than 99% of each measured enzymatic activity. Target analysis of these

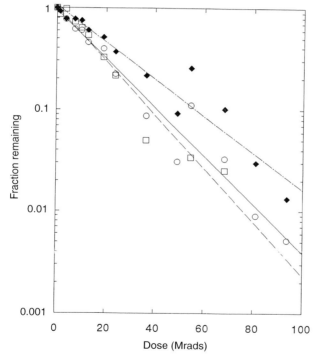

FIG. 1. Surviving enzymatic activities in rat liver cytosol after exposure at −135° to various doses of high-energy electron radiation. Each activity was determined in the same irradiated sample. Cholesteryl ester hydrolase, open circles; retinyl ester hydrolase, open boxes; triglyceride hydrolase, filled diamonds. Data are from a single experiment.

data yielded target sizes of approximately 95–100 kDa. Triolein hydrolase activity determined in the same irradiated samples revealed a slower decrease in activity, which is reflected in a calculated target size about 25% smaller than the other two activities, or about 70 kDa. Importantly, the same quantitative results were obtained in other aliquots of these cytosol samples that had been pretreated with 20 mM cholate (data not shown).

The recent cloning of cDNAs for CEL from a number of species and tissues has now clarified that a confusing number of bile salt-dependent lipases are in fact a single enzyme.[5] This work has also established that the polypeptide MW of the rat protein is approximately 70,000. Although the cloning work has clearly defined the size of the CEL monomer, the enzyme is known to exist in oligomeric forms, and in many cases the oligomerization of the enzyme is induced by trihydroxy bile salts that also activate enzyme activity.

Radiation target analysis is a powerful tool to study the size of the functionally active species of an enzyme. The results of the present experiment involving irradiation of cytosol in the absence of cholate demonstrate a single-size functional unit of CEL (assayed with three different substrates in the presence of cholate) that is roughly the size of the monomer and clearly not the size of a dimer or higher oligomer. This result is then consistent with previous experiments that suggested that the enzyme monomers are active when assayed in the presence of millimolar concentrations of cholate. When aliquots of the cytosol were treated with millimolar concentrations of cholate (i.e., conditions known to induce oligomerization) and then irradiated, the size of the catalytically functional unit did *not* change. This result strongly suggests that the bile salt-induced aggregation of the enzyme monomers and activation of enzyme activity are not functionally related. Thus, cholate-induced oligomerization is catalytically irrelevant. This, in turn, suggests the importance of determining, in detail, the nature of the interaction of bile salts with the enzyme monomer and the mechanism of activation.

Although our preliminary results clearly show that the size of the functionally active unit of CEL is not a multimer, it is surprising that the target size differed depending on the substrates used. The target size obtained with triolein as substrate (about 70 kDa) agrees very closely with the size of the rat CEL polypeptide as determined by cloning, and it is logical to conclude that one such monomer is sufficient to express the triolein lipase activity measured in the cytosols studied here. In contrast, when cholesteryl oleate or retinyl palmitate were used as substrates, target sizes were calculated to be about 25% larger than those for the triolein hydrolase activity, or approximately 95–100 kDa. This could indicate that the structure responsible for the REH and/or CEH activity is a single polypeptide of approximately 95–100 kDa and hence different from CEL, or more than one polypeptide with total mass of 95–100 kDa. The first possibility is unlikely considering (1) evidence that rat CEL does catalyze the hydrolysis of cholesteryl esters and retinyl esters,[30-32] (2) bile salt-dependent hydrolases in the 100,000 molecular weight range have not been reported in the rat, and (3) gel filtration of rat liver cytosols on columns of Sephadex G-100/Sephacryl S200 reveals that activities against all three esters (i.e., cholesteryl esters, retinyl esters, and triglycerides) coelute at a position consistent with the CEL monomer.[33] Thus, we would speculate that the second possibility may be the case, namely, that the CEH and REH activities in whole cytosol represent the activity of the ≈70-kDa CEL monomer and (an)other polypeptide(s) whose total mass is 25–30 kDa. While this suggestion is only speculative at this time, it is interesting to note that a very similar result was obtained previously in a radiation inactivation study of H/K-ATPase,[34] which was known to be composed of a 100-kDa polypeptide. Observed radiation target sizes of ≈140 kDa were difficult to interpret until discovery of a previously unsuspected subunit of 38 kDa.[35] Additional radiation experiments need to be conducted to further delineate the functional size of CEL in the absence and presence of cholate.

Studies on the Acid and Neutral CEHs of Liver Microsomes

Analyses of the microsomal, bile salt-independent CEHs by radiation inactivation have been reported.[27] Specifically, these studies have focused on both an acid (pH 5) CEH and the neutral (pH 7) CEH activities in microsomes.

The loss of the acid CEH activity is shown in Fig. 2. The surviving activity at low radiation doses varies among different preparations, but all preparations show similar losses of activity with radiation doses above 10 Mrad. Data from each experiment were fit with a single exponential function. These curves extrapolated to about 70% at zero dose, and the slope yielded an average target size of 71 ± 14 kDa. The variability at low doses suggests that additional large components may be present in some preparations. Because of their large size, they are rapidly destroyed by radiation and have no effect on measured activity above 10 Mrad. These structures may modulate the CEH activity at pH 5.

[30] C. Erlanson and B. Borgstrom, *Biochim. Biophys. Acta* **167**, 629 (1968).
[31] E. H. Harrison and M. Z. Gad, *J. Biol. Chem.* **264**, 17142 (1989).
[32] K. M. Rigtrup and D. E. Ong, *Biochemistry* **30**, 2920 (1992).
[33] E. H. Harrison, C. J. Rojas, and E. S. Kempner, *Biochim. Biophys. Acta* in press (1997).
[34] E. C. Rabon, R. D. Gunther, S. Bassilian, and E. S. Kempner, *J. Biol. Chem.* **263**, 16189 (1988).
[35] M. A. Reuban, L. S. Lasater, and G. Sachs, *Proc. Natl. Acad. Sci. U.S.A.* **87**, 6767 (1990).

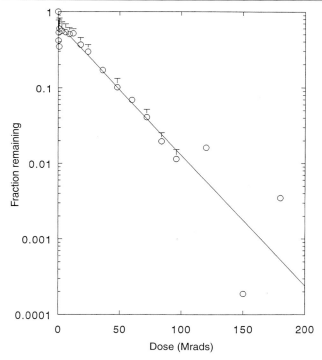

FIG. 2. Remaining acid CEH activity (pH 5) in irradiated rat liver microsomes. (From Harrison et al.[27])

The combined data from six radiation experiments (Fig. 3) show that the radiation-induced loss of the neutral CEH activity in microsomes is clearly not a single exponential. A multicomponent curve suggests that the measured activity is due to two or more structures. If the measured enzymatic activity were due to two independent proteins of different mass, the inactivation curve would be described as the sum of two exponential functions.[21] The experimental data points shown in Fig. 3 can be successfully fit by several different computer-generated curves of this type. The one shown in Fig. 3 is from a model in which a 240-kDa entity contributes 75% of the initial activity and one of 30 kDa contributes 10%.

Radiation target analysis has thus provided new information on the masses of radiation-sensitive structures that are required for the CEH activity in liver microsomes.

The pH 5 CEH activity is due to a structure of about 71 kDa. From radiation alone, it cannot be determined whether this is a single polypeptide

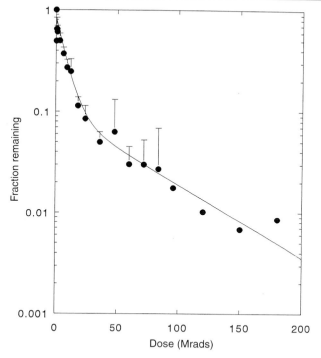

FIG. 3. Remaining neutral CEH activity (pH 7) in irradiated rat liver microsomes. (From Harrison et al.[27])

or an oligomer. However, knowing this target size allows us to make or strengthen several inferences about hepatic enzymes possibly involved in cholesteryl ester metabolism. For example, this target size would also suggest that the pH 5 CEH of microsomes is distinct from the lysosomal acid lipase. Rat liver acid lipase is reported to be a glycoprotein and to have a molecular weight of 58,000.[36] Recent cloning of the human acid lipase indicates a monomer molecular weight of 43,000 for the peptide alone.[37] Thus, our results for the functional target size of the plasma membrane/endosomal pH 5 CEH are not consistent with the known size of acid lipase either as a monomer or as an oligomer.

The microsomal pH 7 CEH activity yielded a multicomponent inactivation curve as shown in Fig. 3. Analysis of these data as the sum of several independent inactivation curves is difficult because each component contri-

[36] R. Klemets and B. Lundberg, *Lipids* **19,** 692 (1984).
[37] R. A. Anderson and G. N. Sando, *J. Biol. Chem.* **266,** 22479 (1991).

butes two variables (the mass of the protein and fraction of the activity) to the equation. One solution for the pH 7 CEH activity data assumed only two components, and suggested that most of the activity was due to a structure considerably larger than that expressing activity at pH 5. This does not mean, however, that a different protein is involved; an oligomer of the pH 5 enzyme is a possibility. This is unlikely, however, given previous experiments involving active site-direted inhibitors, which also pointed to the pH 5 and pH 7 activities being due to different catalytic centers.[19] The target size(s) of the microsomal pH 7 CEH(s) is(are) significantly different than that of the bile salt-dependent neutral CEH (i.e., carboxyl ester lipase) as determined from radiation inactivation target size of the latter enzyme activity as described earlier.

The radiation inactivation results, when taken together with previous observations, are consistent with the suggestion that the microsomal CEH activities are due to distinct enzymes that are different from previously characterized hepatic cholesteryl ester hydrolases such as acid lipase and the bile salt-dependent, neutral CEH. This knowledge, combined with estimates of the sizes of the catalytically active species, is being applied in attempts to solubilize and purify these enzymes.

[7] Immunological Techniques for the Characterization of Digestive Lipases

By Mustapha Aoubala, Isabelle Douchet, Sofiane Bezzine, Michel Hirn, Robert Verger, and Alain De Caro

Introduction

In humans, the digestion of dietary triacylglycerols is mediated by two main enzymes, a gastric lipase, which is secreted in the upper part of the digestive system and acts along the whole gastrointestinal tract, and a pancreatic lipase, which contributes to lipid digestion only in the duodenum.[1–3] Moreover, to overcome the inhibitory effects of the bile salts present in the intestinal lumen, pancreatic lipase specifically requires the presence of a small pancreatic cofactor (colipase), which acts as an anchor for pancreatic

---

[1] Y. Gargouri, H. Moreau, and R. Verger, *Biochim. Biophys. Acta* **1006,** 255 (1989).

[2] M. Hamosh, "Lingual and Gastric Lipases: Their Role in Fat Digestion," CRC Press, Boca Raton, Florida, 1990.

[3] F. Carrière, J. A. Barrowman, R. Verger, and R. Laugier, *Gastroenterology* **105,** 876 (1993).

lipase on bile salt-coated lipid interfaces.[4] Gastric and pancreatic lipases belong to two distinct groups in terms of their primary structure and biochemical properties. The group I family includes pancreatic lipase, lipoprotein lipase, and hepatic lipase.[5] These enzymes have been shown to share common characteristics, such as a molecular weight ranging from 50,000 to 55,000, a high homology of their primary structures (60%), and an optimal activity expressed at alkaline pH. The group II family, defined as the acidic lipase family,[6-8] includes preduodenal lipases and lysosomal lipase and shares no sequence homology with the pancreatic lipase family.

To gain a better understanding of the physiologic functions and structural characteristics of the digestive lipases, human pancreatic lipase (HPL) and human grastric lipase (HGL), monoclonal antibodies (mAbs) may provide useful tools. Such antibodies would also be useful for identification and purification of the enzymes from biological fluids. In this chapter, we describe the preparation and possible application of polyclonal (pAb) and mAbs to HGL and HPL obtained from gastric and pancreatic juices, respectively.

Production and Purification of Monoclonal and Polyclonal Antibodies to Human Gastric and Pancreatic Lipases

Preparation of Antigens

Human Gastric Lipase. HGL is purified to homogeneity from gastric juice[9] from volunteers with the sequential use of chromatographies on Mono S (sulfopropyl cation exchanger, Pharmacia, Uppsala, Sweden) and Mono Q (diethyl aminoethyl anion exchanger, Pharmacia).

Human Pancreatic Lipase. HPL is purified to homogeneity from pancreatic juice according to De Caro et al.[10]

---

[4] C. Erlanson-Albertsson, *Biochim. Biophys. Acta* **1125**, 1 (1992).
[5] S. B. Petersen, and F. Drabløs, in "Lipases: Their Biochemistry, Structure and Application" (P. Woolley and S. Petersen, eds.), p. 23. Cambridge University Press, Cambridge, England, 1994.
[6] R. A. Anderson and G. N. Sando, *J. Biol. Chem.* **266**, 22479 (1991).
[7] D. Ameis, M. Merkel, C. Eckerskorn, and H. Greten, *Eur. J. Biochem.* **219**, 905 (1994).
[8] F. Carrière, Y. Gargouri, H. Moreau, S. Ransac, E. Rogalska, and R. Verger, in "Lipases: Their Structure, Biochemistry and Application" (P. Wooley and S. B. Petersen, eds.), p. 181. Cambridge University Press, Cambridge, England, 1994.
[9] H. Moreau, C. Abergel, F. Carrière, F. Ferrato, J. C. Fontecilla-Camps, C. Cambillau, and R. Verger, *J. Mol. Biol.* **225**, 147 (1992).
[10] A. de Caro, C. Figarella, J. Amic, R. Michel, and O. Guy, *Biochim. Biophys. Acta* **490**, 411 (1977).

Production of Monoclonal Antibodies by Hybridomas

Two young female BALB/c mice were immunized for each antigen with HGL or HPL as follows: the first and the second immunizations were carried out with 100 µg of antigen suspended in 100 µl of 10 mM phosphate buffer (pH 7.4), containing 150 mM sodium chloride (PBS), emulsified in 100 µl of complete Freund's adjuvant (200 µl final volume), and injected subcutaneously at 10-day intervals. Two weeks after the second injection, the mice were bled and the serum tested in a direct binding enzyme-linked immunosorbent assay (ELISA) as described later. The mouse with the highest titre was selected and 100 µg of pure antigen, suspended in 200 µl of PBS, was again injected subcutaneously. Ten days later, three fusion-priming intraperitoneal injections of 50 µg of HGL or HPL each in PBS were given on 3 subsequent days. The fusion procedure followed is that described by Galfré et al.[11] and Kohler and Milstein.[12]

Spleen was harvested 1 day after the last injection and a spleen cell suspension prepared by gently teasing the spleen while holding it with a pair of forceps in a petri dish and fused with nonsecreting mouse myeloma P3 X63 Ag8. 653 (ATCC).[12,13] Cell fusion was performed using polyethylene glycol 1500 (Boehringer Mannheim, Mannheim, Germany). The hybridomas in hypoxanthine-aminopterin-thymidine (HAT) medium were seeded into six microtiter tissue culture plates containing murine peritoneal macrophages as a feeder layer. Ten days after the fusion, 180 µl of culture supernatant was removed from each well to test the antilipase activity in a direct-binding ELISA test.

ELISA Tests for Screening Antilipase MAbs

When not stated otherwise, the following buffers were used for all ELISA procedures. Coating buffer: PBS. Wash medium: PBS containing bovine serum albumin (BSA) (5 g/liter) and Tween-20 (0.5 g/liter). Saturating buffer: PBS containing BSA (5 g/liter). Substrate solution: *o*-phenylenediamine (Sigma, St. Louis, MO) (0.4 g/liter) in 0.05 M sodium phosphate/ citrate, pH 5, containing fresh hydrogen peroxide (0.4%). Stop solution: 1 M sulfuric acid.

The screening of hybridoma supernatants was carried out by performing solid phase immuno assays using 96-well microtiter polyvinyl chloride (PVC) plates (Maxisorb, Nunc, Roskilde, Denmark). For screening anti-HGL mAbs, a simple sandwich ELISA was carried out by coating the

---

[11] G. Galfré, S. C. Howe, C. Milstein, G. W. Butcher, and J. C. Howard, *Nature* **266,** 550 (1977).
[12] G. Kohler and C. Milstein, *Nature* **256,** 495 (1975).
[13] J. F. Kearney, A. Radbruch, B. Liesegang, and K. Rajewki, *J. Immunol.* **123,** 1548 (1979).

plates with 500 ng of pure native HGL per well, in 50 μl coating buffer, overnight at 4°.[14] For screening anti-HPL mAbs, a double sandwich ELISA was performed by coating the plates with 500 ng of pure anti-HPL pAb per well (captor antibody).[15] After saturation of the remaining free sites with saturating buffer (200 μl per well) for 2 hr at room temperature, HPL was added to the PVC-coated pAb (500 ng per well), which, unlike the simple sandwich ELISA, results in the random orientation of the various epitopic regions of HPL. Then hybridoma supernatants (50 μl) were added to each well and incubated for 1 hr at room temperature. mAb-producing hybridomas were detected with peroxidase-conjugated antimouse IgG antibody (Sigma). The substrate solution of peroxidase was used to quantify the positive clones. The reaction was stopped with a 1 M sulfuric acid solution (50 μl per well) and the optical density (OD) was read on an automatic plate reader (Dynatech, Guernsey, UK) at 492 nm. Between each step in the assay, the plates were rinsed three times with wash medium.

Four anti-HPL mAbs (81-23, 146-40, 315-25, and 320-24) were found to react with the simple sandwich ELISA, while mAb 248-31 did not react.[15] All five antibodies, however, interacted with HPL in the double sandwich ELISA. These results can be explained by the fact that the epitope recognized by mAb 248-31 is in a hydrophobic region adsorbed to the PVC plate and is therefore not accessible to the antibody. An alternative explanation might be that a conformational change (denaturation) may have occurred during HPL adsorption to the PVC plate, resulting in the loss of the recognition site. In the case of proteins with functional hydrophobic regions such as lipolytic enzymes, the double sandwich ELISA test involving specific polyclonal antibodies adsorbed as the first layer on PVC plates therefore yields randomly oriented antigenic regions and preserves the second antibody recognition.

MAbs Isotype Identification

A mouse monoclonal antibody isotyping kit (Amersham, Buckinghamshire, UK) was used to determine the antibody class of the various antilipase mAbs. The isotypes of mAbs were set up according to the manufacturer's instructions, with culture supernatants of each clone after limiting dilutions. All the tested mAbs belong to the IgG_1 class with a κ light chain with the exception of the anti-HGL mAb 13-42 and the anti-HPL mAb 248-31, which are of the IgG_{2b} isotype.

[14] M. Aoubala, C. Daniel, A. de Caro, M. G. Ivanova, M. Hirn, L. Sarda, and R. Verger, *Eur. J. Biochem.* **211,** 99 (1993).
[15] M. Aoubala, L. de la Fournière, I. Douchet, A. Abousalham, C. Daniel, M. Hirn, Y. Gargouri, R. Verger, and A. de Caro, *J. Biol. Chem.* **270,** 3932 (1995).

Purification of MAbs

Positive hybridomas were cloned using the limiting dilution technique. For ascites production, 2.5×10^6 hybridoma cells were injected intraperitoneally into BALB/c mice. mAbs were purified from mouse ascitic fluids by precipitation with 50% saturated ammonium sulfate. The protein solution was loaded onto a Protein A-Sepharose CL-4B column (Pharmacia), previously equilibrated with 20 mM borate buffer, pH 8.8, containing 2.5 M sodium chloride. Then the column is extensively washed with the same buffer and the antibodies eluted with 50 mM sodium citrate, pH 4.5 or 5.0, for antibodies of the IgG$_{2b}$ or IgG$_1$ subclasses, respectively. Fractions containing antibodies were dialyzed against PBS containing 0.02% sodium azide, concentrated to 4 mg/ml using Centriprep 30 concentrators (Amicon, Beverly, MA), filtered through 0.22-μM Millipore filters and stored at $-20°$ in small aliquots. The purity of the IgG preparations was checked by SDS–PAGE (7.5%).

Preparation of Anti-HGL and Anti-HPL Polyclonal Antibodies

Rabbits were immunized with native HGL according to the following schedule: 1 mg of HGL in 0.5 ml Freund's complete adjuvant (ICN ImmunoBiologicals, Costa Mesa, CA) was injected subcutaneously on day 0. Three weeks later, 1 mg of HGL, in 0.5 ml Freund's incomplete adjuvant (ICN ImmunoBiologicals), was injected intramuscularly in the footpads every 10 days for 1 month. The last injection included 0.5 mg HGL in PBS and was given intravenously in the ear 1 week before the animals were sacrificed. One week after each booster injection, sera were tested for anti-HGL reactivities by performing ELISA as described earlier.

The rabbit anti-HGL pAb was purified using a column of immobilized HGL. For that purpose, 16 mg of purified HGL was immobilized on 7 ml of swollen Affi-Gel 10 (Bio-Rad, Hercules, CA) equilibrated with PBS. After 4 hr of incubation at $4°$ and under agitation, unreacted gel groups were blocked with a solution of 0.5 M glycinamide (Sigma) (pH 7.5). The gel was poured into a glass column (1.5×3.5 cm) and washed successively with 50 ml of 0.2 M glycine-HCl buffer (pH 2.2) and 100 ml of PBS. Under these conditions, more than 84% of the initial amount of HGL was covalently coupled to the gel. For the purification of specific anti-HGL pAb, rabbit anti-HGL sera were first precipitated with 50% saturated ammonium sulfate. After centrifugation at 10,000g for 30 min, the pellet was solubilized in 25 mM Tris-HCl (pH 7.4) containing 150 mM NaCl (TBS), dialyzed against the same buffer and loaded on the HGL-Affi-Gel 10 column previously equilibrated with TBS. After 4 hr of incubation at $4°$, the column was washed successively with TBS, with 25 mM Tris-HCl (pH 7.4) containing 0.5 M NaCl and 0.2% Triton X-100 and then again with the initial TBS until zero absorbance at 280 nm was reached. Pure antibodies directed against HGL were eluted with a 0.2 M glycine-HCl buffer (pH 2.4). Each fraction (1 ml) was immediately neutralized with 250 μl of 1 mM Tris-HCl (pH 9.0); then the fractions containing antibodies were pooled, dialyzed against PBS, and concentrated to about 3 mg/ml using an Amicon ultrafiltration cell. They are stored at $-20°$ in small aliquots. These antibodies gave a single band on SDS–PAGE.

Characterization and Application of Monoclonal Antibodies

Monoclonal antibodies that have been produced against HGL and HPL are listed in Tables I and II. The mAbs are characterized in terms of affinity to their antigen, capacity to inhibit lipase activity or to bind lipids, epitope mapping, and cross-reactivity with other gastric or pancreatic lipases. These antibodies constitute useful tools for the study of mammalian acidic and pancreatic lipases.

Immunoinactivation Studies

To locate some epitopes at or near the catalytic site of HGL, we tested the ability of each mAb to inhibit lipase activity.[14] A fixed amount of

TABLE I
CHARACTERISTICS OF mAbs TO HGL

	mAb 4.3	mAb 13.42	mAb 25.4	mAb 35.2	mAb 53.27	mAb 83.15	mAb 218.13
Affinity constant ($\times 10^{-7} M^{-1}$)	1.03	ND[a]	1.58	0.19	ND	0.02	8.80
Inhibition of enzymatic activity[b]	Yes	ND	Yes	Yes	ND	Yes	No
Inhibition of lipid binding (monolayer)[c]	Yes	ND	Yes	Yes	ND	Yes	No
Reactivity of HGL by simple sandwich ELISA[d]	+	+	+	+	+	++	+++
Reactivity of HGL by Western blot[e]	Yes	Yes	Yes	Yes	Yes	Yes	Yes
Cross-reactivity to other preduodenal lipases by Western blot							
DGL	Yes	ND	Yes	No	ND	No	No
RGL	Yes	ND	Yes	Yes	ND	No	No

[a] ND, Not determined.
[b] From Aoubala *et al.*[14]
[c] From Ivanova *et al.*[19]
[d] See Fig. 2.
[e] See Fig. 4.

TABLE II
CHARACTERISTICS OF mAbs TO HPL

	mAb 81.23	mAb 146.40	mAb 248.31	mAb 315.25	mAb 320.24
Affinity constant ($\times 10^{-9} M^{-1}$)	7.0	1.9	ND[a]	6.0	4.5
Inhibition of enzymatic activity	Yes	No	Yes	Yes	Yes
Reactivity of HPL by simple sandwich ELISA	+	+	−	+	+
Reactivity of HPL by double sandwich ELISA	+	+	+	+	+
Reactivity of native HPL by Western blot	Yes	Yes	Yes	Yes	Yes
Reactivity of SDS-denatured HPL by Western blot	Yes	Yes	No	Yes	Yes
Cross-reactivity to other pancreatic lipases by Western blot					
Porcine pancreatic lipase	No	No	No	No	No
Dog pancreatic lipase	No	No	No	No	No
Horse pancreatic lipase	Yes	No	No	Yes	Yes
Guinea pig pancreatic lipase	No	No	No	No	No

[a] ND, Not determined.

enzyme (7–30 μg) was incubated with each mAb at various molar ratios (mAb/HGL: 0.5–2). Incubations were performed in PBS (20-μl final volume) for 1 hr at 37°. The residual activity of the HGL–mAb complexes was determined using three substrates differing in their fatty acid chain length: tributyrin and soybean oil emulsions as well as a substrate-forming stable film: 1,2-didecanoyl-sn-glycerol (dicaprin). Assays on tributyrin (Fluka) and soybean oil (commercial grade) emulsified in gumarabic were carried out with the bulk pH-stat method under standard conditions previously described by Gargouri et al.[16] The kinetics of the hydrolysis of 1,2-didecanoyl-sn-glycerol (SRL) films by HGL[17] were recorded, with or without incubation of the enzyme with each mAb, using the barostat technique previously described by Verger and de Haas.[18]

Three classes of mAbs have been identified on the basis of their inhibitory effect: class 1 includes three mAbs (4-3, 25-4, and 35-2) possessing the capability to inhibit HGL on all the substrates tested. The second class, consisting of mAb 83-15, preferentially prevents the lipolytic activity of HGL on long-chain triacylglycerols. Last, mAb 218-13, which belongs to class 3, had no inhibitory effect (Table I).

[16] Y. Gargouri, G. Piéroni, C. Rivière, J.-F. Saunière, P. A. Lowe, L. Sarda, and R. Verger, *Gastroenterology* **91**, 919 (1986).
[17] Y. Gargouri, G. Pièroni, F. Ferrato, and R. Verger, *Eur. J. Biochem.* **169**, 125 (1987).
[18] R. Verger and G. H. de Haas, *Chem. Phys. Lipids* **10**, 127 (1973).

Taking advantage of the low surface activity of IgG molecules,[19] the capacity of the mAbs to inhibit the penetration of native and sulfhydryl-modified HGL (Nbs-HGL) into dicaprin monolayers was investigated using the monomolecular film technique.[19] The HGL–mAb and Nbs-HGL–mAb complexes (molar ratio = 0.5) were therefore obtained by incubating each mAb with HGL or Nbs-HGL exactly as described earlier. All the inhibitory mAbs reduce the lipid-binding capacity of HGL and only the noninhibitory mAb 218-13 does not affect the penetration properties of the enzyme. We have not yet detected any inhibitory mAb that does not affect the lipase penetration. With these experimental data it is not possible to distinguish which of the two extreme situations holds true: the catalytic site and lipid-binding domain are either spatially very close or extremely far apart from each other. Based on the known three-dimensional structures of four lipases,[20–23] it is probable that the two functionally distinct sites of HGL are topographically close.

The potentially inhibitory effects of the five anti-HPL mAbs were also tested.[15] Four mAbs (81-23, 248-31, 315-25, and 320-24) reduced the hydrolysis of trioctanoin emulsion by HPL, whereas mAb 146-40 had no effect (see Table II). It is worth noting that comparable inhibition levels were observed in both the presence and absence of colipase and bile salts, probably because the affinity of HPL for mAb is higher than that for the lipids.[15]

Mapping of Epitopes on the Surface of Native Lipases

To test whether the different mAbs recognized different epitopes on the enzyme, the ELISA double-antibody-binding test (ELISA additivity test) initially developed by Friguet *et al.*[24] was used. Each mAb was first titrated by performing a direct ELISA test in which increasing amounts of each mAb (0.125–2 µg per well) were incubated with a fixed amount of lipase (2–5 ng per well) previously coated on the microplate. The ELISA experiments were performed as described earlier for the screening of mAbs. For each mAb, a titration curve was established, and the mAb concentration

---

[19] M. G. Ivanova, M. Aoubala, A. de Caro, C. Daniel, J. Hirn, and R. Verger, *Colloids Surf.* **B1,** 17 (1993).
[20] L. Brady, A. M. Brzozowski, Z. S. Derewenda, E. Dodson, G. Dodson, S. Tolley, J. P. Turkenburg, L. Christiansen, B. Huge-Jensen, L. Norskov, L. Thim, and U. Menge, *Nature* **343,** 767 (1990).
[21] F. K. Winkler, A. d'Arcy, and W. Hunziker, *Nature* **343,** 771 (1990).
[22] J. D. Schrag, Y. Li, S. Wu, and M. Cygler, *Nature* **351,** 761 (1991).
[23] C. Martinez, P. de Geus, M. Lauwereys, G. Matthyssens, and C. Cambillau, *Nature* **356,** 615 (1992).
[24] B. Friguet, A. F. Chaffotte, L. Djavadi-Ohaniance, and M. E. Goldberg, *J. Immunol. Meth.* **77,** 305 (1985).

that gives the maximal signal corresponds to the saturation of all the accessible epitopes. This mAb concentration was then used for the cotitration of the mAbs in pairs.

Two specific mAbs were mixed prior to incubation with lipase adsorbed to the PVC plate and the competition between antibodies for the antigen was expressed by means of additivity indexes.[24] The different mAbs anti-HGL and anti-HPL were studied in all possible pairs and the additivity index (AI), which characterizes each pair of specific antibodies, was determined using the following formula:

$$AI = 100 \cdot \left(\frac{2 \cdot A_{1+2}}{A_1 + A_2} - 1\right),$$

where A_{1+2} is the absorbance obtained in the ELISA with the mixture of two mAbs at a 1:1 molar ratio, and A_1 and A_2 are the absorbances obtained with each mAb, respectively. This index (expressed as a percentage) makes it possible to evaluate the simultaneous binding of two mAbs to the antigen. According to Friguet et al.,[24] if the two mAbs have the same specificity and bind randomly at the same antigenic site, A_{1+2} will be equal to the mean value of A_1 and A_2, and then AI will be equal to zero (competitive binding). Conversely, if the two mAbs bind independently at two different antigenic sites (additive binding), A_{1+2} will be equal to the sum of A_1 and A_2 and then AI will be equal to 100. Values of AI ranging from 0 to 100 will indicate that the two mAbs bind to closely associated overlapping epitopes (partially additive binding).

From this method, we found that the various epitopes of HGL are partially overlapping and are all located in the same antigenic region.[14] For HPL, the mAbs can be classified into two groups of antibodies directed against two different antigenic determinants: group I includes three mAbs (81-23, 315-25, and 320-24) and group II consists of mAb 146-40.[15] Because mAb 248-31 did not react in a simple ELISA, it was not possible to localize its epitope by this technique.

Mapping of Epitopes along the Primary Sequence of Lipases

An epitope mapping study was also performed using the peptide mapping method described by Wilson and Smith.[25] Briefly, HGL and HPL were subjected to limited digestion with trypsin[26] and chymotrypsin,[27] respec-

[25] J. E. Wilson and A. D. Smith, *J. Biol. Chem.* **260**, 12838 (1985).
[26] M. Aoubala, J. Bonicel, C. Bénicourt, R. Verger, and A. de Caro, *Biochim. Biophys. Acta* **1213**, 319 (1994).
[27] A. Abousalham, C. Chaillan, B. Kerfelec, E. Foglizzo, and C. Chapus, *Protein Eng.* **5**, 105 (1992).

tively. The products were separated by SDS–PAGE and electrophoretically transferred to nitrocellulose or Glassybond membranes for Western blotting and N-terminal amino acid sequence analysis, respectively.

The results of the immunoblotting experiments carried out with the tryptic HGL fragments showed that anti-HGL mAbs can be divided into two groups.[26] The first group comprises five mAbs (4-3, 13-42, 25-4, 35-2, and 83-15), which recognize only the epitopes located in the N-terminal domain of HGL. The second group contains two mAbs (53-27 and 218-13), which immunoreact only with the C-terminal domain of HGL (Fig. 1A).

For HPL, four fragments resulting from the limited chymotryptic cleavage of HPL (14, 26, 30, and 36 kDa) were characterized.[15] MAb 146-40

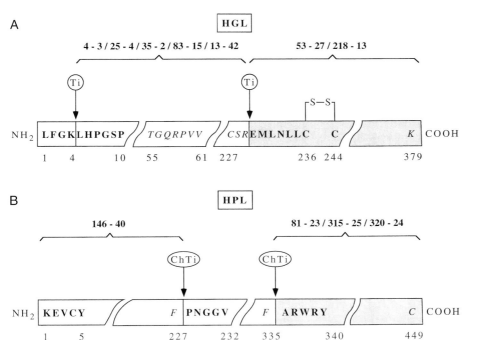

FIG. 1. (A) Schematic representation of the primary sequence of HGL adapted from Bodmer et al.[35] showing the tryptic cleavage sites (Ti) and the N-terminal domain (empty frame) as well as the C-terminal domain (shaded frame). (B) Diagram of the primary sequence of HPL from Winkler et al.[21] The arrow shows the chymotryptic cleavage sites (ChTi). The N-terminal domain (empty frame) as well as the C-terminal domain (shaded frame) are indicated. In both representations, amino acid sequences, determined using the Edman degradation technique, are indicated in boldface letters and those deduced from the cDNA sequence are indicated in italic letters. Numbering over brackets indicates the mAbs that recognize their corresponding fragments.

reacted with the 30- and 36-kDa N-terminal fragments, indicating that the epitope recognized by mAb 146-40 is located in the fragment (Lys-1–Phe-227) (Fig. 1B). In the same study we have reported that three mAbs (81-23, 315-25, and 320-24) did not immunoreact with the C-terminal domain (Ala-336–Cys-449, 14 kDa). More recently, we observed[28] that when HPL was submitted to an extensive chymotryptic cleavage, a positive reactivity of these three mAbs with the 14-kDa fragment. Our initial results[15] can be explained by an insufficient chymotryptic proteolysis generating low amounts of the C-terminal domain. We can now conclude that epitopes recognized by mAbs 81-23, 315-25, and 320-24 are located in the C-terminal domain (Ala-336–Cys-449) (Fig. 1B).

Discrimination of the Pancreatic Lipase Structural Domains by MAbs

We also determined the immunoreactivity of anti-HPL mAbs toward three variants of HPL: N-HPL, HPL(-lid), and a N-GPLRP2/C-HPL chimera, produced in insect cells using the baculovirus expression system.[29] N-HPL consists of the N-terminal domain of HPL only (Lys-1–Phe-335). HPL(-lid) has a mini-lid of 5 amino acid residues, instead of 23 in HPL (Cys-237–Cys-261), and the chimera is made of the N-terminal domain of the guinea pig pancreatic lipase related protein 2 (GPLRP2) and the C-terminal domain of HPL.

Western blot analysis of these different mutants showed that the 146-40 mAb recognized HPL, HPL(-lid), N-HPL but neither GPLRP2 nor the N-GPLRP2/C-HPL chimera nor the purified C-terminal domain of HPL. These results allowed us to conclude that the epitope recognized by the mAb 146-40 is located within the N-terminal domain of HPL but not in the lid (Cys-237–Cys-261). Moreover, the 81-23, 315-25, and 320-24 mAbs recognized HPL, HPL(-lid), the N-GPLRP2/C-HPL chimera, and the purified C-terminal domain of HPL but neither N-HPL nor GPLRP2. We confirmed that these three mAbs are directed against the C-terminal domain of HPL.

Cross-Species Reactivity of Anti-HGL MAbs

Immunological cross-reactivity of the various anti-HGL mAbs with DGL, RGL, and HPL as a control experiment was undertaken by using three different techniques: (1) a simple sandwich ELISA, (2) an indirect competitive assay between two lipases for the same antibody, and (3) the Western blotting technique using SDS-denatured lipases.

[28] S. Bezzine, F. Carrière, J. De Caro, R. Verger, and A. De Caro, in preparation (1997).
[29] F. Carrière, K. Thirstrup, S. Hjorth, F. Ferrato, P. F. Nielsen, C. Withers-Martinez, C. Cambillau, E. Boel, L. Thim, and R. Verger, *Biochemistry* **36,** 239 (1997).

For the simple sandwich ELISA, microtitration plates were coated with 50 ng of each of the pure lipase and incubated with a fixed amount of anti-HGL mAb (500 ng/well). The ELISA procedure was performed as described earlier. Figure 2 shows the cross-reactivity of each mAb with native lipases. None of the five mAbs reacts with HPL. This result is in good agreement with the fact that there is no sequence homology between gastric and pancreatic lipases. These mAbs, however, showed different binding properties toward HGL, DGL, and RGL. Two mAbs (4-3 and 25-4) were found to react with these three lipases. These results show that the epitopes recognized by these two mAbs are conserved for the three enzymes. In contrast, mAbs 35-2 and 83-15 do not show significant reactivity with DGL and RGL adsorbed on the PVC plate. Finally mAb 218-13 presents a significant cross-reactivity with DGL.

Since these three gastric lipases present a high-sequence identity (about 86%), the absence of a cross-reactivity between some anti-HGL mAbs and DGL and RGL could be explained by the fact that the epitopes recognized by these antibodies are not conserved on the surface of these enzymes. An alternative explanation could be that the epitopes are hidden or denatured when the enzymes are adsorbed to the PVC plate.

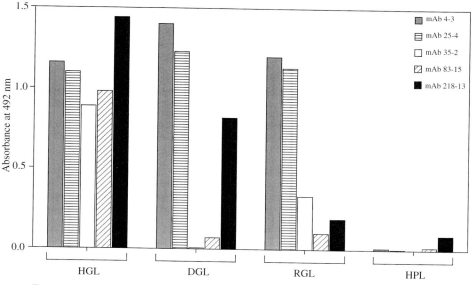

FIG. 2. Cross-species reactivity of anti-HGL mAbs with gastric lipases from human (HGL), dog (DGL), rabbit (RGL), and HPL using a simple sandwich ELISA. The assay was performed by direct coating of the various lipases (50 ng/well) with a constant amount of anti-HGL mAb (500 ng/well).

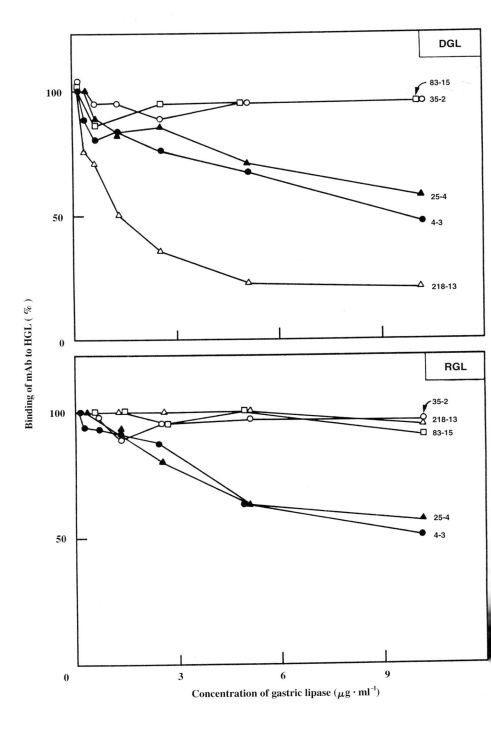

The immunological cross-reactivity of anti-HGL mAbs was also studied using an indirect competitive assay. This technique allowed us to study the interaction between the different mAbs with the antigens in solution. For that purpose, DGL or RGL at various concentrations (0.005–10 μg/ml) was mixed with a constant amount of each mAb (2 μg/ml) in PBS (200-μl final volume). The mixtures were incubated for 2 hr at 37° and then overnight at 4°. After this incubation the presence of unbound mAbs was quantified by performing a simple sandwich ELISA. For that, 100 μl of each mixture was incubated for 2 hr at 37° into the wells of a microtitration plate, previously coated with HGL (20 ng/well). The ELISA procedure was the same as that described earlier. The results (see Fig. 3) confirm that the epitopes recognized by mAbs 4-3 and 25-4 are conserved on the surface of these gastric lipases and correspond to highly conserved structural elements. The epitope of mAb 218-13 is also well conserved only in DGL. In contrast, the epitopes of mAbs 35-2 and 83-15 are not conserved in DGL and RGL.

The reactivity of the preceding mAbs with SDS-denatured lipases was also studied by Western blotting. Results are presented in Fig. 4. First, note that all the anti-HGL–mAbs immunoreact with SDS-denatured HGL, suggesting that the epitopes recognized by these mAbs are probably continuous determinants (see Fig. 4 and Table I). Second, mAbs 4-3 and 25-4 immunoreact with SDS-denatured DGL and RGL. In contrast, mAbs 35-2 and 83-15 show no detectable reactivity toward SDS-denatured DGL, confirming the previous results obtained with the ELISA tests (Figs. 2 and 3).

In contrast, mAb 218-13 loses its immunoreactivity on SDS-denatured DGL (Fig. 4) as compared with the native DGL (see Figs. 2 and 3). These results show that the epitope of mAb 218-13 is a discontinuous antigenic determinant. MAb 35-2 showed a surprising reactivity with SDS-denatured RGL (Fig. 4) in contrast to its recognition of the native enzyme. One hypothesis could be that the epitope of mAb 35-2 is hidden in the core of native RGL and becomes accessible in the SDS-denatured enzyme.

FIG. 3. Determination of the immunoreactivity of the various anti-HGL mAbs toward DGL and RGL using an indirect competitive assay. Gastric lipases, at variable concentrations (10^{-10} M to 0.2×10^{-6} M), were mixed in solution with a fixed amount of each mAb (0.4×10^{-7} M). Then the mixtures were incubated with precoated HGL (20 ng/well) in a microtitration plate. The rest of the procedure is the same as that described earlier for ELISA. Binding values were calculated as percentage reduction in absorbance relative to controls without DGL or RGL (100%).

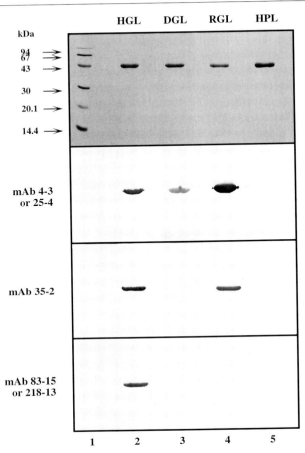

FIG. 4. Upper panel shows SDS–PAGE (12%) patterns of various lipases. Lane 1, molecular mass markers; lane 2, HGL; lane 3, DGL; lane 4, RGL; and lane 5, HPL. Lower panels show immunoblot analysis of the lipases from the upper panel, using anti-HGL mAbs as primary antibody. Alkaline phosphatase-conjugated with goat antimouse IgG (Sigma) was used as secondary antibody.

Immunoaffinity Purification of HGL Using a Monoclonal Antibody

MAb (35-2), with a good affinity to HGL, was produced in large amounts. Purified mAb 35-2 (70 mg) was immobilized on 25 ml of swollen Affi-Gel 10 (Bio-Rad) equilibrated with PBS as described earlier for the case of anti-HGL–pAb. Under these conditions, more than

95% of the initial amount of mAb 35-2 was covalently coupled to the gel.[30]

The first step in the published standard purification of HGL from human gastric juice consisted of a cation exchange chromatography (SP-Sepharose).[9] This step was mostly performed to eliminate components such as mucus and pepsinogens, present in gastric juice, which could affect the immunoaffinity chromatography. Active lipase fractions were concentrated on an Amicon ultrafiltration cell using Diaflo YM-10 membrane, dialyzed against TBS and loaded onto the immobilized mAb column equilibrated with TBS. After a 4-hr incubation period at 4° under rotative agitation (18 runs per minute), the gel was washed several times with TBS until the eluates had no absorbance at 280 nm. Pure HGL was eluted as a single protein peak (see Fig. 5) with 0.2 M glycine-HCl buffer (pH 2.2). The eluted fractions contain a single protein band on SDS–PAGE, corresponding to HGL (insert of Fig. 5). The eluted fractions (1 ml) containing HGL activity were pooled, concentrated to about 2 mg/ml, using an Amicon ultrafiltration cell, dialyzed against 20 mM MES buffer (pH 6), and stored at −20°. As expected from its acidic stability, the elution from the immunoaffinity column under acidic conditions (pH 2.2) does not alter the enzymatic activity of HGL. This procedure offers the advantage of being rapid and reproducible, and it gives a better overall yield (58%) with a 50-fold purification[30] as compared to the previously described methods using conventional chromatographic steps.[9]

Quantitative ELISA for Measuring HGL in Duodenal Contents[30]

The availability of specific antibodies allowed us to set up a sensitive and specific double sandwich ELISA for measuring the HGL in the duodenal contents in which both HGL and HPL are present.[3] The procedure and the buffers used are those described earlier for the screening of mAbs with minor modifications.

Purified anti-HGL pAb was coated in the PVC microplate (125 ng/well) and used as the captor antibody. After saturation of the wells with saturating buffer, 50 μl of standard solution of HGL (0.1–90 ng/ml) or samples from human duodenal contents, appropriately diluted with the washing buffer, were added to each well. Wells containing buffer without HGL or duodenal contents served as controls. Then 50 μl of anti-HGL mAb-biotin conjugate (1 μg/well) (detector antibody) was added to each well. The biotinylation procedure of mAb is described later. HGL was

---

[30] M. Aoubala, I. Douchet, R. Laugier, M. Hirn, R. Verger, and A. de Caro, *Biochim. Biophys. Acta* **1169**, 183 (1993).

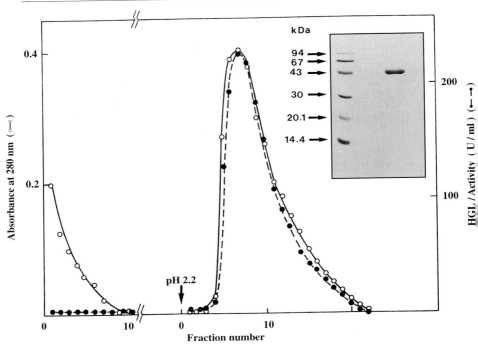

FIG. 5. Purification of HGL by immunoaffinity chromatography using an immobilized mAb (35-2) column. The active fractions obtained by performing SP-Sepharose chromatography were applied to the column and the pure enzyme was eluted with 0.2 M glycine-HCl buffer (pH 2.2). The volume of each fraction was 1 ml. Inset: SDS–PAGE analysis of immunoaffinity purified HGL. Left, Pharmacia standard molecular weight markers; right, 5.8 μg of purified HGL.

measured by using a solution of horse radish peroxidase-labeled streptavidine (Immunotech, Marseilles, France) diluted 20,000-fold according to the manufacturer's instructions.

Since the immunoreactivity of anti-HGL pAb and that of five mAbs are known, all the possible combinations were tested in a double sandwich ELISA.[30] The most sensitive ELISAs able to detect HGL concentrations down to 1 ng/ml were those in which the captor antibody (coating antibody) was a pAb and the detector antibody (antibody–biotin conjugate) was mAb 35-2. Various concentrations of captor antibody (pAb) were tested during the coating procedure. At the optimal concentrations of HGL (60 ng/ml) and biotin-mAb 35-2 conjugate (1 μg/ml), a plateau of absorbance at 492 nm was reached at 2.5 μg/ml (final concentration) of captor antibody corresponding to 125 ng/well.

To study the reproducibility of the assay as well as the interference from the duodenal contents on the ELISA measurements of HGL, two sets of experiments were carried out. First, known concentrations of HGL were prepared in the washing buffer and, second, known concentrations of HGL were added to a complete liquid test meal. Only the ELISA performed with pAb and biotin-labeled mAb 35-2 showed a good correlation between the curves obtained with the known HGL concentrations prepared in the washing buffer and those in a complete liquid test meal. This pair of antibodies (pAb/biotin–mAb 35-2) was therefore selected for measuring the HGL in the duodenal contents. Known amounts of pure HGL were also added to aspirates of duodenal contents containing some endogenous HGL, and then the HGL levels measured by ELISA were compared with those to be expected. The HGL levels measured in these samples ranged between 84 and 110% of the added amounts.[30] In other experiments, various samples from the duodenal contents were assayed using both the ELISA and an enzymatic assay of HGL. A good correlation ($r = 0.95$) was found to exist between the results of the two assays.[30]

We were not able to detect HGL in human sera by using ELISA. Recently, in pre- and postheparin plasma, Bensadoun[31] reported a sensitive double sandwich ELISA for human hepatic lipase (HL) based on the use of two anti-HL mAbs. HL was easily measured in the range of 0.5–5.0 ng/ml.

Interfacial Binding of HGL to Lipid Monolayers, Measured with an ELISA

The ELISA/biotin-streptavidin system has been found to be as sensitive as the use of radiolabeled proteins. Furthermore, biotinylation preserves the biological activities of many proteins. Our aim was to develop a sensitive sandwich ELISA, using the biotin-streptavidin system, and to measure the amount of surface-bound HGL and anti-HGL mAbs adsorbed to monomolecular lipid films.

Protein Biotinylation and Development of an ELISA

The ε-aminocaproic N-hydroxysuccinimide ester D-biotin (ACNHS-biotin) was used to couple biotin moieties to ε-amines of HGL lysine residues using a modified procedure previously described by Guesdon *et al.*[32] To obtain the desired stoichiometric ratio of ACNHS-biotin to total lysine residues of the enzyme, a solution of HGL (1 mg/ml) in 20 mM

---

[31] A. Bensadoun, *Methods Enzymol.* **263**, 333 (1996).
[32] J. L. Guesdon, T. Térnynck, and S. Avrameas, *J. Histochem. Cytochem.* **8**, 1131 (1979).

borate buffer (pH 8.0) containing 150 mM NaCl was mixed with various volumes of a solution of ACNHS-biotin (10 mg/ml) in dimethyl formamide. The reaction mixture was incubated at room temperature for 20 min under stirring and stopped by adding NH$_4$Cl (final concentration 0.1 M). The mixture was then immediately dialyzed at 4° against 10 mM MES buffer (pH 7). The lipolytic activity of HGL was determined both before and after biotin-labeling using the pH-stat and monolayers methods with tributyrin and dicaprin as enzyme substrate, respectively. It was observed for all three ratios used that HGL could be biotin-labeled without any appreciable loss of catalytic activity. Residual activities of 86, 96, and 100% were observed at ratios of ACNHS-biotin to HGL-lysine residues of 0.6, 0.3, and 0.15, respectively. Moreover, biotinylated HGL hydrolyzes dicaprin films similarly as native HGL, indicating that the labeling does not alter the interfacial HGL activity.[33] This is a considerable advantage over previous methods for chemically labeling HGL with radioactive 5,5'-dithiobis(2-nitrobenzoic acid) (DTNB), which led to a catalytically inactive lipase.[34]

Two anti-HGL mAbs were also biotin-labeled as previously described[30] and their capacity to immunoreact with HGL was tested using an ELISA test with native HGL directly coated to the wells of the PVC microplates. No decrease in the antigen-binding capacity of biotinylated mAbs 4-3 and 218-13 was observed.

Determination of the Amount of Protein Adsorbed to the Lipid/Water Interface[33]

We used the monomolecular film technique to study the kinetics of hydrolysis catalyzed by biotinylated HGL as well as its binding to dicaprin films. Measurements were performed with the KSV Barostat equipment (KSV-Helsinki, Helsinki, Finland). Readers are referred to the Chapter [13] by Ransac *et al.* in this volume. After recording of the kinetics and film aspiration, the amounts of biotinylated proteins present in the sample containing the film and a corresponding sample (same volume) containing the bulk phase were measured with an ELISA.

All the ELISA tests were performed as described earlier for the measurement of HGL in duodenal contents, with minor modifications as described by Aoubala *et al.*[33] The OD values at 492 nm were plotted as a function of the concentration of biotinylated protein (Fig. 6). A reference

---

[33] M. Aoubala, M. Ivanova, I. Douchet, A. de Caro, and R. Verger, *Biochemistry* **34**, 10786 (1995).
[34] Y. Gargouri, H. Moreau, G. Piéroni, and R. Verger, *Eur. J. Biochem.* **180**, 367 (1989).
[35] M. W. Bodmer, S. Angal, G. T. Yarranton, T. J. R. Harris, A. Lyons, D. J. King, G. Piéroni, C. Rivière, R. Verger, and P. A. Lowe, *Biochim. Biophys. Acta* **909**, 237 (1987).

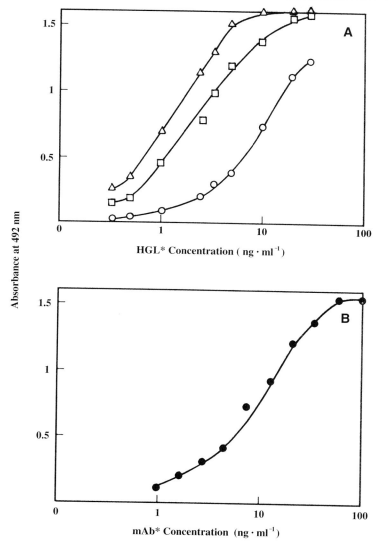

FIG. 6. Reference ELISA curves of (A) biotinylated HGL and (B) biotinylated mAb 4-3. (A) Open triangle, boxes, and circles indicate the titration curves of HGL biotinylated at molar ratios of ACNHS-biotin to HGL lysine amino groups of 0.6, 0.3, and 0.15, respectively. (B) The reference curve of mAb 4-3 biotinylated at one constant molar ratio (0.15) of ACNHS-biotin to mAb lysine amino groups. Biotinylated HGL (HGL*) and biotinylated mAb (mAb*) were dissolved in the washing buffer at concentrations indicated on the abscissa and the immunoreactivity mass of biotinylated proteins was measured with the specific sandwich ELISA.

curve was drawn up for each test and was used to calculate the concentration of biotinylated protein in the aspirated samples recovered from the monomolecular film experiments. Each assay was carried out in duplicate. The detection limit was 25 and 85 pg in the case of HGL and mAb, respectively.[33] The difference between the total amounts of proteins between these two samples was attributed to the surface excess of protein molecules bound to the lipid film.

The volume occupied by the lipid film was not taken into account since it is negligible, with respect to the aspirated subphase. We used the following equation:

$$\Gamma = \frac{[F + B] - [B]}{S} \cdot V_a,$$

where Γ is the surface excess of protein bound to the lipid monolayer, expressed in ng/cm^2; $[F + B]$ is the concentration of protein present in the aspirated film with the aspirated bulk subphase, as determined by ELISA; $[B]$ is the concentration of protein in the bulk sample, also determined by ELISA; V_a is the aspirated volume (ranging from 0.5 to 1.5 ml); and S is the area of the reactional compartment of the trough (39.3 cm^2).

During the monolayer experiments, the validity of the sandwich ELISA for biotinylated HGL was tested in the presence of monomolecular films of either dicaprin or egg phosphatidylcholine (egg PC). The recovery levels of biotinylated HGL injected under lipid monomolecular films were determined after each experiment as

$$\text{Total recovery } [\%] = 100 \cdot \frac{[B] \cdot V_t + \Gamma \cdot S}{T},$$

where $[B]$ is the concentration of biotinylated HGL in the subphase (bulk); V_t is the total volume of the reactional compartment measured after each monolayer experiment (30 ± 2 ml); $\Gamma \cdot S$ is the total amount of biotinylated HGL adsorbed to the monomolecular film, and T is the total amount of biotinylated HGL (200 ng) injected under the monomolecular film. The yields obtained in these conditions were 78 ± 6.2%.

By combining the above sandwich ELISA technique with the monomolecular film technique, it was possible to measure the enzymatic activity of biotinylated HGL on 1,2-didecanoyl-sn-glycerol monolayers as well as to determine the corresponding interfacial excess of the enzyme.[33] Figure 7A shows the amounts of biotinylated HGL in excess at the interface. The amounts of HGL present at or close to the interface increased linearly with increasing surface pressures up to 25 mN/m. Furthermore, a decrease in the amount of adsorbed enzyme was observed above this optimal surface pressure value. The lipolytic activity of biotinylated HGL

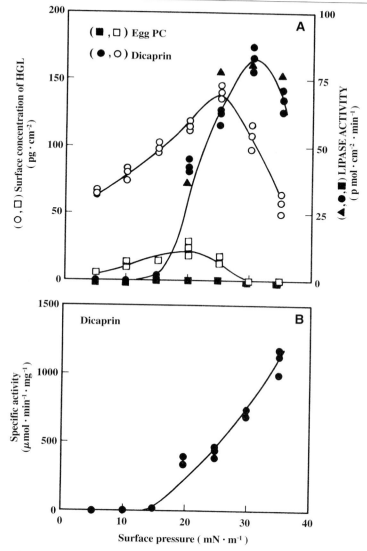

FIG. 7. (A) Variations of the initial velocity of hydrolysis of a dicaprin monolayer with surface pressure (filled circles) by HGL* (200 ng) and (filled triangles) by native nonbiotinylated HGL (200 ng) injected into the subphase (30 ± 2 ml) of a zero-order trough. No enzymatic activity was detected with egg PC films (filled boxes). The interfacial excess of biotinylated HGL was measured 10 min after its injection under a dicaprin (open circles) or egg PC (open boxes) monomolecular film. (B) Variations with surface pressure of the minimal specific activity of biotinylated HGL acting on dicaprin monolayers.

was also measured simultaneously at the corresponding surface pressures. As shown in Fig. 7A, HGL does not significantly hydrolyze dicaprin films at surface pressures of less than 15 mN/m. Above this value the enzymatic activity increases rapidly, reaching a maximum value at 30 mN/m. The ratio of observed enzyme activity to the amount of adsorbed protein, as determined by the specific sandwich ELISA, allows one to calculate the specific activity of the enzyme acting on a monomolecular film of dicaprin. As shown in Fig. 7B, the specific activity increased continuously from 15 up to 35 mN/m. At the latter value, the maximal specific activity reached was 1100 ± 97 U/mg whereas at 15 mN/m, its value was zero. The specific activities determined at 35 mN/m were found to be in the range of the values measured under optimal bulk assay conditions, using tributyrin emulsion as substrate (i.e., 1000 μmol/min per milligram of enzyme).

At a given lipase concentration in the water subphase, the interfacial binding of biotinylated HGL to the nonhydrolyzable egg PC monolayers was found to be 10 times lower than in the case of dicaprin monolayers[33] (Fig. 7A). This observation suggests that the weak affinity of biotinylated HGL for egg PC is probably due to repulsive forces originating from the zwitterionic phospholipid headgroups; whereas the strong interactions observed between the enzyme and the electrically neutral dicaprin monomolecular films in fact reflect the occurrence of strong enzyme/substrate interactions.

It is worth noting that the surface-bound enzyme includes not only those enzyme molecules directly involved in the catalysis but also an unknown amount of protein present close to the monolayer. These enzyme molecules were not necessarily involved in the enzymatic hydrolysis of the film.

Given the low tensioactivity of the mAbs of the IgG isotype,[19] we also investigated the effects of five anti-HGL mAbs (4-3, 25-4, 35-2, 83-15, and 218-13) on the catalytic activity as well as on the interfacial binding of biotinylated HGL to lipid/water interfaces.[33] Four out of these five mAbs (4-3, 25-4, 35-2, and 83-15) were found to reduce significantly the lipolytic activity of HGL. Moreover, three of the four inhibitory mAbs (4-3, 25-4, and 35-2) were found to reduce the specific activity of HGL, whereas mAb 83-15 had no effect on the specific activity. These results clearly indicate that the latter mAb (83-15) complexed with biotinylated HGL mainly affects the binding of the enzyme to the lipid/water interface, whereas the other three inhibitory mAbs (4-3, 25-4, and 35-2) affect both the binding and the catalytic steps of HGL. Therefore, it was possible for the first time to correlate the lipase activity with the surface excess of the enzyme present at the interface.

Acknowledgments

This work was supported by the BRIDGE-T-Lipase Program of the European Communities under contract BIOT-CT 910274 (DTEE) and by BIOTECH G-Program BIO 2-CT94-3041.

Section II

Substrate and Inhibitor Characterization

[8] Physical Behavior of Lipase Substrates

By DONALD M. SMALL

Introduction

Lipases including phospholipases, triacylglycerol lipases, cholesterol ester hydrolases, and wax ester hydrolyses are a diverse and evolutionarily ancient group of proteins with very different primary structures whose general catalytic site resembles a serine protease. Many lipases have some affinity for hydrophobic surfaces and surfaces of emulsion particles or membranes. Most lipases, especially triacylglycerol lipases, cholesterol ester hydrolases, and wax ester hydrolases, probably encounter their substrates in partially or fully emulsified particles since the long acyl chain members of triacylglycerols, cholesterol esters, and wax esters are virtually insoluble in aqueous systems. Phospholipases encounter their substrates at the interface of bilayer membranes or on the emulsifying layer of emulsion particles. Long-chain phospholipids are also highly insoluble and thus encounter their enzymes at the lipid/water interface. In 1968, studies from Desneuelle's lab,[1] clearly indicated the importance of an interface for the catalytic action of pancreatic lipase. In the same year a similar conclusion was reached by Olive and Dervichian[2] concerning the hydrolysis of phospholipid. The interface between a monolayer of lipid and water is probably the site of reaction of most of the lipases. Of course, lipid and water are the substrates for the reaction. In this chapter, we consider some of the known orientations of lipid substrates at the oil or phospholipid water interface.

Phospholipids have been known for 70 years to be interfacial molecules.[3,4] Most long-chain phospholipids have two acyl chains of 12 to 22 carbons and a highly hydrophilic headgroup anchored to one of the two primary carbons of the glycerol.[5] These groups, such as phosphorylcholine, phosphorylethanolamine, phosphoryl serine, phosphoryl inositol, and phosphoryl glycerol, all lend a strong hydrophilic part to the phospholipid molecule that prevents it from being soluble in oil (hydrocarbon). The two hydrocarbon chains, however, partition strongly into oil. In fact, these acyl chains gain 0.85 K calories per mole of $-CH_2-$ when the hydrocarbon

---

[1] B. Entressangels and P. Desnuelle, *Biochim. Biophys. Acta* **159,** 285 (1968).
[2] J. Olive and D. G. Dervichian, *Bull. Soc. Chim. Biol.* **50,** 1409 (1968).
[3] J. B. Leathes, *Lancet* 853; 957; 1019 (1925).
[4] E. Gorter and F. Grendel, *J. Exp. Med.* **41,** 439 (1925).
[5] G. Cevc, ed., "Phospholipids Handbook," Marcel Dekker, New York, 1993.

chain is transferred from water to oil.[6] Thus, these schizophrenic molecules are destined to find hydrophobic/aqueous interfaces in which to reside. There the hydrophobic acyl chains enter the oily side and the water-soluble groups enter the aqueous side. Because the solubility of phospholipids is virtually nil in both the aqueous and the oil phase, if only an aqueous phase is present then phospholipids will form their own surface by making bilayers or other related structures[5] in which the chains of the lipid are segregated with each other and separated from the polar groups, and the polar groups interact with water.

Glycerol and Glycerol-Ester Conformation of Acyl Glycerol Substrates

The conformation around the glycerol region, that is, the site of action in lipases, has been discussed for lipids in the crystalline state by Pascher *et al.* in detail.[7,8] However, in most of the biological systems in which lipases act, the lipids are not in the crystalline state. Phospholipids and other lipids are usually in some liquid or liquid crystalline state. Do specific conformations occur in liquid or liquid crystals? Recently molecular modeling,[9,10] nuclear magnetic resonance (NMR),[10,11] and Fourier transform infrared spectroscopy (FTIR)[12] suggest that a single specific crystalline conformation probably does not exist at the liquid crystal interface but rather that several conformations around the glycerol region of similar energy may rapidly oscillate. It is not clear whether increased pressure (decreased surface area) or increased surface viscosity in the liquid state above the gel (crystalline)–liquid transition might favor a specific conformation, but that is a possibility. Certain conformations observed by NMR in monoglactosyl diglyceride liquid crystals appear to be favored.[10]

Using Pascher's terminology,[8] glycerol conformation is described by specifying the mutual orientation of the three glycerol oxygens, $O(1)$, $O(2)$, $O(3)$, as expressed by torsion angles around the glycerol $C(1)$ and $C(2)$ bond, $O(1)-C(1)-C(2)-O(2)$, and the C2-C3 bond, $O(2)-C(2)-C(3)-O(3)$. These two torsion angles called Θ_4 and Θ_2 taken along with the way the acyl chain extends from the glycerol oxygens defines the glycerol ester conformation. Torsion angles of $0°$ and $\pm 120°$ are not allowed because they produce an eclipsed conformation that is sterically hindered. The

---

[6] C. Tanford, "The Hydrophobic Effect," John Wiley, New York, 1980.
[7] I. Pascher, M. Lundmark, P. G. Nyholm, and S. Sundel, *Biochem. Biophys. Acta.* **1113,** 339 (1992).
[8] I. Pascher, *Curr. Opin. Struct. Biol.* **6,** 439 (1996).
[9] H. Hauser, I. Pascher, and S. Sundell, *Biochem.* **27,** 9166 (1988).
[10] K. P. Howard and J. Prestegard, *J. Am. Chem. Soc.* **117,** 5031 (1995).
[11] A. Sanson, M. A. Monck, and J. M. Neumann, *Biochem.* **34,** 5938 (1995).
[12] W. Hübner, H. H. Mantsch, F. Paltauf, and H. Hauser, *Biochem.* **33,** 320 (1994).

conformations that are possible are + and −60 and 180°, called +synclinal (+sc), −synclinal (−sc), and antiplanar (+ap). Deviations from these angles of plus or minus 30° are allowed, but are usually much less. According to Pascher,[8] there are nine possible orientations around the glycerol. Of the nine possible conformers of glycerol, seven are found in crystalline structures of acylglycerols,[8] and these are, according to Θ_4/Θ_2, 60°(sc)/60°(sc), 60°(sc)/−60°(−sc), −60°(−sc)/−60°(−sc), 60°(sc)/180°(ap), −60°(−sc)/180°(ap), 180°(ap)/60°(sc), and 180°(ap)/−60°(−sc).

The acyl chains esterified to the glycerol oxygens start with a fairly rigid carboxyl ester group, which has an antiplanar conformation where the torsion angles of the C–O—C(O)–C are 180°. If the chain conformation were all *trans* then the acyl chains could stick out from the glycerol in a number of different directions, depending on the Θ_4/Θ_2 torsion angles. However, acyl chains pack together in parallel layers in crystals,[7] and thus there must be distortions in the CH_2–CH_2 bonds of some of the chains to allow parallel chain packing. The zigzag conformation of the glycerol O–CH_2–CH–CH_2–O is often continued down one of the two primary acyl chains. Some minor rotations around the primary carbon–oxygen bond are allowed from the 180°, but as a rule the carbonyl oxygen points to the two hydrogens on the primary glycerol carbon. However, the acyl chain esterified to the oxygen of the two carbon (secondary glycerol carbon) is arranged so that the carbonyl oxygen points toward the single hydrogen atom of the middle glycerol carbon. Deviations of the torsion angle between the secondary glycerol carbon and the oxygen of the attached chain are small and the torsion angle is fairly fixed at about 120°. Because the chains pack together in parallel layers in crystals, parallel chain packing is usually accomplished by skewing one of the CH_2–CH_2 bonds near the carbonyl groups of one of the chains to bend it back so that it lies next to an adjacent chain. Pascher has classified the acyl chains as all-*trans* extending from the ester group or bent so that parallel chain packing is possible.[8] Extrapolations of the glycerol conformations to monolayers at the air/water interface in which the acyl chains must lie above and perpendicular to the water surface show several conformations are possible, including sc/−sc, −sc/sc, and ap/sc.[8] Although a specific conformation may predominate when the surface is solid, several conformations could alternate rapidly in liquid monolayers.

Are glycerol conformations maintained in the active site of lipolytic enzymes? Using a variety of substrate inhibitors, transition state analogues, and nonideal substrates, two glycerol conformations have been identified in the active site,[8,13] sc/ap in pancreatic phospholipase A_2[14] and two other

---

[13] C. Camballiau, A. Nicolas, S. Longhi, and C. Martinez, *Curr. Opin. Struct. Biol.* **6,** 449 (1996).
[14] B. van den Berg, M. Tessari, R. Boelens, R. Dijkman, R. Kaptein, G. H. De Haas, and H. M. Verheij, *J. Biomol NMR* **5,** 110 (1995).

phospholipase A_2s,[15,16] and ap/sc, the natural "tuning fork" conformation of crystalline triacylglycerols[17,18] in the pancreatic lipase–colipase complex.[19,20] Other conformations have been modeled into cobra venom phospholipase A_2 and *Candida rugosa* lipase (see Pascher[8]). Although the glycerol conformations are generally preserved, the chain packing is greatly disturbed in the active site of *Candida antartica* lipase B[13] where the chains are bent and splayed away from each other due to interactions of the chains with the hydrophobic pockets of the enzyme.

Conformations of Substrates at the Air/Water Interface

The general conformation at interfaces, that is without defining the specific torsion angles around the glycerol, are known for a number of substrates including triacylglycerols, diacylglycerols, monoacylglycerols, cholesterol esters, and wax esters (see Small[21] for general review).

The conformation of saturated single-chain triacylglycerols in their most stable crystalline state (β form) shows that trilaurin and tridecanoin have a typical tuning fork conformation.[17,18] The Θ_4/Θ_2 mean angles of the two structures are $-175°/68°$, that is, ap/sc conformation (Fig. 1). The acyl chains on one of the primary carbons and the secondary carbon form an almost all-*trans* line through the glycerol extending in opposite directions without kinks. The chain on the other primary position bends backward at the number two acyl carbon so that the chain lies parallel to the other primary chain. It is clear that while trilaurin and tridecanoin are not chiral molecules, the conformations of the chains at the two primary positions differ explicitly around the number two acyl carbon, one being bent, the other being straight. The extended tuning fork conformation of the molecule appears to be present in the melted liquid state.[22] However, the specific conformation around the glycerol and its esters have not been defined. When melted triglyceride comes in contact with an aqueous interface a major conformational change occurs such that all three chains must now lie roughly parallel to one another and extend away from the water side

[15] D. L. Scott and P. B. Sigler, *Adv. Protein Chem.* **45**, 53 (1994).
[16] M. M. G. M. Thunnissen, E. Ab, K. H. Kalk, J. Drenth, B. W. Dijkstra, O. P. Kuipers, R. Dijkman, G. H. De Haas, and H. M. Verheij, *Nature* **347**, 689 (1990).
[17] K. Larrson, *Ark. Kemi.* **23**, 1 (1964).
[18] L. H. Jensen and A. J. Mabis, *Acta Crystallogr.* **21**, 770 (1966).
[19] H. van Tilbeurgh, M. P. Egloff, C. Martinez, N. Rugani, R. Verger, and C. Cambillau, *Nature (London)* **362**, 814 (1993).
[20] M. P. Egloff, F. Marguet, G. Buono, R. Verger, C. Cambillau, and H. van Tilbeurgh, *Biochem.* **34**, 2751 (1995).
[21] D. M. Small, "The Physical Chemistry of Lipids: From Alkanes to Phospholipids," Handbook of Lip. Res. Series (D. Hanahan, ed.), Vol. 4, pp. 1–672. Plenum Press, New York, 1986.
[22] K. Larsson, *Fette Seifen Austrichm.* **74**, 136 (1972).

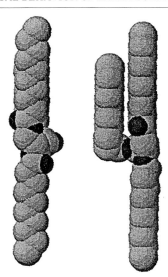

FIG. 1. Views of the crystalline structure tridecanoin (Tricaprin) drawn from the crystalline coordinates.[18] One primary chain extends up in an all-*trans* configuration from the glycerol. The chain coming off the 2 glycerol carbon (β position) is a linear extension through the glycerol in the opposite direction. The other primary chain comes off the glycerol oxygen and bends at the 2 acyl carbon so that the rest of the chain lies parallel to the other primary chain. It is clear from this picture that while the 2 primary acyl groups are chemically identical they have different conformations. Packing of the chains is triclinic parallel (β).

of the interface. A number of possible conformations have been suggested[8]; some may be present in a stable form in solidified monolayers, that is, monolayers compressed to high pressures where the chains solidify.[21,23,24] However in the liquid state it is more likely that different conformations of similar energy fluctuate at the interface. How the surface pressure or packing density affects the conformation in the fluid chain melt state is not known. From the enzymatic activity of lipoprotein lipase on racemic 1,2-dicaprin at different surface pressures it might appear that there was a conformational change at about 30 dyn/cm.[25] The selectivity of the enzyme for one of the antipodes of racemic dicaprin from 5 to 30 dyn/cm is extremely high suggesting that a single conformation may be recognized. Between 30 and 40 dyn/cm that selectivity greatly decreases and the other antipode is also hydrolyzed effectively. However, it is known that the pressure area

---

[23] G. Gains, "Insoluble Monolayers at the Liquid-Gas Interface," Vol. 14, pp. 1–386. Interscience Publishers, New York, 1966.

[24] D. G. Dervichian, "Progress in the Chem. of Fats and Other Lipids," Vol. 2, pp. 193–242. Pergamon Press, New York, 1954.

[25] E. Rogalska, S. Ransac, R. Verger, *J. Biol. Chem.* **268**, 792 (1993).

isotherm of dicaprin has no obvious transitions and forms a liquid monolayer from very low pressures to its collapse pressure at just above 40 dyn/cm. Furthermore, other lipases do not show a break in the selectivity, but rather a continuing decrease in the selectivity of antipodes as the pressure is increased. No molecular explanation has yet been given for this decrease in selectivity as surface pressure increases, but it does not appear to be related to a phase change in the lipid substrate.

The area occupied by long-chain triacylglycerols at their collapse pressure depends on the physical state of the monolayer.[21,24] If the chains are solid then the area is about 20 Å^2 per chain very similar to the area of aliphatic chains packed in a hexagonal lattice.[21] If the monolayer is fluid, the area is 38 Å^2 or greater, and if condensed, the area is between 23 and 38 Å^2.[21,24] The collapse pressure is low for triacylglycerols since only the three ester groups anchor the triglyceride to the interface. At ambient temperature the collapse pressure ranges between 12 and 15 dyn/cm for most long-chain triacylglycerols.

An interesting phenomenon is observed if one of the chains, for instance, the fatty acid on the *sn*-3 position, is considerably shorter than that on the 1 and 2 position as occur in milk fat.[26] For instance, in the case of 1,2-dipalmitoyl-*sn*-3 acyl glycerols (PPX series), the surface and bulk properties have been compared.[27–29] The bulk and surface melting points for the PPX series where the acyl group at the *sn*-3 position increases from 2 to 16 is shown in Fig. 2. The surface and bulk melting points for *sn*-3 acetate (PP2), proprionate (PP3), and butyrate (PP4) are similar, but as the chain lengthens the surface melting point becomes much lower than the bulk so that by eight carbons (PP8) the surface melting point is below 0°. However, as the chain *sn*-3 lengthens further and begins to approach the length of palmitate (PP16), the surface melting point rises, but it never gets as high as the bulk melting point. An explanation for this divergence in surface melting was given by analysis of pressure/area isotherms of the PPX series[28,29] (see Fig. 3). As the monolayer is compressed from large areas (<1 dyn/cm surface pressure), a short chain in the -*sn*-3 position (PP3, PP4, PP5, PP6) will at first lie flat, parallel to the air/water interface, and then be pushed beneath the surface into the aqueous phase.[28] The energy required to push the chain into the water is approximately 500 calories per $-CH_2-$ group.[28] This is less than the 850 cal/$-CH_2-$ it takes to push a $-CH_2-$ group from oil to water[6] and indicates that about 2/5 of the chain is already in water, that is, floating

---

[26] W. C. Breckinridge, *in* "Fatty Acids and Glycerides," Handbook of Lipid Res. Series (D. Hanahan, ed.), Vol. 1, p. 197, Plenum Press, New York, 1978.
[27] D. R. Kodali, D. Atkinson, T. G. Redgrave, and D. M. Small, *J.A.O.C.S.* **61,** 1078 (1984).
[28] D. A. Fahey and D. M. Small, *Biochem.* **25,** 4468 (1986).
[29] D. A. Fahey and D. M. Small, *Langmuir* **4,** 589 (1988).

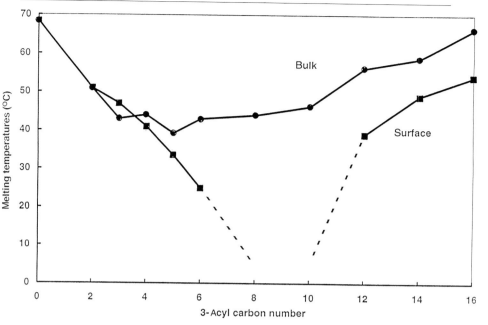

Fig. 2. Surface and bulk melting points of 1,2-dipalmitoyl, 3-acyl-sn-glycerols (PPX series). The melting point is on the vertical axis and the length of the sn-3 acyl chain is on the horizontal axis. The melting point of the bulk phrase is of the most stable crystalline form and the melting point of the surface is taken from a monolayer at the collapse point. Note that there is an odd/even effect of PP3 and PP5 in the bulk melting. The odd/even effect is lost in the surface melting. There is a marked decrease in surface melting in the medium-chain series PP8 and PP10 so that the melting point is below the freezing point of water. See text for further explanation. (Data from Kodali et al.[27] and Fahey and Small.[28,29])

on it. Once the shorter chain has been pushed into the aqueous phase, then the two palmitates can be pushed together to the area, about 20 Å² per chain, where they solidify on the surface. The surface pressure then abruptly rises until the collapse point is reached. Surface melting is the temperature at which the solid dipalmitoyl part liquefies. Some decrease from bulk melting is observed in PP5 and PP6. However, in PP8 the energy required to push the octanoate chain into water is greater than the energy necessary to collapse the film. Therefore, the octanoate chain cannot be submerged and bends up and lies parallel to the two palmitates and prevents them from solidification even at temperatures near 0° (see Fig. 3). The same effect occurs with PP10 but as the acyl chain in the sn-3 position gets longer it approaches the length of palmitate and permits solidification albeit at temperatures lower than bulk melting.[29] Note that tripalmitin bulk melting

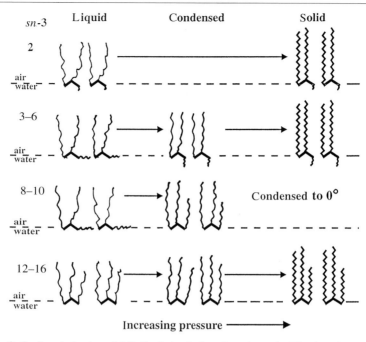

FIG. 3. Surface behavior of 1,2 dipalmitoyl, 3-acyl-*sn*-glycerols. The length of the acyl group in the 3 position is given as *sn*-3. When PP2 is spread on the air/water surface, the acetate is already in the aqueous phase. As PP2 is compressed the two palmitate chains go directly from a liquid to a solid monolayer. Each chain occupies 20 Å^2 at collapse. With PP3-6 the short fatty acid chain lies partly submerged on the surface and a force of 500 cal/–CH$_2$– is necessary to push the chain into the water. When this occurs, the two palmitate chains first form a condensed monolayer and then a solid monolayer, which collapses at high surface pressure at an area of about 40 Å^2, that is, 20 Å^2 per chain. The melting point of this monolayer is shown in Fig. 2. When the chain at the *sn*-3 position is intermediate, 8–10 carbons, then the collapse pressure of the monolayer becomes less than the pressure necessary to force the medium chain into the aqueous phase. Therefore, the chain folds upward and prevents the palmitates from solidfying at the interface, that is, they act as a disrupting component in the surface. Thus, the layer does not solidify even at the freezing point of water. When the chain becomes longer PP12-PP16, it is able to form a solid at the surface when compressed to about 60 Å^2 (20 Å^2 per chain), and this is reflected in the increase in the surface melting points shown in Fig. 2. (Data from Kodali *et al.*[27] and Fahey and Small.[28,29])

is melting of the stable β form and this relates directly to the differing conformation in the β crystal (tuning fork) and at the interface (all three chains lying parallel). Triacylglycerols containing one short and two long chains (PP2-PP6) clearly present a different substrate conformation than a triglyceride with three long chains. The fact that gastric lipase prefers the short chain at the *sn*-3 position may be related to this physical property (see Fahey and Small[28] and references therein).

Esters of long chain fatty acids behave in a manner similar to that of the PPX series. If the alcohol esterified is short, for example, buteryl stearate, then the four-carbon alcohol can be forced into water. If the alcohol is long, it forms a hairpin conformation.[30] Long-chain diacylglycerols are considerably more surface active and collapse above 25 dyn/cm,[28] and long-chain monoacylglycerols are very surface stable molecules collapsing at even higher pressures.

Cholesterol esters of long-chain fatty acids (see Small,[21] Chapter 11, for a review of physical properties) do not spread at an air/water interface to form a monolayer, but rather form a lens on the surface. The conformation of cholesterol esters in an oily phase is also extended,[31] however, cholesterol esters on meeting an air/water interface also change their conformation to be bent like a hairpin at the interface. Thus although cholesterol esters do not spread, they do expose the ester group to the aqueous interface, and this is reflected in a decrease in the interfacial tension compared to a pure hydrocarbon. Therefore, when highly nonpolar substrates such as triacylglycerols and cholesterol esters are broken into small globules by kinetic action, the interfaces presented are those that will be present at the oil/water interface and are primarily rich in the ester groups. A similar orientation occurs when these molecules are solubilized by membrane lipids.

Solubility and General Conformation of Substrates in Phospholipid Interfaces

Because most nonpolar or weakly polar lipids such as cholesterol esters and triglycerides exist in tissue fluids as microemulsions (lipoproteins) or in cells as droplets that are emulsified by phospholipids and proteins, it is important to understand how they partition into phospholipids. The first indications that long-chain triglycerides partition into phospholipids came from the surface chemistry experiments.[32] When egg phosphotidylcholine was spread with triolein on an air/water interface and compressed, a small amount of triolein (~5%) persisted in the monolayer even at the collapse pressure of the phospholipid. These experiments were repeated, confirmed, and extended by Brockman and colleagues (see Chapter 14), who showed that a few percent of triacylglycerols and cholesterol esters were trapped

[30] N. K. Adam, *Proc. Roy. Soc. Ser. A* **126,** 366 (1929).
[31] C. Burks and D. Engelman, *Proc. Natl. Acad. Sci. U.S.A.* **78,** 6863 (1981).
[32] P. Desnuelle, J. Molines, and D. Dervichian, *Bull. Soc. Chim. Biol.* **18,** 197 (1951).

TABLE I
Solubility of Enzyme Substrate Molecules in Phosphatidylcholine Bilayers

Substrate molecule	Mol%	Wt%	Technique	PC preparation	T (°C)	R
Cholesterol esters						
Cholesteryl octanoate	5.9	3.9	^{13}C NMR	SUV[b]	35	4
Cholesteryl octanoate	4.5	2.9	^{13}C MAS NMR	MLV[b]	25	4
Cholesteryl myristate	3.6–6.1	2.8–5	Light microscopy, DSC, and X-ray diffraction	MLV	37	3
Cholesteryl myristate	0	0	Light microscopy, DSC, and X-ray diffraction	MLV(<Tm)[a]	<23	3
Cholesteryl palmitate	3.7	2.9	^{13}C NMR	SUV	35	4
Cholesteryl palmitate	1.9	0.6	^{13}C MAS NMR	MLV	25	4
Cholesteryl stearate	3.3	2.7	^{13}C NMR	SUV	35	4
Cholesteryl stearate	1.5	1.2	^{13}C MAS NMR	MLV	25	4
Cholesteryl oleate	3.2–3.7	2.6–3.0	^{13}C NMR	SUV	35	4
Cholesteryl oleate	2.2	1.8	^{13}C MAS NMR	MLV	25	4
Cholesteryl oleate	2.1	1.6	^{13}C NMR	SUV	35	4
Cholesteryl oleate	0.7–2.0	0.5–1.5	^{13}C MAS NMR	MLV	35	4
Cholesteryl oleate	2.9	2.4	^{13}C NMR	SUV	37	4
Cholesteryl oleate	2.3	1.9	^{13}C NMR	SUV + 33%C	37	4
Cholesteryl oleate	1.2	1.0	^{13}C NMR	SUV + 50%C	37	4
Cholesteryl linolenate	3–4.3	2.4–3.5	Light microscopy, DSC, and X-ray diffraction	MLV	23	3

in the monolayers at high surface pressures. Phase equilibrium studies carried out with cholesterol esters, egg phosphotidylcholine (PC), and water[33] indicated that the bilayer phase can solubilize about 3% by weight of cholesteryl linolenate (Table I). Cholesteryl myristate is similarly soluble in liquid crystalline dimyristoyl PC but not in the quasi-crystalline gel phase.[34] Emulsions made of egg PC and triolein were broken ultracentrifugally and the surface and core phases collected.[35] The surface phase contained about 2% by weight of triolein. Similar studies on mixtures containing triolein and cholesterol esters[36] and reconstituted chylomicron lipids[37,38]

---

[33] M. J. Janiak, C. R. Loomis, G. G. Shipley, and D. M. Small, *J. Mol. Biol.* **86**, 325 (1974).
[34] M. J. Janiak, G. G. Shipley, and D. M. Small, *J. Lip. Res.* **20**, 183 (1979).
[35] K. W. Miller and D. M. Small, *J. Colloid Interface Sci.* **89**, 466 (1982).
[36] K. W. Miller and D. M. Small, *Biochem.* **22**, 443 (1983).
[37] K. W. Miller and D. M. Small, *J. Biol. Chem.* **258**, 13772 (1983).
[38] K. W. Miller and D. M. Small, *New Comp. Biochem., Plasma Lipoproteins,* Elsevier Sci. Pub. **14**, 1 (1987).

TABLE I (continued)

Substrate molecule	Mol%	Wt%	Technique	PC preparation	T (°C)	Ref.
Triacylglycerols						
Trioctanoin	~14	~9	^{13}C NMR	SUV		41
Trioctanoin	10	6.1	^{13}C NMR	SUV	30	44
Trioctanoin	7.0	4.2	^{13}C NMR	SUV + 3%TO	30	44
Trioctanoin	2.8	2.0	^{13}C NMR	SUV + 30%C	30	44
Trioctanoin	8.5	5.2	^{13}C NMR	SUV + 2%CO	30	44
Tripalmitin	~2	~2	^{13}C NMR	SUV	35	41
Tripalmitin	3	3.2	^{13}C NMR	SUV	40–50	41
Triolein	3	3.3	^{13}C NMR	SUV	35	41
Triolein	3	2.7	^{13}C NMR	SUV	40–50	41
Triolein	2.2	2.5	^{13}C NMR	SUV	35	47
Triolein	1.1	1.2	^{13}C NMR	SUV + 33%C	35	47
Triolein	0.15	0.17	^{13}C NMR	SUV + 50%C	35	47
Triolein	2.5	2.8	^{13}C NMR	SUV	35	39
Triolein	~2.7	~3	^{13}C MAS NMR	MLV	35	42
Triolein	1.8	2.0	Centrifugation and analysis	MLV	23	35
Triolein	3.0	3.3	^{13}C NMR	SUV	30	44
Triolein	1.1	1.3	^{13}C NMR	SUV + 10%Toct	30	44
Triolein	0.6	0.7	^{13}C MAS NMR	MLV + 10%Toct	30	44
Diacylglycerols						
Dilauroyl-*sn*-glycerol	>20	>12.5	^{13}C NMR	SUV	10–55	50
Monoacylglycerols						
All long chain	Infinitely soluble					
Wax esters						
Ethyl oleate	25	~9.6	^{13}C NMR	SUV	35	51

^{a} Tm, Chain transition from La (melted chains) to gel (frozen chains).
^{b} SUV, Small unilamellar vesicles; MLV, multilamellar vesicles.

confirmed the small amount of triglyceride solubilized in the surface of emulsions and chylomicrons.[37,38]

The use of ^{13}C carbonyl carbon enrichment of cholesterol esters, triacylglycerols, and other molecules allowed us to identify by NMR molecules, which partitioned into the surface of a bilayer and those which partitioned into small emulsion particles that separated from the bilayer without disrupting the system.[39–41] These results are summarized in Table I. By using increasing amounts of cholesterol ester[40] or triacylglycerol[39] in PC vesicles, the limiting solubility could be achieved where the concentration of triacylglycerol or cholesterol ester in the surface phase remained constant even

[39] J. A. Hamilton and D. M. Small, *Proc. Natl. Acad. Sci. U.S.A.* **78,** 6878 (1981).
[40] J. A. Hamilton and D. M. Small, *J. Biol. Chem.* **257,** 7318 (1982).
[41] J. A. Hamilton, *Biochem.* **28,** 2514 (1989).

though more was added. The original studies[39–41] were done on small unilamellar vesicles (SUV) of PC but were later expanded to use solid-state magic angle NMR (MAS NMR) techniques and multilamellar vesicles (MLV), which showed very similar solubilities[42,43] although slightly lower in some cases (Table I). Several things can be noted from Table I. First the ordinary cholesterol esters, such as cholesteryl oleate, palmitate, and stearate, have solubilities of about 2 to 3 mol% by several of the techniques studied. When the cholesterol ester side chain is shortened to eight carbons (cholesteryl octanoate) then the solubility increases.

The long-chain triglycerides, triolein and tripalmitin, are also soluble to about 2 to 3 mol% in bilayers. The medium-chain triglycerides, such as trioctanoin, are soluble to somewhat larger amounts, about 10 mol%. Interestingly, the incorporation of both triolein and trioctanoin appears to be competitive with both solubilities falling some so that the sum of the solubilities of the two is between 3 and 10 mol%.[44] Cholesterol is known to have a rigidifying effect on bilayers and appears to increase the surface pressure of emulsions as a function of cholesterol phospholipid ratio.[45] When free cholesterol is included in small unilamellar vesicles it displaces cholesteryl oleate and triolein from the bilayer.[46,47] It also appears to push the remaining cholesteryl oleate[46] and triolein[47] somewhat deeper in the bilayer so that the carbonyl group is less hydrated. When cholesterol is at a 1:1 molar ratio with phospholipid, cholesteryl oleate solubility is reduced to about 1.2 mol% and triolein solubility is almost abolished (0.15 mol%). When cholesteryl oleate and triolein are both included, they partly compete for positions in the bilayer,[48] however, the maximum total of both does not exceed 4 mol%.[48] Like cholesterol, monopalmitin forces triolein out of the interface.[49]

From the chemical shift difference of the carbonyl groups compared to fully hydrated carbonyls, the general conformations of the molecules in the bilayer interface are shown in Fig. 4. The conformation of triacylglycerol in the phospholipid monolayer of a bilayer vesicle was about half hydrated compared to aqueous triacetin. The two primary carboxyl groups are more exposed to the aqueous interface than the secondary or β carboxyl sug-

[42] J. A. Hamilton, D. A. Fujita, and C. F. Hammer, *Biochem.* **30,** 2894 (1991).
[43] A. Salmon and J. A. Hamilton, *Biochem.* **35,** 16065 (1996).
[44] J. A. Hamilton, J. Vural, and R. Deckelbaum, *J. Lipid Res.* **37,** 773 (1996).
[45] D. M. Small and M. C. Phillips, *Adv. Colloid Interface Sci.* **41,** 1 (1992).
[46] P. J. R. Spooner, D. Gantz, J. A. Hamilton, and D. M. Small, *Biochim. Biophys. Acta.* **860,** 345 (1986).
[47] P. J. R. Spooner and D. M. Small, *Biochem.* **26,** 5820 (1987).
[48] J. A. Hamilton, K. W. Miller, and D. M. Small, *J. Biol. Chem.* **258,** 12821 (1983).
[49] E. Boyle, D. M. Small, D. Gantz, J. A. Hamilton, and J. B. German, *J. Lipid Res.* **37,** 764 (1996).

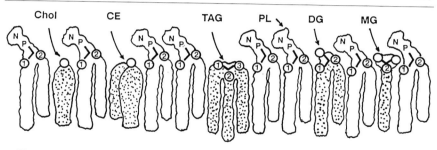

Fig. 4. The general conformation of weakly polar molecules in a phosphatidylcholine bilayer. One monolayer of the bilayer is shown. The phospholipids are shown with the phosphorous (P) and choline (N) in the standard conformation in which the sn-1 chain extends in the linear fashion from the glycerol and the sn-2 chain bends at the second acyl carbon to lie parallel to the sn-1 chain. PL, Phosphotidylcholine; Chol, cholesterol; CE, cholesterol ester; TAG, triacylglycerol; DG, diacylglycerol; MG, monoacylglycerol. 1 and 2 refer to the sn-1 and -2 positions, respectively. Free cholesterol interdigitates between the fatty acyl chains of PL causing a condensation of liquid monolayers. Its maximum solubility is one molecule of cholesterol for one molecule of phosphatidylcholine. Cholesterol esters form a hairpin configuration with the acyl chain and the steroid lying side by side and the ester group protruding into the aqueous space. Triacylglycerols have all three chains lying parallel and roughly perpendicular to the aqueous phospholipid interface. If one chain were short (less than eight carbons), it would protrude into the aqueous space as in Fig. 3. The sn-1 and -3 ester groups are more exposed to the water than the group in the 2 position. Diacylglycerols have a conformation very similar to phospholipids with the sn–3–OH in the aqueous interface. The sn-2 is bent over and the sn-1 acyl chain extends straight down into the bilayer. The 2 monoacylglycerols would have both –OH groups in the interface hydrogen bonded to water.

gesting that the β ester was buried in the membrane somewhat deeper than the two primary ester groups.[39] Cholesteryl oleate had a party hydrated ester group indicating that the molecule has a hairpin-like conformation in the bilayer with the oleate lying next to the steroid.[40]

Diacylglycerols were shown to be quite soluble (>20 mol%) and to have a glycerol conformation very much like that of phosphotidylcholine in bilayer. That is, the sn-1 chain forms a linear all-*trans* conformation with the glycerol oxygens and carbons and the unesterified primary alcohol protrudes into the aqueous space to hydrogen bond with water.[50] This is quite different from crystalline diacylglycerol in which the axis of the glycerol molecule is roughly parallel to the bilayer plane and the free hydroxyl group hydrogen bonds with the sn-2 carbonyl of the neighboring diacylglycerol.[7,8,21] The solubility of 1,2-dilauryl-sn-glycerol, is at least 20 mol% and probably higher. Long-chain monoacylglycerols are bilayer-forming molecules themselves and are thus infinitely soluble in bilayers of phospholipids.

[50] J. A. Hamilton, S. P. Bhamidipati, D. R. Kodali, and D. M. Small, *J. Biol. Chem.* **266**, 1177 (1991).

However, high concentrations of glycerol monooleate and lower concentrations of saturated monoacylglycerols displace triglycerides from surfaces.[49] The wax ester, ethyl oleate, has been studied[51] by ^{13}C NMR and found to be soluble to about 25 mol%. It appears that the carbonyl is partly hydrated, which suggests that it is in the interface with the oleate lying parallel to the phospholipid chains. The ethyl group is probably in the water.

The Trans Bilayer Movements of Lipase Substrates

The *trans* bilayer movement of molecules such as cholesterol esters and triglycerides is important because if fast, it allows such molecules to be formed on one side of a membrane and to flip rapidly to the other side. Furthermore, the movement between a core-located triglyceride and a surface-located triglyceride, if rapid, would allow the enzyme rapid access to core molecules. If, however, the exchange rate between core and surface were slow, then the surface concentrations might be rate limiting. Using NMR and small concentrations of ^{13}C-labeled molecules in small unilamellar vesicles, the chemical shift difference between the inner layer and the outer layer can be used to distinguish the internally located pool from an externally located pool, provided the exchange rate is slow. Therefore, in the case of molecules such as bile acids[52,53] and 1,2-dilauryl-*sn*-glycerol it was shown that there were two pools of these molecules, one existing on the external surface and the other on the internal surface whose exchange rate was of the order of milliseconds. When the temperature increases, the exchange rate increases and therefore the two peaks will merge and appear as one single peak. Using these analyses, the exchange rates of 1,2-dilauryl-*sn*-glycerol have been measured at several temperatures.[50] At 38° the exchange rate was 60/sec ($T_{1/2} \sim 10$ msec). However, only a single peak for cholesterol ester or triglyceride is seen even when the temperature is cooled down near 0° or when cholesterol is added to make the bilayer more viscous.

In general, the flip rate is related to the mass of the molecule, to the fluidity of the bilayer, and to the energy necessary to dehydrate hydrated groups as they pass through the hydrophobic core of the membrane. Cholesterol esters with a single ester group should have a small barrier to flipping since they hydrate only modestly. Triacylglycerols probably require about three times the energy since they have three ester groups. Nevertheless, only one peak is observed under all the conditions in which the triglycerides and cholesterol esters have been studied (Table I) and, therefore, it is our

---

[51] D. A. Bird, M. Laposata, and J. Hamilton, *J. Lip. Res.* **37**, 1449 (1996).
[52] D. J. Cabral, J. A. Hamilton, and D. M. Small, *J. Lip. Res.* **27**, 334 (1986).
[53] D. J. Cabral, D. M. Small, H. S. Lilly, and J. A. Hamilton, *Biochem.* **26**, 1801 (1987).

(unproved) assumption that the flip-flop rate is fast on an NMR timescale, that is, microseconds or faster.

A similar dilemma was present for long-chain protonated fatty acids. However, studies using radioactive fatty acids in multilamellar liposomes suggested that not only was flip-flop rapid, but also movement between bilayers was relatively rapid.[54] The use of fluorescent indicator dye to measure pH changes inside unilamellar vesicles showed that the flip rate was indeed very rapid for protonated fatty acids.[55] Thus, a single peak for a carbonyl appears to indicate very rapidly mixing pools that appears by NMR to be a single pool.

A related matter is the rate of exchange between the core of a droplet and the surface. This is pertinent to the hydrolysis of triglyceride by lipoprotein lipase. If there were two pools (one in the surface and the other in the core) for which there is good evidence,[35-38] and they were in rapid equilibrium, then the rapidity of the exchange would allow lipoprotein lipase to see all of the triglyceride, that is, the substrate concentration would be the total triglyceride concentration. If, however, the exchange rates were exceedingly slow, and lipase could only hydrolyze the surface-located pool then the surface concentration would become a rate-limiting factor in the reaction. Although there is no direct evidence, the fact that only a single peak of ^{13}C-labeled cholesteryl oleate is present in microemulsions of ^{13}C-cholesteryl oleate and that the chemical shift of that peak is consistent with a rapidly changing surface and core pools (unpublished results) makes it probable that the surface and core pools of cholesterol ester rapidly exchange. The interaction of enzymes with membranes might also be influenced by the flip rate of a molecule from one side of the membrane to the other. If the substrate could flip rapidly then the enzyme on one side could react with all of the substrate in the membrane. Thus a molecule, such as diacylglycerol, which has a relatively rapid but measurable flip rate, could be generated on one side of the membrane, flip to the other, where it could interact with another enzyme. However, bilayers of asymmetric phospholipids may be prevented from interacting with enzymes because the *trans* bilayer movement of phospholipids is so slow. A protein catalyzed *trans* bilayer movement (flipase) might overcome this so that phospholipids primarily found on one side of the membrane could be utilized by enzymes on the other side once they were catalytically transferred across the membrane.

---

[54] P. Brescher, R. Saouaf, J. Sugerman, and A. V. Chobanian, *J. Biol. Chem.* **259**, 13395 (1984).
[55] F. H. S. Kamp and J. A. Hamilton, *Proc. Natl. Acad. Sci. U.S.A.* **89**, 11367 (1992).

[9] Phospholipase A_2 Activity and Physical Properties of Lipid-Bilayer Substrates

By Thomas Hønger, Kent Jørgensen, Deborah Stokes, Rodney L. Biltonen, and Ole G. Mouritsen

A. Introduction

It has been known for a long time[1,2] that the activity of phospholipases depends on the properties of the lipid substrate, such as the form of aggregation (i.e., the lamellar symmetry), the thermodynamic state, and the chemical composition of the lipid bilayer. An in-depth study of this relationship is highly important because the substrate, over the course of time, changes composition and possibly morphology due to accumulation of hydrolysis products in the bilayer and because the binding and the activity of the enzyme is controlled by small-scale local properties of the bilayer, which may be very different from the global properties of the bilayer substrate and therefore difficult to detect.

In this chapter, we propose a rational strategy for a methodological study of phospholipase activity by combining insight from a parallel set of theoretical calculations and experimental measurements, with the aim of unraveling details of the relationship between the physical properties on different scales of the substrate, on the one hand, and the activity of particularly PLA_2,* on the other.

Hydrolysis of lipid-bilayer substrates by PLA_2 is a convenient and important model system for studying generic aspects of the relationship between lipase biocatalytic activity and the physical and thermodynamic properties of the lipid substrate. It is a convenient model system because it is sufficiently simple and well controlled such that it holds a promise for unraveling, in a quantitative manner, the physical determinants of phospholipase activity. It is a biologically important model system because lipid bilayers are models of the fluid bilayer component of biological membranes.[3] PLA_2 activity is omnipresent in cells and there seem to be some

---

[1] J. A. F. Op den Kamp, J. De Gier, and L. L. M. van Deenen, *Biochim. Biophys. Acta* **345**, 235 (1974).

[2] R. L. Biltonen, *J. Chem. Thermodynamics* **22**, 1 (1990).

* Abbreviations: $DC_{14}PC$, dimyristoyl phosphatidylcholine; $DC_{16}PC$, dipalmitoyl phosphatidylcholine; $DC_{18}PC$, distearoyl phosphatidylcholine; DC_nPC, diacylphosphatidylcholine with n carbon atoms in each saturated acyl chain; PLA_2, phospholipase A_2.

[3] M. Bloom, E. Evans, and O. G. Mouritsen, *Quart. Rev. Biophys.* **24**, 293 (1991).

common determinants of their mode of action at the lipid bilayer site. Therefore, despite the fact that phospholipases constitute a very heterogeneous group of enzymes, a study of PLA_2/substrate interactions is important for understanding generic aspects of the interaction between biological membranes and peripheral proteins. The full perspective of this statement may become clear when one considers the accumulating evidence that lipids play a functional role in a variety of cellular processes, for example, growth and signaling.[4]

The viewpoint we want to stress in this chapter is that, despite the fact that the biochemistry and the kinetics of PLA_2 activity[5] are very complex and to some extent poorly understood, the physical properties of the lipid-bilayer substrate pose some specific constraints on the system that lead to certain correlations between the lipid-bilayer physical properties and the phospholipase activity. In particular, the activity is controlled not only by the chemical structure of the individual lipid molecules and their gross interfacial properties (i.e., the fact that they are aggregated) but also to a large extent by the lateral organization and the cooperative dynamic characteristic of an assembly of many molecules.

To clarify our methodological approach designed to investigate correlations between PLA_2 activity and the physical properties of lipid-bilayer substrates, this chapter is organized as follows: In Section B we review the relevant physical properties of lipid bilayers because these are controlled by the underlying thermodynamics and phase equilibria. Particular attention is paid to in-plane phase transitions and dynamic lateral bilayer organization in terms of nanoscale lipid domains and microheterogeneity. This section contains the key information on the substrate structure and organization that can be obtained from theoretical and computational considerations. The effects of non-bilayer-forming lipids are described with reference to the possible influence on the substrate of the products of the hydrolysis. Section C provides the necessary background information on the hydrolysis of lipids by PLA_2 and how it can be monitored. This section sets the stage for Section D, which describes the data for the dependence of the PLA_2 activity on the composition of the substrate and it presents a comparison between experimental and theoretical results that provides strong evidence for a correlation between lipid-bilayer dynamic microheterogeneity and enzyme activity. Section E discusses how solutes, such as sterols and drugs, can modify PLA_2 activity via modulating the physical properties of the substrate. This section discusses the potential of our proposed combined

[4] O. G. Mouritsen and P. J. K. Kinnunen, *in* "Biological Membranes" (K. Mertz and B. Roux, eds.), p. 465, Birkhäuser, Boston, 1996.
[5] M. Jain, M. H. Gelb, J. Rogers, and O. G. Berg, *Methods Enzymol.* **249,** 567 (1995).

theoretical and experimental methodological approach to study PLA_2 activity as being guided by an insight into which factors control the physical properties of the lipid-bilayer substrate. Finally, in Section F we discuss the future prospects of our methodological approach, considering both its strengths and weaknesses.

B. Physical Properties of Lipid-Bilayer Substrates

Lateral Bilayer Organization and Lipid Domains

In its capacity of being a many-particle system, the large assembly of lipid molecules that form a lipid bilayer displays cooperative phenomena such as phase transitions and phase equilibria. These phase equilibria determine not only the macroscopic bilayer behavior, such as bilayer area, thickness, and mechanical moduli, but they also underlie the formation of lipid domains and small-scale lipid structures within the bilayer.[6-8] We are concerned here with the main phase transition, which takes the bilayer from a solid (gel phase) structure composed of conformationally ordered lipid acyl chains to a fluid phase with predominantly disordered lipid acyl chains. Numerous results from both theoretical and experimental investigations of well-defined model bilayer systems have provided insight into the nature of the main transition and how its cooperativity may give rise to a dynamic heterogeneous lateral bilayer structure composed of fluctuating lipid domains in the nanometer range. The lipid domains correspond to spontaneous fluctuations in lateral structure or composition.[9]

Whereas substantial information exists on lipid-bilayer structure and dynamics on the molecular, mesoscopic, and macroscopic scales, information derived predominantly from spectroscopic and scattering studies,[3] there is a conspicous lack of information on nanoscopic scales, that is, in the range of 10–1000 Å. This range is very difficult to access experimentally, and possibly only via further developments of ultrasensitive surface-probe techniques, such as atomic force microscopy, will this regime become directly accessible. We would like to claim that this nanoscopic regime is of paramount interest for quantitative studies of phospholipase activity.

We can investigate the nanoscale regime by means of computer simulation techniques using statistical mechanical models of well-defined lipid-

---

[6] O. G. Mouritsen and K. Jørgensen, *Chem. Phys. Lipids* **73,** 3 (1994).
[7] O. G. Mouritsen and K. Jørgensen, *Mol. Membr. Biol.* **12,** 15 (1995).
[8] S. Pedersen, T. Baekmark, K. Jørgensen, and O. G. Mouritsen, *Biophys. J.* **71,** 554 (1996).
[9] For a recent selection of papers on lipid domains in membranes, see *Mol. Membr. Biol.* **12,** 1 (1995).

bilayer systems. Computer simulation of lipid-bilayer systems, including the molecular interaction between PLA_2 and lipid bilayers,[10] is increasingly being used in the study of the biophysics of lipid membranes.[11] However, due to limitations in today's computational power, only rather simplified bilayer models can be studied when lipid-phase transitions and lipid-domain formation are at issue, simply because these phenomena require calculations on very large arrays containing typically of the order of 10^4 lipid molecules.[12] Such simplified models include certain lattice models,[12] which contain enough detail of the molecular interactions that a faithful picture of the cooperative phenomena can be reproduced.

Figure 1 shows schematic top-view snapshots of the lateral lipid organization of bilayers composed of $DC_{14}PC$, $DC_{16}PC$, and $DC_{18}PC$ lipids undergoing the gel-to-fluid acyl chain-melting transition. These snapshots were obtained from Monte Carlo computer simulation calculations on simple lattice models of the phase transition.[6] In the transition region a pronounced heterogeneous lateral lipid-domain structure develops due to strong thermally induced structural fluctuations. The fluctuations are manifested in the formation of dynamic fluid domains in the bulk gel phase below the transition temperature, T_m, and dynamic gel domains in the bulk fluid phase above T_m as illustrated in Fig. 1. The coexisting dynamic gel and fluid domains in the transition region imply the appearance of interfacial regions that are likely to have poor packing properties. Hence the bilayer possesses dynamic microheterogeneity in the transition region.[13] Figure 1 moreover demonstrates that the dynamic heterogeneity becomes more pronounced and prevails over a wider temperature region the shorter the acyl chains are.[14] We return to this observation later in relation to PLA_2 activity (Section D).

The lateral organization of the lipid species in a mixed lipid bilayer is basically determined by the interactions between the different lipids as well as the thermodynamic conditions such as temperature, pressure, and ionic strength. Binary mixtures of lipids are subject to phase separation phenomena resulting in coexisting static gel and fluid phases due to the

---

[10] S. T. Jones, P. Alström, H. J. Berendsen, and R. W. Pickersgill, *Biochim. Biophys. Acta* **1162**, 135 (1993).

[11] O. G. Mouritsen, in "Molecular Description of Biological Membrane Components by Computer Aided Analysis" (R. Brasseur, ed.), p. 3, CRC Press, Boca Raton, Florida, 1990.

[12] B. Dammann, H. C. Fogedby, J. H. Ipsen, C. Jeppesen, K. Jørgensen, O. G. Mouritsen, J. Risbo, M. C. Sabra, M. M. Sperotto, and M. J. Zuckermann, in "Nonmedical Applications of Liposomes" (Y. Barenholz and D. Lasic, eds.), p. 85, CRC Press, Boca Raton, Florida, 1996.

[13] O. G. Mouritsen and K. Jørgensen, *BioEssays* **14**, 129 (1992).

[14] J. H. Ipsen, K. Jørgensen, and O. G. Mouritsen, *Biophys. J.* **58**, 1099 (1990).

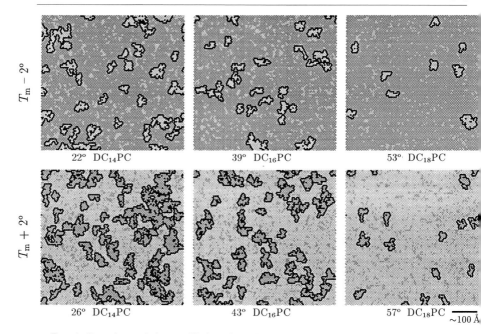

FIG. 1. Snapshots of the equilibrium lateral organization of lipid bilayers composed of one-component $DC_{14}PC$, $DC_{16}PC$, and $DC_{18}PC$ saturated phospholipids in the gel-to-fluid main transition temperature regions. The heterogeneous bilayer structure is composed of dynamic coexisting gel and fluid phases. The snapshots show typical bilayer configurations in the fluid and gel phase 2° above and 2° below the characteristic gel-to-fluid phase transition temperature, T_m, for the three different lipid bilayers. Gel and fluid regions of the bilayer are denoted by dark gray and light gray areas, respectively. The interfacial regions between the dynamic coexisting gel and fluid regions are highlighted in dark. The data were obtained by computer simulation calculations on systems corresponding to 5000 lipid molecules.

nonideal mixing properties.[15] The phase diagram shown in Fig. 2 for a binary mixture composed of $DC_{14}PC$ and $DC_{18}PC$ lipids shows that a fairly nonideally behaving mixture gives rise to a broad gel–fluid phase coexistence region.[16] A visualization of the lateral lipid organization in the different parts of the phase diagram for this mixture as obtained from computer simulation is provided by Fig. 3. In the one-phase fluid and gel regions, a pronounced heterogeneous nonuniform distribution of the lipids is seen. The appearance of a heterogeneous bilayer structure

[15] A. G. Lee, *Biochim. Biophys. Acta* **472**, 285 (1977).
[16] K. Jørgensen, M. M. Sperotto, J. H. Ipsen, O. G. Mouritsen, and M. J. Zuckermann, *Biochim. Biophys. Acta* **1152**, 135 (1993).

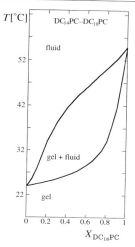

Fig. 2. Phase diagram for a lipid mixture composed of $DC_{14}PC$–$DC_{18}PC$ as determined from computer simulation calculations. The phase diagram reveals a nonideal phase behavior for the mixture, which is manifested as a broad gel–fluid phase coexistence region.

in the one-phase regions is caused by compositional fluctuations that give rise to the formation of dynamic lipid domains characterized by a local concentration of each of the lipid species that is different from the corresponding global concentration. This type of lipid-domain formation becomes more pronounced near phase boundaries as quantified in Fig. 4 by the lipid-acyl-chain pair correlation function, $g_{ff}(DC_{18}PC)$, in the case of an equimolar $DC_{14}PC$–$DC_{18}PC$ mixture for two different temperatures in the fluid one-phase region. This correlation function, which provides a quantitative measure of the local structure of the fluid, clearly reveals that the compositional fluctuations and the appearance of a dynamic heterogeneous bilayer structure become more pronounced as the liquidus phase boundary is approached.[16] The development of a heterogeneous bilayer structure in the vicinity of phase boundaries may play an important role for the activity of membrane-associated phospholipases. In particular, PLA_2 lag-time measurements of $DC_{14}PC$–$DC_{18}PC$ multilamellar vesicles have shown an increase in the activity of the enzyme in the temperature regions of the liquidus and solidus phase boundaries,[17] which may be relevant for the interpretation of PLA_2 activity in the vicinity of phase coexistence regions.

[17] E. Goormaghtigh, M. Van Campenhoud, and J.-M. Ruysschaert, *Biochim. Biophys. Res. Commun.* **101**, 1410 (1981).

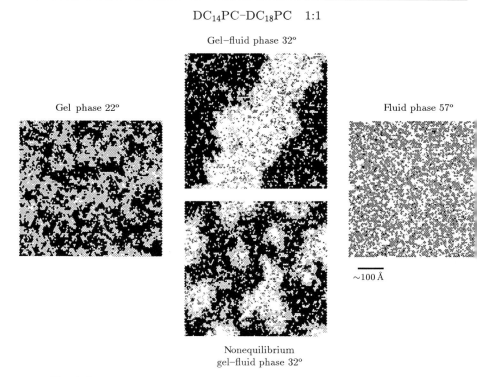

FIG. 3. Snapshots of the equilibrium and nonequilibrium microscopic lateral organization of an equimolar $DC_{14}PC$–$DC_{18}PC$ mixture in the one-phase fluid region, $T = 57°$, in the gel–fluid phase coexistence region, $T = 32°$, and in the one-phase gel region, $T = 22°$. The snapshots, which are obtained by computer simulation calculations, represent bilayer systems corresponding to 5000 lipid molecules. The symbols for the conformational states of the acyl chains are gel–$DC_{14}PC$ (light gray), fluid–$DC_{14}PC$ (blank), gel–$DC_{18}PC$ (black), fluid–$DC_{18}PC$ (dark gray).

Inside the gel–fluid phase coexistence region it is still a controversial issue as to what extent the lipid-bilayer system displays fully developed macroscopic phase separation as dictated by the equilibrium phase diagram, or the lateral structure instead is characterized as a nonequilibrium percolative bilayer structure[18] composed of coexisting gel and fluid domains as shown in the nonequilibrium snapshot in Fig. 3 of the $DC_{14}PC$–$DC_{18}PC$ lipid mixture.[19] Nonequilibrium phenomena and the formation of long-

[18] P. F. F. Almeida, W. L. C. Vaz, and T. E. Thompson, *Biochemistry* **31,** 7198 (1992).
[19] K. Jørgensen and O. G. Mouritsen, *Biophys. J.* **95,** 942 (1995).

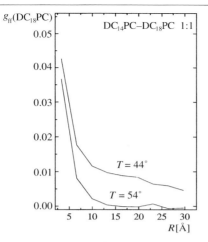

FIG. 4. Pair correlation function, $g_{ff}(DC_{18}PC)$, for the fluid–$DC_{18}PC$ lipids in an equimolar $DC_{14}PC$–$DC_{18}PC$ lipid mixture at two different temperatures in the fluid phase, $T = 44$ and $54°$, as a function of the acyl chain separation, R. The data were derived from computer simulation calculations.

living lipid domains may be of particular interest in relation to a deeper understanding of the dynamic behavior of the lipid bilayer substrate during the PLA$_2$ hydrolysis time course. It has been proposed that product formation of lysolipids and free fatty acids, which at a critical mole fraction can give rise to phase separation phenomena and the formation of a heterogeneous bilayer structure composed of long-living nonequilibrium lipid domains, is closely related to the sudden increase in the activity of PLA$_2$.[20] A recent study of nonequilibrium phenomena in lipid mixtures has revealed a characteristic relaxation time on the order of hours for the dynamic nonequilibrium lipid-domain formation inside gel–fluid phase coexistence regions.[21] A similar slow relaxation behavior for the lateral bilayer structure might be of importance for unraveling the dynamic aspects of PLA$_2$ activity during the hydrolysis time course, which leads to a time-dependent change in the concentration of immiscible reaction products.

Dynamic lipid-domain formation caused by compositional fluctuations that prevail in lipid mixtures, for example, in the one-phase fluid regions, depends on the degree of nonideality of the mixture. If the nonideality of the mixture is substantial, as is the case for a $DC_{12}PC$–$DC_{20}PC$ lipid mixture,

[20] W. R. Burack, Q. Yuan, and R. L. Biltonen, *Biochemistry* **32**, 583 (1993).
[21] K. Jørgensen, A. Klinger, M. Braiman, and R. L. Biltonen, *J. Phys. Chem.* **100**, 2766 (1996).

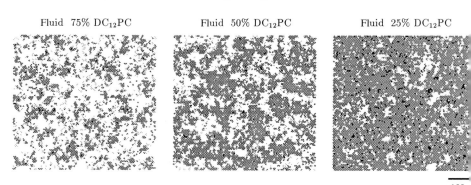

FIG. 5. Snapshots of the equilibrium microscopic lateral organization of a binary $DC_{12}PC$–$DC_{20}PC$ lipid mixture in the one-phase fluid region at a fixed temperature, $T = 55°$, as a function of composition. The snapshots, which were obtained by computer simulation calculations, show typical bilayer configurations corresponding to 5000 lipid molecules. The symbols for the conformational states of the acyl chains are gel–$DC_{12}PC$ (light gray), fluid–$DC_{12}PC$ (blank), gel–$DC_{20}PC$ (black), fluid–$DC_{20}PC$ (dark gray).

the compositional fluctuations become very pronounced. This effect is seen in snapshots shown in Figs. 3 and 5 of equimolar $DC_{14}PC$–$DC_{18}PC$ and $DC_{12}PC$–$DC_{20}PC$ mixtures in the one-phase fluid regions. Increasing the hydrophobic mismatch of the lipid-acyl chains of the two lipid species clearly leads to a more pronounced dynamic bilayer heterogeneity. Such effects may have functional consequences for the activity of phospholipases toward different fluid bilayer substrates, which macroscopically are considered to exist in a fluid homogeneous state but on a shorter length scale can carry a lot of local structure corresponding to lipid domains bounded by interfaces with special molecular packing characteristics. The lipid-acyl-chain pair correlation function and the degree of lipid-domain formation display a maximum at equimolar composition as shown by the lipid acyl chain pair correlation function, $g_{ff}(DC_{20}PC)$ in Fig. 6 for the $DC_{12}PC$–$DC_{20}PC$ mixture. Moreover, the pair correlation function shown in Fig. 6 reveals a compositional dependence of the degree of lipid-bilayer heterogeneity that becomes less pronounced when only small amounts of one of the lipid species is incorporated into the lipid bilayer. The variation of bilayer heterogeneity with composition is clearly visualized on the snapshots in Fig. 5 for 25, 50, and 75 mol% of $DC_{12}PC$ in the binary mixture. It certainly would be of interest to test to what extent the dynamic heterogeneity in a one-phase fluid region, which can be systematically varied by simply

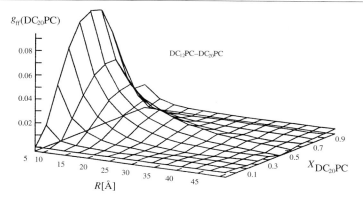

Fig. 6. Pair correlation function, $g_{ff}(DC_{20}PC)$, for the fluid–$DC_{20}PC$ lipids in the one-phase fluid region of a binary $DC_{12}PC$–$DC_{20}PC$ lipid mixture as a function of composition, $X_{DC_{20}PC}$, and acyl-chain separation, R. The data were derived from computer simulation calculations.

changing the composition of the lipid mixture, would influence the overall activity of membrane-associated enzymes like phospholipases.

In summary, bilayer substrates of lipids and lipid mixtures, even in one-phase fluid regions of their phase diagram, display a substantial degree of small-scale lateral structure. This lateral structure, which can be described as dynamic lipid domains, is caused by density and compositional fluctuations and is a basic consequence of the fact that the bilayer substrate is a many-particle system.

Bilayer Curvature and Morphologic Instability to Non-Bilayer-Forming Lipids

A lipid bilayer has lamellar symmetry and would adopt a planar configuration if it were not for boundary conditions or spontaneous curvature caused by possible differences in the chemical composition of its two monolayer leaflets. However, being composed of two juxtaposed monolayers that may each have an intrinsic spontaneous curvature, the lipid bilayer may furthermore suffer from a distributed curvature stress field.[22] This curvature stress field implies that the bilayer has a certain degree of frustration. This frustration may be relieved and new nonlamellar lipid phases of different symmetries can form,[22] such as hexagonal and cubic phases, when thermodynamic conditions change or if certain lipids that promote nonla-

---

[22] M. W. Tate, E. F. Eikenberry, D. C. Turner, E. Shyamsunder, and S. M. Gruner, *Chem. Phys. Lipids* **57**, 147 (1991).

mellar packing symmetry are incorporated into the bilayer. Even if the frustration is not released globally and the overall bilayer symmetry is preserved, the built-in curvature stress and the propensity for forming nonbilayer phases may change fusiogenity of the bilayer membrane and its susceptibility to interact locally with proteins in a way that depends specifically on the local curvature stress. It is well known that biological membranes contain a substantial fraction of lipids that support nonbilayer phases, and it has been suggested that these lipids are required for biological viability of the cell.[2,4,23] In particular, the propensity for forming inverse hexagonal H_{II} lipid phases[23] has been discussed as important, for example, to lipid/protein interactions and cell growth.[2,4]

The tendency of specific amphiphiles to promote nonlamellar phases can to a large extent be understood in terms of simple geometric packing considerations.[24] From such considerations one is to expect that the products from the PLA_2 catalyzed hydrolysis, lysolipids, and fatty acids, when incorporated into bilayer-forming diacyl phospholipids, induce a finite curvature stress field leading to considerable frustration: the lysolipids have micellar M_I propensity and the fatty acids support the formation of a inverse hexagonal H_{II} phase. Thus there are two conflicting demands that on the average appear to cancel. However, local fluctuations in composition may lead to a significant propensity for nonbilayer symmetry. We return to this in connection with PLA_2 activity in Section D.

C. Hydrolysis of Lipids by PLA_2

Phospholipase A_2 catalyzes the hydrolysis of the ester linkage of the sn-2 acyl chain of phospholipids, yielding fatty acid and 1-acyl-lysophospholipid.[25] Although small molecular weight, water-soluble PLA_2 enzymes are capable of catalyzing the hydrolysis of monomeric short-chain substrates, they are much more efficient in the catalysis of aggregated substrates. For example, using diheptanoylphosphatidylcholine as the substrate, it was found that the activity of the porcine pancreatic enzyme increased by more than one order of magnitude when the substrate was increased above its critical micelle concentration.[26] This enhancement of catalytic activity in the presence of an extensive water/lipid interface led to the concept of interfacial activation.[27] Using mixed micelles containing nonhydrolyzable

[23] R. M. Epand, *Chem.–Biol. Interact.* **63**, 239 (1991).
[24] J. Israelachvili, "Intermolecular and Surface Forces," 2nd ed., Academic Press, New York, 1992.
[25] D. L. Scott and P. B. Sigler, *Adv. Prot. Chem.* **45**, 53 (1994).
[26] W. A. Peterson, J. C. Vidal, J. J. Voliverk, and G. H. de Haas, *Biochemistry* **13**, 1455 (1974).
[27] R. Verger, *Methods Enzymol.* **64**, 340 (1980).

detergents and phospholipid substrates, Dennis and coworkers[28] demonstrated that the activity of PLA_2 from cobra venom was related to the surface concentration and not the bulk concentration of substrate. The term *surface dilution* was coined to describe this effect. A model for the activation that involved either two distinct binding sites (the activation site and the catalytic site) on the same molecule or the existence of a dimer as the active species has been suggested.[29] PLA_2 may exhibit spontaneously high activity toward small unilammellar vesicles of phosphatidylcholine[30] and negatively charged vesicles in the gel state.[31] The hydrolysis time courses for these systems are approximately hyperbolic. Attempts have been made to analyze the kinetics in terms of Michaelis–Menten type kinetics. Jain and coworkers[5] have argued that a mechanism involving binding, feedback inhibition of the reaction products, and the existence of the enzyme in a monomeric form is sufficient to describe the data, but it is not clear if the proposed mechanism is either unique or general.

An important observation regarding PLA_2 activation was the appearance of a lag period in activity[32,33] as shown in Fig. 7A. This lag period, defined as the time, τ, required after the addition of enzyme to the onset of rapid hydrolysis, is evident in phosphatidylcholine large vesicles, and its length depends on substrate concentration, enzyme concentration, and calcium concentration. In addition, the lag time is strongly influenced by the physical properties of the lipid-bilayer substrate as determined by temperature, type of lipid-bilayer material, and composition of the mixed bilayer substrate (see Burack and Biltonen[34] and Sections D and E). The modulation of this lag period by the physical nature of the aggregated substrate may provide clues to the mechanism of PLA_2 activation.

The lag period can readily be monitored using a setup with pH-stat titration, fluorescence spectroscopy, and light scattering (Fig. 7A), as described in details in a previous volume of *Methods in Enzymology*.[35] Although the experimental results described in this chapter apply primarily to the PLA_2 from *A. piscovorous piscovorous*, they appear to be general for water-soluble PLA_2 enzymes. These results are used in conjunction

---

[28] R. A. Deems, B. R. Eaton, and E. A. Dennis, *J. Biol. Chem.* **250,** 9013 (1975).
[29] M. F. Roberts, R. A. Deems, and E. A. Dennis, *Proc. Natl. Acad. Sci. U.S.A.* **74,** 1950 (1977).
[30] M. Menashe, G. Romero, R. L. Biltonen, and D. Lichtenberg, *J. Biol. Chem.* **261,** 5328 (1986).
[31] M. K. Jain, J. Rogers, D. V. Jahagerdar, J. F. Marecek, and F. L. Ramirez, *Biochim. Biophys. Acta* **860,** 435 (1986).
[32] R. Apitz-Castro, M. K. Jain, and G. H. De Haas, *Biochim. Biophys. Acta* **688,** 349 (1982).
[33] G. Romero, K. Thompson, and R. L. Biltonen, *J. Biol. Chem.* **262,** 13476 (1987).
[34] W. R. Burack and R. L. Biltonen, *Chem. Phys. Lipids* **73,** 209 (1994).
[35] J. D. Bell and R. L. Biltonen, *Methods Enzymol.* **197,** 249 (1991).

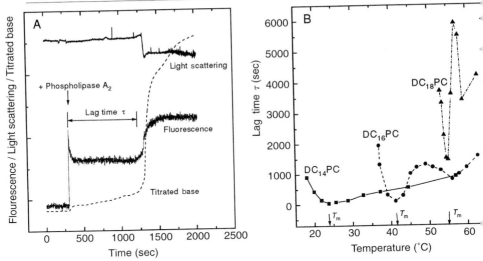

Fig. 7. (A) Characteristic reaction time course for PLA$_2$ catalyzed hydrolysis of initially large unilamellar one-component diacyl phosphatidylcholine vesicles. The curves represent a typical hydrolysis reaction time course obtained for DC$_{16}$PC unilamellar vesicles at 39°. The PLA$_2$ hydrolysis reaction is monitored by pH-stat titration, intrinsic fluorescence from PLA$_2$ emitted at 340 nm on excitation at 285 nm, and 90° light scattering from the suspension at 285 nm. All data are given in arbitrary units. The lag time, τ, is defined as the onset in the increase in the intrinsic PLA$_2$ fluorescence intensity. (B) Lag time, τ, as a function of the reaction temperature for the hydrolysis of unilamellar vesicles of either DC$_{14}$PC, DC$_{16}$PC, or DC$_{18}$PC. PLA$_2$ was purified from venom of *Agkistrodon piscivorus piscivorus*.

with theoretical model studies to describe approaches to answer possible implications of bilayer structure in the modulation of PLA$_2$ activity.

D. Dependence of PLA$_2$ Activity on Lipid-Bilayer Organization

Dependence on Substrate/Product Composition

The origin of the lag phase in the PLA$_2$ hydrolysis time course for vesicular substrates has been argued to be due to accumulation of a so-called "critical mole fraction of hydrolysis products" in the bilayer.[32] The pH-stat titration of the fatty acid production in combination with fluorescence energy transfer experiments has been used to demonstrate that products, specifically fatty acid molecules, segregate in the plane of the bilayer at the end of the lag period.[20,36] PLA_2 has high affinity for such fatty acid segregated domains due to a favorable electrostatic interaction between the positively charged docking region of PLA_2 and the negatively charged fatty acid domain.[5,37,38] Depending on the surface potential of the vesicles, fatty acid domain formation could promote additional binding of PLA_2 to the bilayer surface and accelerate the rate of hydrolysis.[5] The presence of a lag phase is, however, not merely a result of low enzyme binding affinity to the substrate vesicles. Hydrolysis experiments with vesicles of mixtures of zwitterionic and anionic lipids, where all PLA_2 initially binds to the vesicles, indeed exhibit a lag phase.[34] It has been hypothesized that the crossover from slow to fast hydrolysis is related to the relief of "local product inhibition" in the lag phase due to the local attraction of hydrolysis products to the surface-bound enzyme.[34] As commented on later, the termination of the lag phase also results in global structural changes in the vesicle suspension, which require consideration of the impact of morphologic stability of the bilayer on the PLA_2 hydrolysis time course.

Fluorescence probe studies of the polarity of the surface region of the bilayer have been used to support lysophospholipid-induced "disruption of the bilayer surface" at the end of the lag phase.[39] Thus, both product molecules are considered to be important for the lateral changes in lipid organization. Ternary codispersions of substrate and product molecules have been used extensively in attempts to analyze the physicochemical events leading to the burst in PLA_2 activity at the end of the lag phase (see, e.g., Refs. 20, 32, and 39). However, it was pointed out some time ago[32] that the impact of the product molecules depends critically on the equilibration conditions for the lipid mixture. The polymorphism of ternary lipid mixtures of a phospholipid and its PLA_2 hydrolysis products is complex and may include various nonvesicular topologies. When the PLA_2 reaction approaches the burst region, most likely the hydrolysis proceeds with a speed far above that allowed for a compositional equilibration of the lipid morphology.[20,32] No models have so far accounted for the nonequilibrium dynamic nature of the PLA_2 reaction time course. As mentioned in Section B along with the presentation of Fig. 3, a nonequilibrium distribution of components in the plane of the bilayer may induce transient ordering processes that could influence the PLA_2 activity.[34]

[36] M. K. Jain, B.-Z. Yu, and A. Kozubek, *Biochim. Biophys. Acta* **980**, 23 (1989).
[37] D. L. Scott, A. M. Mandel, P. B. Sigler, and B. Honig, *Biophys. J.* **67**, 493 (1994).
[38] K. M. Maloney, M. Grandbois, D. W. Grainger, C. Salesse, K. A. Lewis, and M. F. Robers, *Biochim. Biophys. Acta* **1235**, 395 (1995).
[39] J. D. Bell, M. L. Baker, E. D. Bent, R. W. Ashton, D. J. B. Hemming, and L. D. Hansen, *Biochemistry* **34**, 11551 (1995).

Dependence on Microheterogeneity and Fluctuations

It has been suggested that so-called lipid-packing defects in the bilayer may promote high PLA_2 activity.[1,17,32,40–43] Recently, we proposed[44] that such packing defects may be related in a systematic way to the structural microheterogeneity of the unhydrolysed lipid bilayer as quantified from computer simulation calculations on microscopic molecular interaction models of the type described in Section B. Figure 7B shows the variation in the characteristic lag time, τ, for the typical PLA_2 reaction time course (Fig. 7A) as a function of temperature and chain length for large unilamellar vesicles of $DC_{14}PC$, $DC_{16}PC$, and $DC_{18}PC$. The three curves of $\tau(T)$ each display a minimum at the respective gel-to-fluid phase transition temperature, T_m, which is deeper and wider the shorter the lipid chain is. For $DC_{16}PC$ and $DC_{18}PC$, $\tau(T)$ has in addition a local maximum above T_m. A similar local maximum is resolvable for $DC_{14}PC$ when the reaction conditions are altered such that τ becomes longer at all temperatures. This local maximum in $\tau(T)$ is sharper and closer to T_m, the longer the lipid chain. Eventually, $\tau(T)$ increases at reaction temperatures further above T_m for all three series of data possibly due to an augmented fraction of denatured PLA_2 at high temperatures. From data covering a less extended temperature range, it has been stated that $\tau(T)$ is of parabolic-like form with a minimum at T_m,[1,17,32,40,41] reflecting the time required to accumulate the so-called critical mole fraction of products in the lipid suspension. However, the data in Fig. 8A demonstrate a highly asymmetric, lipid-chain-length-dependent behavior for $\tau(T)$. This set of data strongly indicates that the physical properties of the lipid-bilayer substrate and the degree of microheterogeneity in the gel-to-fluid transition regions play important roles for the activity of PLA_2. As visualized by the Monte Carlo computer simulation results shown in Fig. 1, the closer to the phase transition and the shorter the lipid-chain length, the stronger the microheterogeneity. The degree of heterogeneity can readily be quantified as the amount of bilayer area occupied by the interfaces between gel and fluid lipid domains. Figure 8B shows the temperature and lipid-chain-length dependence of the inverse interfacial area as obtained from simulations. The experimental data for $\tau(T)$ in Fig. 8A in combination with the simulated data for the inverse interfacial area in Fig. 8B demonstrate a strong qualitative correlation between $\tau(T)$ and the degree of dynamic bilayer heterogeneity both with respect to tempera-

---

[40] C. R. Kensil and E. A. Dennis, *J. Biol. Chem.* **254**, 5843 (1979).
[41] D. Lichtenberg, G. Romero, M. Menashe, and R. L. Biltonen, *J. Biol. Chem.* **261**, 5334 (1986).
[42] N. E. Gabriel, N. V. Agman, and M. F. Roberts, *Biochemistry* **26**, 7409 (1987).
[43] J. Y. A. Lehtonen and P. K. J. Kinnunen, *Biophys. J.* **68**, 1888 (1995).
[44] T. Hønger, K. Jørgensen, R. L. Biltonen, and O. G. Mouritsen, *Biochemistry* **35**, 9003 (1996).

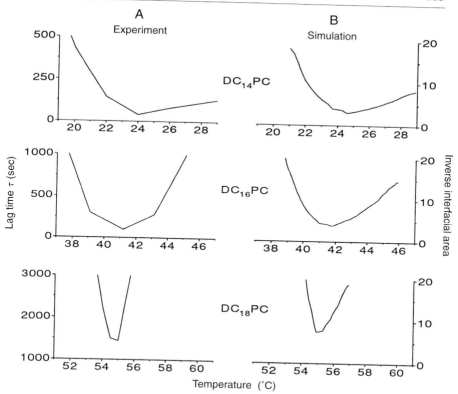

Fig. 8. (A) Enlargement of the lag time curves shown in Fig. 7B for the PLA_2 catalyzed hydrolysis of unilamellar vesicles in the characteristic temperature regions of the gel-to-fluid phase transition for the $DC_{14}PC$, $DC_{16}PC$, and $DC_{18}PC$ lipid bilayers. (B) Monte Carlo computer simulation calculations of the temperature dependence of the inverse interfacial area of lipid bilayers composed of $DC_{14}PC$, $DC_{16}PC$, and $DC_{18}PC$ lipids in the gel-to-fluid phase transition regions. The heterogeneous bilayer structure is divided into the bulk area, the domain area, and the interfacial area, the latter being determined by the interfaces between the bulk and domain regions composed of dynamic coexisting gel and fluid phases as shown in Fig. 1.

ture and lipid-chain-length variations. This correlation is valid for the width, the depth, and the asymmetry of the $\tau(T)$ function in the transition region.

The presence of a local maximum in $\tau(T)$ above T_m (Fig. 7B) is quite interesting and may, in fact, provide a new lead to the interpretation of the PLA_2 hydrolysis reaction of vesicular substrates. It is possible that the general behavior of $\tau(T)$, including this local maximum, could be rationalized effectively in terms of the temperature- and chain-length dependence of a mesoscopic or macroscopic physical property of the bilayer. The bilayer bending rigidity, κ, is a candidate for such a property. Recent work shows that κ displays a maximum above T_m,[45] and κ has been found to become anomalously low in the transition region due to a coupling between bilayer curvature and density fluctuations.[46]

The correlation between PLA_2 activity and dynamic bilayer microheterogeneity corroborates to the general hypothesis that the function of membrane-associated proteins can be modulated by collective physical properties of the lipid bilayer, specifically the dynamic bilayer heterogeneity in the nanometer range. As pointed out in Section B, it would be interesting to examine the response in PLA_2 activity to the nature of the bilayer heterogeneity by exploring other regions of the phase space for lipid bilayers, for example, close to phase boundaries and in relation to nonbilayer phases or as a function of the concentration of membrane-active drugs and detergents. We elaborate on this last issue in Section E.

Dependence on Bilayer Curvature and Morphology

The influence of the large-scale structure of the aggregated lipid substrate, for example, curvature and polymorphism, on the activity of phospholipases has been an ongoing subject of interest.[47] In the case of PLA_2, the most intensively examined lipid model systems are monolayers[38] micelles, and vesicles of different curvature.[5,34]

It has been found that the PLA_2 activity may be raised by supplying the phospholipid substrate in the form of bilayers with increasing curvature.[41] This observation was attributed to an increasing number of "lipid packing defects" in high-curvature vesicles, which could ease the assessibility of PLA_2 to the substrate molecules. Similar arguments have been used to explain the sensitivity of the PLA_2 activity to osmotic stressing of vesicles.[43] The nature of such mechanically induced bilayer defects is only partially understood.[48] Combined experimental and theoretical studies of supported bilayers of varying curvature suggest that curvature-induced stress is an important parameter in controlling the effective lateral tension of the lipid bilayer and the corresponding phase behavior of the system.[49]

---

[45] L. Fernandez-Puente, I. Bivas, M. D. Mitov, and P. Méléard, *Europhys. Lett.* **28**, 181 (1994).
[46] T. Hønger, K. Mortensen, J. H. Ipsen, J. Lemmich, R. Bauer, and O. G. Mouritsen, *Phys. Rev. Lett.* **72**, 3911 (1994).
[47] M. Waite, in "Biochemistry of Lipids, Lipoproteins and Membranes" (D. E. Vance and J. Vance, eds.), p. 269, Elsevier, New York, 1991.
[48] B. L.-S. Mui, P. R. Cullis, E. A. Evans, and T. D. Madden, *Biophys. J.* **64**, 443 (1993).
[49] T. Brumm, K. Jørgensen, O. G. Mouritsen, and T. M. Bayerl, *Biophys. J.* **70**, 1373 (1996).

Only little attention has been paid to the dynamic interplay between PLA$_2$ activity and lipid morphology.[50–52] As the composition of the lipid suspension is changing, several global (thermodynamic) and local energetic effects come into play. The elastic free energy of the individual vesicles will be altered due to the PLA$_2$-induced changes in composition. First, the spontaneous curvature of the lipid bilayer will change as a mere consequence of creating a compositional asymmetry between the two leaflets of the bilayer (see, e.g., Ref. 53). Second, both the lysophospholipid and the fatty acid products have different "intrinsic packing properties" as compared to the diacyl phospholipid substrate molecule[24] (see Section B). The lysophospholipid molecule will favor positive curvature of the lipid aggregate, whereas the fatty acid molecule has the opposite propensity. Changing the lipid composition of a vesicle due to PLA$_2$ activity may consequently promote a coupling between the local lateral composition and the local curvature of the lipid bilayer. This phenomenon has been termed *curvature instability*[54] and was proposed to explain the destabilization of the vesicle structure observed in the burst region of the PLA$_2$ reaction time course.[52] Upon PLA$_2$ hydrolysis of the external side of lipid vesicles, the bilayer may also relax tension by a trans-bilayer motion of material. Lipids with weak acid–base properties may easily translocate over the bilayer and this has indeed been observed for the fatty acid component during PLA$_2$ hydrolysis of vesicles.[55] These effects have been considered in the interpretation of the PLA$_2$ reaction time course data[51,52] (and in Hønger et al., in preparation).

Globally, the hydrolysis of vesicular substrates may induce phase transitions in the lipid suspension. Only a small region of the phase diagrams of aqueous codispersions of a phospholipid and its PLA$_2$ hydrolysis products has been systematically studied.[56] Both lamellar and isotropic phases have been detected. Our results on the mesomorphism of these systems show that drastic changes in the lipid state of aggregation may occur when the fraction of hydrolysis products exceeds what corresponds to approximately 10% hydrolysis of the substrate. In general, our data imply a very rich mesomorphism of these ternary lipid mixtures (Hønger et al., unpublished).

[50] W. R. Burack, A. Dibble, M. Allietta, T. Hønger, and R. L. Biltonen, *Biophys. J.* **66,** A58 (1994).
[51] T. Hønger, Ph.D. thesis, The Technical University of Denmark (1994).
[52] T. Hønger, A. Dibble, W. R. Burack, M. Allietta, and R. L. Biltonen, *Biophys. J.* **68,** A464 (1995).
[53] E. Sackmann, *Can. J. Phys.* **68,** 999 (1990).
[54] S. Leibler, *J. Physique* **47,** 507 (1986).
[55] F. Kamp and J. A. Hamilton, *Biochemistry* **32,** 11074 (1993).
[56] I. Brentel, G. Arvidson, and G. Lindblom, *Biochim. Biophys. Acta* **904,** 401 (1987).

Hence, the observations of a complex phase behavior of the substrate/product lipid suspensions together with the accumulating knowledge about the nonequilibrium character of the PLA_2 hydrolysis time course call for some care in the interpretation and modeling of the PLA_2 hydrolysis time course.

E. Modulation of PLA_2 Activity Via Changes of the Substrate by Membrane-Soluble Compounds

PLA_2 activity toward aggregated substrates can be modulated by the lipid composition of the bilayer substrate as well as by membrane-interacting drugs.[5,34] In this section we attempt to relate the effects of cholesterol, cholate, alcohols, and anesthetics on PLA_2 activity to their effects on the structure of the aggregated lipid-bilayer substrate. In an early work, van Deenen and coworkers[57] demonstrated that cholesterol reduced the activity of porcine pancreatic PLA_2 toward $DC_{16}PC$ bilayers in the gel–fluid transition region, until no activity was observed above 35 mol%. Gheriani-Gruszka et al.[58] investigated the influence of the bile salt cholate on PLA_2-catalyzed hydrolysis of egg lecithin unilamellar vesicles at 38°. If the vesicles contain sufficient cholate, a burst in activity is observed within a relatively short time. These experiments suggested that the lag phase is inversely related to the cholate concentration in the vesicles. In addition, the amount of hydrolysis at the burst decreased and the maximum rate of hydrolysis at the burst increased with cholate concentration in the vesicles. The authors[41] interpreted these results in terms of cholate-enhancing fluctuations, or dynamic heterogeneity, in the bilayer. Jain and Cordes[59] investigated the effects of several alcohols on the rate of bee venom PLA_2 catalyzed hydrolysis of egg lecithin vesicles. They observed a rather complex dependence of activity on alcohol concentration. The maximum PLA_2 activity increased with increasing temperature. However, for vesicles of $DC_{16}PC$ a maximum in the alcohol-induced activity was observed near the gel-to-fluid phase transition temperature.[60] These types of effects are likely to be the result of alcohol/bilayer interactions because it was recently shown that octanol has no effect on PLA_2 activity toward monomolecular substrates and does not affect its binding to gel state $DC_{16}PC$ vesicles.[61] A detailed study of the temperature dependence of the effect of octanol on the duration of

---

[57] J. A. Op den Kamp, M. T. Kauerz, and L. L. van Deenen, *Biochim. Biophys. Acta* **406**, 169 (1975).
[58] N. Gheriani-Gruszka, S. Almog, R. L. Biltonen, and D. Lichtenberg, *J. Biol. Chem.* **263**, 11808 (1988).
[59] M. K. Jain and E. H. Cordes, *J. Membr. Biol.* **14**, 101 (1973).
[60] M. K. Jain and E. H. Cordes, *J. Membr. Biol.* **14**, 119 (1973).
[61] R. L. Biltonen, D. Stokes, and B. Lathrop, *Prog. Anes. Mech.* **3**, 162 (1995).

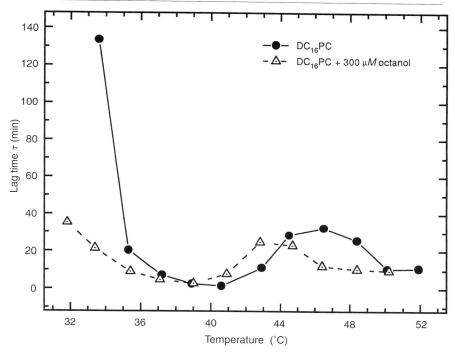

FIG. 9. Lag time measurements of PLA$_2$ catalyzed hydrolysis of 0.26 mM unilamellar DC$_{16}$PC vesicles in the absence and presence of 300 μM octanol. The PLA$_2$ concentration was 232 nM (D. Stokes, Ph.D. thesis, in preparation).

the lag phase as shown in Fig. 9 (D. Stokes, Ph.D. thesis, in preparation) revealed that even in the presence of the alcohol, the temperature dependence in the vicinity of the transition temperature is quite similar to that shown in Fig. 7B, except that it was shifted to lower temperature in a dose-dependent manner.[61] The shift in temperature corresponding to the minimum in τ is, however, greater than the octanol-induced shift in the gel-to-fluid transition temperature of the unhydrolyzed bilayers. Further studies in the fluid phase at 47° indicated that chloroform and several alcohols including octanol reduced the lag phase in a monotonic fashion correlating with their water/lipid partition coefficient. These results suggest that alcohols may affect the dynamic bilayer heterogeneity in the vicinity of the phase transition.

The potential of anesthetics to modulate lipid-bilayer microheterogeneity is illustrated in the snapshots of Fig. 10, which were obtained from computer simulation calculations on a generic molecular model developed

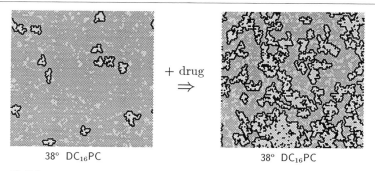

Fig. 10. Effects on the dynamic lipid-bilayer microheterogeneity by a water-soluble membrane-active drug like octanol. The results are obtained by computer simulations on a system corresponding to 5000 lipid molecules. The symbols denoting the different bilayer regions are explained in Fig. 1.

to describe the interaction of water-soluble drugs with lipid bilayers.[62] The drug molecule has a high affinity to kink-like acyl-chain conformations that have a high occurrence at the boundary between the gel and fluid regions. This leads to an enhancement of the dynamic microheterogeneity of the lipid bilayer and a much more ramified domain structure as seen in Fig. 10. The interfacial lipid-packing defects, which in Section D were suggested to promote high PLA_2 activity, might at the same time be altered due to a tendency of the drug molecules to accumulate in the boundary regions. In fact, the anesthetic molecules may have the capacity of both enhancing the dynamic microstructure of the lipid bilayer and making the interfacial lipids less susceptible to PLA_2 catalyzed hydrolysis. A related study of the effect of the insecticide lindane on the $DC_{16}PC$ lipid-bilayer microheterogeneity and the associated anomalously high permeability properties in the transition region has revealed a sealing effect, although lindane at the same time increased the dynamic microheterogeneity.[63] This effect could be rationalized in terms of the ability of lindane to change the interfacial molecular packing characteristics resulting in a concomitant decreased transmission through such leaky interfaces.

Alcohols may also affect the duration of the lag phase by altering the fraction of hydrolysis at the burst or through a modulation of the bilayer bending rigidity (Section D). It is noteworthy that Gruner and Shyamsunder[64] have suggested that changes in curvature stress may be the basis of anesthetic action. In studies such as those just described, it is necessary

---

[62] K. Jørgensen, J. H. Ipsen, M. J. Zuckermann, and O. G. Mouritsen. *Chem. Phys. Lipids* **65**, 205 (1992).
[63] M. Sabra, K. Jørgensen, and O. G. Mouritsen, *Biochim. Biophys. Acta* **1282**, 85 (1996).
[64] S. M. Gruner and E. Shyamsunder, *Ann. N.Y. Acad. Sci.* **625**, 685 (1991).

to understand the effect of the added constituents on lipid structure and their transition characteristics. For example, high concentrations of alcohols can promote interdigitation of lipids in the opposing monolayers in the gel state.[65] Thus, care must be taken to evaluate these effects. To develop a better understanding of alcohol-induced alteration in PLA_2 catalysis it is necessary first to understand the effects of alcohols on lipid structure and phase behavior. This can rationally be done by a combination of theoretical and experimental approaches.

Obviously, the situation as to which way membrane-soluble compounds affect PLA_2 activity via modulation of the lipid-bilayer substrate is at present fairly unclear, although there are some striking observations reported that indicate that the physics of the substrate is an important determinant. It is also unclear whether there are any universal effects in play. The situation may be further complicated by the possibility that membrane solutes may affect the activity qualitatively differently at different solute concentrations. Cholesterol is a particularly interesting compound in this respect since it has been predicted both by theory and experiment, on the basis of the lecithin-cholesterol phase diagram,[66-68] that whereas cholesterol in amounts above about 5 mol% suppresses the bilayer fluctuations, cholesterol in small amounts has the opposite effect, leading to enhanced dynamic heterogeneity. It would be interesting to investigate if there is a nonmonotonous dependence of cholesterol concentration on PLA_2 activity, that is, whether small amounts of cholesterol enhance the PLA_2 activity whereas high concentrations have the opposite effect.

F. Outlook

We have in this chapter proposed a methodological and generic study of phospholipase activity and its dependence on the physical properties of the lipid-bilayer substrate using a combined experimental and theoretical approach. The main result is that it is possible to correlate the activity of PLA_2 with the physical properties, in particular, the microheterogeneity, of the bilayer substrate in the absence of the enzyme. The theoretical calculations are based on computer simulation, which reveals both the global thermodynamic bilayer behavior as well as the lateral microscopic

---

[65] E. S. Rowe and J. M. Campion, *Biophys. J.* **67,** 1888 (1994).
[66] L. Cruzeiro-Hansson, J. H. Ipsen, and O. G. Mouritsen, *Biochim. Biophys. Acta* **979,** 166 (1989).
[67] E. Corvera, O. G. Mouritsen, M. A. Singer, and M. J. Zuckermann, *Biochim. Biophys. Acta* **1107,** 261 (1992).
[68] J. Lemmich, T. Hønger, K. Mortensen, J. H. Ipsen, R. Bauer, and O. G. Mouritsen, *Eur. Biophys. J.* **25,** 61 (1996).

organization of the lipid bilayer in the planar configuration. In its present state, such model calculations can provide guidelines and suggest rational ways for future experimental studies on relevant bilayer substrate properties that are considered candidates for modulation of the activity of membrane-associated enzymes. Future development of this type of modeling would have to come to terms with the deviations from planar symmetry as well as being able to account for the propensity of the hydrolysis products for forming nonbilayer phases.

On the experimental side, future extensions along the lines suggested in this chapter would involve a careful and quantitative determination of the actual composition of the lipid substrate during the time course of hydrolysis in the lag phase as well as a characterization of the morphologic structure of the substrate subsequent to the burst in activity.

Finally, the combined theoretical and experimental methodological approach to phospholipase activity described here would meet its ultimate test in studies of the effect of various membrane solutes on lipase activity as being controlled by specific changes in the physical properties of the substrate.

Acknowledgments

Work in the authors' laboratories is supported by The Danish Natural Science Research Council, The Danish Technical Research Council, the Carlsberg Foundation, NIH, and NSF. Dr. Mouritsen is a Fellow of the Canadian Institute for Advanced Research.

[10] Covalent Inactivation of Lipases

By STÉPHANE RANSAC, YOUSSEF GARGOURI, FRANK MARGUET, GÉRARD BUONO, CHRISTOPH BEGLINGER, PIUS HILDEBRAND, HANS LENGSFELD, PAUL HADVÁRY, and ROBERT VERGER

1. Introduction

In higher animals, the intestinal absorption of dietary triacylglycerols requires, first, their enzymatic conversion into the more polar fatty acids and monoacylglycerols by digestive lipolytic enzymes. Lipolysis is catalyzed by preduodenal and pancreatic lipases. In humans, the hydrolysis of alimentary triacylglycerols begins in the stomach and is catalyzed by human gastric lipase (HGL),[1–3] which is able to hydrolyze short- and long-chain triacyl-

[1] Y. Gargouri, H. Moreau, and R. Verger, *Biochim. Biophys. Acta* **1006**, 255 (1989).

glycerols at comparable rates. Under acidic pH conditions, HGL has been shown to be remarkably stable and active, whereas pancreatic lipase irreversibly loses its lipolytic capacity. The optimum pH for HGL activity is around 5.4, which is close to the pH of the gastric content during a test meal,[4] compared to 8 to 9 in the case of pancreatic lipase. The partial hydrolysis of alimentary triacylglycerols that occurs at the pH levels prevailing in the stomach rapidly triggers pancreatic lipase activity in the intestine.[5,6] Conventional treatments for obesity have focused largely on strategies to control energy intake, however, the long-term efficacy of such approaches is limited.[7] A reduction of dietary fat adsorption by an inhibitor of digestive lipases holds great promise as an antiobesity agent.

Bodmer et al.[8] have cloned, sequenced, and expressed active HGL in yeast. The amino acid sequence obtained from the cDNA consists of 379 residues as compared to 449 residues constituting porcine pancreatic lipase (PPL). No amino acid sequence homology exists between HGL and PPL, except for short regions of six residues containing serine in a position analogous to that of the essential serine 152. The cDNA sequence of HGL shows the presence of three cysteine residues in comparison to PPL, which contains 14 cysteine residues (6 disulfide bridges and 2 free sulfydryl groups).

1.1 Structure of Human Pancreatic Lipase

The determination of the 3D structure of human pancreatic lipase (HPL) by Winkler et al.[9] confirmed the existence of two distinct domains in pancreatic lipase: a larger N-terminal domain comprising residues 1–336 and a smaller C-terminal domain made up of residues 337–449. The high degree of amino acid sequence homology observed within the lipase gene family

---

[2] F. Carrière, Y. Gargouri, H. Moreau, S. Ransac, E. Rogalska, and R. Verger, in "Lipases: Their Structure, Biochemistry and Application" (P. Wooley and S. B. Petersen, eds.), p. 181. Cambridge University Press, Cambridge, England, 1994.
[3] M. Hamosh, "Lingual and Gastric Lipases: Their Role in Fat Digestion." CRC Press, Boca Raton, Florida, 1990.
[4] F. Carrière, J. A. Barrowman, R. Verger, and R. Laugier, *Gastroenterology* **105,** 876 (1993).
[5] Y. Gargouri, G. Piéroni, C. Rivière, P. A. Lowe, J.-F. Saunière, L. Sarda, and R. Verger, *Biochim. Biophys. Acta* **879,** 419 (1986).
[6] S. Bernbäck, L. Bläckberg, and O. Hernell, *Biochim. Biophys. Acta* **1001,** 286 (1989).
[7] W. Bennet, *Ann. N.Y. Acad. Sci.* **449,** 250 (1987).
[8] M. W. Bodmer, S. Angal, G. T. Yarranton, T. J. R. Harris, A. Lyons, D. J. King, G. Piéroni, C. Rivière, R. Verger, and P. A. Lowe, *Biochim. Biophys. Acta* **909,** 237 (1987).
[9] F. K. Winkler, A. d'Arcy, and W. Hunziker, *Nature* **343,** 771 (1990).

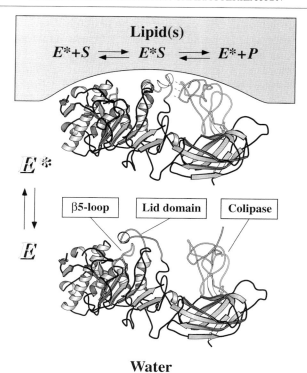

Fig. 1. Structure of human pancreatic lipase and conformational change induced by the adsorption to a lipid interface. (From Carrière et al.,[116] adapted from van Tilbeurgh et al.[25,27])

supports the view that this particular architecture is also common to lipoprotein lipase and to hepatic lipase.[10–14]

As illustrated in Fig. 1, and discussed in a review by Winkler and Gubernator,[15] the large N-terminal domain is a typical α/β hydrolase fold

[10] T. G. Kirchgessner, J. C. Chuat, C. Heinzmann, J. Etienne, S. Guilhot, K. Svenson, D. Ameis, C. Pilon, L. d'Auriol, A. Andalibi, M. C. Schotz, F. Galibert, and A. J. Lusis, *Proc. Natl. Acad. Sci. U.S.A.* **86**, 9647 (1989).
[11] B. Persson, G. Bentsson-Olivecrona, S. Bernbäck, T. Olivecrona, and H. Jörnvall, *Eur. J. Biochem.* **179**, 39 (1989).
[12] Z. S. Derewenda and C. Cambillau, *J. Biol. Chem.* **266**, 23112 (1991).
[13] W. A. Hide, L. Chan, and W. H. Li, *J. Lipid Res.* **33**, 167 (1992).
[14] H. van Tilbeurgh, A. Roussel, J. M. Lalouel, and C. Cambillau, *J. Biol. Chem.* **269**, 4626 (1994).
[15] F. K. Winkler and K. Gubernator, in "Lipases. Their Structures, Biochemistry and Application" (P. Wooley and S. B. Petersen, eds.), p. 139. Cambridge University Press, Cambridge, England, 1994.

dominated by a central parallel β-sheet.[16] It contains the active site with a catalytic triad formed by serine 152, aspartate 176, histidine 263, all of which are conserved in lipoprotein lipase and hepatic lipase. This catalytic triad is chemically analogous to that originally described in serine proteases such as chymotrypsin[17] but is structurally distinct. The nucleophilic elbow, β-strand/εSer/α-helix, including the Gly-X-Ser-X-Gly consensus sequence, has been detected mostly in lipases and esterases.[18-21] *Bacillus* and *Candida antartica* B lipases have a variation to Ala-X-Ser-X-Gly and Thr-X-Ser-X-Gly, respectively. This consensus sequence is, however, homologous to that found in other enzymes such as haloalkane dehalogenase, bromoperoxidase, and acyltransferase. The structure of HPL clearly demonstrated that serine 152 is the nucleophilic residue essential for catalysis, in agreement with the chemical modification of serine 152 in porcine pancreatic lipase,[22] and in contradiction with results suggesting a function of serine 152 in interfacial recognition.[23] The hydrolytic mechanism is directed by a nucleophilic attack by the serine 152, and the tetrahedral intermediate formed is stabilized by an oxyanion hole. An acyl enzyme is then formed, and the alcohol moiety leaves the active site. The action of a water molecule permits the regeneration of the enzyme and the liberation of the acyl moiety.

In the structure resolved by Winkler *et al.*,[9] the active site is covered by a surface loop between the disulphide-bridge cysteine 237 and 261. This surface loop includes a short one-turn α-helix with a tryptophan residue (Trp 252) completely buried and sitting directly on top of the active site serine 152. Under this closed conformation, found mainly in water solution, the "lid" prevents the dispersed substrate from having access to the active site. Spectroscopic studies of tryptophan fluorescence have shown that large spectral changes are induced by acylation of pancreatic lipase with the

---

[16] D. L. Ollis, E. Cheah, M. Cygler, B. Dijkstra, F. Frolow, S. M. Franken, M. Harel, S. J. Remington, I. Silman, J. Schrag, J. L. Sussman, K. H. G. Verschueren, and A. Goldman, *Protein Eng.* **5,** 197 (1992).
[17] D. Blow, *Nature* **351,** 444 (1991).
[18] Z. S. Derewenda and U. Derewenda, *Biochem. Cell Biol.* **69,** 842 (1991).
[19] Z. S. Derewenda, U. Derewenda, and G. G. Dodson, *J. Mol. Biol.* **227,** 818 (1992).
[20] H. W. Anthonsen, A. Baptista, F. Drablos, P. Martel, S. B. Petersen, M. Sebastiao, and L. Vaz, in "Biotechnology Annual Review" (M. R. Elgewely, ed.), Vol. 1, p. 315. Elsevier, Amsterdam, 1995.
[21] K.-E. Jaeger, S. Ransac, B. W. Dijkstra, C. Colson, M. Vanheuvel, and O. Misset, *FEMS Microbiol. Rev.* **15,** 29 (1994).
[22] A. Guidoni, F. Benkouka, J. de Caro, and M. Rovery, *Biochim. Biophys. Acta* **660,** 148 (1981).
[23] C. Chapus, M. Sémériva, C. Bovier-Lapierre, and P. Desnuelle, *Biochemistry* **15,** 4980 (1976).

inhibitor tetrahydrolipstatin (THL) in the presence of bile salt micelles.[24] By crystallizing the pancreatic lipase–procolipase complex in the presence of mixed lipid micelles, it was shown that the "lid" was shifted to one side, exposing both the active site and a larger hydrophobic surface.[25] This motion is induced when the binding to the lipid occurs and is probably the structural basis for "interfacial activation" of pancreatic lipase. For more information, the "interfacial activation" phenomenon is described and discussed in Chapter [16] of this volume.[26]

The β-sandwich C-terminal domain of pancreatic lipase is necessary for colipase binding to occur, as shown in the 3D structure of the HPL–porcine procolipase complex.[27] Procolipase is a "three-finger" protein that is topologically comparable to snake toxins, known to bind phospholipids, even though these proteins do not share any sequence homology. In the 3D structure of the pancreatic lipase–procolipase complex, the N-terminal pentapeptide (activation peptide or "enterostatin") was not visible in the electronic density map.[27] This is why we consider that pro and colipase have basically identical 3D structures. Colipase lacks any well-defined secondary structural elements. This small protein seems to be stabilized mainly by an extended network of five disulfide bridges that runs throughout the flatly shaped molecule, reticulating its finger-like loops. The colipase surface can be divided into a rather hydrophilic part, interacting with lipase, and a more hydrophobic part, formed by the tips of the fingers, which are very mobile and constitute the lipid interaction surface. The interaction between colipase and the C-terminal domain of lipase is stabilized by eight hydrogen bonds and about 80 van der Waals contacts. Upon the opening of the lid three more hydrogen bonds and about 28 van der Waals contacts are added, explaining in part the higher apparent affinity in the presence of a lipid/water interface. The fact that colipase and lipase are colocalized at the interface may contribute to a higher affinity constant. In the absence of an interface, no conformational change in the lipase molecule is induced by the binding of procolipase. The structure of the open form of the HPL–procolipase complex in the presence of mixed lipid micelles revealed, however, that the "lid" binds to the procolipase N-terminal domain when the complex is activated at an interface.[25] The open structure of the lipase–procolipase complex illustrates how colipase might anchor the lipase at the interface in the presence of bile salts: colipase binds to the noncatalytic β-

---

[24] Q. Lüthi-Peng and F. K. Winkler, *Eur. J. Biochem.* **205,** 383 (1992).
[25] H. van Tilbeurgh, M.-P. Egloff, C. Martinez, N. Rugani, R. Verger, and C. Cambillau, *Nature* **362,** 814 (1993).
[26] F. Ferrato, F. Carrière, L. Sarda, and R. Verger, *Methods Enzymol.* **286,** [16], 1997 (this volume).
[27] H. van Tilbeurgh, L. Sarda, R. Verger, and C. Cambillau, *Nature* **359,** 159 (1992).

sheet of the C-terminal domain of HPL and exposes the hydrophobic tips of its fingers at the opposite side of its lipase-binding domain. This hydrophobic surface, in addition to the hydrophobic back side of the lid, helps to bring the catalytic N-terminal domain of HPL into close contact with the lipid/water interface, as shown in Fig. 1.

Horse pancreatic lipase[28] 3-D structure has been solved recently and is almost identical to that of the human. Unfortunately, the 3D structure of gastric lipase is not yet available. Furthermore, these acid lipases have no sequence homology with the microbial or the other mammalian lipases known.[2]

1.2 Methods for Lipase Inhibition[29–31]

To describe the kinetics of a lipolytic enzyme acting at an interface, a simple and versatile model has been proposed by Verger et al.[32] This model consists basically of two successive equilibria. The first describes the reversible penetration of a water-soluble enzyme into an interface ($E \rightleftarrows E^*$). This is followed by a second equilibrium in which one molecule of penetrated enzyme binds a single substrate molecule, forming the enzyme–substrate complex (E^*S). This is the two-dimensional equivalent of the classical Michaelis–Menten equilibrium. Once the complex (E^*S) has been formed, the catalytic steps take place, regenerating the enzyme in the form E^* and liberating the lipolysis products. An extension of the previous kinetic model was proposed by Ransac et al. for depicting the competitive inhibition,[33] as well as the covalent inhibition of lipolytic enzymes at a lipid/water interface (Fig. 2).[34]

Achieving specific and covalent inhibition of lipolytic enzymes is a difficult task, because of nonmutually exclusive processes such as interfacial denaturation, changes in "interfacial quality,"[35] and surface dilution phenomena.[36] Furthermore, the interfacial enzyme binding and/or the catalytic

---

[28] Y. Bourne, C. Martinez, B. Kerfelec, D. Lombardo, C. Chapus, and C. Cambillau, *J. Mol. Biol.* **238,** 709 (1994).
[29] A. Moulin, J.-D. Fourneron, G. Piéroni, and R. Verger, *Biochemistry* **28,** 6340 (1989).
[30] Y. Gargouri, H. Chahinian, H. Moreau, S. Ransac, and R. Verger, *Biochim. Biophys. Acta* **1085,** 322 (1991).
[31] C. Cudrey, H. van Tilbeurgh, Y. Gargouri, and R. Verger, *Biochemistry* **32,** 13800 (1993).
[32] R. Verger, M. C. E. Mieras, and G. H. de Haas, *J. Biol. Chem.* **248,** 4023 (1973).
[33] S. Ransac, C. Rivière, C. Gancet, R. Verger, and G. H. de Haas, *Biochim. Biophys. Acta* **1043,** 57 (1990).
[34] S. Ransac, Y. Gargouri, H. Moreau, and R. Verger, *Eur. J. Biochem.* **202,** 395 (1991).
[35] R. Verger, and G. H. de Haas, *Ann. Rev. Biophys. Bioeng.* **5,** 77 (1976).
[36] E. A. Dennis, *Drug. Dev. Res.* **10,** 205 (1987).

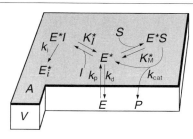

FIG. 2. Kinetic model illustrating the covalent inhibition of a lipolytic enzyme at a lipid/water interface. Symbols and abbreviations are as follows: A, Total interfacial area (surface); V, total volume (volume); E, free enzyme concentration (molecule/volume); E^*, interfacial enzyme concentration (molecule/surface); S, interfacial concentration of substrate (molecule/surface); I, interfacial concentration of inactivator (molecule/surface); P, product concentration (molecule/volume); E^*S, interfacial enzyme–substrate complex concentration (molecule/surface); E^*I, interfacial enzyme–inactivator complex concentration (molecule/surface); E_i^*, covalently inactivated enzyme concentration (molecule/surface); k_d, desorption rate constant (time^{-1}); k_p, penetration rate constant (volume · surface^{-1} · time^{-1}); k_{cat}, catalytic rate constant (time^{-1}); k_i, inhibition rate constant (time^{-1}); K_M^*, interfacial Michaelis–Menten constant (molecule/surface); K_I^*, interfacial dissociation constant for the enzyme–inactivator complex (molecule/surface). (From Ransac et al.[34])

turnover can be diversely affected by the presence of potential amphipathic inhibitors.[37,38]

In this chapter, we present and discuss results chiefly concerning the covalent inhibition of gastric and pancreatic lipases. Rather than presenting an exhaustive list of compounds tested so far with lipases of animal and microbial origin, we have selected experimental data illustrating the specific problems encountered during the covalent inhibition of digestive lipases. We do not review the noncovalent "inhibition" observed with proteins and tensioactive agents that usually do not interact specifically with lipases but generally adsorb at the lipid/water interface, affecting the "interfacial quality"[39] and hence the lipase activity.

Figure 3 depicts the chemical structures of all inhibitors we discuss: 4,4'-dithiopyridine (4-PDS), 5,5'-dithiobis(2-nitrobenzoic acid) (NbS$_2$), dodecyl-dithio-5-(2-nitrobenzoic acid) (C$_{12:0}$-S-NbS), Ajoene, diethyl p-nitrophenyl

[37] R. Verger, in "Lipases" (B. Borgström, and H. L. Brockman, eds.), p. 83. Elsevier, Amsterdam, 1984.
[38] B. I. Kurganov, L. G. Tsetlin, E. A. Malakhova, N. A. Chebotareva, V. Z. Lankin, G. D. Glebova, V. M. Berezovsky, A. V. Levashov, and K. Martinek, J. Biochem. Biophys. Methods 11, 177 (1985).
[39] G. Piéroni, Y. Gargouri, L. Sarda, and R. Verger, Adv. Colloid Interface Sci. 32, 341 (1990).

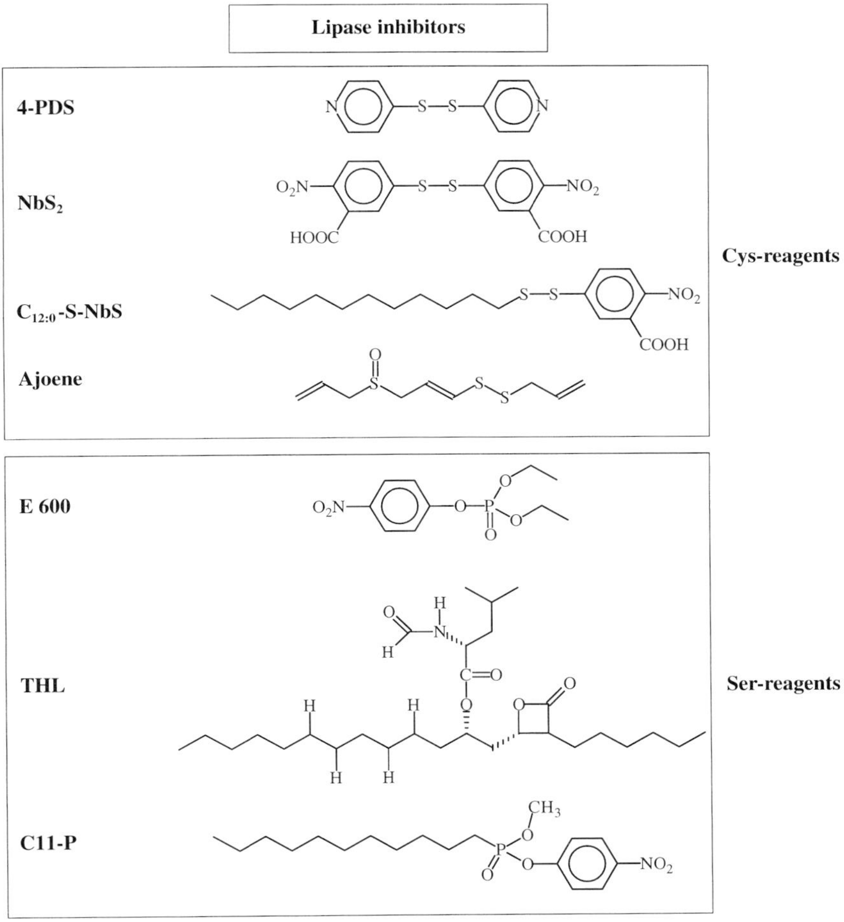

FIG. 3. Structures of various lipase inhibitors. (Adapted from Egloff et al.[74]) Cysteine reagents: 4-PDS, 4,4'-dithiopyridine; NBS$_2$, 5,5'-dithiobis(2-nitrobenzoic acid); C$_{12:0}$-S-NbS, dodecyldithio-5-(2-nitrobenzoic acid); and ajoene. Serine reagents: E$_{600}$, diethyl p-nitrophenyl phosphate; THL, tetrahydrolipstatin; C11-P, O-methyl O-(p-nitrophenyl) n-undecyl phosphonate.

phosphate (E$_{600}$), tetrahydrolipstatin (THL), O-methyl-O-(p-nitrophenyl) n-undecylphosphonate (C11-P). To study the effects of these reagents, which can in principle modify covalently several essential amino acid residues of the lipases, we have proposed four different methods,[29–31] depending on the order of addition of lipase, substrate, and inactivator (Fig. 4).

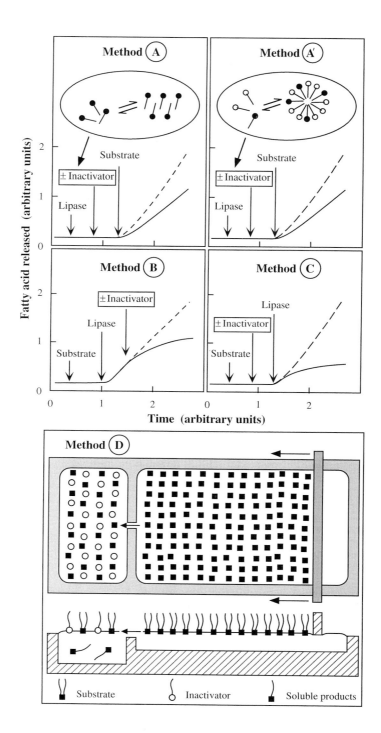

1.2.1 Method A: Lipase/Inactivator Preincubation Method (Fig. 4A). This method was set up to test, in aqueous medium and in the absence of substrate, the possible reactions between lipases and all the reagents to be tested. Residual lipase activity was measured separately using an emulsified substrate. Sometimes the inhibitor is incorporated into a solution of a tensioactive agent to create a mixed micellar interface, containing the incorporated inactivator, and described as method A' by Cudrey et al.[31]

1.2.2 Method B: Inhibition During Lipolysis (Fig. 4B). This method was designed to test whether any inhibition reaction occurred in the presence of a water-insoluble substrate during lipase hydrolysis. Lipase was injected into the reaction vessel of a pH-stat containing the emulsified substrate maintained under vigorous stirring. The inhibitor was injected a few minutes after lipase addition. Lipase activity was then continuously recorded.

1.2.3 Method C: Poisoned Interface Method (Fig. 4C). The inactivator was injected before the enzyme addition at various concentrations into a lipolysis assay containing the emulsified substrate. Lipase activity was then continuously recorded.

1.2.4 Method D: Monolayer Technique (Fig. 4D). With conventional emulsified systems, it is not possible to control the "interfacial quality" and to assess easily the distribution of soluble versus adsorbed amphiphilic molecules. This prompted us to use the monolayer technique, based on surface pressure decrease due to Lipid-film hydrolysis.[40,41] The technique is applicable to those cases in which the lipid forms a stable monomolecular film at the air/water interface and in which reaction products are freely soluble and diffuse away rapidly into the aqueous phase.

The monolayer technique was also used to study the covalent inhibition of lipases, and this approach can also be considered a poisoned interface method (Method C). Covalent inhibition of lipases is studied using mixed monomolecular films of substrate containing amphiphilic inhibitor molecules such as THL or $C_{12:0}$-S-NbS and using a zero-order trough with a reaction compartment containing a mixed substrate/inhibitor film, whereas

[40] D. G. Dervichian, *Biochimie* **53**, 25 (1971).
[41] G. Zografi, R. Verger, and G. H. de Haas, *Chem. Phys. Lipids* **7**, 185 (1971).

FIG. 4. Methods used to study the effects of inactivators on lipase activities. Three methods, A, B, and C, were used. With each method, arrows indicate the order of successive injections of inhibitor, lipases, and substrate. (From Gargouri et al.[30]) Method D: Principle of the method for studying lipase inhibition using mixed substrate/inactivator monomolecular films. (Adapted from Piéroni et al.[42,117])

the reservoir compartment was covered with a film of pure substrate, as shown in Fig. 4D.[34,42]

2. Inhibition of Lipases by Diethyl-*p*-nitrophenyl Phosphate (E_{600}) and Various Phosphonates

In previous studies, it has been shown that organophosphorus compounds such as diethyl *p*-nitrophenyl phosphate (E_{600}) are able to inactivate PPL stoichiometrically.[43-45] This chemical inhibition of pancreatic lipase occurred exclusively with mixed E_{600}/bile salt micelles and in the presence of colipase.[45] Guidoni *et al.*[22] have shown that serine 152 in PPL can be specifically labeled with E_{600}. Furthermore, it was reported by Chapus and Sémériva[46] that E_{600}-modified PPL or diethylphospholipase (DP-lipase) had a decreased interfacial binding capacity onto siliconized glass beads but was still able to hydrolyze a water-soluble substrate such as *p*-nitrophenyl acetate (*p*-NPA). These authors extrapolated their results obtained with siliconized glass beads to lipid/water interfaces and speculated that the essential serine 152 of PPL was involved in the lipid binding domain, and not in the catalytic site. This interpretation was probably due to the implicit assumption that siliconized glass beads were a good substitute and a valid model for the "interfacial quality" of triacylglycerol/water interfaces. Finally, the first 3D crystallographic structures of human pancreatic[9] as well as several fungal and bacterial[47-54] lipases have recently been solved. From

---

[42] G. Piéroni and R. Verger, *J. Biol. Chem.* **254**, 10090 (1979).

[43] P. Desnuelle, L. Sarda, and G. Ailhaud, *Biochim. Biophys. Acta* **37**, 570 (1960).

[44] M. F. Maylié, M. Charles, and P. Desnuelle, *Biochim. Biophys. Acta* **276**, 162 (1972).

[45] M. Rouard, H. Sari, S. Nurit, B. Entressangles, and P. Desnuelle, *Biochim. Biophys. Acta* **530**, 227 (1978).

[46] C. Chapus and M. Sérémiva, *Biochemistry* **15**, 4988 (1976).

[47] L. Brady, A. M. Brzozowski, Z. S. Derewenda, E. Dodson, G. Dodson, S. Tolley, J. P. Turkenburg, L. Christiansen, B. Huge-Jensen, L. Norskov, L. Thim, and U. Menge, *Nature* **343**, 767 (1990).

[48] J. D. Schrag, Y. Li, S. Wu, and M. Cygler, *Nature* **351**, 761 (1991).

[49] C. Martinez, P. de Geus, M. Lauwereys, G. Matthyssens, and C. Cambillau, *Nature* **356**, 615 (1992).

[50] P. Grochulski, Y. Li, J. D. Schrag, F. Bouthillier, P. Smith, D. Harrison, B. Rubin, and M. Cygler, *J. Biol. Chem.* **268**, 12843 (1993).

[51] M. E. M. Noble, A. Cleasby, L. N. Johnson, M. R. Egmond, and L. G. J. Frenken, *FEBS Lett.* **331**, 123 (1993).

[52] D. Lang, L. Haalck, B. Hofmann, H.-J. Hecht, F. Spener, R. D. Schmid, and D. Schomburg, *Acta Cryst.* **50**, 225 (1994).

[53] D. M. Lawson, A. M. Brzozowski, S. Rety, C. Verma, and G. G. Dodson, *Protein Eng.* **7**, 543 (1994).

[54] J. Uppenberg, M. T. Hansen, S. Patkar, and T. A. Jones, *Structure* **2**, 293 (1994).

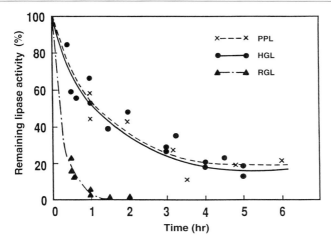

FIG. 5. Time course of inhibition of PPL, HGL, and RGL during incubation with radiolabeled E_{600} (E_{600} to lipase molar ratio of 84). Lipases (25 nmol) were incubated with 2.1 μmol of E_{600} (200 μCi) at 25° in 50 mM acetate buffer, pH 6.0, 50 mM NaCl, 25 mM $CaCl_2$, and 3 mM NaTDC. Residual lipase activity was measured on tributyrin as substrate. (From Moreau et al.[57])

these lipase structures, it is now clear that essential serine 152 of the pancreatic lipase is part of the classical Asp-His-Ser triad and constitutes the nucleophilic residue essential for catalysis. The spatial arrangement of the proposed catalytic triad in these lipases is very similar to the catalytic triad of classical serine esterases, such as trypsin. The widespread distribution of Gly-X-Ser-X-Gly pentapeptide in the primary structures of all the known lipases and the apparent contradiction between previous interpretations of the interfacial binding data and the 3D structure of pancreatic lipase have brought the question of the potential role of serine 152 very much to the forefront. We therefore reinvestigated the inhibition of PPL using radiolabeled E_{600} and extended this study to gastric lipases such as HGL and rabbit gastric lipase (RGL) purified according to published methods.[55,56]

2.1 Inhibition of PPL, HGL, and RGL by Radiolabeled E_{600}[57]

Figure 5 gives the time course of inhibition of PPL, HGL, and RGL incubated in pH 6.0 with radiolabeled E_{600}/NaTDC mixed micelles at an

---

[55] C. Tiruppathi and K. A. Balasubramanian, *Biochim. Biophys. Acta* **712**, 692 (1982).
[56] H. Moreau, Y. Gargouri, D. Lecat, J.-L. Junien, and R. Verger, *Biochim. Biophys. Acta* **960**, 286 (1988).
[57] H. Moreau, A. Moulin, Y. Gargouri, J.-P. Noël, and R. Verger, *Biochemistry* **30**, 1037 (1991).

E_{600} to lipase molar ratio of 84. PPL and HGL showed the same inhibition rates, reaching a plateau at a 20% remaining activity level after 5 hr of incubation. This partial inhibition of PPL by E_{600} was previously reported by Rouard *et al.*[45] By contrast, RGL inhibition proceeded at a faster rate, and the lipolytic activity was completely abolished after 60 min of incubation. From previous data on gastric lipase inhibition by sulphydryl reagents, RGL can be said to have shown a faster reaction rate than HGL.[58,59] This difference in reaction velocity might reflect a difference in active site accessibility.

After removing the excess micellar radiolabeled E_{600}, the labeling stoichiometry of the inactivated lipases was determined and correlated with the percentage of lipase inhibition.[57] Note the existence of a direct correlation between the inhibition levels of the three digestive lipases tested and the stoichiometric labeling obtained with [^{14}C]E_{600}. In the case of PPL, it has been clearly shown that the classical serine reagent (E_{600}) inactivates the enzyme after reacting with the essential serine 152.[22] Site-directed mutagenesis has confirmed the involvement of serine 152 in the catalytic activity.[60] Furthermore, we showed for the first time that gastric lipases were also stoichiometrically inactivated by mixed E_{600}/bile salt micelles under identical experimental conditions to those used with PPL.

Chemical modification of the single free sulfydryl group of gastric lipases, with either 4-PDS or NbS_2,[58,59] induced a complete loss of activity on both water-soluble *p*-NPA and emulsified tributyrin used as substrates.[57] Both catalytic activities were also abolished by treatment with a serine reagent such as E_{600}, suggesting that *p*-NPA and tributyrin hydrolysis might take place at the same catalytic site of gastric lipases in sharp contrast with PPL. Previous investigations[61,62] showed that, after a limited chymotrypsin treatment of PPL, two domains were cleaved with a concomitant loss of tributyrin activity. The isolated *C*-terminal domain of PPL (12 kDa) was still able to hydrolyze *p*-NPA and not tributyrin. The rate of hydrolysis of *p*-NPA was no longer affected by the presence of an interface. The recently published 3D structure of HPL[9] shows that the essential serine 152 is located in the larger *N*-terminal domain, at the *C*-terminal edge of a doubly wound, parallel β-sheet- and that it is part of an Asp-His-Ser triad. This putative hydrolytic site is covered by a surface loop and therefore rendered inaccessible to solvents. Thus interfacial activation is likely to involve a substantial

---

[58] Y. Gargouri, H. Moreau, G. Piéroni, and R. Verger, *J. Biol. Chem.* **263**, 2159 (1988).
[59] H. Moreau, Y. Gargouri, G. Piéroni, and R. Verger, *FEBS Lett.* **236**, 383 (1988).
[60] M. E. Lowe, *J. Biol. Chem.* **267**, 17069 (1992).
[61] M. Bousset-Risso, J. Bonicel, and M. Rovery, *FEBS Lett.* **182**, 323 (1985).
[62] J. D. de Caro, P. Rouimi, and M. Rovery, *Eur. J. Biochem.* **158**, 601 (1986).

conformational change during adsorption at the lipid/water interface. A second potential "catalytic site," located in the C-terminal domain and containing histidine and aspartate residues, can be hypothesized to be responsible for the hydrolysis of p-NPA by pancreatic lipases. It should be stressed, however, that the specific activity of digestive lipases on p-NPA can be considered negligible since its amounts to less than one-thousandth of the catalytic activity measured on tributyrin as substrate.[57] The significance of p-NPA hydrolysis by pancreatic lipases has been overestimated in previous investigations, and there is no justification for extrapolating triacylglycerol hydrolysis mechanisms from the p-NPA ones.[46,63]

2.2 Interfacial Binding to Tributyrin Emulsion of Native and Chemically Modified Digestive Lipases[57]

Several proteins (bearing lipolytic activity or otherwise) were incubated with a tributyrin emulsions. After centrifugation of the emulsified system, it emerged that the nonlipolytic proteins were entirely located in the aqueous supernatants, whereas each protein bearing lipolytic activity was entirely located in the tributyrin phase. These data show that the experimental binding protocol in which a tributyrin emulsion is used and the oil and water phases are separated by centrifugation is an appropriate means of distinguishing between proteins able specifically to bind lipid/water interfaces such as lipolytic enzymes. With this experimental protocol, the binding of E_{600}-modified lipases was found to be comparable to that of native lipases.[57] Furthermore, when DP-PPL was used in the presence of a tributyrin emulsion, the interface-mediated labeling reaction with [^{3}H] sulfobenzoic cyclic anhydride was still possible,[29] which also indicates that DP-PPL was present at the lipid/water interface.

All in all, the present results suggest that in gastric, as in pancreatic, lipases an essential serine residue, which was stoichiometrically labeled with the organophosphorus reagent (E_{600}), is involved in catalysis and not in lipid binding.

2.3 Inhibition of Lipases by Phosphonates and 3D Structures of Lipase-Inhibitor Complexes

2.3.1 Synthesis of New Chiral Organophosphorus Compounds, Analogous to Triacylglycerols. New synthetic methods to obtain lipase inactivators

---

[63] B. Kerfelec, E. Foglizzo, J. Bonicel, P. E. Bougis, and C. Chapus, *Eur. J. Biochem.* **206**, 279 (1992).

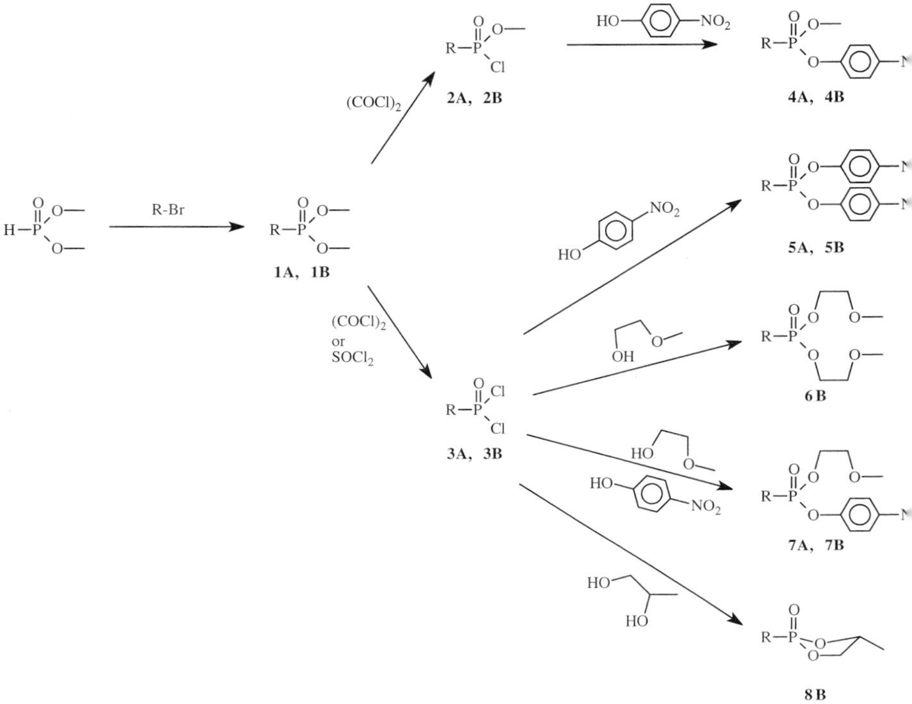

FIG. 6. Structure of various phosphonates and synthesis pathways. $R = n\text{-}C_5H_{11}$ for A products, $R = n\text{-}C_{11}H_{23}$ for B products. (From Marguet et al.[65])

have been described,[64–68] replacing the carbonyl of the hydrolyzable ester bonds by a phosphonate group including a good leaving group, as shown in Fig. 6. These compounds, mimicking in both their charge distribution and configuration the transition state that occurs during carboxyester hydrolysis, were synthesized and investigated as potential inactivators of HPL and

[64] S. Patkar and F. Björkling, in "Lipases. Their Structure, Biochemistry and Application" (P. Wooley and S. B. Petersen, eds.), p. 207. Cambridge University Press, Cambridge, England, 1994.
[65] F. Marguet, C. Cudrey, R. Verger, and G. Buono, Biochim. Biophys. Acta **1210,** 157 (1994).
[66] F. Marguet, "Synthèse d'inhibiteurs chiraux organophosphores: Contribution à l'étude du mécanisme d'action des lipases digestives." Thèse d'Universitè, University of Aix-Marseille II, 1994.
[67] M. L. M. Mannesse, J. W. P. Boots, R. Dijkman, A. J. Slotboom, H. T. W. V. Vanderhijden, M. R. Egmond, H. M. Verheij, and G. H. De Haas, Biochim. Biophys. Acta **1259,** 56 (1995).
[68] P. Stadler, G. Zandonella, L. Haalck, F. Spener, A. Hermetter, and F. Paltauf, Biochim. Biophys. Acta **1304,** 229 (1996).

HGL lipases. Their efficiency as inactivators was studied on the basis of the alkyl chain length, the nature of the leaving group, and the influence of the ester substituent.[65] The released p-nitrophenol to enzyme ratio indicates that a 1:1 complex was formed. In the absence of substrate, the most powerful inactivator was O-methyl O-(p-nitrophenyl) n-pentylphosphonate, which has a short alkyl chain, a small methoxy substituent, and a good leaving group.

Stadler *et al.*[69] studied the influence of substrate hydrophobicity and steric hindrance by variation of the alkyl and acyl chain length at the sn-2 position of the glycerol backbone. Hydrolysis of these synthetic substrates demonstrated that minor structural variations at this sn-2 position of triacylglycerol strongly affect the stereoselectivity of the lipases tested. These authors concluded that the ester carbonyl in the (nonhydrolyzed) sn-2 position of a triacylglycerol was responsible for correct positioning of the substrate in the binding site of lipases. Taking into consideration the above-mentioned results, 1,2-diacyl-3-phosphonoglycerides, which are true glyceride analogues, were therefore synthesized by Marguet.[66] The author intentionally conserved the two carboxyl ester linkages to keep as closely as possible to the structure of acylglycerols (natural substrates). This choice turned out *a posteriori* to be rational, since no or only negligible hydrolysis of these compounds occurred during the experiments with digestive lipases.

These organophosphorus compounds have been prepared in an optically pure form. The final step in the synthesis was the introduction of the phosphonate group leading to a pair of diastereoisomers that were separated by performing liquid chromatography. In this way, four pure stereoisomers, $S_C R_P$, $S_C S_P$, $R_C S_P$, and $R_C R_P$, were obtained as shown in Fig. 7. Marguet *et al.*[66,70] recently studied the inhibition of HGL and HPL by the monomolecular film technique using mixed films of these chiral organophosphorus compounds, which are true triacylglycerol analogues, and dicaprin. Interfacial lipase binding has been evaluated by means of ELISA tests with biotinalyted lipases, with which it was possible to measure the surface density of enzymes in the nanogram range. With both enzymes, kinetic experiments were performed at various molar ratios of dicaprin premixed with each of the four chiral inhibitors. All four stereoisomers investigated reduced the hydrolysis of dicaprin by HGL and HPL. With HPL, the four stereoisomers exhibited a rather weak inhibition capacity, and no significant differences were observed among them. With each inhibitor tested, interfacial binding experiments using ELISA tests showed no significant difference in the surface density of HPL, which confirmed the low stereoselectivity

---

[69] P. Stadler, A. Kovac, L. Haalck, F. Spener, and F. Paltauf, *Eur. J. Biochem.* **227,** 335 (1995).
[70] F. Marguet, I. Douchet, R. Verger, and G. Buono, *J. Am. Chem. Soc.* (1997).

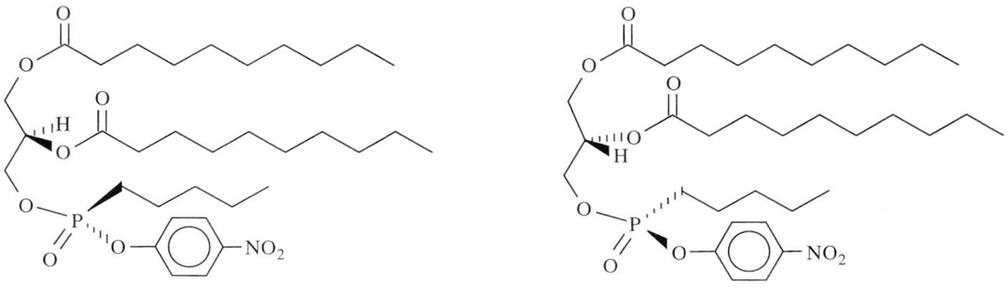

A (R_c, R_p)

A' (S_c, S_p)

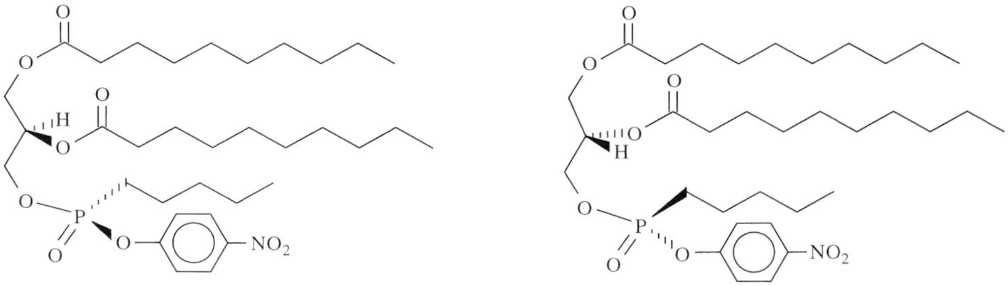

B (R_c, S_p)

B' (S_c, R_p)

FIG. 7. Structure of organophosphorus inhibitors that are triacylglycerol analogues. (From Marguet et al.[66,74])

of pancreatic lipases using either triacylglycerols[71] or triacylglycerol analogues.[72] With respect to gastric lipase, however, the enzyme adsorbed less on each stereoisomeric inhibitor that on the dicaprin substrate. Furthermore, the various organophosphorus enantiomers displayed differential inhibitory effects. The inhibition was very dependent on the chirality on the sn-2 carbon of the glycerol backbone, while the chirality on the phosphorus atom had no influence. The $R_C S_P$ and $R_C R_P$, which both contain the

---

[71] E. Rogalska, C. Cudrey, F. Ferrato, and R. Verger, *Chirality* **5**, 24 (1993).
[72] S. Ransac, E. Rogalska, Y. Gargouri, A. M. T. J. Deveer, F. Paltauf, G. H. de Haas, and R. Verger, *J. Biol. Chem.* **265**, 20263 (1990).

phosphorus moiety at the *sn*-3 position, were found to be the best inhibitors. This latter finding correlates well with the *sn*-3 preference, during the hydrolysis of triacylglycerols catalyzed by gastric lipases.[71-73] Moreover, the levels of surface density of gastric lipase differed significantly with each enantiomeric inhibitor used. A clear correlation was observed between the molar ratio (α_{50}) of inhibitor leading to half inhibition and the surface concentration of gastric lipase: the highest enzymatic inhibition was observed with films containing the enantiomeric inhibitor to which the human gastric lipase was best adsorbed.[66,74]

2.3.2 The 2.46-Å Resolution Structure of the Pancreatic–Procolipase Complex Inhibited by C_{11} Alkyl Phosphonate.[75] In an attempt to further characterize the active site and catalytic mechanism, a C_{11} alkyl phosphonate compound has been synthesized. This compound is an effective inactivator of pancreatic lipase. The crystal structure of the pancreatic lipase–colipase complex inhibited by this compound was determined at a resolution of 2.46 Å.[75] As was observed in the case of the structure of the ternary pancreatic lipase–colipase–phospholipid complex,[25] the binding of the ligand induces rearrangements of two surface loops in comparison with the closed structure of the enzyme.[27] The inhibitor, which could be clearly observed in the active site, was covalently bound to the active serine 152. A racemic mixture of the inhibitor was used in the crystallization, and evidence exists that both enantiomers are bound at the active site. The C_{11} alkyl chain of the first enantiomer fits into a hydrophobic groove and is believed to mimic the interaction between the leaving fatty acid of a triacylglycerol substrate and the protein (Fig. 8A). The alkyl chain of the second enantiomer also has an elongated conformation and interacts with hydrophobic patches on the surface of the open amphipathic lid (Fig. 8B). This may indicate the location of a second alkyl chain of a triacylglycerol substrate. The alkyl portions of the two C_{11} alkyl phosphonate enantiomers superimpose well with the two fatty acyl chains of the phospholipid (Fig. 8C) observed in the ternary phospholipid–lipase–colipase complex.[25] Some of the detergent molecules, needed for the crystallization, were also observed in the crystal. Some of them were located at the entrance of the active site, bound to the hydrophobic part of the lid. On the basis of this crystallographic study, a hypothesis about the binding mode of real substrates and the organization of the active site was proposed.[75] After

---

[73] E. Rogalska, S. Nury, I. Douchet, and R. Verger, *Chirality* **7,** 505 (1995).
[74] M.-P. Egloff, S. Ransac, F. Marguet, E. Rogalska, H. van Tilbeurgh, G. Buono, C. Cambillau, and R. Verger, *Oléagineux, Corps Gras et Lipides* **2,** 52 (1995).
[75] M.-P. Egloff, F. Marguet, G. Buono, R. Verger, C. Cambillau, and H. van Tilbeurgh, *Biochemistry* **34,** 2751 (1995).

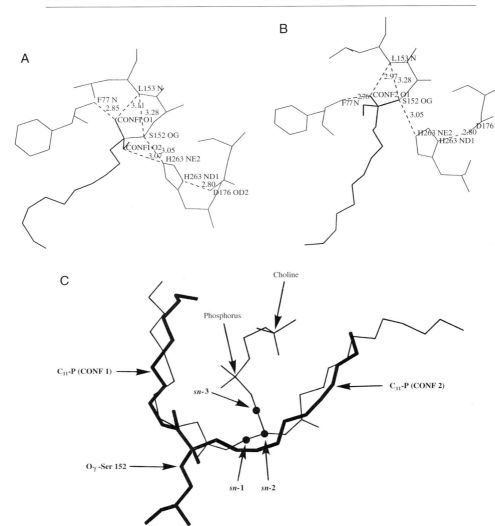

FIG. 8. (A) Polar interactions between the active site residues with the first enantiomer of C_{11}-P (CONF1) and (B) with the second enantiomer of C_{11}-P (CONF2). (C) Phosphatidylcholine molecule (thin lines) as refined in the ternary lipase–colipase–mixed micelle complex[25] superimposed on the two enantiomers of the C_{11}-P inhibitor (thick lines). (From Egloff et al.[75]) Carbon atoms of the glycerol moiety of the phosphatidylcholine are marked as sn-1, sn-2, and sn-3. The phosphorus atom and the choline moiety are indicated by arrows. Both C_{11}-P inhibitors (CONF 1 and CONF 2) are shown covalently linked to the $O\gamma$ of the active site serine 152.

partly leaving the lipid particle, the scissile acyl chain (sn-1 or sn-3) of the substrate may bind to the hydrophobic groove, implying that the substrate adopts a "fork" conformation at the active site.

2.3.3 *Inhibition of Microbial Lipases by Phosphonates.*[64,67,68] Patkar and Björkling[64] recently reviewed several families of lipase inhibitors including boronic acids, phosphorus-containing inhibitors and β-lactone-containing inhibitors. Ethyl hexylchlorophosphonate and analogues thereof were investigated as inhibitors of lipases. Both microbial and mammalian lipases were irreversibly inhibited. The inhibition could be monitored by p-nitrophenol release from ethyl p-nitrophenyl hexylphosphonate. Quantitative analysis of the data indicated that a 1:1 lipase–inhibitor complex was formed during inhibition. Enantioselective inhibition was found for the lipases from *Candida antarctica* and *Rhizomucor miehei* using pure enantiomers of ethyl p-nitrophenyl hexylphosphonate as inhibitors. Using the same inhibitors, reversed enantioselectivity was found for the protease α-chymotrypsin as compared to the two lipases.[76]

1,2-Dioctylcarbamoylglycero-3-0-p-nitrophenyl alkylphosphonates, where the alkyl is a methyl or octyl group, were synthesized and their activity was tested as irreversible inhibitors of cutinase from *Fusarium solani pisi* and *Staphylococcus hyicus* lipase by Mannesse *et al.*[67] Rapid inhibition of these enzymes occurred with a concomitant release of 1 mol of p-nitrophenol per mole of enzyme. Both lipases show great selectivity toward the chirality of these compounds at the glycerol and phosphorus sites. Rapid inhibition at an inhibitor concentration of 0.1 mol% in 100 mM NaTDC ($t_{1/2} < 60$ min) occurred when the glycerol moiety had the (R) configuration, while inhibitors of the (S) configuration reacted 4- to 10-fold more slowly. The isomer with the p-nitrophenyl octylphosphonate attached to the secondary hydroxyl group of glycerol hardly inhibited ($t_{1/2} > 1$ day) the lipases. These results reflect the known position and stereopreference of these enzymes, which preferentially release the fatty acid at sn-3 of natural triacylglycerols. The enzymes showed an even greater selectivity toward the chirality at the phosphorus site, since the differences in reactivity between the faster and slower reacting isomers were as high as about 250-fold for the case of the methylphosphonates and about 60-fold for that of the octylphosphonates. These phosphonates can be regarded as true active site-directed inhibitors. The inhibited enzymes can be said to be analogues of the tetrahedral intermediate in the acylation step that occurs during triacylglycerol hydrolysis.

Stadler *et al.*[68] synthesized 1,2(2,3)-diradylglycero O-(p-nitrophenyl) n-hexylphosphonates, with the diradylglycerol moiety being di-O-octylglycerol,

---

[76] F. Björkling, A. Dahl, S. Patkar, and M. Zundel, *Bioorganic Medicinal Chem.* **2**, 697 (1994).

1-*O*-hexadecyl-2-*O*-pyrenedecanoylglycerol, or 1-*O*-octyl-2-oleoylglycerol, and tested for their ability to inactivate lipases from *Chromobacterium viscosum* and *Rhizopus oryzae*. The experimental data indicate the formation of stable, covalent 1:1 enzyme-inhibitor adducts with the di-*O*-alkylglycerophosphonates. Both lipases exhibited the same preference for the chirality at the phosphorus that was independent from the absolute configuration at the glycerol backbone. However, the inhibitors with the phosphonate linked at position *sn*-1 of the glycerol moiety reacted significantly faster than the corresponding *sn*-3 analogues, reflecting the *sn*-1 stereopreference of the enzymes toward triacylglycerol analogues with a *sn*-2 *O*-alkyl substituent. In contrast, the phosphonates based on the 1-*O*-octyl-2-oleoylglycerol did not significantly inactivate *Chromobacterium viscosum*. Unexpectedly, these substances were hydrolyzed in the presence of lipase.

2.3.4 The 3-D Structure of Inhibited Microbial Lipases. The active center of *R. miehei*, like the human pancreatic lipase, contains a structurally analogous Asp-His-Ser triad (characteristic of serine proteases), which is buried completely beneath a similar short helical segment, or "lid," as HPL. The crystal structure (at 3-Å resolution) of a complex of *R. miehei* lipase with ethyl hexylphosphonate ester reveals that the enzyme's active site is exposed by the movement of the helical lid.[77] This movement also increases the nonpolarity of the surface surrounding the catalytic site. The structure of the enzyme in this complex is probably equivalent to the activated state generated by the oil/water interface (see Fig. 9).[77]

Another fungal lipase from *Candida rugosa* has been crystallized with and without inhibitor, revealing a similar structural modification: the helical lid has to be shifted to unmask a buried active site.[78–80] Furthermore, using both enantiomers of menthyl hexylphosphonate as inhibitors of *C. rugosa* lipase, Cygler *et al.*[80] established a structural basis for the chiral preferences of lipases. These authors determined X-ray crystal structures of covalent complexes of *C. rugosa* lipase with transition-state analogues for the hydrolysis of menthyl esters. One structure contains (1*R*)-menthyl hexylphosphonate (1*R*) derived from the fast-reacting enantiomer of menthol; the other contains (1*S*)-menthyl hexylphosphonate (1*S*) derived from the slow-

---

[77] A. M. Brzozowski, U. Derewenda, Z. S. Derewenda, G. G. Dodson, D. M. Lawson, J. P. Turkenburg, F. Bjorkling, B. Huge-Jensen, S. A. Patkar, and L. Thim, *Nature* **351**, 491 (1991).

[78] P. Grochulski, F. Bouthillier, R. J. Kazlauskas, A. N. Serreqi, J. D. Schrag, E. Ziomek, and M. Cygler, *Biochemistry* **33**, 3494 (1994).

[79] P. Grochulski, Y. Li, J. D. Schrag, and M. Cygler, *Protein Sci* **3**, 82 (1994).

[80] M. Cygler, P. Grochulski, R. J. Kazlauskas, J. D. Schrag, F. Bouthillier, B. Rubin, A. N. Serreqi, and A. K. Gupta, *J. Am. Chem. Soc.* **116**, 3180 (1994).

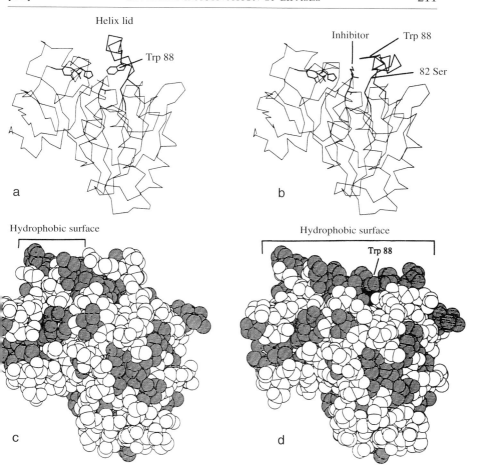

FIG. 9. Diagrammatic representation of the conformational change in *R. miehei* lipase. (a) The native enzyme and (b) the complex are drawn with their Cα backbone. The catalytic triad, the tryptophan 88 and the lid, and the inhibitor are drawn with all atoms and highlighted with thick bonds. The complete atomic structures are drawn with (c) their van der Waals' radii of the native molecule and (d) the complex. The nonpolar atoms are shaded. The increase in the extent of the nonpolar surface is readily seen in this view. (From Brzozowski et al.[77])

reacting enantiomer. These high-resolution 3D structures show first that the empirical rule determined on the basis of substrate mapping is an accurate low-resolution description of the alcohol binding site in *C. rugosa* lipase. Second, interactions between the menthyl ring of the slow-reacting enantiomer and the histidine of the catalytic triad disrupt the hydrogen bond between Nε2 of the imidazole ring and the menthol oxygen atom, which probably explains why the reaction of the (1S) enantiomer of menthol is slower than that of the (1R) enantiomer. Third, the enantiopreference of *C. rugosa* lipase toward secondary alcohols is set not by a separate alcohol-binding site from the catalytic site but by the same loops as those that assemble the catalytic machinery. The common orientation of these loops among many lipases and esterases accounts for their common enantiopreference toward secondary alcohols.

Recently, Longhi *et al.*[81] have crystallized a cutinase from *F. solani* with a triacylglycerol analogue: (R)-1,2-dibutyl carbamoylglycero-3-O-p-nitrophenylbutylphosphonate. This inhibitor is covalently linked to the active site serine, mimicking the first tetrahedral intermediate along the reaction pathway. This structure provides a more realistic model for a complex between a lipolytic enzyme and a triacylglycerol.[81]

3. Inhibition of Digestive Lipases by Sulphydryl Reagents and a Natural Disulphide Compound, Ajoene

3.1 Inhibition of Human Pancreatic Lipase by Sulphydryl Reagents

3.1.1 Reaction of Human Pancreatic Lipase with NbS_2 or $C_{12:0}$-S-NbS (Method A). As shown in Fig. 10A, during incubation of HPL with NbS_2 at a molar ratio of NbS_2 to HPL of 16, 1 mol of NbS ion per mole of enzyme was measured. The reaction proceeded much faster, however, when incubation was carried out in the presence of EDTA (see Fig. 10A) probably due to chelation of divalent ions that interfere with the thiol–disulfide interchange reaction. In both cases, pancreatic lipase activity was unaffected. Likewise, when HPL was incubated with $C_{12:0}$-S-NbS at a $C_{12:0}$-S-NbS to lipase molar ratio of 2, one -SH group was titrated without any loss in enzymatic activity. Incubating HPL with $C_{12:0}$-S-NbS at a $C_{12:0}$-S-NbS to HPL molar ratio of 16 resulted however in the release of about 1.75 mol of NbS ions per mole of HPL (Fig. 10B). The presence of sodium taurodeoxycholate (final concentration 5 mM) with or without colipase in the incubation medium decreased the rate of NbS ion liberation (Fig. 10B). After 2 hr of incubation, only 1.4 sulfydryl group was modified per mole of HPL.

Figure 10C shows that after a complete modification of -SH_1 with NbS_2, further addition of $C_{12:0}$-S-NbS (molar excess of 16) led to a further release

[81] S. Longhi, M. Mannesse, H. M. Verheij, G. H. de Haas, M. Egmond, E. Knoops-Mouthuy, and C. Cambillau, *Protein Sci.* **6**, 275 (1997).

of about 0.5 mol of NbS ion per mole of HPL, which was accompanied by a concomitant loss of lipase activity.

The diagram in Fig. 11 summarizes schematically the various sulphydryl modification states of HPL. The triangles and squares represent catalytically active and inactive enzyme forms, respectively. The numbers between brackets are the molar excesses of sulfydryl reagents to lipase that were used.[82]

To study more closely and quantitatively the reactivity of $C_{12:0}$-S-NbS to the SH_{II} group of HPL, NbS-HPL, and $C_{12:0}$-S-HPL derivatives were prepared. NbS-HPL was obtained after incubating HPL in the presence of NbS_2 (at a NbS_2 to HPL molar ratio of 16). After being incubated for 30 min, the mixture was applied to a Superose 12 column. Using a FPLC system (Pharmacia, Sweden), NbS-HPL was completely separated from the free NbS ions and excess NbS_2. We checked that after addition of excess mercaptoethanol to NbS-HPL, 1 mol of NbS ion per mole of HPL was released as expected. To prepare $C_{12:0}$-S-HPL, we used conditions similar to those described earlier; namely, 100 min incubation of HPL with $C_{12:0}$-S-NbS at a $C_{12:0}$-S-NbS to HPL molar ratio of 2. No optical absorption was observed at 412 nm after adding mercaptoethanol to the gel filtered preparation of $C_{12:0}$-S-HPL, which indicated that no adsorption of $C_{12:0}$-S-NbS to $C_{12:0}$-S-HPL occurred. With the above -SH_I-modified HPL, no significant loss of lipase activity was observed when the modified enzyme was tested on emulsified tributyrin.

After -SH_I modification of HPL by NbS_2 or $C_{12:0}$-S-NbS, a further 1-hr incubation in the presence of $C_{12:0}$-S-NbS (at a molar excess of 16) induced the release of about 0.9 mol of NbS ion per mole of either NbS-HPL or $C_{12:0}$-S-HPL. In the latter case, the $(C_{12:0}$-S$)_2$-HPL obtained was purified by gel filtration under conditions identical to those described earlier for preparing $C_{12:0}$-S-HPL. Furthermore, a good anticorrelation was observed to exist between the time dependency of HPL inhibition and the number of modified sulfydryl groups per mole of HPL. It is worth noting that NbS-HPL and $C_{12:0}$-S-HPL are inactivated at about the same rate using $C_{12:0}$-S-NbS, regardless of the assay system used, that is, emulsified tributyrin in the presence or absence of 6 mM sodium taurodeoxycholate and colipase.

3.1.2 Mechanisms Involved. To determine whether the modification of sulfydryl groups in HPL affects the interfacial binding step or only the hydrolysis step, we performed interfacial binding of $(C_{12:0}$-S$)_2$-HPL to tributyrin emulsion. We showed that $(C_{12:0}$-S$)_2$-HPL was still able to bind to a tributyrin emulsion.[82] We took advantage of the "open" 3D structure of the

---

[82] Y. Gargouri, C. Cudrey, H. Mejdoub, and R. Verger, *Eur. J. Biochem.* **204**, 1063 (1992).

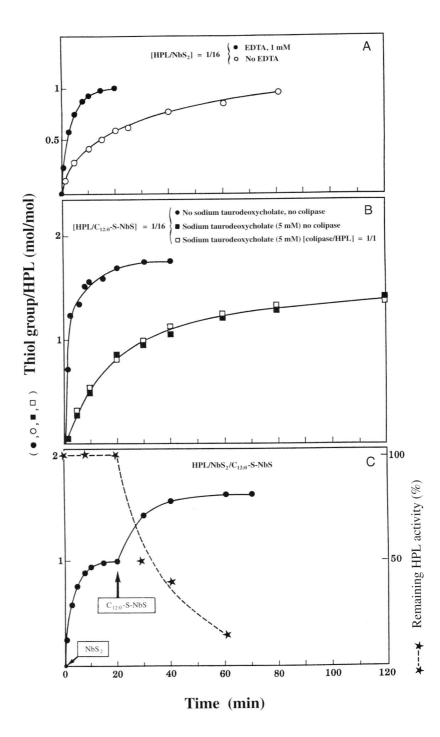

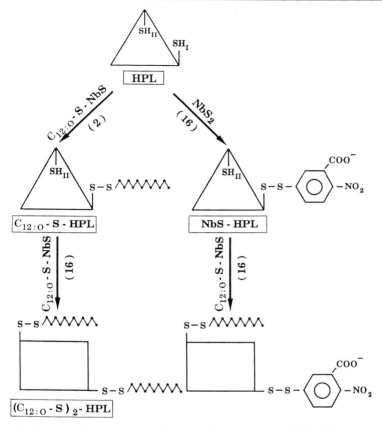

Fig. 11. Diagram of the various sulfhdryl modification states of HPL. The triangles and squares represent catalytically active and inactive enzyme forms, respectively. The numbers between brackets are the molar excesses of sulfhdryl reagents to lipase that were used. (From Gargouri et al.[82])

Fig. 10. Time course of modification of sulfydryl groups of HPL induced by NbS$_2$ or C$_{12:0}$-S-NbS. HPL (20 nmol) was incubated at pH 8.0 and 30°. (A) NbS$_2$ (320 nmol) in the presence (●) or absence (○) of 1 mM EDTA. (B) C$_{12:0}$-S-NbS (320 nmol) in the presence of 1 mM EDTA (●) or 1 mM EDTA and 5 mM NaTDC (■) or 1 mM EDTA, 5 mM NaTDC and colipase (20 nmol) (□). (C) Same conditions as in (A) in the presence of EDTA 1 mM, followed 20 min later by the addition to the incubation medium of C$_{12:0}$-S-NbS (320 nmol). The NbS ions released were measured spectrophotometrically at 412 nm (E^M_{1cm} = 13,600). The number of sulphydryl groups modified per lipase molecule was calculated at various incubation times (●). The residual lipase activity was estimated in parallel by performing tributyrin hydrolysis (★). (From Gargouri et al.[82])

new pancreatic lipase–procolipase complex determined by van Tilbeurgh et al.,[25] which has a phospholipid molecule located at the enzyme's active site. Drastic structural changes in the pancreatic lipase architecture occurred on lipid binding. For instance, a considerable conformational-induced fit led to the creation of the oxyanion hole, which is known to be essential for serine esterase activity. The $\beta 5$ loop (residues 76–85), partly covering the active site in the "closed" HPL structure, moved toward the β-strand containing cysteine 103 to allow the formation of the oxyanion hole. Although it is spatially removed from the catalytic triad (16.1 Å is the distance between the γ-sulfur of cysteine 103 and the γ-oxygen of the active site serine 152), the SH_{II} group of pancreatic lipases, when chemically labeled, was found to be responsible for the loss of lipolytic activity. The presence of a bulky dodecyl chain linked by a disulfide bond to the SH_{II} (cysteine 103) may have hampered the $\beta 5$ loop movement due to steric hindrance. The oxyanion hole could therefore not be formed, and the modified lipase therefore remains inactive on monomeric as well as emulsified substrates. A steric conflict was found by Cudrey et al.[31] to exist between glycine 81 of the $\beta 5$ loop and cysteine 103 after modification of the latter by a thiododecyl substituent, whereas this substituent can be accommodated into the "closed" structure of HPL. Pancreatic lipase inhibition by amphiphilic sulfydryl reagent ($C_{12:0}$-S-NBS) can thus be said to be due to the prevention of a productive induced fit.[31]

3.2 Reaction of NbS_2 and 4-PDS with the Sulfydryl Group of Gastric Lipases (Method A)

We have determined the number of free sulfydryl groups in HGL and RGL and established their contribution to the expression of lipase activity.[1] Our data show that during the incubation of HGL or RGL with classical sulfydryl reagents, such as NbS_2 or 4-PDS, only one sulfydryl group was titrated in HGL or RGL (Fig. 12). Further addition of a denaturing agent (SDS or urea) did not increase the number of modified SH groups. We calculated that the accessible sulfydryl group in RGL reacts three times faster than the corresponding residue in HGL (see Fig. 12). Using NbS_2 as an SH reagent, RGL reacted 30 times more slowly than PPL (SH_I). In view of the hydrophilic character of NbS_2, it can be tentatively concluded that the immediate spatial surroundings of the sulphydryl group in HGL or RGL may be more hydrophobic than the microenvironment of SH_I in PPL. Similar results are obtained using dog gastric lipase (DGL).[83] It can

---

[83] F. Carrière, H. Moreau, V. Raphel, R. Laugier, C. Bénicourt, J.-L. Junien, and R. Verger, *Eur. J. Biochem.* **202**, 75 (1991).

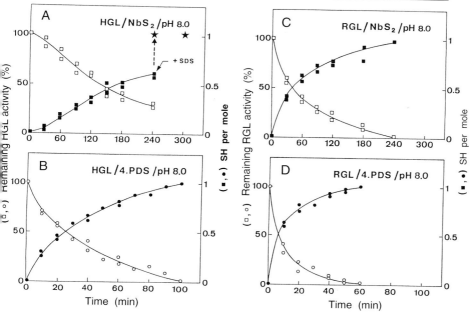

FIG. 12. Variation in gastric lipase activity and sulfhydryl group modification during incubation with sulfhydryl reagents. (A) HGL (24 nmol) and (C) RGL (20 nmol) were incubated at pH 8.0 at 25° with NbS$_2$ (1333 nmol). The liberated NbS was measured spectrophotometrically at 412 nm (E^M_{1cm} = 13,600). The number of sulfhydryl groups modified per lipase molecule was calculated at different incubation times (■). In parallel, residual lipase activity was measured by tributyrin hydrolysis (□). The arrow in (A) indicates the injection of 3% SDS and the star indicates the value reached on SDS addition. (B) HGL (20 nmol) and (D) RGL (12 nmol) were incubated at pH 8.0 and 25° with 4-PDS (1000 nmol with HGL and 833 nmol with RGL). The liberated 4-thiopyridone was measured spectrophotometrically at 324 nm (E^M_{1cm} = 19,800). The number of sulfhydryl groups modified per lipase molecule was calculated at various incubation times (●). Residual lipase activity was measured on tributyrin (○). (Panels A and B are from Gargouri et al.[58]; panels C and D from Moreau et al.[59])

thus be concluded that RGL, HGL, and DGL possess a sulfydryl group that is essential for the expression of their catalytic activity.[58,59]

3.3 Reaction of $C_{12:0}$-S-NbS with the Sulfydryl Group of Gastric Lipases (Method A)

An incubation period (several hours) with NbS$_2$ was needed to modify the essential sulfhydryl. To increase this reaction rate, we synthesized and tested a new sulphydryl reagent, $C_{12:0}$-S-NbS, which bears a hydrophobic moiety.

The incubation of HGL or RGL at pH 6.0 or 8.0 with $C_{12:0}$-S-NbS in the absence of NaTDC (Method A) leads to the rapid and complete inhibition of enzymes.[30] This inhibition parallels the release in the incubation medium of one ion per lipase molecule. Further addition of SDS did not increase the number of NbS ions released per molecule of enzyme.

3.4 Effect of a Natural Disulfide Compound (Ajoene) on Gastric Lipases

Garlic has been used for dietetic purposes for thousands of years by many different people and cultures.[84,85] Bordia[86] has reported that an ether garlic extract inhibits platelet aggregation. Twenty different components have been isolated from ethanolic garlic extract. Apitz-Castro *et al.*[87] then isolated an odorless nonvolatile antiplatelet compound, named "ajoene" (see Figs. 3 and 13). The antiplatelet action of ajoene has been studied at physiological, pharmacological, and biochemical levels.[87,88] Diallyl disulfide (see Fig. 13A), which is given off by garlic when heated, shows no antiplatelet activity.[87]

The inhibitory effect of ajoene was checked on HGL (Fig. 13A) and RGL (Fig. 13B). For purpose of comparison, PPL was used in both cases as a reference. It is clear that pancreatic lipase activity was not affected by incubation with ajoene, diallyl disulfide, or NbS_2 for 4 hr at pH 8.0 at a molar ratio of 2000, whereas HGL and RGL activity was inhibited by treatment with ajoene and to a lesser extent with diallyl disulfide. A 20% reduction in HGL activity was observed after a 4-hr incubation with a 1500 molar excess of diallyl disulfide. Under similar experimental conditions, a 1-hr incubation with ajoene resulted in a drop in the HGL and the RGL activity levels of 70 and 80%, respectively. A 4-hr incubation resulted in 90% inhibition. Ajoene was also able to inactivate gastric lipases under acidic conditions (0.1 M acetate buffer, pH 5.4): a decrease of about 50% in HGL activity was observed after a 2-hr incubation at an ajoene to enzyme molar ratio of 2000:1. To compare the inactivating effects of garlic compounds with classical chemical sulfhydryl reagents, we used NbS_2 as a control. As shown in Fig. 13, NbS_2 and ajoene were found to have comparable inhibition properties.

[84] E. Block, S. Ahmad, J. L. Catalfamo, M. K. Jain, and R. Apitz-Castro, *J. Am. Chem. Soc.* **108**, 7045 (1986).
[85] L. J. Harris, "The Book of Garlic." Aris Books, 1979.
[86] A. Bordia, *Atherosclerosis* **30**, 355 (1978).
[87] R. Apitz-Castro, S. Cabrera, M. R. Cruz, E. Ledezma, and M. K. Jain, *Throm. Res.* **32**, 155 (1983).
[88] R. Apitz-Castro, E. Ledezma, J. Escalante, A. Jorquera, F. M. Pinate, J. Moreno-Rea, O. Carrilo Leal, and M. K. Jain, *Drug. Res.* **38**, 901 (1988).

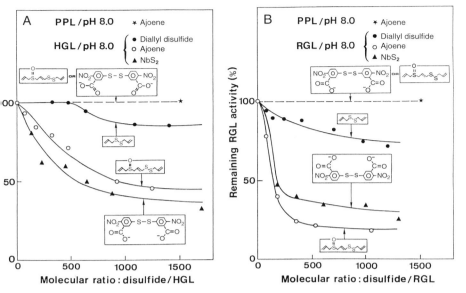

FIG. 13. Effects of ajoene, diallyl disulfide, and NbS$_2$ on (A) HGL and (B) RGL activity. HGL or RGL (5 μM) were incubated in the presence of variable amounts of ajoene (○) diallyl disulphide (●) or NbS$_2$ (▲) in 0.1 M Tris-HCl buffer (pH 8.0) at 37°. After a 1-hr incubation, the remaining lipase activity was measured titrimetrically at pH 6.0 under optimal assay conditions using tributyrin as substrate. Ajoene and diallyl disulfide are synthetic compounds that have been purified by HPLC. Control experiments without adding disulfide showed no reduction in HGL or RGL activity. Further addition of a reducing agent, such as mercaptoethanol (up to 43 mM), failed to restore the lost HGL or RGL activities. (From Gargouri et al.[118])

Patients with congenital pancreatic lipase[89] or exocrine pancreatic insufficiency[90] are able to absorb more than 50% of the ingested fat, despite the complete absence of pancreatic lipase. This is indirect evidence that HGL plays a relatively important physiological role in the general gastrointestinal lipid absorption process. It has been shown by means of *in vitro* experiments that gastric lipolysis is of prime importance for lipid digestion in humans. Hence, inhibition of the first step in gastrointestinal lipolysis may induce a decrease in the overall dietary fat absorption. The present *in vitro* data show that HGL can be selectively inactivated by preincubation with ajoene,

[89] D. P. R. Muller, J. P. K. McCollum, R. S. Tompeter, and J. T. Harries, *Gut* **16,** 838 (1975).
[90] C. K. Abrams, M. Hamosh, S. K. Dutta, V. S. Hubbard, and P. Hamosh, *Gastroenterology* **92,** 125 (1987).

whereas diallyl disulfide is a much less effective inhibitor. Disulfide, however, has no such inhibitory effect on pancreatic lipase activity. The selective HGL inactivating properties of nontoxic, odorless disulfides, such as ajoene, may therefore have promising applications in the field of hypolipemia and antiobesity therapy.

4. Inhibition of Digestive Lipases by THL

It has been shown that tetrahydrolipstatin (THL), derived from lipstatin that is produced by *Streptomyces toxytricini*, acts *in vitro* as a potent inhibitor of pancreatic and gastric lipases as well as cholesterol ester hydrolase.[30,34,91,92] It has been suggested that a stoichiometric enzyme-inhibitor complex of an acyl–enzyme type is formed that is slowly hydrolyzed, with water as the final acceptor, leaving an intact enzyme and the inactive form of THL.[93,94] A "reactivation" of inactivated HPL was observed, as evidenced by kinetics showing a lag phase, and probably due to a slow deacylation in the pH-stat assay system of the inhibited enzyme.

THL is thus an amphiphilic inactivator reacting with the essential serine of the active site. Comparisons on the partitioning between the micellar and oil phases of THL and $C_{12:0}$-S-NbS were made to estimate the hydrophobic–lipophilic balance of each inactivator. Its preferential micellar partitioning makes $C_{12:0}$-S-NbS inefficient in the presence of sodium taurodeoxycholate. In contrast, THL is associated chiefly with the triacylglycerol phase, even in the presence of bile salts.[31] The latter physicochemical property is probably a requirement for prototypic lipase inactivators to be effective under physiological conditions, that is, in the presence of bile and lipids.

Furthermore, HPL loses more than 80% of its activity when incubated with THL in a buffer containing bile salts. During the inhibition process, Lüthi-Peng and Winkler[24] reported large changes in intrinsic tryptophan fluorescence and in the near-ultraviolet circular dichroism. The rate of chemical inhibition is highly comparable to that determined from the time dependence of the spectral changes. It is concluded that HPL undergoes a conformational transition on inhibitor binding, resulting in a change in the microenvironment of tryptophan residues. Bile salts are needed in this

[91] P. Hadvàry, H. Lengsfeld, and H. Wolfer, *Biochem. J.* **256**, 357 (1988).
[92] B. Borgström, *Biochim. Biophys. Acta* **962**, 308 (1988).
[93] P. Hadvàry, W. Sidler, W. Meister, W. Vetter, and H. Wolfer, *J. Biol. Chem.* **266**, 2021 (1991).
[94] Q. Lüthi-Peng, H. P. Maerki, and P. Hadvary, *FEBS Lett.* **299**, 111 (1992).

system for effective inhibition of the enzyme by THL, and a large increase in the inhibition rate takes place at about the critical micellar concentration (CMC) of bile salts. The inhibited enzyme can be reactivated by reducing the bile salt concentration to below the CMC, and the changes in tryptophan fluorescence induced by acylation with THL are thereby reversed. This suggests that bile salts above their CMC stabilize the acyl–enzyme complex.[24]

4.1 Inhibition of Digestive Lipases During Lipolysis (Method B)

The hydrolysis of tributyrin was recorded at pH 6.0 with HGL and RGL and pH 8.0 with PPL. Two minutes after enzyme addition, THL was injected at variable final concentrations. As can be seen from the kinetic curves in Fig. 14, the rate of tributyrin hydrolysis by PPL (Fig. 14A), RGL (Fig. 14B), and HGL (Fig. 14C) decreased to different extents. The lipase hydrolysis rate decreased more rapidly at the higher THL concentrations. With all the lipases tested, more than 90% of the enzyme inhibition was observed to occur with final THL concentration of 200 μM.

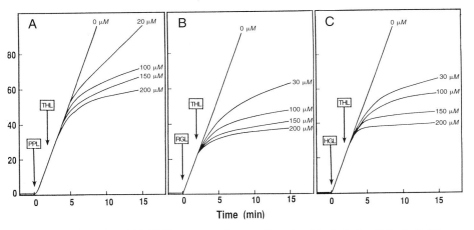

FIG. 14. Effect of variable THL concentrations on the rate of hydrolysis of tributyrin by (A) PPL (5 nM) + colipase (100 nM), (B) RGL (24 nM), and (C) HGL (24 nM). Assays were carried out at 37°, pH 8.0 with PPL and pH 6.0 with HGL and RGL using 0.2 ml tributyrin emulsified in 10 ml of 150 mM NaCl, 2 mM NaTDC, 1.5 μM bovine serum albumin. Each kinetic represents one typical experiment. (From Gargouri et al.[30])

4.2 Effect of Inactivator Concentration on the Rate of Tributyrin Hydrolysis by Digestive Lipases (Method C: Poisoned Interface)

The rate of tributyrin hydrolysis by HGL, RGL, or PPL was determined as a function of the THL concentration (Fig. 15). In all cases, THL was injected into a tributyrin emulsion 1 min prior to the enzyme addition and the initial lipase hydrolysis rate was measured. As shown by the curve in Fig. 15, the rate of tributyrin hydrolysis by HGL, RGL, and PPL decreased according to a similar pattern, with increasing THL concentrations. The inactivator concentration that reduced the lipase activity to 50% (I_{50}) of its initial value was around 10 μM.

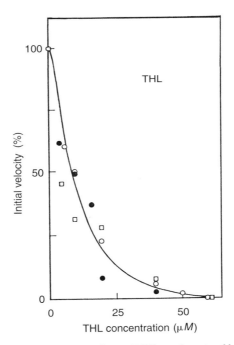

FIG. 15. Effect of increasing concentrations of THL on the rate of hydrolysis of tributyrin by PPL (5 nM) in the presence of colipase (100 nM) (□), or RGL (24 nM) (●), or HGL (24 nM) (○). Experiments were performed at 37°, pH 8.0, with PPL and pH 6.0 with HGL and RGL using 0.2 ml tributyrin emulsified in 10 ml of 150 mM NaCl, 2 mM NaTDC, 1.5 μM bovine serum albumin. In all cases, THL, from a 20 mM ethanolic solution, was injected 1 min prior to the enzyme addition. Initial velocity measurements are given. Each point represents an independent assay. (From Gargouri et al.[30])

4.3 Gastric and Pancreatic Lipase Activity on Mixed Films Containing THL

To study the effects of THL on the gastric and pancreatic lipase activity, we first determined the film stability of THL and its interfacial properties at the air/water interface at pH 4.0 and 8.0.[34] The force/area curves were not significantly different when the experiments were carried out on a subphase at pH 4.0 or 8.0. The collapse point (25.2 mN/m) was reached at a molecular area of about 0.7 nm^2/molecule. This collapse pressure was identical to that reported previously by Borgström.[92] All the mixed films consisting of the previous compounds plus dicaprin were stable with time (no significant decrease in the surface pressure was observed to occur within 1 hr).

Lipase activities were measured as a function of the inactivator surface concentration, as indicated in Fig. 16. With all lipases tested, the hydrolysis of dicaprin decreased sharply as the surface molar fraction of THL increased. The surface concentrations of THL (α_{50}), which reduces lipase activity to 50%, are 0.013, 0.025, and 0.25% with PPL, RGL, and HGL, respectively. On the one hand, it should be remembered that the turnover rates of gastric and pancreatic lipases are quite different.[95] Consequently, the amounts of each lipase injected into the reaction compartment of the zero-order trough were different (see legend for Fig. 16), leading to various THL/lipase molar ratios. However, we know that in monolayer systems characterized by a low specific surface (around 1 cm^2/ml), a small fraction of the total amount of enzyme actually binds to the monomolecular film. In the particular cases of PPL and HGL, using pure dicaprin films, it was found that with both lipases around 1.5% of the total amount of injected enzyme was recovered after film aspiration and measuring radioactivity.[95,96] Judging from this percentage of lipid-bound enzyme, the effective (surface) molar excesses of THL to film-bound lipase inducing a 50% reduction in catalytic activity were estimated as 10-fold and 20-fold with PPL, RGL, and HGL, respectively. One can say that the THL is an amphipathic molecule reacting at its lactone ring. THL is tightly bound and slowly hydrolyzed by lipases and esterases to give a long-lived transient acyl–enzyme complex.[92] Small amounts of THL are therefore required to reduce lipase activity. The stoichiometry of the interfacial situation can be described as follows: one lipase molecule embedded among 10^5 substrate molecules will be inactivated to half its initial velocity by the presence of 10 THL molecules.

---

[95] Y. Gargouri, H. Moreau, G. Piéroni, and R. Verger, *Eur. J. Biochem.* **180**, 367 (1989).
[96] Y. Gargouri, G. Piéroni, C. Rivière, L. Sarda, and R. Verger, *Biochemistry* **25**, 1733 (1986).

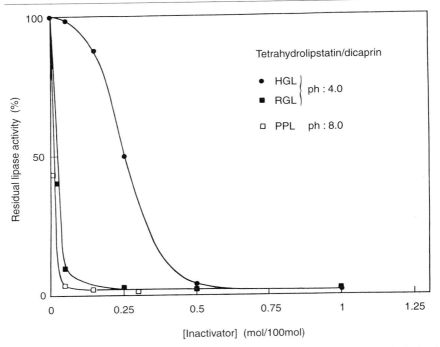

FIG. 16. Effect of increasing surface concentration of THL on the rate of hydrolysis of dicaprin monolayers at a constant surface pressure (25 mN/m) by HGL (final concentration 2 nM) (●), RGL (final concentration 0.4 nM) (■), and PPL (final concentration 0.2 nM) (□). The subphase was 50 mM glycine-HCl, pH 4.0, 100 mM NaCl, and 10 mM CaCl$_2$ in the HGL and RGL assays, and 10 mM Tris-HCl, pH 8.0, 100 mM NaCl, 21 mM CaCl$_2$, and 1 mM EDTA in the PPL assays. (From Ransac et al.[34])

4.4 Experimental Design to Evaluate Inhibition of Pancreatic Lipase in Humans

4.4.1 Measurement of Pancreatic Enzyme Secretion and Intraduodenal Hydrolysis of Triacylglycerols. Several publications, illustrating the clinical application of THL in the treatment of human obesity have recently appeared.[97–107] Accurate measurement of human pancreatic enzyme secretion

---

[97] C. Güzelhan, H. J. M. J. Crijns, P. A. M. Peeters, J. H. G. Jonkman, and D. Hartmann, *Int. J. Obes.* **15,** 29 (1991).
[98] M. K. Meier, D. Blum-Kaelin, K. Bremer, et al., *Int. J. Obes.* **15,** 31 (1991).
[99] J. Hauptman, F. Jeunet, and D. Hartmann, *Am. J. Clin. Nutr.* **55,** 309S (1992).
[100] M. L. Drent and E. A. Vanderveen, *Int. J. Obes.* **17,** 241 (1993).
[101] C. Güzelhan, J. Odink, J. J. N. Jansenzuidema, and D. Hartmann, *J. Int. Med. Res.* **22,** 255 (1994).

and intraduodenal hydrolysis of triacylglycerols requires collection of duodenal juice. Indirect methods as measurement of serum levels of exocrine pancreatic enzymes and measurement of the function of enzymes such as stool fat determination do not specifically quantify exocrine pancreatic secretion. A direct approach to measure pancreatic secretion (including pancreatic lipase) after a meal in humans is to aspirate duodenal contents via a tube at a distal duodenal site while infusing a nonabsorbable marker at a more proximal site to allow for determination of recovery and thus for calculation of total output of each component.[4,108–110] Measurement of concentration of pancreatic secretory components in duodenal fluid is meaningless for determining secretory rates, since the volume of duodenal contents may vary markedly due to gastric emptying, bile flux, pancreatic secretion, and possibly mucosal bicarbonate secretion.[110] Further considerations such as intraduodenal activation of enzymes[108–110] or adherence to food particles[110] are relevant for interpreting meal-induced pancreatic secretion in humans and pancreatic enzyme content in duodenal aspirates obtained during intestinal perfusion.[110]

Tetrahydrolipstatin is an inhibitor of many lipases including gastrointestinal lipases. It inhibits hydrolysis and absorption of triacylglycerols in the duodenum and is being developed as an antiobesity agent.[30,91,92,100,111] We measured the effect of THL on secretion and activity of pancreatic lipase and on fat hydrolysis in duodenal aspirates of human volunteers using polyethyleneglycol 4000 (PEG 4000) as a nonabsorbable recovery marker. THL was infused in an oil/egg emulsion in parallel with ingestion of a meal.

---

[102] Y. Hussain, C. Guzelhan, J. Odink, E. J. Vanderbeek, and D. Hartmann, *J. Clin. Pharmacol.* **34,** 1121 (1994).

[103] J. Zhi, A. T. Melia, R. Guerciolini, J. Chung, J. Kinberg, J. B. Hauptman, and I. H. Patel, *Clin. Pharmacol. Ther.* **56,** 82 (1994).

[104] M. L. Drent, I. Larsson, T. Williamolsson, F. Quaade, F. Czubayko, K. Vonbergmann, W. Strobel, L. Sjostrom, and E. A. Vanderveen, *Int. J. Obes.* **19,** 221 (1995).

[105] D. Isler, C. Moeglen, N. Gains, and M. K. Meier, *Br. J. Nutr.* **73,** 851 (1995).

[106] J. G. Zhi, A. T. Melia, H. Eggers, R. Joly, and I. H. Patel, *J. Clin. Pharmacol.* **35,** 1103 (1995).

[107] M. L. Drent and E. A. Vanderveen, *Obes. Res.* **3,** S623 (1995).

[108] C. Beglinger, M. Fried, I. Whitehouse, J. B. Jansen, C. B. Lamers, and K. Gyr, *J. Clin. Invest.* **75,** 1471 (1985).

[109] J. R. Malagelada, E. P. Di Magno, W. H. J. Summerskill, and V. L. W. Go, *J. Clin. Invest.* **58,** 493 (1976).

[110] T. E. Solomon, *in* "Physiology of the Gastrointestinal Tract" (L. R. Johnson, ed.), p. 1499. Raven Press, New York, 1994.

[111] E. K. Weibel, P. Hadvàry, E. Hochuli, E. Kupfer, and H. Lengsfeld, *J. Antibiotics* **40,** 1081 (1987).

Duodenal juice was aspirated at the ligament of Treitz for determination of exocrine pancreatic secretory products, triglyceride emulsions, and their hydrolysis products.

4.4.2 Inhibition of Pancreatic Lipase and Triglyceride Hydrolysis by THL in the Duodenum in Humans. After an overnight fast, subjects were intubated by the nasal route with a polyvinyl 3-lumen tube (outer diameter 4 mm), the tip being placed at the ligament of Treitz under fluoroscopic control. Throughout each experiment, PEG 4000 dissolved in saline (1.2 g/liter) was perfused by a roller pump to the duodenum via openings near the papilla Vateri (3.4 ml/min; 4.1 mg PEG 4000/min). The PEG 4000 solution was weighed before and after infusion to control pumping volumes.

The duodenal juice was gently aspirated by a syringe to fill the tube and let drip by gravity into a glass vessel on ice below the volunteer. Aspirates were collected for 10 min within 20-min intervals. After taking aliquots the remainder of the aspirate was reinjected through the aspiration tube into the duodenum. Aliquots were stored at −20° until analyzed for PEG, pancreatic lipase, total and free fatty acids, and sitostanol. Duodenal samples for determination of lipase activity were diluted 50-fold with PBS/5% BSA for lipase stabilization during storage.

Following a 40-min basal collection period, subjects ingested a standardized meal consisting of 100 g mashed potatoes (water and potatoe flakes corresponding to 13 g carbohydrate and 1 g protein) and 100 g Béarnaise sauce (butter/egg emulsion with 10 g butter) prepared by a cook. The butter was labeled with 300 mg sitostanol as nonabsorbable recovery marker. Simultaneously, with the start of the meal intake, the intraduodenal saline infusion was changed to an egg/sunflower oil/olive oil emulsion containing PEG 4000 (same amount infused as without oil). The oil emulsion contained 300 mg tri(undecanoyl) glycerol as a recovery marker for triacylglycerols and free fatty acids. The emulsion was weighed at the beginning and at the end of the infusion period of 4 hr.

The meal emulsion was prepared essentially according to Hernell *et al.*[112] In 200 ml saline, 18.75 g olive oil and 18.75 g sunflower oil were homogenized gently, to avoid foaming, with a polytron at RT for 5 min with 375 mg tri(undecanoyl) glycerol, 40 ml mixed raw egg, 12 g sucrose, 1200 mg PEG 4000, and the appropriate doses of THL. After filling up to 1000 ml with tap water 800 ml of the emulsion (30 g fat) were infused.

From the concentrations of PEG in the infusate [PEG_i] and in the duodenal samples [PEG_s] and the volume perfused during the sampling

---

[112] O. Hernell, J. E. Staggers, and M. Carey, *Biochemistry* **29,** (1990).

interval, V_i, the total intraduodenal volume, V_t, during the sampling interval can be calculated:

$$V_t = V_i([PEG_i]/[PEG_s])$$

Enzyme output, EO, during the sampling interval is calculated by multiplying V_t with enzyme concentrations $[E]$, that is,

$$EO = V_t[E]$$

4.4.3 Analytical Procedures. PEG 4000 determination was modified from Hydén et al.[113] Pancreatic lipase activity was measured according to Imamura et al.,[114] and lipase protein by a sandwich ELISA test. Total fat was extracted according to Bligh and Dyer[115] and weighed. Free fatty acids were methylated in the fat extract with diazomethane and quantitated by GLC. Esterified fatty acids in the fat extract were transmethylated with sodium methanolate and determined by GLC.

4.4.4 Results and Discussion. When THL ingestion increased, secretion of pancreatic lipase decreased (Fig. 17A). This, however, does not represent a direct effect of THL but instead is, via a feedback regulation mechanism, a consequence of the diminished availability of free fatty acids. The specific activity of pancreatic lipase in the controls remained constant during the whole experiment and was reduced up to 50% during THL infusion (Fig. 17B). No dose relationship was observed. The hydrolysis of the infused egg/oil emulsion and of the ingested meal butter fat was followed separately by measuring free and esterified undecanoic and lauric acids. Free fatty acid formation was dose-dependently inhibited by THL. The highest dose completely suppressed lipolysis from both fat sources (Fig. 17C). Although under these experimental conditions triglyceride hydrolysis was strongly suppressed, a significant level of active lipase was observed. Possibly, pancreatic lipase on THL containing fat droplets is inactivated, whereas lipase not associated with fat droplets remains active. Gastric emptying of the meal was finished at 140 min after start of ingestion, that is, only very low levels of butterfat were present in the duodenal aspirates then. This was

[113] S. Hydén, *Kungliga Lantbrukshögskolans Annaler* **22,** 139 (1955).
[114] S. Imamura, T. Hirayama, T. Arai, K. Takao, and H. Misaki, *Clin. Chem.* **35,** 1126 (1989).
[115] E. G. Bligh and W. J. Dyer, *Can. J. Biochem. Physiol.* **37,** 911 (1959).
[116] F. Carrière, R. Verger, A. Lookene, and G. Olivecrona, in "Interface between Chemistry and Biochemistry" (H. Jörnvall and P. Jollès, eds.), Birkhäuser Verlag, Germany, 1995.
[117] G. Piéroni and R. Verger, *Eur. J. Biochem.* **132,** 639 (1983).
[118] Y. Gargouri, H. Moreau, M. K. Jain, G. de Haas, and R. Verger, *Biochim. Biophys. Acta* **1006,** 137 (1989).

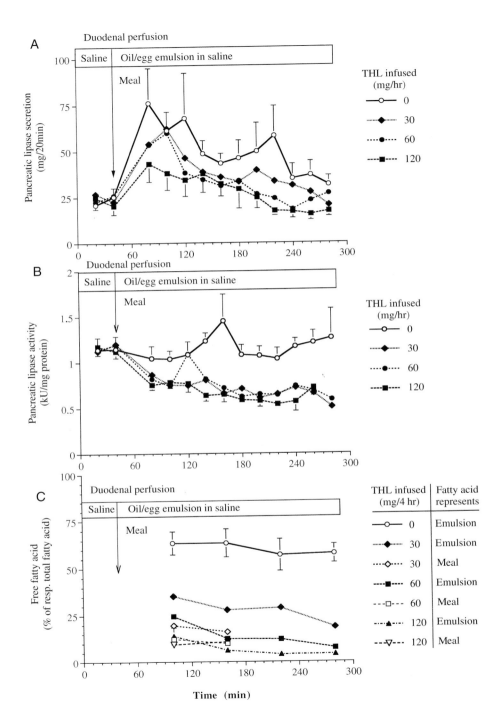

clearly visible also from the complete absence of sitostanol at these time intervals (not shown).

These experiments demonstrate that in humans intraduodenal infusion of THL strongly reduces activity of pancreatic lipase, both by reducing its specific activity and pancreatic secretion. In parallel, hydrolysis of triacylglycerols from a test meal containing a butter emulsion can be completely suppressed as well as lipolysis of an oil emulsion infused into the duodenum.

5. Kinetic Model Illustrating the Covalent Inhibition of a Lipolytic Enzyme at a Lipid/Water Interface[34]

The kinetic model presented in Fig. 2 can be used to depict the covalent inhibition of lipolytic enzymes at a lipid/water interface. In this model, we only consider cases where the regeneration of enzyme form, E, form E_i^* is very slow, and therefore does not take place during our experimental procedure. The kinetic treatment of this model is derived from our analogous treatment described by Ransac et al.[33] In the present kinetic treatment, the inhibitor is assumed to be in large molar excess compared to the adsorbed enzyme, giving the following sigmoidal expression of the time dependence of the products released:

$$P = k_{cat} \cdot E_0 \cdot \frac{1}{k_i} \cdot \frac{S}{K_M^*} \cdot \frac{K_I^*}{I} \cdot \frac{\tau_2(e^{-t/\tau_2} - 1) - \tau_1(e^{-t/\tau_1} - 1)}{\tau_1 - \tau_2}$$

where τ_1 and τ_2 represent the time constants of the combined partial reactions $E \rightleftarrows E^*$, $E^* \rightleftarrows E^*S$, $E^* \rightleftarrows E^*I$, and the irreversible reaction $E^*I \rightarrow E_i^*$, respectively. E^*I is the interfacial enzyme–inhibitor complex and E_i^* the interfacial covalently inhibited enzyme. The corresponding equation of the enzymatic velocity is

$$v = \frac{dP}{dt} = k_{cat} \cdot E_0 \cdot \frac{1}{k_i} \cdot \frac{S}{K_M^*} \cdot \frac{K_I^*}{I} \cdot \frac{e^{-t/\tau_1} - e^{-t/\tau_2}}{\tau_1 - \tau_2}$$

FIG. 17. (A) Effect of increasing doses of THL on secretion of pancreatic lipase in volunteers (mean ± SE; $n = 10$). Spot sampling of duodenal juice at the ligament of Treitz. After a baseline period of meal (egg/butter emulsion with potatoes) was ingested and infusion of an egg/oil emulsion was started. THL was dissolved in the emulsified oil. (B) Reduction of specific activity of pancreatic lipase during infusion of increasing doses of THL into the duodenum in volunteers. (C) Effect of increasing doses of THL on free fatty acid formation in duodenal aspirates from a meal of potatoes with Béarnaise sauce (an egg/butter emulsion identified by myristic acid), and from an intraduodenally infused egg/oil emulsion [labeled by tri(undecanoyl) glycerol]. THL was dissolved in the emulsified oil.

As can be seen from this equation, the enzyme velocity possesses a maximal value ($v_{\text{inflection}}$) at a given time ($t_{\text{inflection}}$) which corresponds to the inflection point of the experimental sigmoidal curve describing the product release as a function of time.

$$v_{\text{inflection}} = k_{\text{cat}} \cdot E_0 \cdot \frac{1}{k_i} \cdot \frac{S}{K_M^*} \cdot \frac{K_I^*}{I} \cdot \frac{\left(\frac{\tau_2}{\tau_1}\right)^{\frac{\tau_2}{\tau_1 - \tau_2}} - \left(\frac{\tau_2}{\tau_1}\right)^{\frac{\tau_1}{\tau_1 - \tau_2}}}{\tau_1 - \tau_2}$$

$$t_{\text{inflection}} = \frac{\ln \tau_1 - \ln \tau_2}{\tau_1 - \tau_2} \tau_1 \tau_1$$

One can see that both $v_{\text{inflection}}$ and $t_{\text{inflection}}$ are independent of time.

In our experiments, when using mixed inactivator/substrate films the enzyme activity ($v_{\text{inflection}}$) was always measured at the inflection time ($t_{\text{inflection}}$).

If we normalize the inhibitor concentration using its molar fraction [$\alpha = (I/I + S)$], $v_{\text{inflection}}$ can be rewritten:

$$v_{\text{inflection}} = k_{\text{cat}} \cdot E_0 \cdot \frac{1}{k_i} \cdot \frac{K_I^*}{K_M^*} \cdot \frac{1 - \alpha}{\alpha} \cdot \frac{\left(\frac{\tau_2}{\tau_1}\right)^{\frac{\tau_2}{\tau_1 - \tau_2}} - \left(\frac{\tau_2}{\tau_1}\right)^{\frac{\tau_1}{\tau_1 - \tau_2}}}{\tau_1 - \tau_2}$$

then we can deduce the molar fraction of inactivator (α_{50}) that reduces the enzyme activity (measured at the inflection time) to 50% of its value in the absence of inactivator.

Assuming a large excess of inactivator relative to the adsorbed enzyme, the expression of α_{50} (not shown) is independent of the enzyme concentration. By contrast, in an experimental setting, we observed such a dependence using mixed THL/dicaprin. Thus, we have to conclude that, contrary to our initial assumption, the inactivator and the bound enzyme are probably in the same concentration range. Consequently, the previously estimated (10- to 20-fold) of the inhibitor/film-bound-enzyme ratio inducing a 50% reduction in catalytic activity was probably an overestimate.

In the presence of small amounts of THL, the kinetics did not present a sigmoidal pattern. In contrast, inhibition experiments using the alkyl-S-NbS analogues showed a marked sigmoidal nature in good agreement with the preceding kinetic model. These observations can be explained by different values of the inhibition rate constant (k_i) and the interfacial dissociation constant (K_I^*) for the two types of inactivator. For the THL, the interfacial dissociation constant (K_I^*) and the inhibition rate constant (k_i) are probably very low, we can therefore consider it to act as a very good competitive inhibitor. In contrast, the alkyl-S-NbS inactivator seems to show a lower affinity for the enzyme, but once complexed with the enzyme it reacts quickly to give the covalently inactivated enzyme.

If one compares the α_{50} values obtained between THL and the disulfide compounds [$C_{12:0}$-S-NbS, $C_{16:0}$-S-NbS, $C_{18:1(9)}$-S-NbS], it is obvious that with the "poisoned interface" system, THL is the most potent inactivator.

Acknowledgment

Our thanks are due to Drs. F. Carrière, H. Moreau, C. Cudrey, and A. Moulin (CBBM, Marseilles) for stimulating discussions. Research was carried out with financial support from the BRIDGE-T-Lipase (contract BIOT-CT 91-0274 DTEE) and the BIOTECH G-lipase (contracts BIO2-CT94-3041 and BIO2-CT94-3013) programs of the European Union as well as the CNRS-IMABIO project.

[11] Mechanism-Based Inhibitors of Mammalian Cholesterol Esterase

By Shawn R. Feaster and Daniel M. Quinn

Introduction

Cholesterol esterase (CEase,[1] sterol ester hydrolase; EC 3.1.1.13) is a lipolytic enzyme that is synthesized in the pancreas and secreted into the duodenum in response to an oral fat load.[2–4] The enzyme is also found in human, feline, canine, and gorilla milk, wherein its digestive action allows alimentary absorption of lipids in suckling infants.[5–8] CEase has long been suspected of playing a role in intestinal absorption of dietary lipids, including cholesterol,[9,10] although there is considerable controversy on this is-

---

[1] Abbreviations: CEase, cholesterol esterase; CMC, critical micelle concentration; PNPB, p-nitrophenyl butyrate; SSV, summed-squared-variance.
[2] H. Brockeroff and R. G. Jensen, in "Lipolytic Enzymes" p. 176. Academic Press, New York, 1974.
[3] D. Kritchevsky and H. V. Kothari, *Adv. Lipid Res.* **16,** 221 (1978).
[4] E. A. Rudd and H. L. Brockman, in "Lipases" (B. Borgström and H. L. Brockman, eds.), p. 185. Elsevier, Amsterdam, 1984.
[5] S. Bernbäck, L. Bläckburg, and O. Hernell, *J. Clin. Invest.* **85,** 1221 (1990).
[6] C.-S. Wang, M. E. Martindale, M. M. King, and J. Tang, *Am. J. Clin. Nutr.* **49,** 457 (1989).
[7] L. M. Freed, C. M. York, M. Hamosh, J. A. Sturman, and P. Hamosh, *Biochem. Biochim. Acta* **878,** 209 (1986).
[8] E. Freudenberg, *Experientia* **22,** 317 (1966).
[9] S. G. Bhat and H. L. Brockman, *Biochem. Biophys. Res. Commun.* **109,** 486 (1982).
[10] L. L. Gallo, S. B. Clark, S. Myers, and G. V. Vahouny, *J. Lipid Res.* **25,** 604 (1985).

sue.[9-11] Different laboratories have characterized the effect of removal of endogenous CEase activity on lipid absorption in bolus-fed, duodenally cannulated rats.[11] Contradictory results have been reported: CEase either has no role[11] or is a major player[12] in cholesterol absorption.

As described elsewhere in Chapter [3], Hui and co-workers have addressed the role of CEase in the absorption of dietary lipids by developing a CEase gene knockout mouse model, and report that a loss of pancreatic CEase activity has no effect on cholesterol absorption, but is a decisive element in cholesteryl ester absorption.[13] Of considerable interest are recent reports from the same group that plasma cholesterol and plasma CEase levels are correlated,[14] and that plasma CEase is involved in LDL metabolism,[14] may function as a lipid transfer protein,[15] and is involved in uptake of cholesteryl esters by liver cells in culture.[16] Therefore, inhibition of CEase may provide a novel approach to the development of hypolipidemic agents. Accordingly, this chapter summarizes the various classes of CEase inhibitors and describes in some detail methods for characterizing inhibition by carbamates. The following two sections outline features of CEase structure and function that facilitate the rational design of inhibitors.

Catalytic Mechanism

Physiological lipid substrates of CEase have very low water solubilities, and consequently are contained in supramolecular aggregates, such as micelles or lipoproteins.[4] Accordingly, physiological CEase catalysis is interfacial biocatalysis, wherein substrate binding to the active site and consequent catalytic turnover are preceded by reversible binding of the enzyme to the lipid/aqueous interface, as outlined in Fig. 1. The mechanistic complexity of interfacial biocatalysis makes characterization of the inhibition mechanisms of CEase inhibitors a nontrivial task. A steady-state derivation for the interfacial mechanism of Fig. 1 gives Eqs. (1)–(3), in which A, I_L, V_L, and $[E]_T$ are, respectively, the substrate concentration per interface area, the total lipid interface area, the total volume of lipid, and the analytical enzyme concentration[17]; additional terms are illustrated in Fig. 1. As shown in Eq. (2), the apparent V_{max} monitors both catalytic turnover (k_{cat}) and substrate binding at the active site (K_m^*), while Eq. (3) shows that the

---

[11] S. M. Watt and W. J. Simmonds, *J. Lipid Res.* **22**, 157 (1981).
[12] C. R. Borha, G. V. Vahouny, and C. Treadwell, *Am. J. Physiol.* **206**, 223 (1960).
[13] P. N. Howles, C. P. Carter, and D. Y. Hui, *J. Biol. Chem.* **271**, 7196 (1996).
[14] J. Brodt-Eppley, P. White, S. Jenkins, and D. Y. Hui, *Biochim. Biophys. Acta* **1272**, 69 (1995).
[15] S. C. Meyers-Payne, D. Y. Hui, and H. L. Brockman, *Biochemistry* **34**, 3942 (1995).
[16] F. Li, Y. Huang, and D. Y. Hui, *Biochemistry* **35**, 6657 (1996).
[17] R. Verger and G. H. de Haas, *Annu. Rev. Biophys. Bioeng.* **5**, 77 (1976).

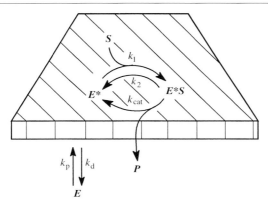

FIG. 1. Mechanism of interfacial lipolytic enzyme catalysis.

apparent K_m depends both on enzyme binding to the interface (k_d/k_p) and on subsequent substrate binding to the active site. Consequently, a traditional Michaelis–Menten and Lineweaver–Burk analysis[18] of inhibition of CEase-catalyzed hydrolysis or synthesis of lipid esters is worthless. Fortunately, CEase catalyzes the hydrolysis of water-soluble substrates, which provides assays that avoid the vagaries of interfacial biocatalysis.

$$v_i = \frac{V_{\max}^{\mathrm{app}} \left(A \cdot \frac{I_L}{V_L} \right)}{K_m^{\mathrm{app}} + \left(A \cdot \frac{I_L}{V_L} \right)} \quad (1)$$

$$V_{\max}^{\mathrm{app}} = \frac{k_{\mathrm{cat}}[E]_T A}{K_m^* + A} \quad (2)$$

$$K_m^{\mathrm{app}} = \frac{k_d}{k_p} \frac{K_m^* A}{K_m^* + A} \quad (3)$$

The catalytic mechanism of CEase, outlined in Fig. 2, utilizes a hydrogen-bonded triad of active site residues, namely, Ser-194-His-435-Asp-320. After substrate binding, the hydrolytic reaction proceeds in successive acylation and deacylation stages.[19,20] This mechanism describes events that follow binding of CEase to substrate-containing lipid interfaces, and inclusively

---

[18] I. R. Segel, in "Enzyme Kinetics" p. 100. John Wiley & Sons, New York, 1975.
[19] D. Lombardo and O. Guy, Biochem. Biophys. Acta **657**, 425 (1981).
[20] J. S. Stout, S. D. Sutton, and D. M. Quinn, Biochem. Biophys. Acta **837**, 6 (1985).

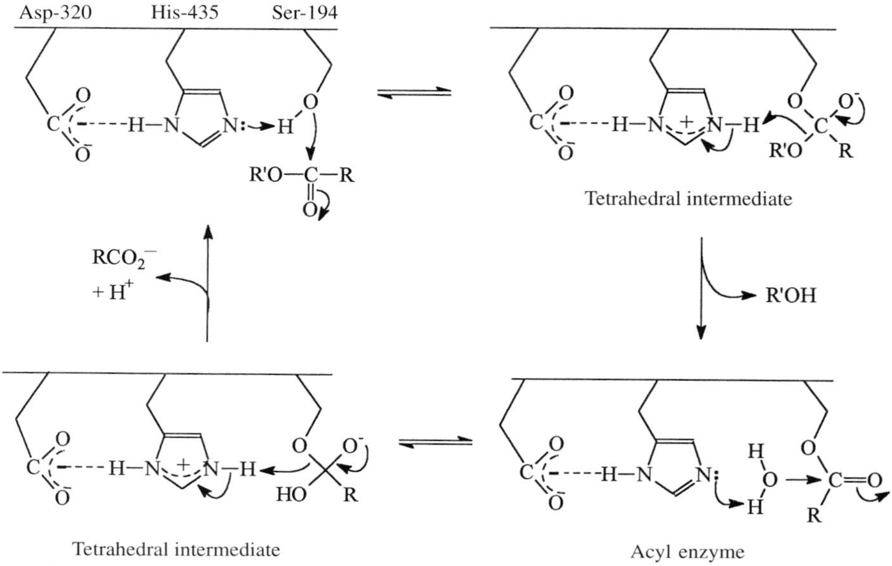

FIG. 2. Chemical mechanism of CEase analysis.

describes the events in each catalytic cycle for turnover of water-soluble substrates. This acyl enzyme mechanism is reminiscent of the mechanisms of serine proteases[21] and acetylcholinesterase,[22] which suggests that the classes of agents that inhibit these enzymes will inhibit CEase as well.

Substrate Specificity

Although it is the only pancreatic enzyme with high hydrolytic activity for cholesteryl esters, CEase nonetheless catalyzes hydrolytic and synthetic reactions of a wide range of substrates.[2,4,23,24] In Volume 71 of *Methods in Enzymology,* Gallo outlined radiochemical methods for assaying hydrolysis of cholesteryl oleate, and synthesis of the same ester from oleic acid and cholesterol, in mixed micelles that contained bile salts.[25] Rudd and Brockman have tabulated the various classes of substrates that are hydrolyzed by CEase.[4] These include cholesteryl esters that have saturated and unsatu-

[21] J. Kraut, *Annu. Rev. Biochem.* **46,** 331 (1977).
[22] P. Taylor and Z. Radić, *Annu. Rev. Pharmacol. Toxicol.* **34,** 281 (1994).
[23] D. Kritchevsky and H. V. Kothari, *Adv. Lipids Res.* **16,** 221 (1978).
[24] J. Hyun, H. Kothari, E. Herm, J. Mortenson, C. R. Treadwell, and G. V. Vahouny, *J. Biol. Chem.* **244,** 1937 (1969).
[25] L. L. Gallo, *Methods Enzymol.* **71,** 664 (1981).

rated fatty acyl chains; short acyl chain synthetic cholesteryl esters; tri-, di-, and monoacylglycerols; phospholipids; vitamin esters[26]; and phenyl esters. This broad substrate specificity suggests that CEase possesses an extensive molecular surface, contiguous with the locus of Ser-194 of the catalytic triad, that recognizes the functional groups of lipid substrates. Recent molecular modeling studies, outlined in the next section, suggest the regions of CEase that provide this extended active site.[27] Understanding the interactions that occur in this extended active site is crucial in the design of CEase inhibitors of maximum potency and specificity.

Enzyme Structure

No X-ray crystal structure of a mammalian CEase has been reported. However, human, rat, and bovine CEases are members of a supergene family referred to as the α/β hydrolase-fold family.[28,29] Enzymes of this family share secondary and tertiary structural features, even when sequence similarity is slight.[29] Of the 65 members of this supergene family,[30] 61 are hydrolytic enzymes and 55 of these utilize the type of Ser–His–carboxylate triad mechanism outlined in Fig. 2.[31,32] X-ray structures of several hydrolytic enzymes in this family are available on the Internet from the Protein Data Bank in Brookhaven, NY. Of these, the structures of *Torpedo californica* acetylchoalinesterase and the lipases from *Candida rugosa* and *Geotrichum candidum* have been used as templates to formulate a model of rat CEase by homology modeling.[27] The rat CEase model that resulted from this process is shown in Fig. 3, and features of the model that enlighten the development and evaluation of inhibitors are discussed below.

The active site of CEase is noted by the positions of the catalytic triad residues in Fig. 3 and is situated at the bottom of a wide cleft in the center of the view. Three topological hydrophobic domains converge at Ser-194 of the triad and are proposed to be involved in recognition of active site ligands.[27] The domain denoted by ST, just to the left of Ser-194, is the

---

[26] H. A. Zahalka, P. J. Dutton, B. O'Doherty, T. A.-M. Smart, J. Phipps, D. O. Foster, G. W. Burton, and K. U. Ingold, *J. Am. Chem. Soc.* **113**, 2797 (1991).
[27] S. R. Feaster, B. L. Barnett, and D. M. Quinn, *Protein Sci.* **6**, 71 (1997).
[28] D. L. Ollis, E. Cheah, M. Cygler, B. Dijkstra, F. Frolow, S. M. Franken, M. Harel, S. J. Remington, I. Silman, J. Schrag, J. Sussman, K. H. G. Verschueren, and A. Goldman, *Protein Eng.* **5**, 197 (1992).
[29] M. Cygler, J. D. Schrag, J. L. Sussman, M. Harel, I. Silman, M. K. Gentry, and B. P. Doctor, *Protein Sci.* **2**, 366 (1993).
[30] The Web address for the Esther homepage is www.montpellier.inra.fr:7010/cholinesterase. This site contains sequence comparisons for proteins of the α/β hydrolase-fold family.
[31] D. M. Blow, *Acc. Chem. Res.* **9**, 145 (1976).
[32] R. M. Stroud, *Sci. Am.* **231**, 74 (1974).

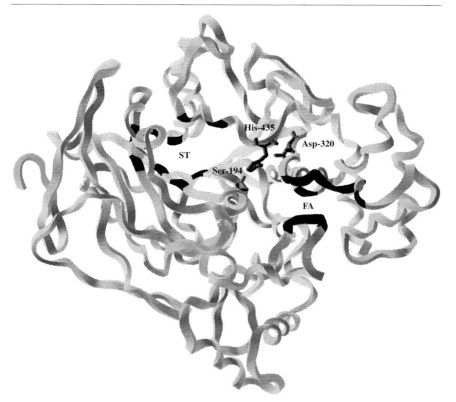

FIG. 3. Ribbon representation of the rat CEase model. In this figure, ST and FA represent the steroid and fatty acyl binding domains, respectively. For clarity, segments of the ribbon corresponding to the residues defining these domains have been colored black. The members of the catalytic triad (Ser-194-His-435-Asp-320) appear as capped sticks.

steroid binding locus. The localization of ST in the structure is similar to that of the steroid-binding domain of a fungal CEase from *Candida cylindracea,* identified by X-ray crystallography.[33] Radiating from Ser-194 toward the right in the view is one of the two putative fatty acid binding sites.[27]

The convergence of three hydrophobic domains at Ser-194 is consistent with the high catalytic activity of CEase for diverse lipid substrates, such as triacylglycerols and cholesteryl esters. A final and notable feature of the structure is that the active site cleft is not occluded by surface loops, as

[33] D. Ghosh, Z. Wawrzak, V. Z. Pletnev, N. Li, R. Kaiser, W. Pangborn, H. Jörnvall, M. Erman, and W. L. Duax, *Structure* **3,** 279 (1995).

previously noted for lipases from *Rhizomucor meihei*,[34,35] *G. candidum*,[36,37] *C. rugosa*,[38,39] and pancreatic lipase.[40–43] Consequently, the evaluation of inhibition kinetics for CEase is not complicated by the interfacial activation phenomena that are reported for many lipases.[34,40,42]

Structural Classes of CEase Inhibitors

This section briefly outlines the classes of agents that have been used to inhibit CEase, and where appropriate describes methods used to characterize the inhibitors. The next section describes in detail the methods used to characterize inhibition by carbamates, the most studied class of CEase inhibitors.

Figure 4 outlines the various classes of active site-directed CEase inhibitors that have been reported to date. Among them are boronic and borinic acids,[44,45] aryl haloketones,[46] aryl phosphates and phosphonates,[47,48] aryl carbamates,[49–53] and cholesteryl carbamates.[54] One of the first reported

[34] A. M. Brzozowski, U. Derewenda, Z. S. Derewenda, G. G. Dodson, D. M. Lawson, J. P. Turkenburg, R. Bjorkling, B. Huge-Jensen, S. A. Patkar, and L. Thim, *Nature* **351**, 491 (1992).
[35] U. Derewenda, A. M. Brzozowski, D. M. Lawson, and Z. S. Derewenda, *Biochemistry* **31**, 1532 (1992).
[36] J. D. Schrag, Y. Li, S. Wu, and M. Cygler, *Nature* **351**, 761 (1991).
[37] J. D. Schrag and M. Cygler, *J. Mol. Biol.* **230**, 575 (1993).
[38] P. Grochulski, Y. Li, J. D. Schrag, F. Bouthillier, P. Smith, D. Harrison, B. Rubin, and M. Cygler, *J. Biol. Chem.* **268**, 12,843 (1993).
[39] P. Grochulski, F. Bouthillier, R. J. Kazlauskas, A. N. Serreqi, J. D. Schrag, E. Ziomek, and M. Cygler, *Biochemistry* **33**, 3494 (1994).
[40] H. van Tilbeurgh, L. Sarda, R. Verger, and C. Cambillau, *Nature* **359**, 159 (1992).
[41] H. van Tilbeurgh, M.-P. Egloff, C. Martinez, N. Rugani, R. Verger, and C. Cambillau, *Nature* **362**, 814 (1993).
[42] M. P. Egloff, L. Sarda, R. Verger, C. Cambillau, and H. van Tilbeurgh, *Protein Sci.* **4**, 44 (1995).
[43] F. K. Winkler, A. D'Arcy, and W. Hunziker, *Nature* **343**, 771 (1990).
[44] L. D. Sutton, J. L. Lantz, T. Eibes, and D. M. Quinn, *Biochem. Biophys. Acta* **1041**, 79 (1990).
[45] L. D. Sutton, J. S. Stout, L. Hosie, P. S. Spencer, and D. M. Quinn, *Biochem. Biophys. Res. Commun.* **134**, 386 (1986).
[46] J. Sohl, L. D. Sutton, D. J. Burton, and D. M. Quinn, *Biochem. Biophys. Res. Commun.* **151**, 554 (1988).
[47] D. M. Quinn, L. D. Sutton, J. S. Stout, T. Calogeropoulou, and D. F. Wiemer, *Phosphorous Sulfur Silicon* **51/52**, 43 (1990).
[48] S. K. Murthy and J. Ganguly, *Biochem. J.* **83**, 460 (1962).
[49] G. Lin and C.-Y. Lai, *Tetrahedron Lett.* **34**, 6117 (1995).
[50] S. R. Feaster, K. Lee, N. Baker, D. Y. Hui, and D. M. Quinn, *Biochemistry* **35**, 16723 (1996).
[51] D. M. Quinn, U.S. Patent 5066674A (1991).
[52] T. J. Commons and D. P. Strike, U.S. Patent 5169844A (1992).
[53] R. E. Mewshaw, T. J. Commons, and D. P. Strike, E.P. 42835 A1 (1991).
[54] L. Hosie, L. D. Sutton, and D. M. Quinn, *J. Biol. Chem.* **262**, 260 (1987).

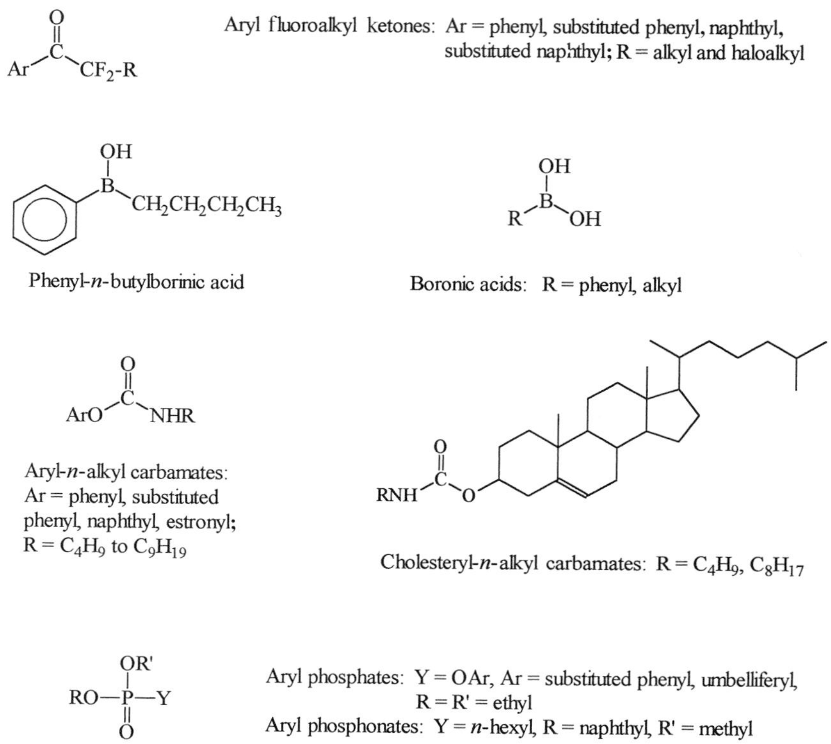

Aryl fluoroalkyl ketones: Ar = phenyl, substituted phenyl, naphthyl, substituted naphthyl; R = alkyl and haloalkyl

Phenyl-*n*-butylborinic acid

Boronic acids: R = phenyl, alkyl

Aryl-*n*-alkyl carbamates:
Ar = phenyl, substituted phenyl, naphthyl, estronyl;
R = C_4H_9 to C_9H_{19}

Cholesteryl-*n*-alkyl carbamates: R = C_4H_9, C_8H_{17}

Aryl phosphates: Y = OAr, Ar = substituted phenyl, umbelliferyl, R = R' = ethyl
Aryl phosphonates: Y = *n*-hexyl, R = naphthyl, R' = methyl

FIG. 4. Structural classes of CEase inhibitors.

inhibitions of CEase was by the phosphorylating agent diethyl *p*-nitrophenyl phosphate,[48] which presumably reacts with the active site serine. Although this inhibitor was shown to block 50% of CEase activity in only 15 min even at low concentrations (e.g., 200 nM), no details of the kinetics of inhibition were provided.

Measurement of inhibitions of porcine pancreatic CEase by boronic and borinic acids[44,45] and haloketones[46] utilized the chromogenic, water-soluble substrate *p*-nitrophenyl butyrate (PNPB). The assay and reaction conditions are outlined in Scheme 1. Reactions are typically run in sodium phosphate buffers and the corresponding time courses are monitored by following the production of *p*-nitrophenoxide in a UV-visible spectrophotometer at 400 nm. The high surface activity of CEase[4] can be a source of irreproducibility of kinetics measurements, probably because the enzyme is sticking to the walls of the cells in which the reaction is monitored. This situation can be avoided by including micellar concentrations of Triton

p-Nitrophenyl butyrate (PNPB) → (CEase, H₂O) → p-Nitrophenoxide + CH₃CH₂CH₂CO₂⁻

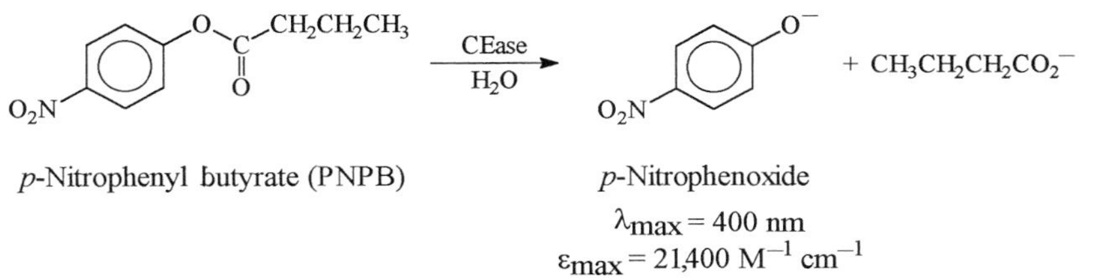

p-Nitrophenyl butyrate (PNPB)

p-Nitrophenoxide
$\lambda_{max} = 400$ nm
$\varepsilon_{max} = 21{,}400$ M⁻¹ cm⁻¹

SCHEME 1. Spectrophotometric assay for cholesterol esterase. Typical reaction conditions: T, 25°; pH, 7.0 (0.1 M sodium phosphate buffer, 0.1 M NaCl); [PNPB], 50–200 μM; [CEase], 0.3–3 μg ml⁻¹.

X-100 [critical micellar concentration (CMC) = 0.2 mM][55] in enzyme stock solutions. Dilution of the stock enzyme on injection into the reaction mixture (e.g., by 100-fold) also dilutes Triton X-100 to well below its CMC value. Rates measured in this way are similar in magnitude to but far more reproducible than those measured in the absence of Triton X-100. Additional reaction conditions that are germaine to particular classes of inhibitors are provided in figure legends.

Boronic and borinic acids and haloketones are rapid, linear competitive inhibitors of CEase. This means that inhibition patterns are easily discerned by plotting rates as a function of varied substrate and inhibitor concentrations according to the Lineweaver–Burk transform [Eq. (5)] of the Michaelis–Menten equation [Eq. (4)][18]:

$$\nu_i = \frac{V_{max}^{app}[A]}{K_m^{app} + [A]} \quad (4)$$

$$\frac{1}{\nu_i} = \frac{K_m^{app}}{V_{max}^{app}} \frac{1}{[A]} + \frac{1}{V_{max}^{app}} \quad (5)$$

where V_{max}^{app} and K_m^{app} depend on the concentration of reversible inhibitors in a way that is diagnostic of the inhibition mechanism.[18] For example, $V_{max}^{app} = V_{max}$ and $K_m^{app} = K_m(1 + [I]/K_i)$ for linear competitive inhibition. An alternative and less labor-intensive method for characterizing reversible inhibition is to follow reaction time courses to completion, beginning with $[A] > K_m$. The time courses are consequently fit to the integrated form of the Michaelis–Menten equation [Eq. (6)][20]:

$$t = \frac{K_m^{app}}{V_{max}^{app}} \ln \frac{[A]_0}{[A]} + \frac{1}{V_{max}^{app}} ([A]_0 - [A]) \quad (6)$$

[55] "Rohm & Haas Surfactants and Dispersants—Handbook of Physical Properties." Rohm & Hass, Philadelphia, 1978.

A linear transform of the integrated Michaelis–Menten equation can be used, as the Lineweaver–Burk transform in Eq. (5) of the differential Michaelis–Menten equation is used, to analyze visually the inhibition mechanism [Eq. (7)]:

$$\frac{\Delta t}{\Delta [A]} = \frac{K_m^{app}}{V_{max}^{app}} \frac{\ln([A]_i/[A]_{i+10})}{\Delta[A]} - \frac{1}{V_{max}^{app}} \qquad (7)$$

In Eq. (7) Δt is the time increment between observations at times t_i and t_{i+10}, and $\Delta[A]$ is the corresponding substrate concentration change (i.e., $[A]_{i+10} - [A]_i$). Sutton et al. show such an analysis for inhibition of porcine pancreatic CEase by hexylboronic acid,[44] and Table I gives results so obtained for boronic acids and haloketones.[44–46]

Phenyl-n-butylborinic acid is a sufficiently potent inhibitor of porcine pancreatic CEase that, in its characterization, inhibitor concentrations are comparable to the assay concentrations of the enzyme. In this situation inhibitor depletion occurs, and consequently Henderson tight-binding anal-

TABLE I
BORONIC ACID AND HALOKETONE INHIBITORS OF CEASE

Inhibitor	K_i
Phenylboronic acid[a]	36 μM
Phenyl-n-butylborinic acid[a]	2.9 nM
n-Butylboronic acid[a,b]	223 μM
Methylboronic acid[a]	No inhibition
Pentylboronic acid[a]	187 μM
Hexylboronic acid[a]	13 μM
Heptylboronic acid[a]	2.5 mM
Octylboronic acid[a]	7 mM
Chlorodifluoroacetophenone[c]	36 nM
Trifluoroacetophenone[c]	340 nM
Pentafluoropropiophenone[c]	4.6 mM
Heptafluorobutyrophenone[c]	32 mM

[a] L. D. Sutton, J. L. Lantz, T. Eibes, and D. M. Quinn, *Biochem. Biophys. Acta* **1041**, 79 (1990).
[b] L. D. Sutton, J. S. Stout, L. Hosie, P. S. Spencer, and D. M. Quinn, *Biochem. Biophys. Res. Commun.* **134**, 386 (1986).
[c] J. Sohl, L. D. Sutton, D. J. Burton, and D. M. Quinn, *Biochem. Biophys. Res. Commun.* **151**, 554 (1988).

ysis is employed.[56] For tight-binding competitive inhibitors the following interaction [Eq. (8)] mechanism applies:

$$E + A \underset{K_m}{\rightleftarrows} EA \xrightarrow{k_{cat}} E + P$$
$$I \updownarrow \Big\downarrow K_i$$
$$EI$$

$$\frac{\nu_i}{\nu_0} = \frac{[E]_T - [EI]}{[E]_T} \tag{8}$$

In Eq. (8), ν_i and ν_0 are rates of CEase-catalyzed hydrolysis of PNPB in the presence and absence of inhibitor, and $[E]_T$ is the total enzyme concentration. Equation (9) describes the enzyme-inhibitor equilibrium for the interaction mechanism just given:

$$K_i = \frac{[E][I]}{[EI]} = \frac{([E]_T - [EI])([I]_0 - [EI])}{[EI]\left(1 + \frac{[A]}{K_m}\right)} \tag{9}$$

Solving Eq. (8) for $[EI]$, inserting the result into Eq. (9), and rearranging gives Eq. (10):

$$\nu_i = \frac{\nu_0}{2[E]_T}\{[E]_T - [I]_0 - K_i^a$$
$$+ \sqrt{[I]_0^2 + [E]_T^2 + (K_i^a)^2 + 2K_i^a([E]_T + [I]_0) - 2[I]_0[E]_T}\} \tag{10}$$

This equation is a nonlinear version of the linear steady-state rate equation derived by Henderson.[56] The K_i value for phenyl-n-butylborinic acid was determined by fitting ν_i versus $[I]_0$ data to Eq. (10), with the CEase activity in the absence of inhibitor, ν_0, the apparent inhibition constant, $K_i^a = K_i(1 + [A]/K_m)$, and the total enzyme concentration, $[E]_T$, as adjustable parameters. The fit is displayed in Fig. 5.

Inhibition by Carbamates

Carbamates comprise the most studied class of CEase inhibitors, both in a mechanistic sense[49,50] and for their potential as hypocholesterolemic agents.[51,57] These compounds interact with the CEase active site according

---

[56] P. J. F. Henderson, *Biochem. J.* **127,** 321 (1972).
[57] D. E. Clark, T. J. Commons, D. H. Prozialeck, M. L. McKean, and S. J. Adelman, *Fed. Am. Soc. Exp. Bio. J.* **6,** A1388 (1992).

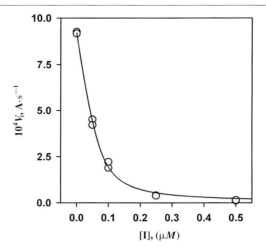

FIG. 5. Henderson tight-binding analysis for inhibition of pig CEase by phenyl-*n*-butylborinic acid. Reaction conditions are as described in Scheme 1 except that the buffer contained 1.0 mM sodium taurocholate and the pH was 7.05. The solid line is a least-squares fit to Eq. (10); the parameters of the fit are $[E]_T = 72 \pm 6$ nM, $\nu_0 = 9.2 \pm 0.1 \times 10^{-4}$ A·s^{-1} (i.e., A = absorbance), and $K_i^a = 11 \pm 2$ nM. Since $1 + [A]/K_m$ was 1.5, $K_i = 8 \pm 1$ nM.

to the mechanism in Scheme 2. This mechanism is *pro forma* that for substrate turnover, and therefore carbamates are best described as pseudosubstrate inhibitors whose potencies derive in part from the slowness of the decarbamylation phase of the reaction. This section details methods used in the laboratory of the authors for kinetically resolving the equilibrium and rate constants of this mechanism, $K_C = k_2/k_1$, k_3, and k_5. This information can be utilized in structure-activity studies of reversible binding to the active site (i.e., K_C), and of the carbamylation and decarbamylation transition states, monitored by k_3 and k_5, respectively. In this way, one can maximize inhibitor potency by optimizing reversible affinity and carbamylenzyme lifetime.

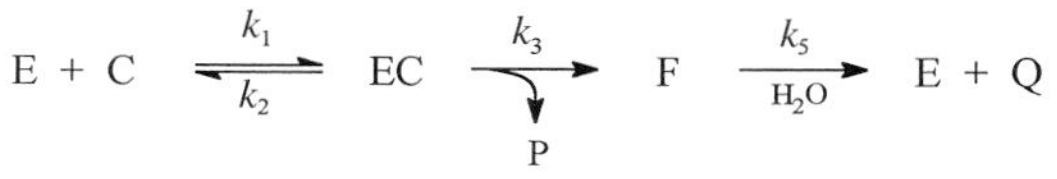

$$E + C \underset{k_2}{\overset{k_1}{\rightleftharpoons}} EC \overset{k_3}{\underset{P}{\searrow}} F \overset{k_5}{\underset{H_2O}{\longrightarrow}} E + Q$$

SCHEME 2. Carbamyl enzyme mechanism of CEase inhibition. C, carbamate inhibitor; F, carbamyl enzyme intermediate; P, Q, alcohol and carbamic acid products; EC, noncovalent complex.

Assay Methods for Kinetic Characterization of Carbamylation Events

For arylcarbamates, the carbamylation phase of the reaction is considerably more rapid than the decarbamylation phase (i.e., $k_3 \gg k_5$) and, therefore, the successive phases can be characterized separately. The more rapid carbamylation phase can be kinetically characterized in two ways. One is a traditional stopped-time assay in which CEase and inhibitor are mixed in the assay buffer mentioned in Scheme 1 and, at various times, aliquots are removed to a spectrophotometer cell for determination of residual activity of CEase-catalyzed hydrolysis of PNPB. A parallel control (i.e., no inhibitor in the mixture from which aliquots are taken) allows one to adjust activities measured at various times for spontaneous loss of CEase activity. These control corrections are not needed for stopped-time assays of 2 hr or less duration, since the enzyme maintains >95% of its initial activity over this period of time. Stopped-time assays are needed when inhibitor potency is not great, or when low concentrations of an inhibitor, whether potent or not, are being assayed. One obtains values of the observed pseudo first-order inhibition rate constant, k_{obs}, by fitting data to Eq. (11):

$$R = R_0 e^{-k_{obs} t} + R_\infty \tag{11}$$

where R, R_0, and R_∞ are ratios of the inhibited activity to the control activity at times t, 0, and ∞, respectively. To obtain reliable estimates of R_0, R_∞, and k_{obs}, inhibition reactions should be followed for at least three half-lives.

An alternative continuous assay method can be used for inhibitions that are complete in 2 hr or less. In this method, CEase, inhibitor, and substrate are mixed in the assay buffer and the reaction time course is monitored continuously, for example, at 400 nm with a UV-visible spectrophotometer when PNPB is used as the substrate (cf. Scheme 1). An initial CEase concentration is determined by trial and error, such that <10% substrate depletion occurs during the time course of the inhibition reaction. This is necessary so that time-dependent decreases in instantaneous activity reflect progressive inhibition only. When inhibition reactions are so monitored, a biphasic time course such as that shown in Fig. 6 is observed that is described by the following function [Eq. (12)]:

$$A = A_0 + v_{ss} t + \frac{v_0 - v_{ss}}{k_{obs}} (1 - e^{-k_{obs} t}) \tag{12}$$

In Eq. (12) A and A_0 are absorbances at times t and 0, respectively; v_0 and v_{ss} are instantaneous velocities at time 0 and in the linear, steady-state phase of the reaction, respectively; and k_{obs} is again the observed pseudo first-order rate constant. Other methods of continuously monitoring reac-

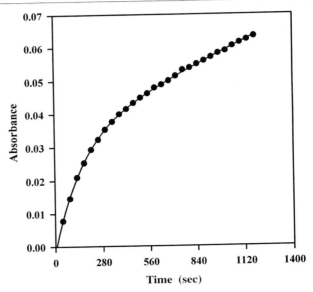

FIG. 6. Biphasic time course for inhibition of rat CEase by 2-(5,6,7,8-tetrahydronaphthyl)-N-hexyl carbamate. The reactions were run as described in Scheme 1 in buffers that also contained 0.160 N NaCl, 100 μM sodium taurocholate, 20 μM Triton X-100, 50 μM PNPB, 500 nM carbamate, 2.7 nM RCEase, and 2% acetonitrile (v/v). The solid line is a least-squares fit to Eq. (12); the parameters of the fit are $A_0 = -0.0017 \pm 0.0001$, $k = 6.16 \pm 0.03 \times 10^{-3}$ s^{-1}, $v_{ss} = 2.563 \pm 0.008 \times 10^{-5}$ A·s^{-1}, and $v_0 = 2.418 \pm 0.006 \times 10^{-4}$ A·s^{-1}.

tions, such as fluorescence or the pH-stat, could also be used. Reactions must be followed for at least six half-lives (>98% completion) to obtain reliable estimates of the parameters, especially v_{ss} and k_{obs}.

Once k_{obs} values have been determined at various inhibitor concentrations, the resulting data are fit to Eq. (13):

$$k_{obs} = \frac{k_3[I]}{K_C^{app} + [I]} \quad (13)$$

This equation is analogous *pro forma* to the Michaelis–Menten equation [cf. Eq. (4)], and gives values of the carbamylation rate constant k_3 and the thermodynamic dissociation constant K_C, per the mechanism in Scheme 2. If k_{obs} values were determined by the stopped-time method, then $K_C^{app} = K_C$; if k_{obs} values were determined by the continuous assay method, then $K_C^{app} = K_C(1 + [A]/K_m)$ is used to calculate K_C, where $[A]$ and K_m are the initial substrate concentration and its Michaelis constant, respectively. Figure 7A provides an example of data fit to Eq. (13) for inhibition of rat CEase by 2-(5,6,7,8)-tetrahydronaphthyl-N-hexyl carbamate.[50]

The use of Eq. (12) and the continuous assay method affords an additional means for determining K_C. If the formation of EC in Scheme 2 is a rapid equilibrium process, then v_0 values determined from fits of inhibition time courses to Eq. (12) will be a function of $[I]$ [Eq. (14)]:

$$v_0 = \frac{v_0^0 K_C^{app}}{[I] + K_C^{app}} \tag{14}$$

Here v_0 is the initial velocity of the biphasic inhibition timecourse and v_0^0 is the corresponding least-squares extrapolation to $[I] = 0$ on fitting data to Eq. (14). Hence, v_0^0 is always in close agreement with the control CEase activity. A fit of data to Eq. (14) for inhibition of rat CEase by 2-(5,6,7,8)-tetrahydronaphthyl-N-hexyl carbamate is plotted in Fig. 7B. Values of K_C determined by this approach agree well with those that come from fitting data to Eq. (13).

Residual Velocity Methods for Kinetic Characterization of Decarbamylation

Residual velocity in time courses for inhibition of CEase by carbamates is measured by R_∞ and v_{ss} in Eqs. (11) and (12) for the stopped-time and continuous assay methods, respectively. This residual velocity consists of two components: (1) the rate of nonenzymic, buffer-catalyzed hydrolysis of the substrate, which can be measured in blank reactions that contain substrate and inhibitor but no CEase; and (2) the rate of steady-state turnover of the carbamate, that is, $k_3[EC] = k_5[F]$ in Scheme 2. This steady-state turnover is analogous to that for substrates in an acyl enzyme mechanism, and provides a way of measuring the decarbamylation rate constant k_5. The rate of steady-state turnover is calculated by subtracting the rate of an appropriate substrate blank from the residual velocity.

The initial velocity of a control CEase assay is, via the Michaelis–Menten equation [Eq. (4)], a measure of $[E]_T$, the analytical enzyme concentration. Correspondingly, the steady-state turnover rate of an inhibition assay provides a measure of the concentration of free CEase that is continually replenished by decarbamylation [Eq. (15)]:

$$v_{ss}^c = \frac{v_0 K_m^I + v_I[I]}{K_m^I + [I]} \tag{15}$$

The rate v_{ss}^C is the measured blank-corrected steady-state rate, and the least-squares adjustable parameters v_0, v_I, and K_m^I are, respectively, the rate extrapolated to zero inhibitor concentration, the asymptotic rate at

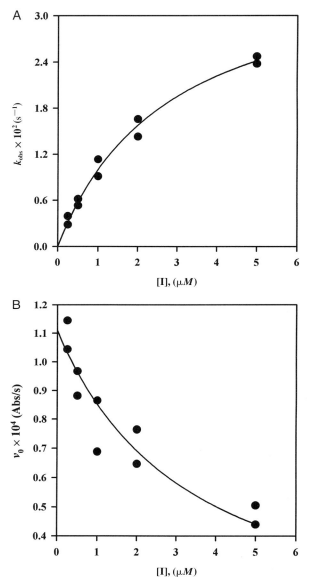

FIG. 7. Inhibition of rat CEase by 2-(5,6,7,8-tetrahydronaphthyl)-N-hexyl carbamate. The reactions were run as described in Scheme 1 in buffers that also contained 0.160 N NaCl, 100 μM sodium taurocholate, 20 μM Triton X-100, 50 μM PNPB, 0.25–5.00 μM 2-(5,6,7,8-tetrahydronaphthyl)-N-hexyl carbamate, 1.2–9.4 nM RCEase, and 2% acetonitrile (v/v). (A) Nonlinear dependence of k_{obs} on [I]. The solid line is a least-squares fit to Eq. (13); the parameters of the fit are $k_3 = 3.7 \pm 0.3 \times 10^{-2}$ s^{-1} and $K_i = 2.7 \pm 0.4$ μM. (B) Nonlinear dependence of v_0 on [I]. The solid line is a least-squares fit to Eq. (14); the parameters of the fit are $v_0^0 = (1.18 \pm 0.06) \times 10^{-4}$ A · s^{-1} and $K_i = 3.3 \pm 0.7$ μM.

infinite inhibitor concentration, and the Michaelis constant for CEase-catalyzed hydrolysis of the carbamate pseudosubstrate. As for the K_m value of a substrate that is hydrolyzed via an acyl enzyme mechanism, K_m^1 is given by Eq. (16).

$$K_m^1 = K_C \frac{k_5}{k_3 + k_5} \qquad (16)$$

Values of ν_0 and K_m^1 are calculated by least-squares fits to Eq. (15), and k_5 is subsequently calculated from Eq. (16) by using values of K_C and k_3 determined as described previously.

The asymptotic value of the turnover rate at saturating carbamate concentration, ν_1, also provides a measure of the decarbamylation rate constant k_5 [Eq. (17)]:

$$\nu_1 = \frac{k_5 \nu_0}{k_3 + k_5} \qquad (17)$$

When decarbamylation is much slower than carbamylation (i.e., $k_5 \ll k_3$), then $\nu_1 \approx (k_5/k_3)\nu_0$. In this situation ν_1 is usually less than 2% of ν_0, and therefore k_5/k_3 cannot be determined accurately. However, for inhibitors that carbamylate the active site of CEase slowly, ν_1 provides a convenient second check of k_5.

Kinetic Characterization of Decarbamylation by Numerical Integration

When the initial inhibitor concentration is just greater than the CEase concentration (typically two- to threefold), stopped-time assays show rapid loss of CEase activity and subsequent slow but full activity return, as illustrated in Fig. 8. The time course in this figure therefore monitors full pseudosubstrate turnover of the carbamate inhibitor, an observation that is described by the following simplified version of Scheme 2:

$$E + C \xrightarrow[P]{k_3/K_C} F \xrightarrow{k_5}_{H_2O} E + Q$$

This simplified version is obtained when $[C]_0$, the initial carbamate concentration, is $\ll K_C$. The following set of differential equations [Eqs. (18)–(20)] describes the time dependencies of $[E]$, $[C]$, and $[F]$ for this version of the mechanism in Scheme 2:

$$\frac{d[E]}{dt} = k_5[F] - \frac{k_3}{K_C}[E][C] \qquad (18)$$

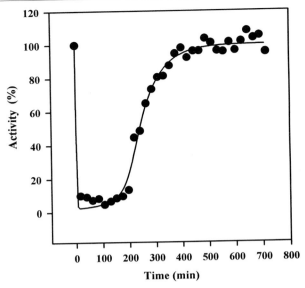

FIG. 8. Stopped-time assay for transient inhibition of rat CEase by 2-naphthyl-N-octyl carbamate. The reaction contained 64 nM CEase in 0.1 M sodium phosphate buffer, pH 7.00, that contained 0.16 N NaCl, 6 mM sodium taurocholate, 2% acetonitrile (v/v), and 150 nM 2-naphthyl-N-octyl carbamate. The incubation temperature was 25.0 ± 0.1°. At the indicated times, 40 μl of the reaction mixture were added to 2.00 ml of the buffer in Scheme 1 that also contained 0.16 N NaCl, 6 mM sodium taurocholate, and 40 μl of 5.0 mM PNPB dissolved in acetonitrile. Initial rates were calculated for each assay by linear least-squares analysis. The solid line is the best fourth-order Runge–Kutta numerical simulation of the data; the parameters for this simulation are k_3/K_C = 88,000 ± 2000 $M^{-1} \cdot s^{-1}$ and k_5 = (2.828 ± 0.003) × 10^{-4} s^{-1}.

$$\frac{d[C]}{dt} = -\frac{k_3}{K_C}[E][C] \tag{19}$$

$$\frac{d[F]}{dt} = \frac{k_3}{K_C}[E][C] - k_5[F] \tag{20}$$

There is no analytical solution for this set of differential equations, which consequently must be solved by numerical methods. This is done by approximating the differential equations in terms of finite time intervals, Δt, and species concentrations. The resulting approximate versions of the differential equations are integrated numerically by using fourth-order Runge–Kutta integration.[58,59] The requisite expressions for the species concen-

[58] B. K. Carpenter, in "Determination of Organic Reaction Mechanisms" p. 76. John Wiley & Sons, New York, 1984.
[59] H.-C. Shin and D. M. Quinn, *Biochemistry* **31**, 811 (1992).

trations at time j of the CEase activity time course are given by Eqs. (21)–(23):

$$[E]_j = [E]_{j-1} + \frac{e_{1j} + 2e_{2j} + 2e_{3j} + e_{4j}}{6} \tag{21}$$

$$[C]_j = [C]_{j-1} + \frac{c_{1j} + 2c_{2j} + 2c_{3j} + c_{4j}}{6} \tag{22}$$

$$[F]_j = [F]_{j-1} + \frac{f_{1j} + 2f_{2j} + 2f_{3j} + f_{4j}}{6} \tag{23}$$

The concentration increments, e_{ij}, in Eq. (21) are calculated in the following set of equations [Eqs. (24)–(27)]:

$$e_{1j} = \left\{ k_5[F]_{j-1} - \frac{k_3}{K_C}[E]_{j-1}[C]_{j-1} \right\} \Delta t \tag{24}$$

$$e_{2j} = \left\{ k_5\left([F]_{j-1} + \frac{f_{1j}}{2}\right) - \frac{k_3}{K_C}\left([E]_{j-1} + \frac{e_{1j}}{2}\right)\left([C]_{j-1} + \frac{c_{1j}}{2}\right) \right\} \Delta t \tag{25}$$

$$e_{3j} = \left\{ k_5\left([F]_{j-1} + \frac{f_{2j}}{2}\right) - \frac{k_3}{K_C}\left([E]_{j-1} + \frac{e_{2j}}{2}\right)\left([C]_{j-1} + \frac{c_{2j}}{2}\right) \right\} \Delta t \tag{26}$$

$$e_{4j} = \left\{ k_5([F]_{j-1} + f_{3j}) - \frac{k_3}{K_C}([E]_{j-1} + e_{3j})([C]_{j-1} + c_{3j}) \right\} \Delta t \tag{27}$$

Equation (24) is the finite time interval analog of differential Eq. (18), in which the time step is Δt; Eqs. (25)–(27) are similar in form except that the concentration increments are increased by using concentration increments calculated in the respective preceding equations. Similar sets of equations are formulated for c_{ij} and f_{ij} by referring to differential Eqs. (19) and (20). Typical finite time intervals used in integrations of CEase activity time courses are in the range of 0.5–1.0 sec, as specified by the experimenter.

Equations (21)–(23) illustrate that species concentrations at time j are sums of their respective concentrations at time $j - 1$ and weighted averages of corresponding concentration increments. The concentrations of E, C, and F are set at $[E]_T$, $[C]_0$, and 0 at time 0, and concentrations of these species are then calculated as a function of reaction time by using Eqs. (21)–(23). The experimental data that are input for Runge–Kutta integration are CEase activities, measured at various times as initial rates of PNPB hydrolysis, versus control activities, measured under the same conditions in the absence of the carbamate inhibitor. The rate of spontaneous hydrolysis of PNPB (i.e., in the absence of CEase) is measured separately, and subtracted from the overall rate at each time of the CEase inhibition time course. The relationship between active enzyme concentration, $[E]$, and activity is given by Eq. (28):

$$[E] = \frac{\nu_t^I}{\nu_t^c}[E]_T \tag{28}$$

The rates ν_t^I and ν_t^c are those for the inhibited and control reactions, respectively, at time t; $[E]_T$ is the analytical enzyme concentration.

A computer program that performs the numerical integration has been written in the C++ programming language. This program, which is available on request from the authors,[60] performs a grid search in parameter space for the combination of parameters k_3/K_C, k_5, and $[E]_T$ that minimizes the summed-squared-variance [Eq. (29)],

$$\text{SSV} = \sum_{j=1}^{n} [\nu_i^t(\text{observed}) - \nu_i^t(\text{calculated})]^2 \tag{29}$$

The index j is over the n observations of time-dependent CEase activity. The experimenter specifies initial values of the parameters and the initial carbamate concentration $[C]_0$. A simulation routine in the program can be used to generate initial parameter estimates. An 11 × 11 × 11 grid is computed about the user-input parameter estimates, with the parameter estimate being the central (sixth) value along each dimension of the grid. The grid tolerance (i.e., the space between parameter values along each grid dimension) is computed by dividing each parameter by 5. This grid of 1331 parameter estimates is searched for the set that has the least SSV, per Eq. (29). The grid is then compressed by a factor of 5 along each coordinate about this new, albeit optimized, set of parameter estimates, and the entire process is repeated. The user specifies the number of grid searches for each parameter, and hence the number of grid compressions, for each numerical integration of an inhibition time course. When the final parameter set estimate has been calculated, uncertainties in the parameters are calculated by assuming that the variation of χ^2 with respect to each parameter is independent of the values of the other parameters. In this case the following expression gives the uncertainty of the kth parameter a_k[61]:

$$\sigma_{a_k} = \sqrt{\frac{2}{\partial^2 \chi^2/\partial a_k^2}}. \tag{30}$$

---

[60] Send inquiries to Daniel M. Quinn, Dept. of Chemistry, University of Iowa, Iowa City, IA 52242, or to Shawn R. Feaster, Division of Biochemistry, Building 40, Walter Reed Army Institute of Research, Washington, DC 20307-5100.

[61] P. R. Bevington, in "Data Reduction and Error Analysis for the Physical Sciences," p. 242. McGraw-Hill, New York, 1969.

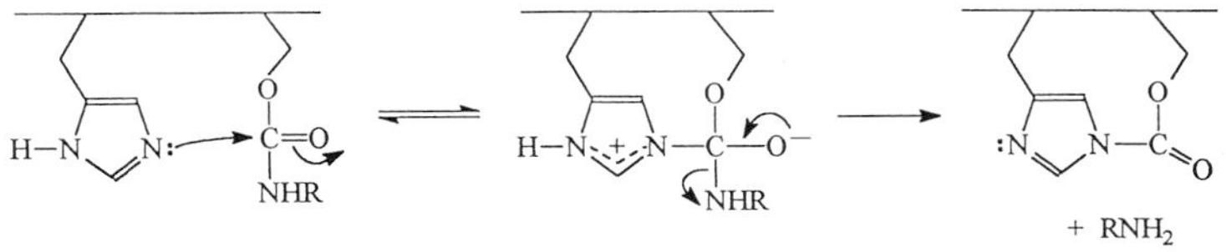

SCHEME 3. Irreversible inhibition of CEase by carbonyl cross-linking of the active site.

Various additional processes can accompany pseudosubstrate turnover by CEase of carbamate inhibitors. These include spontaneous loss of CEase activity, spontaneous hydrolysis of the carbamate, and diversion of F in Scheme 2 to an irreversibly inactivated form. These processes can be incorporated into differential Eqs. (18)–(20) and accounted for in the numerical integration in Eqs. (21)–(23). The program written in the authors' laboratory allows inclusion of side reactions as a user option. Generally the best way to handle spontaneous CEase inactivation and carbamate hydrolysis is to determine the appropriate first-order rate constants in separate control experiments, and subsequently account for these reactions in numerical integrations by introducing the rate constant as constrained parameters, rather than as parameters that are optimized by grid searching. This cannot be done for diversion of F to an irreversibly inactivated form of CEase. However, it is easy to detect this process, since in the activity recovery phase of the time course the final activity falls below that of the control time course. This situation has been encountered only rarely in the authors' laboratory, and then only for inhibition by structurally diminutive carbamates, such as 2-naphthyl-N-butyl carbamate in the absence of sodium taurocholate. The constitution of the diversion reaction of F to an inactivated form is not known. However, a good possibility is nucleophilic displacement of the alkylamine fragment of F by the imidazole sidechain of His-435, as shown in Scheme 3. Although this process has not been characterized in detail, the carbonyl cross-linking of the active site of the serine protease α-chymotrypsin on interaction with diarylcarbonate inhibitors provides a literature precedent.[62] An experiment that would be consistent with the mechanism in Scheme 3 is to resurrect the activity of CEase by treating the inhibited enzyme with NH_2OH. This experiment has not yet been done.

---

[62] T. H. Fife, J. E. Hutchins, and D. M. McMahon, *J. Am. Chem. Soc.* **94**, 1316 (1972).

Conclusion

This chapter describes the procedures for following the rapidly reversible and time-dependent inhibition of CEase. An effort was made to not only outline assay procedures for characterizing inhibition kinetics, but to provide detailed information on data analysis methods and software. Hopefully the information provided will prove useful to researchers who are undertaking investigations of the inhibition of CEase or related lipolytic enzymes.

[12] On the Inhibition of Microbial Lipases by Tetrahydrolipstatin

By LUTZ HAALCK and FRITZ SPENER

Introduction

Among the various serine-specific inhibitors available today, for example, organophosphates,[1] phosphonates,[2-4] alkyl, or arylsulfonates,[5] tetrahydrolipstatin (THL) represents the first known selective and irreversible inhibitor for triacylglycerol lipases. THL has been reported to inhibit digestive lipases, namely, pancreatic lipase from several species, but also carboxyl ester lipase (cholesterol esterase) of pancreatic origin and human gastric lipase.[6-8] The inhibitor is generated by catalytic hydrogenation of lipstatin,[9]

---

[1] A. M. Brzozowski, U. Derewenda, Z. S. Derewenda, G. G. Dodson, D. M. Lawson, and J. P. Turkenburg, *Nature* **351,** 491 (1991).

[2] U. Derewenda, A. M. Brzozowski, D. M. Lawson, and Z. S. Derewenda, *Biochemistry* **31,** 1532 (1992).

[3] P. Stadler, G. Zandonella, L. Haalck, F. Spener, A. Hermetter, and F. Paltauf, *Biochim. Biophys. Acta* **1304,** 229 (1996).

[4] M. L. Mannesse, J. W. Boots, R. Dijkman, A. J. Slotboom, H. T. van der Hijden, M. R. Egmond, H. M. Verheij, and G. H. de Haas, *Biochim. Biophys. Acta* **1259,** 56 (1995).

[5] A. J. G. Horrevoets, T. M. Hackeng, H. M. Verheij, R. Dijkman, and G. H. de Haas, *Biochemistry* **28,** 1139 (1989).

[6] B. Borgström, *Biochim. Biophys. Acta* **962,** 308 (1988).

[7] P. Hadváry, W. Sidler, W. Meister, W. Vetter, and H. Wolfer, *J. Biol. Chem.* **266,** 2021 (1991).

[8] S. Ransac, Y. Gargouri, F. Marguet, G. Buono, H. Lengsfeld, and R. Verger, *Methods Enzymol.* **286,** [10], 1997 (this volume).

[9] E. K. Weibel, P. Hadváry, E. Hochschuli, E. Kupfer, and H. Lengsfeld, *J. Antibiot. (Tokyo)* **40,** 1081 (1987).

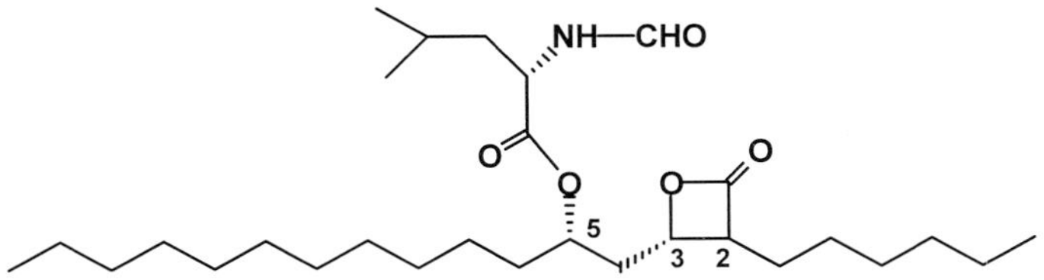

FIG. 1. Chemical structure of THL.

a lipid isolated from *Streptomyces toxytricini* or by chemical synthesis.[10] Tetrahydrolipstatin as well as esterastin[11] and valilacton[12] belong to a family of esterase inhibitors derived from mycolic acids, which are hydroxy fatty acids displaying a 2-alkyl substitution, essential for the inhibitory potency. The common structural feature of these inhibitors is the reactive β-lactone ring (Fig. 1) leading to an ester with the serine hydroxyl group of the catalytic triad of the lipase.[7]

Few data are available on the inhibition of microbial lipases by THL. Borgström[6] found no inhibition of the lipases from *Staphylococcus aureus* and *Rhizopus arrhizus* by THL with emulsified tributyrin as substrate. We investigated the inhibitory potency of THL toward the microbial lipases from *Chromobacterium viscosum* and *Rhizopus oryzae* (formerly denoted *R. arrhizus*) and found a strong dependence of the inhibitor's action on substrates and media applied.

Experimental Procedures

Materials

Purified porcine pancreatic lipase (PPL) was a generous gift from R. Verger (Laboratoire de Lipolyse Enzymatique du CNRS, Marseille, France). Tetrahydrolipstatin (M_r 496) and [^{14}C]THL (specific activity: 20.3 mCi/mmol) were kindly provided by Hoffmann-La Roche (Basel, Switzerland). Tetrahydrolipstatin is soluble in organic solvents, such as heptane, chloroform, and dimethyl sulfoxide, and stable in these solvents at 4° for months. Purified colipase from porcine pancreas is purchased from Boehringer (Mannheim, Germany) (lot 13618222-12). The specific activity of

---

[10] P. Barbier and M. P. Schneider, *Helv. Chim. Acta* **70**, 196 (1987).
[11] H. Umezawa, T. Aoyagi, T. Hazato, K. Uotani, F. Kojima, M. Hamada, and T. Takeuchi, *J. Antibiot.* (*Tokyo*) **31**, 639 (1978).
[12] M. Kitahara, M. Asano, H. Naganawa, K. Maeda, M. Hamada, T. Aoyagi, H. Umezawa, Y. Iitaka, and H. Nakamura, *J. Antibiot.* (*Tokyo*) **40**, 1647 (1987).

PPL is 1700 U/mg and is determined using olive oil as substrate in the presence of colipase.[13]

Olive oil, tributyrin, p-nitrophenyl palmitate (p-NPP) and dimethyl sulfoxide (DMSO) are purchased from Sigma (Munich, Germany); all other chemicals are from Merck (Darmstadt, Germany).

Lipase Purification

Chromobacterium viscosum lipase (CVL), also known as *Pseudomonas glumae* lipase,[14,15] is purchased as crude fermentation extract from Toyo Jozo (Shizuoka, Japan) (lot LP 251 S). Forty grams of the extract are dissolved in 300 ml starting buffer (10 mM Tris-HCl, pH 6.5). The clear solution is applied to anion exchange chromatography on Fractogel EMD-TMAE 650(M) (Merck, Darmstadt, Germany; 5 × 20 cm) equilibrated in starting buffer at a flow rate of 4 ml/min. This purification step removes brown colorants present with the lipase, whereas the lipase does not interact with the chromatographic material and elutes directly in the void volume. After subsequent washing with 400 ml starting buffer, the column is regenerated by applying 1.0 M NaCl in starting buffer. The lipolytic active fractions of the void volume are pooled, NaCl is added to a final concentration of 2 M, and the pH is adjusted to 8.4. This solution is applied to hydrophobic interaction chromatography on Phenyl-Sepharose CL-4B (Pharmacia, Uppsala, Sweden; 4 × 30 cm) at 6°. The lipase binds strongly to the column and usually cannot be removed without the use of detergents. Exploiting the extreme stability of this enzyme in various organic solvents (e.g., methanol, ethanol, 2-propanol, acetone, or acetonitrile), the lipase can be eluted with a decreasing NaCl gradient (2–0 M) and a concomitantly increasing 2-propanol gradient (0–75%, w/w) at a flow rate of 3 ml/min. The lipolytic active fractions elute at 55% 2-propanol and are pooled, desalted, and applied to size exclusion chromatography for final polishing. Best results are achieved using Superdex 75 (XK 26/60) with 10 mM calcium acetate and 200 mM NaCl as buffer. With this procedure 280 mg of >96% pure CVL is obtained within 3 days. The pure lipase (32.9 kDa) was crystallized[16] and the 3D structure was solved recently at high resolution by Lang *et al.*[17]

---

[13] R. Ruyssen and A. Lauwers, "Pharmaceutical Enzymes, Properties and Assay Methods," p. 210. E. Story-Scientia, Gent, 1978.

[14] L. G. Frenken, M. R. Egmond, A. M. Batenburg, J. W. Bos, C. Visser, and C. T. Verrips, *Appl. Environ. Microbiol.* **12,** 3787 (1992).

[15] M. E. Noble, A. Cleasby, L. N. Johnson, M. R. Egmond, and L. G. Frenken, *Protein Eng.* **7,** 559 (1994).

[16] D. Lang, B. Hofmann, L. Haalck, H.-J. Hecht, F. Spener, R. D. Schmid, and D. Schomburg, *Acta Cryst.* **50,** 272 (1994).

[17] D. Lang, B. Hofmann, L. Haalck, H.-J. Hecht, F. Spener, R. D. Schmid, and D. Schomburg, *J. Mol. Biol.* **259,** 704 (1996).

The protein profile is monitored spectrophotometrically at 280 nm and the lipase activity is determined using p-NPP.[18] Unless stated otherwise all experiments are performed at room temperature.

Rhizopus oryzae lipase (ROL) (formerly denoted *R. arrhizus* lipase) is purified as described[19] from a crude extract obtained from Gist-Brocades (Delft, The Netherlands) (lot 80.000 ST).

The specific activity of the purified CVL and ROL is 3200 U/mg and 12,600 U/mg, respectively, when tested with olive oil as substrate.[13]

Inhibition of Lipase-Catalyzed Hydrolysis of Emulsified Substrates

The hydrolysis of emulsified substrate is monitored by the titration of fatty acids released at pH 8.0 (olive oil) and pH 7.0 (tributyrin), respectively, for 15 min at 37° using a recording pH-stat (Metrohm, Herisau, Switzerland). The lipase activity obtained after inhibition by THL is calculated by determining the slopes of the titration curves during the first 4 min and is expressed as $k/k_0 \times 100\%$, where k represents the zero-order reaction rate constant of the substrate hydrolysis in the presence and k_0 in absence of THL.

Reagents

Solution A:	THL (1 mg/ml, 2 mM) in DMSO
Solution B:	10% (w/v) gum arabic and 1% (w/v) CaCl$_2$ × 2H$_2$O (25.3 mM) in bidestilled water
Solution C:	0.1% sodium taurocholate (18.6 mM) in bidestilled water
Buffer D:	1 mM Tris-maleate, pH 8.0, 1 mM CaCl$_2$, and 150 mM NaCl
Solution E:	0.01 μM lipase in buffer D
Solution F:	0.5 μM THL in tributyrin
Buffer G:	1 mM Tris-maleate, pH 7.0, 1 mM CaCl$_2$, and 150 mM NaCl
Solution H:	0.1 μM lipase in buffer G

In addition, 0.025 and 0.1 M NaOH are necessary as titrant.

Procedure. The olive oil substrate emulsion is prepared by mixing 1 ml olive oil intimately with 100 μl solution A and emulsified subsequently in 4 ml solution B using a high-speed homogenizer (IKA Ultra Turrax T 25, Staufen, Germany) at 10,000 rpm for 10 min. Different concentrations of THL ranging from 0.05 to 20 μM are obtained by dilution of aliquots taken from solution A with DMSO. Upon addition of 750 μl solution C, and 5

---
[18] U. K. Winkler and M. Stuckmann, *J. Bacteriol.* **138**, 663 (1979).

[19] L. Haalck, F. Paltauf, J. Pleiss, R. D. Schmid, F. Spener, and P. Stadler, *Methods Enzymol.* **284**, 353 (1997).

ml bidistilled water, the pH is adjusted to 8.0 and the reaction is started by addition of 100 μl of the lipase containing solution E. In the presence of PPL, the lipase solution also contains 0.01 μM colipase.

The tributyrin emulsion is obtained according to the method described by Borgström.[6] Five hundred microliters of solution F are added to 10 ml buffer G giving rise to a final substrate concentration of 162 μM. Certain concentrations of THL ranging from 0.31 to 15.6 μM are obtained by dilution of aliquots taken from solution F with tributyrin. The suspension is stirred vigorously and the pH is adjusted to 7.0. The reaction is started by the addition of 100 μl solution H. In the presence of PPL, the lipase solution also contains 0.1 μM colipase.

THL has a low hydrophilicity and distributes in an oil/aqueous two-phase system in favor of the lipid phase, so it can be concluded that the inhibitor is present at the oil/water interface of the substrate. Thus most effective inhibition is achieved when the inhibitor is mixed with the oil prior to emulsification, avoiding diffusion from the water to the interface.[6]

The liberation of fatty acids remains constant over time as observed with *C. viscosum* lipase and PPL (data not shown). Therefore, the lipase activity can be expressed by a zero-order reaction rate constant (k) when the concentration x of liberated fatty acids equals $k \cdot t$, then dx/dt equals k. For ROL, however, the curves reveal a hyperbolic shape,[20] yet we assume that hydrolysis followed a zero-order reaction up to 4 min. With increasing reaction progress the change of reaction order seen can be due to product inhibition.

With olive oil as substrate, a rapid increase in inactivation of CVL, ROL, and PPL with increasing amounts of THL is observed (Fig. 2A). In presence of a 30-fold molar excess of THL, CVL loses more than 50% of its lipolytic activity, whereas a 50% inhibition of ROL is only achieved using a 100-fold molar excess. In all experiments, the inhibition of PPL is more effective as compared to the microbial lipases. Surprisingly, in the presence of tributyrin the sensitivity of the microbial lipases toward THL dramatically decreases (Fig. 2B). Compared to the olive oil system, CVL is inhibited by approximately 50% in the presence of a 150-fold molar excess of THL, whereas the lipase from *R. oryzae* is not inhibited at all. For PPL results similar to those for the olive oil system are obtained.

Due to its amphiphilic character, it is reasonable to assume that the C_{11} acyl chain of THL, with the higher hydrophobic character as compared to the 2-hexyl substituent, is located in the oil phase, whereas the leucine, lactone, and 2-hexyl residues are located in the oil/water interface, being

---

[20] A. Potthoff, L. Haalck, and F. Spener, *Biochim. Biophys. Acta,* submitted.

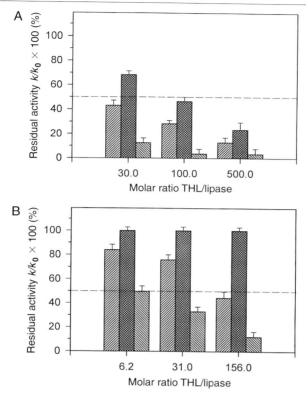

FIG. 2. Inhibition of lipase-catalyzed hydrolysis by THL: (A) substrate emulsion, 1 ml olive oil in 10% (w/w) gum arabic; lipase, 0.01 μM; and (B) substrate emulsion, 162 mM tributyrin in 1 mM Tris-maleate buffer, pH 7.0; lipase, 0.1 μM. Lipase activity is determined by the pH-stat method. Bars represent the means of three measurements. ▨, CVL; ▦, ROL; ▧, PPL.

accessible to the lipases. Thus the emulsified triacylglycerols serve as substrate for the lipases as well as solvent for the inhibitor.

Effect of 2-Propanol on the Inhibition of CVL

The lipases tested here exhibit no activity toward water-insoluble, nonemulsified substrates in aqueous phase. It is well established, however, that lipase activity is significantly increased at the lipid/water interface, which is known as interfacial activation and is related to a conformational change in the enzyme on interaction with the oil/aqueous interface.[21] In the course

[21] L. Sarda and P. Desnuelle, *Biochim. Biophys. Acta* **30**, 513 (1958).

of this event, an α-helical lid moves and the active center becomes fully accessible (open conformation). To transform the lipase into such active conformation without presenting an interface, 2-propanol is added to the aqueous buffer to lower the dielectric constant of the medium. In similar experiments the covalent modification of *Rhizomucor miehei* lipase with phenylmethylsulfonyl fluoride was enhanced significantly, which is consistent with the finding that the active site triad is buried under the lid in an aqueous environment.[22] The experiment here is performed with CVL alone, because ROL and PPL are not stable in the presence of higher concentrations of 2-propanol. In contrast to the inhibitor studies in the presence of substrates as described earlier, the following experiments are carried out in the absence of a substrate.

Reagents

Solution A: THL (1 mg/ml, 2 mM) in DMSO
Buffer B: 10 mM NaH$_2$PO$_4$, pH 7.0, and 50% 2-propanol (v/v)
Solution C: 1 mg/ml CVL (30.4 μM) in buffer B
Solution D: 30 mg p-NPP in 10 ml 2-propanol
Buffer E: 207 mg sodium deoxycholate and 100 mg gum arabic in 50 mM sodium phosphate buffer, pH 8.0
Solution F: prior to analysis solution D and buffer E are mixed 1 : 9 (v/v)

Procedure. CVL (200 μl buffer C, 1 μM) is incubated in 700 μl 10 mM NaH$_2$PO$_4$, pH 7.0, in the absence and presence of 50% 2-propanol (buffer B), respectively, with different amounts of THL (100 μl buffer A). The concentrations of THL ranging from 0.02 to 200 μM (1 ml final volume) are attained by dilution of aliquots taken from solution A with DMSO. Due to the absence of substrates, the residual lipase activity is monitored by hydrolysis of p-NPP[18] after taking 10 μl aliquots at certain times. The aliquots are mixed with 900 μl of the self-emulsifying substrate in a cuvette and the formation of the yellow p-nitrophenol is monitored at room temperature in a two-channel UV spectrophotometer by plotting the absorbance at 410 nm. The lipase activity can be calculated from the slope of the curve.[18]

In the absence of 2-propanol THL does not affect the lipase up to a 150-fold molar excess of THL; beyond this excess an 80% inhibition is attained (data not shown). In contrast, in the presence of 50% 2-propanol a 3.5-fold molar excess leads to a 50% inhibition, and a 30-fold molar excess abolishes the lipase activity completely after 5 min of incubation (Fig. 3).

[22] L. Brady, A. M. Brzozowski, Z. S. Derewenda, S. Tolley, J. P. Turkenburg, L. Christiansen, B. Huge-Jensen, L. Norskov, L. Thim, and U. Menge, *Nature* **343**, 767 (1990).

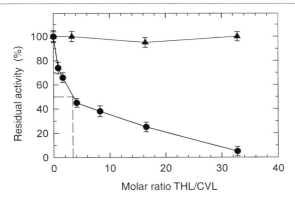

FIG. 3. The inhibition of CVL by THL affected by 2-propanol. CVL (6.1 μM) is incubated with THL in 10 mM sodium dihydrogenphosphate buffer, pH 7.0. Lipase activity is determined by the pH-stat method. —▲—, without 2-propanol; incubation, 2 hr; —●—, 50% 2-propanol; incubation, 5 min. Data represent the mean ± S.D. for 3 measurements.

As stated before, the addition of 50% 2-propanol decreases the dielectric constant of the buffer system and should allow for a conformational change, that is, opening of the α-helical lid to give full access to the active site. However, the poor water-soluble THL becomes more soluble. It is noteworthy that under these conditions the molar excess of inhibitor necessary to attain 50% inhibition of CVL within the same period of time is reduced 10-fold as compared to the experiments performed in the presence of substrate emulsions (Figs. 2 and 3).

Binding Experiments with [^{14}C]THL

For PPL and human pancreatic lipase, Hadváry et al.[7] provided evidence that the inhibition is caused by covalent binding of [^{14}C]THL to the active site serine. Adopting the procedure of these authors, investigations are carried out to elucidate the mechanism of inhibition of the microbial lipases. Subsequently, reactivation of the enzymes is studied and a possible cleavage of the inhibitor extracted from the enzymes is investigated.

Reagents

Solution A: THL (1 mg/ml, 2 mM) in DMSO
Solution B: 10% (w/v) gum arabic and 1% (w/v) $CaCl_2 \times 2H_2O$ (25.3 mM) in bidistilled water
Solution C: 0.1% sodium taurocholate (18.6 mM) in bidistilled water

Buffer D: 1 mM Tris-maleate, pH 8.0, 1 mM CaCl$_2$, and 150 mM NaCl
Solution E: 0.01 μM lipase in buffer D
Solution F: chloroform-methanol (1:3.5, v/v)

Procedure. The inhibition of the lipases (6.1 μM each) in the presence of the olive oil emulsion is carried out as described in the section "Inhibition of Lipase-Catalyzed Hydrolysis of Emulsified Substrates" except that unlabeled THL (solution A) is mixed with 18.7 μM (0.6 μCi) [^{14}C]THL 1:10 (w/w) giving rise to a 30-fold molar excess of the inhibitor. The [^{14}C]THL–lipase complex then formed is extracted with 7.2 ml solution F to precipitate the protein and to separate unbound inhibitor in the organic phase. The precipitate is removed by centrifugation at 10,000g and resuspended in 1 mM Tris-maleate buffer, pH 8.0. Upon addition of 3 ml Aquasafe 500 (Zinsser Analytik, Frankfurt, Germany) and 3 ml Lipoluma (Baker, Deventer, Amsterdam) to 500-μl aliquots of the aqueous suspension and the organic phase, respectively, the radioactivity is counted by scintillation spectrometry.

In the case of a covalent modification of the active site serine by [^{14}C]THL the precipitate formed should remain radioactively labeled as found for PPL, yet the absence of radioactivity in the precipitates of the microbial lipases excludes such a mechanism for CVL and ROL (Fig. 4).

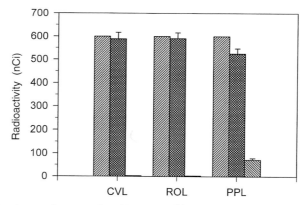

FIG. 4. Covalent and noncovalent binding of [^{14}C]THL to the lipases. The lipases (6.1 μM) are incubated for 1 hr with a 30-fold molar excess of [^{14}C]THL emulsified in 1 ml olive oil. Unbound THL is removed by vortexing with 7.2 ml chloroform-methanol (1:3.5, v/v) and the protein precipitate is removed by centrifugation. The precipitate is dissolved in 1 mM Tris-maleate buffer, pH 8.0, and the radioactivity is counted in the aqueous solution as well as in the organic solvent. ▨, radioactivity supplied; ▩, radioactivity in the organic phase; ▧, radioactivity in the protein pellet. Data represent the mean ± S.D. for 3 measurements.

The high stability of CVL in organic solvents offers the opportunity to investigate whether this lipase regains activity on extraction of the inhibitor.

Reagents

Solution A:	THL (2.5 mg/ml, 5 mM) in DMSO
Buffer B:	10 mM NaH$_2$PO$_4$, pH 7.0, and 50% 2-propanol (v/v)
Solution C:	1 mg/ml CVL (30.4 μM) in buffer B
Solution D:	chloroform-methanol (1:3.5, v/v)

Procedure. For the reactivation studies, 250 μl solution C (CVL, 26 μM) is incubated with 45 μl solution A to give a 30-fold molar excess of THL (780 μM). As checked with the *p*-NPP assay, the lipase activity is completely abolished within 10 min. Then 20-μl aliquots are withdrawn, extracted with increasing volumes of solution D and the lipase activity again is determined by the *p*-NPP assay.

Once the complex between CVL and THL is formed it is stable for months, however, the lipolytic activity of CVL can partially be restored by extracting the THL–CVL complex with chloroform-methanol (Fig. 5).

Next it is interesting to know whether the THL bound to CVL and then extracted is inactivated by cleavage of the β-lactone ring of the inhibitor. To answer this question the inactivation and reactivation of CVL are repeated and the extracts containing the inhibitor removed from the lipase are analyzed by TLC.

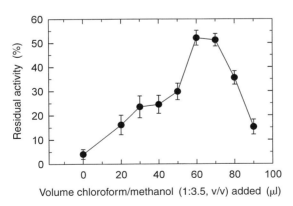

FIG. 5. Reactivation of THL inhibited CVL by chloroform-methanol extraction. CVL (26 μM) is inhibited with 780 μM THL in 10 mM NaH$_2$PO$_4$, pH 7.0, in the presence of 50% 2-propanol. On 1-hr incubation, 20 μl of this solution is extracted with increasing volumes of chloroform-methanol (1:3.5, v/v) and lipase activity in the aqueous phase is determined using the *p*-NPP test.[18] The data represent the means of three measurements.

Reagents

Solution A:	THL (1 mg/ml, 2 mM) in 2-propanol
Buffer B:	10 mM NaH$_2$PO$_4$, pH 7.0, and 30% 2-propanol (v/v)
Solution C:	chloroform-methanol (1:3.5, v/v)
TLC plates:	silica gel 60, 0.2 mm (Merck, Darmstadt, Germany)
Solvent system 1	chloroform-acetone-acetic acid (96:4:0.5, by vol)
Solvent system 2	chloroform-acetone-acetic acid (70:30:0.5, by vol)
Detection reagent:	10% (w/w) phosphomolybdate in ethanol

In contrast to the procedure described before, the inhibitor is dissolved in 2-propanol because DMSO, even after dialysis, interferes with the mobile phase used in TLC, causing unreproducible R_f values. The lower concentration of 2-propanol (30%, v/v) is necessary due to the reduced stability of the ultrafiltration membrane toward 2-propanol.

Procedure. CVL (2.72 mg, 8.27 μM) is dissolved in 8.79 ml buffer B and incubated for 2 hr with 1.21 ml solution A to ensure a 30-fold molar excess of THL (0.248 mM). The complete loss of lipase activity after 30 min is confirmed by the p-NPP assay.[18] The lipase is dialyzed extensively against buffer B using ultrafiltration membranes (Amicon, Witten, Germany; cutoff 10 kDa) to remove unbound inhibitor. After 48 hr, no THL can be detected in the supernatant as judged by TLC using solvent system 1 as mobile phase. The remaining solution (10 ml) is divided into two parts, subsequently extracted (3 × 2 ml) with chloroform or solution C. As control, THL is treated identically without lipase present. In the TLC analysis of the THL extracts β-hydroxypalmitic acid serves as reference for the cleavage of THL.[6] For visualization of spots the dried TLC plates are sprayed with detection reagent and charred at 180°.

The more unpolar solvent system, system 1, gives better resolution for the active uncleaved THL (R_f = 0.43), whereas β-hydroxypalmitic acid remains near the origin (R_f = 0.12). In solvent system 2 both substrates migrate, and the resolution for THL and β-hydroxypalmitic acid is R_f = 0.87 and R_f = 0.38, respectively. This analysis reveals no difference between the chloroform and chloroform-methanol extracts and the respective controls. In fact, unmodified THL is seen on all plates accompanied by two faint spots in solvent system 1 (R_f = 0.73 and 0.28) and 2 (R_f = 0.92 and 0.62), respectively. From this observation we conclude that in contrast to the results obtained with the human carboxyl ester lipase (HCEL)[6] the inhibitor is not consumed by CVL.

Conclusions

These experiments, on the one hand, confirm that *R. oryzae* lipase is not inhibited by THL in tributyrin emulsions.[6] On the other hand, THL inhibits ROL as well as CVL in the presence of emulsified olive oil as substrate, the latter with a higher efficiency. Obviously olive oil and tributyrin not only serve as substrates, but also as solvents for THL, thus mediating the interaction between lipase and the inhibitor. Due to the different hydrophilic/lipophilic balance of the two triacylglycerols, one can assume that THL accumulates at higher concentrations at the interface of emulsified olive oil as compared to emulsified tributyrin, resulting in a more efficient inhibition of lipolysis.

CVL is most sensitive toward THL in the presence of 50% 2-propanol without the emulsifying power of triacylglycerol substrates. We explain this phenomenon in terms of the better solubility of the inhibitor and the activation of the lipase in the homogeneous system.

The binding experiments with [^{14}C]THL proved that THL and the microbial lipases do not form a covalent THL–lipase complex. In addition, it is shown that CVL is reactivated from the stable THL–lipase complex to more than 50% of its initial activity by extracting the THL with chloroform-methanol. The fact that the removed inhibitor remains chemically unchanged indicates that inhibition of CVL by THL is achieved by noncovalent interaction.

[13] Monolayer Techniques for Studying Lipase Kinetics

By STÉPHANE RANSAC, MARGARITA IVANOVA, ROBERT VERGER, and IVAN PANAIOTOV

1. Introduction: Why Use Lipid Monolayers as Lipase Substrates?

Lipids have been classified by Small[1] depending on how they behave in the presence of water (Fig. 1). This makes it possible to distinguish between polar (e.g., acylglycerols) and apolar lipids (e.g., hydrocarbon, carotene). The polar lipids can be further subdivided into three classes. Class I consists of those lipids that do not swell in contact with water and form stable monomolecular films (these include triacylglycerols, diacylglycerols, phytols, retinols, vitamins A, K, and E, waxes, and numerous sterols). The class II lipids (which include phospholipids, monoacylglycerols, and

[1] D. M. Small, *J. Am. Oil Chemists Soc.* **45,** 108 (1968).

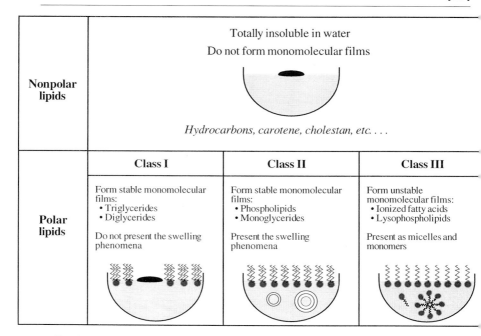

FIG. 1. A classification of biological lipids based on their interaction in aqueous systems. (From Small.[1])

fatty acids) spread evenly on the surface of water, but since they become hydrated, they swell up and form well-defined lyotropic (liquid crystalline) phases such as liposomes. The class III lipids (such as lysophospholipids and bile salts) are partly soluble in water and form unstable monomolecular films, and beyond the critical micellar concentration level, micellar solutions.

Lipases act on acylglycerol substrates. Because the plane of symmetry of the glycerol molecule is a prochiral plane, the two primary hydroxyl groups are stereochemically distinct (i.e., they are enantiotopic groups). A Fischer projection of a glycerol molecule is drawn with the secondary alcohol chain branching off to the left of the main hydrocarbon chain (Fig. 2). The carbon atoms are numbered 1, 2, and 3, working downward. With this system of numbering, glycerol becomes sn-glycerol (i.e., stereospecifically numbered glycerol). This makes it possible to obtain unambiguous expressions for the stereoisomeric forms of (phospho)glycerides. Natural phospholipids, for example, all belong to the sn-glycero-3-phosphate series. In the case of natural triacylglycerols, the fatty acids that esterify positions

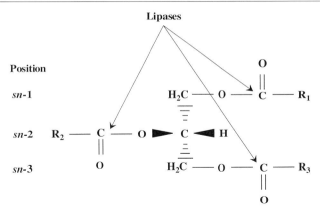

FIG. 2. Fisher representation of a triacylglycerol molecule. Identification of potentially hydrolyzable ester bonds.

sn-1, sn-2, and sn-3 are often different, which results in various chiral substrates.

The monolayer is a suitable model system to study the enzymatic reactions carried out at the interfaces in a heterogeneous medium. A new field of investigation was opened in 1935 when Hughes[2] used the monolayer technique for the first time for this purpose. He observed that the rate of the phospholipase A-catalyzed hydrolysis of a lecithin film, measured in terms of the decrease in surface potential, decreased considerably when the number of lecithin molecules per square centimeter increased. Since this early study, several laboratories have used the monolayer technique to monitor lipolytic activities, mainly with glycerides and phospholipids as substrates.

There are at least five major reasons for using lipid monolayers as substrates for lipolytic enzymes. The reader is referred to previous reviews for details.[3-7]

1. It is easy to follow the course of the reaction monitoring one of several physicochemical parameters characteristic of the monolayer film: surface pressure, potential, density, etc.

[2] A. Hughes, *Biochem. J.* **29**, 437 (1935).
[3] R. Verger and G. H. de Haas, *Annu. Rev. Biophys. Bioeng.* **5**, 77 (1976).
[4] R. Verger, *Methods Enzymol.* **64**, 340 (1980).
[5] R. Verger and F. Pattus, *Chem. Phys. Lipids* **30**, 189 (1982).
[6] S. Ransac, H. Moreau, C. Rivière, and R. Verger, *Methods Enzymol.* **197**, 49 (1991).
[7] G. Piéroni, Y. Gargouri, L. Sarda, and R. Verger, *Adv. Colloid Interface Sci.* **32**, 341 (1990).

2. Probably the most important basic reason is that it is possible with lipid monolayers to vary and control the *interfacial quality,* which depends on the nature of the lipids forming the monolayer, the orientation and conformation of the molecules, the molecular and charge densities, the water structure, the viscosity, etc. One further advantage of the monolayer technique as compared to bulk methods is that with the former, it is possible to transfer the film from one aqueous subphase to another.
3. Using the surface barostat balance, the lipid packing of a monomolecular film of substrate can be kept constant during the course of hydrolysis, and it is therefore possible to obtain accurate pre-steady-state kinetic measurements with minimal perturbation caused by increasing amounts of reaction products.
4. The monolayer technique is highly sensitive and very little lipid is needed to obtain kinetic measurements. This advantage can often be decisive in the case of synthetic or rare lipids. Moreover, a new phospholipase A_2 has been discovered using the monolayer technique as an analytical tool.[8]
5. Inhibition of lipase activity by water-insoluble substrate analogues can be precisely estimated using a zero-order trough and mixed monomolecular films in the absence of any synthetic, nonphysiological detergent.

The monolayer technique is therefore suitable for modeling *in vivo* situations.

2. Kinetic Models for Interfacial Enzymatic Lipolysis

2.1 Kinetics in the Presence of Substrate Only

One of the main assumptions implicitly underlying the classical Michaelis–Menten model is the fact that the enzymatic reaction must take place in an isotropic medium (i.e., both the enzyme and the substrate must be in the same phase). This model therefore cannot be used as it stands to study lipolytic enzymes acting mainly at the interface between a water phase and an insoluble lipid phase. In principle, the mechanism of the chemical reactions carried out at the interfaces in a heterogeneous medium depend strongly on the interfacial organization, steric coordination, and physical interaction between the reacting molecules. The chemical interactions are coupled with processes of adsorption, desorption, convection, and

---

[8] R. Verger, F. Ferrato, C. M. Mansbach, and G. Piéroni, *Biochemistry* **21**, 6883 (1982).

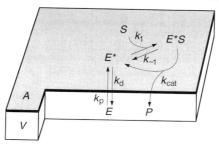

FIG. 3. Proposed model for lipase kinetics at the interfaces. (From Verger and de Haas.[3])

molecular diffusion of the reacting molecules and products of the reaction, etc. One or any combination of these processes may be rate determining.

To adapt the Michaelis–Menten kinetic scheme to the interfacial lipolysis, various models were proposed. In the simplest one, described by Verger et al.,[9] an instantaneous solubilization of the products of the reaction is assumed (Fig. 3). It is based on the idea that an enzyme–substrate complex might be formed at the interface. The model consists of two successive steps. The first one is a reversible penetration of a water-soluble enzyme (E) into the lipid substrate at the interface. The enzyme is fixed at the interface by an adsorption–desorption molecular mechanism. One formal consequence of this stage is the dimensional change in the enzyme concentration. The penetration stage (adsorption) sometimes, but not always, involves enzyme activation (e.g., via the opening of the amphiphilic lid covering the active site). This penetration step, leading to a more favorable energy state of the enzyme (E^*), is followed by a two-dimensional Michaelis–Menten kinetic scheme. The enzyme in the interface E^* binds a substrate molecule S to form the E^*S complex followed by its decomposition. The products of reaction P are soluble in the water phase, desorb instantaneously, and induce no change with time in the physicochemical properties of the interface. An important conceptual detail should be emphasized here: to be consistent with the fact that the enzyme-catalyzed reaction occurs at the interface, the concentration of E^*, E^*S, and S must be expressed as surface concentration units.

Resolving the kinetic equations corresponding to the model presented in Fig. 3 together with the equation of mass conservation at steady-state conditions, the following expression [Eq. (1)] for the concentration P of the product (expressed as molecules per volume) released with time t is obtained:

[9] R. Verger, M. C. E. Mieras, and G. H. de Haas, *J. Biol. Chem.* **248**, 4023 (1973).

$$P = k_{\text{cat}} E_0 \frac{S}{K_M^*} \frac{t + \dfrac{\tau_1^2}{\tau_1 - \tau_2}(e^{-t/\tau_1} - 1) - \dfrac{\tau_2^2}{\tau_1 - \tau_2}(e^{-t/\tau_2} - 1)}{1 + \dfrac{k_d}{k_p(A/V)} + \dfrac{S}{K_M^*}} \quad (1)$$

where $K_M^* = (k_{\text{cat}} + k_{-1})/k_1$ is the interfacial Michaelis–Menten constant; τ_1 and τ_2 are the induction times describing the establishment of the penetration–desorption and the interfacial Michaelis–Menten steady states, respectively; and A and V are the total interfacial area and volume of the system, respectively.

If we assume a rapid equilibrium between E^* and E^*S as compared to E and E^*, then Eq. (1) can be simplified as [Eq. (2)]

$$P = k_{\text{cat}} E_0 \frac{S}{K_M^*} \frac{t + \tau(e^{-t/\tau} - 1)}{1 + \dfrac{k_d}{k_p(A/V)} + \dfrac{S}{K_M^*}} \quad (2)$$

The induction time τ essentially reflects the establishment of the penetration–desorption equilibrium[9] and is the intercept of the asymptote with the time axis. We have to note, however, that some experimental data do not fit with this model.[10,11] Some lag times were called *catastrophic lag times*. They are characterized by the fact that after a certain time of low hydrolytic rate, a short accelerating phase leads to a high reaction rate. This sudden increase in hydrolytic rate is accompanied by a concomitant increase in lipase binding to the interface. Furthermore, these lag times are dependent on the presence of products in the interface.[10]

The rate of product release, in the stationary state, can be written as follows:

$$v = \frac{dP}{dt} = k_{\text{cat}} E_0 \frac{S}{K_M^* \left(1 + \dfrac{k_d}{k_p(A/V)}\right) + S} \quad (3)$$

For definitions of the parameters in these equations, see Refs. 9, 12, and 13.

[10] T. Wieloch, B. Borgström, G. Piéroni, F. Pattus, and R. Verger, *J. Biol. Chem.* **257**, 11523 (1982).
[11] S. Ransac, E. Rogalska, Y. Gargouri, A. M. T. J. Deveer, F. Paltauf, G. H. de Haas, and R. Verger, *J. Biol. Chem.* **265**, 20,263 (1990).
[12] S. Ransac, C. Rivière, C. Gancet, R. Verger, and G. H. de Haas, *Biochim. Biophys. Acta* **1043**, 57 (1990).
[13] S. Ransac, "Modulation des activités (phospho)lipasiques. Mise en œuvre de la technique des films monomoléculaire pour l'étude d'inhibiteur spécifiques et la détermination de la stéréo-sélectivité des enzymes lipolytiques." Thesis, University of Aix-Marseille II, 1991.

A more complex kinetic model, taking into account the interfacial accumulation and reorganization of the insoluble products of the reaction, was also proposed.[14] An important simplification can be made in the monolayer condition. In fact A/V has a very low value (around 1 cm^{-1}) and hence a limited number of all enzyme molecules present in bulk solution are adsorbed at the interface $[E \gg (E^* + E^*S)(A/V)]$. With this assumption and from Eq. (3), we obtain the following simplified expression of the rate of hydrolysis at steady state:

$$v_\mathrm{m} = \frac{dP}{dt} = \frac{k_\mathrm{cat}}{K_\mathrm{M}^* \dfrac{k_\mathrm{d}}{k_\mathrm{p}(A/V)}} E_0 S = Q_\mathrm{m} E_0 S \tag{4}$$

where v_m is the rate of hydrolysis of the monolayer, and Q_m is a global kinetic constant, called *interfacial quality*,[9] taking into account the influence of the physicochemical properties of the interface on the enzyme activity.

In the case of the hydrolysis of particles (emulsion for lipases or liposomes for phospholipase A_2) where all the enzyme is bound to these particles $[E \ll (E^* + E^*S)(A/V)]$, Eq. (3) can be simplified[9]:

$$v_\mathrm{p} = \frac{dP}{dt} = \frac{k_\mathrm{cat}}{K_\mathrm{M}^* + S} E_0 S = Q_\mathrm{p} E_\mathrm{p} S \tag{5}$$

where v_p is the rate of hydrolysis of the particles, and Q_p is an overall kinetic constant.

2.2 Kinetics in the Presence of a Competitive Inhibitor

Kinetic models accounting for the competitive inhibition in the presence and absence of detergent[12] and for irreversible inactivation[15] have also been developed. With the monolayer technique, we usually work without detergent. The rate of product formation, in the stationary state, in the presence of a competitive inhibitor (see Fig. 4) can be written as follows:

$$v = \frac{dP}{dt} = k_\mathrm{cat} E_0 \frac{S}{K_\mathrm{M}^* \left[1 + \dfrac{k_\mathrm{d}}{k_\mathrm{p}(A/V)} + \dfrac{I}{K_\mathrm{I}^*}\right] + S} \tag{6}$$

where I is a lipase competitive inhibitor.

As stated in Section 3.4, the minimal change that can be made at increasing inhibitor concentrations is to maintain constant the sum of inhibi-

---

[14] V. Raneva, T. Ivanova, R. Verger, and I. Panaiotov, *Colloids Surf.* **B3**, 357 (1995).
[15] S. Ransac, Y. Gargouri, H. Moreau, and R. Verger, *Eur. J. Biochem.* **202**, 395 (1991).

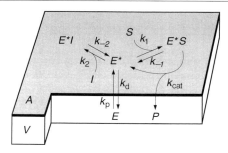

FIG. 4. Proposed model for competitive inhibition at interfaces. (From Ransac et al.[12])

tor and substrate $(I + S)$ when varying the inhibitor molar fraction $\alpha = I/(I + S)$, that is, to progressively substitute a molecule of substrate by a molecule of inhibitor. Under these conditions, the above equation becomes

$$v = \frac{k_{cat}E_0(1 - \alpha)}{\alpha\left(\frac{K_M^*}{K_I^*} - 1 + K_M^*(A_I - A_S)\right) + \frac{K_M^*}{I + S}\left[1 + \frac{k_d}{k_p(A/V)} + I(A_S - A_I)\right] + 1} \quad (7)$$

With the hypothesis of identical molecular area for the substrate and the inhibitor, this rate can be simplified as follows:

$$v = k_{cat} E_0 \frac{1 - \alpha}{\alpha\left(\frac{K_M^*}{K_I^*} - 1\right) + \frac{K_M^*}{I + S}\left[1 + \frac{k_d}{k_p(A/V)}\right] + 1} \quad (8)$$

This equation fits perfectly with the experimental data obtained in the presence of competitive inhibitors.[16] The variations of the relative velocity as a function of the inhibitor molar fraction are described in Fig. 5A. Ransac et al.[12] proposed to plot a velocities ratio defined as

$$R_v = \frac{v_1}{v_2}$$

$$= \frac{\text{velocity in the presence of an inhibitor with } K_I^* = K_M^* \text{ and } A_I = A_S}{\text{velocity in the presence of an inhibitor with } K_I^* \neq K_M^* \text{ and } A_I \neq A_S} \quad (9)$$

[16] S. Ransac, A. M. T. J. Deveer, C. Rivière, A. J. Slotboom, C. Gancet, R. Verger, and G. H. de Haas, *Biochim. Biophys. Acta* **1123**, 92 (1992).

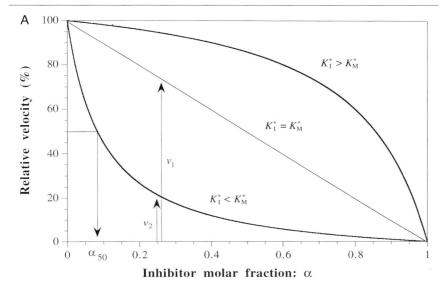

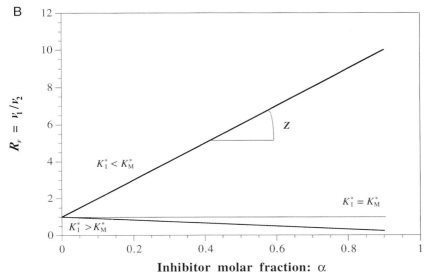

Fig. 5. (A) Relative enzyme velocity as a function of the interfacial molar ratio of inhibitor (α).[12] (B) Ratio between velocities v_1 and v_2 from part A as a function of the interfacial molar ratio of inhibitor (α).[12]

FIG. 6. Schematic representation of two extreme models of interfacial catalysis of liposomes: (A) scooting mode and (B) hopping mode.

This ratio (R_v) varies linearly with the inhibitor molar fraction (α) (Fig. 5B) and, under the same hypothesis of Eq. (8), can be written as

$$R_v = 1 - \alpha \frac{\frac{K_M^*}{K_I^*} - 1}{\frac{K_M^*}{I + S}\left[1 + \frac{k_d}{k_p(A/V)}\right] + 1} = 1 - \alpha Z \tag{10}$$

The slope of these lines (Fig. 5B) is a function of the ratio K_I^*/K_M^*, and has been coined Z by Ransac et al.[12] and named the *inhibitory power* of the molecule tested. The molar fraction of inhibitor that reduces the enzyme activity to 50%, α_{50}, can be written as

$$\alpha_{50} = \frac{1}{2 + Z} \tag{11}$$

2.3 Bilayer Liposomes as Substrate for Lipolytic Enzymes

To account for the interface catalysis of liposomes, two extreme and theoretical kinetic situations have been considered by Jain and Berg,[17] as shown in Fig. 6. In the so-called *scooting mode* of catalysis, the enzyme molecules are irreversibly bound to vesicles. A bound enzyme molecule E^* remains adsorbed at the interface between the catalytic cycles and hydrolyzes the substrate molecules from the outer monolayer of one single vesicle. In the pure *hopping mode* of catalysis, the enzyme molecules are exchanged between vesicles. The binding ($E \rightarrow E^*$) and the desorption ($E^* \rightarrow E$) of a bound enzyme molecule occur after each catalytic cycle. Hopping from one vesicle to the other, ultimately it will hydrolyze available substrate molecules in the outer monolayer of all vesicles. However, we have to realize that such a hopping model, if applicable, would be extremely

[17] M. K. Jain and O. G. Berg, *Biochim. Biophys. Acta* **1002**, 127 (1989).

inefficient in terms of catalysis, due to a rate-limiting step in the adsorption–desorption process.

2.3.1 Scooting Mode of Interfacial Catalysis

2.3.1.1 LOW ENZYME/VESICLES RATIO (E/V). If the enzyme to vesicle ratio E/V is low, there is an excess of vesicles without any bound enzyme and a small number of vesicles with only one enzyme. Assuming that all enzyme molecules are randomly distributed between the available vesicles, and follow a Poissonian distribution with an average number E/V of enzymes per vesicle:

$$p_j = \frac{\left(\frac{E}{V}\right)^j}{j!} e^{-E/V} \tag{12}$$

where p_j is the probability that a vesicle has j bound enzymes.

The probability for vesicles without any bound enzyme is

$$p_0 = e^{-E/V} \tag{13}$$

At a low E/V ratio, $p_0 = 1 - (E/V)$ and $p_1 \approx (E/V)$, that is, the number of vesicles containing one enzyme molecule is nearly the same as the number of enzyme molecules in the liposomal suspension.

In the scooting mode, the kinetics of the catalytic reaction are determined by the two-dimensional Michaelis–Menten step. According to Jain and Berg's nomenclature,[17] $N(t)$ is the number of substrate molecules per one enzyme-containing vesicle that have been hydrolyzed after time t and N_T is the total number of phospholipid molecules initially present in the outer monolayer of one vesicle. Hence, $n(t) = [N_T - N(t)]/N_T$ corresponds to the fraction of substrate molecules remaining on the outer surface of one liposome.

According to Jain and Berg's nomenclature,[17] the corresponding normalized Michaelis–Menten equation is

$$\frac{dN}{dt} = \frac{k_{cat} n(t)}{K_M^* + n(t)} \tag{14}$$

Assuming a high normalized Michaelis–Menten constant ($K_M^* \gg 1$), Eq. (14) can be written

$$\frac{dN}{dt} = \frac{k_{cat}}{K_M^*} \left(1 - \frac{N}{N_T}\right) \tag{15}$$

and after integration:

$$N(t) = N_T(1 - e^{-k_{cat}t/K_M^*N_T}) \tag{16}$$

To compare the results obtained with monolayers and liposomes, it is necessary to express the surface concentration of enzyme and substrate consistently. In the kinetic Eq. (4) and Eq. (5), S, which is expressed in molecules per square centimeter, could be expressed in substrate molecules per liposome containing only one enzyme molecule as in Eq. (15). In fact $N(t) = P\dfrac{V}{A}a$ and $N_T - N(t) = Sa$, where a is the surface area of one liposome. Then, from Eq. (4), one obtains Eq. (17), which is analogous to Eq. (15):

$$\frac{dN}{dt} = \frac{k_{cat}N_T E_0}{(k_d/k_p)K_M^*}\left(1 - \frac{N}{N_T}\right) \tag{17}$$

and after integration:

$$N(t) = N_T(1 - e^{-k_{cat}E_0 t/(k_d/k_p)K_M^*}) \tag{18}$$

2.3.1.2 HIGH ENZYME/VESICLES RATIO. When the enzyme to vesicle ratio E/V is high, many enzyme molecules can be bound to one vesicle. The probability that a vesicle has at least one enzyme is given by

$$p_{j\neq 0} = 1 - e^{-E/V}. \tag{19}$$

Under these conditions, the concentration of lipolytic product at time t is given by[17]

$$N(t) = N_T(1 - e^{-E/V})(1 - e^{-k_{cat}t/(K_M^*N_T)}). \tag{20}$$

When E/V is very high, Eq. (20) is reduced to Eq. (16).

2.3.2 Hopping Mode of Interfacial Catalysis. In the hopping mode, when the lipase is not irreversibly bound, the average number of substrate molecules, remaining on a bound or unbound vesicle, can change in two ways: either by hydrolysis at a bound vesicle or by exchange of an enzyme from a bound to an unbound vesicle. Assuming the same conditions (low E/V and $K_M^* \gg 1$) as in Eq. (16), the number of hydrolyzed substrate molecules $N(t)$ can be expressed as follows[17]:

$$N(t) = \frac{K_i N_T}{K_i + k_d}\left(k_d t + \frac{K_i}{K_i + k_d}\right)(1 - e^{-(K_i + k_d)t}), \tag{21}$$

where $K_i \equiv k_{cat}/K_M^* N_T$. For $k_d \ll K_i$, the scooting mode is restored and Eq. (21) can be simplified to Eq. (16).

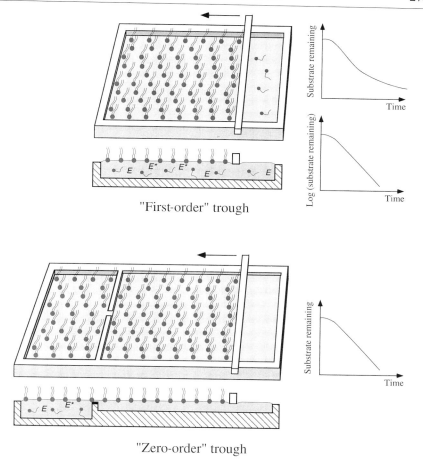

FIG. 7. Comparison of lipase kinetic obtained with a first-order trough and a zero-order trough.

3. Monolayer Technique for Studying Lipase Kinetics

3.1 Zero-Order Trough

Several types of troughs have been used to study enzyme kinetics. The simplest of these is made of Teflon that is rectangular in shape (Fig. 7, top) but gives nonlinear kinetics.[18,19] To obtain rate constants, a semilogarithmic

[18] D. G. Dervichian, *Biochimie* **53**, 25 (1971).
[19] G. Zografi, R. Verger, and G. H. de Haas, *Chem. Phys. Lipids* **7**, 185 (1971).

transformation of the data is required. This drawback was overcome by a new trough design (zero-order trough, Fig. 7, bottom) consisting of a substrate reservoir and a reaction compartment containing the enzyme solution.[20] The two compartments are connected by a narrow surface canal made of etched glass. The kinetic recordings obtained with this trough are linear, unlike the nonlinear plots obtained with the usual one-compartment trough. The surface pressure can be kept constant automatically by the surface barostat method described elsewhere.[20] Fully automated monolayer systems of this kind are now commercially available (KSV, Helsinki, Finland), and have been found to have many advantages.

3.2 Pure Lipid Monolayers as Lipase Substrates

3.2.1 Medium-Chain Lipid as Substrates. To use the monolayer technique for lipase kinetics, one necessary condition is that the substrate must be insoluble in the aqueous subphase and must form stable monomolecular films. The enzymatic hydrolysis of short- or medium-chain lipid substrates yielding water-soluble products can be easily interpreted on the basis of Eqs. (1) and (3). A typical example is the didecanoylglycerol, which forms perfectly stable film up to 32 mN/m, and both products, monoacylglycerol and fatty acid, desorb rapidly into the aqueous subphase. In the case of the trioctanoylglycerol, the dioctanoylglycerol, formed during the first lipolysis step, could transitively remain in the film and be either further hydrolyzed or solubilized.

In the case of short- or medium-chain lipid substrate when the reaction products are water soluble, the enzymatic hydrolysis can also be followed at constant area A by measuring the fall of surface density (Γ) detected by the radioactive method[21,22] or of surface pressure (π).[23] The main objection is related to the fact that the state of the monolayer changes during the reaction, and the kinetic constants (k_d/k_p, K_M^*, and k_{cat}) are not true constants.

3.2.2 Long-Chain Substrates. In the case of long-chain lipid substrates such as diolein, for which the reaction products are insoluble and remain in the monolayer, the rate of the barrier movement is much slower than the corresponding rates obtained with short- and medium-chain lipid such as dicaprin. The main reason for such a difference is certainly that oleoyl products are hardly water soluble, and hence the rate of the barrier movement does not directly reflect the enzymatic activity. Other factors are the

---

[20] R. Verger and G. H. de Haas, *Chem. Phys. Lipids* **10,** 127 (1973).
[21] A. D. Bangham and R. M. C. Dawson, *Biochem. J.* **75,** 133 (1960).
[22] R. M. C. Dawson, *Biochem. J.* **98,** 35c (1966).
[23] J. W. Jagocki, J. H. Law, and F. J. Kézdy, *J. Biol. Chem.* **248,** 580 (1973).

possible molecular reorganization of the lipolytic products at the interface ($P^* \rightarrow P^{**}$) together with product inhibition. The complete kinetic treatment of such a model will require the introduction of additional kinetic equations describing the slow desorption process ($P^* \rightarrow P$), the possible reorganization process ($P^* \rightarrow P^{**}$), and product inhibition. The target seems rather illusory, however, because even in the simplest situation of water-soluble lipolytic products, the intrinsic interfacial hydrolysis kinetic constant K_M^* cannot yet be experimentally determined. This is the reason why we prefer to use the initial model (Fig. 3) and Eq. (4) to estimate the global kinetic constant Q_m.

We have developed a method in our laboratory for studying the hydrolysis of long-chain lipid monolayers involving controlled surface density.[5,6] A large excess of serum albumin has to be present in the aqueous subphase to solubilize the lipolytic products. This step was found not to be rate limiting. The linear kinetics obtained with natural long-chain phospholipids were quite similar to those previously described in the case of short-chain phospholipids using a zero-order trough with the barostat technique.[20]

Another way of using long-chain lipid monolayers is based on the unique properties of β-cyclodextrin (β-CD). A study on the desorption rate of insoluble monomolecular films of oleic acid, monoolein, 1,2-diolein, 1,3-diolein, and triolein at the argon/water interface by the water-soluble β-CD has been published.[24] The desorption of the water-insoluble reaction products (oleic acid and monoolein) probably involves the complexation of the single acyl chain into the β-CD cavity and desorption into the aqueous phase of the soluble oleic acid/β-CD and monoolein/β-CD complexes.[24] In the case of monolayers of multiple acyl chain molecules such as diolein and triolein, no detectable change in the surface pressure occurred after β-CD injection.

The surface rheological dilatational properties of the monolayers of diolein and triolein in the presence of β-CD in the subphase were studied. The elasticity of the diolein monolayer remained unchanged, whereas the decrease in the surface elasticity of the triolein film was attributed to the formation of a water-insoluble triolein/β-CD complex. With the "tuning fork" model, one acyl chain of triolein can be included in the β-CD cavity. Schematic models have been proposed to attempt to explain the complexation of these lipids by β-CD at the argon/water interface (see Fig. 8). In addition to the preceding results, the presence of β-CD in the water subphase makes it possible for the first time to perform kinetic measurements

---

[24] S. Laurent, M. G. Ivanova, D. Pioch, J. Graille, and R. Verger, *Chem. Phys. Lipids* **70**, 35 (1994).

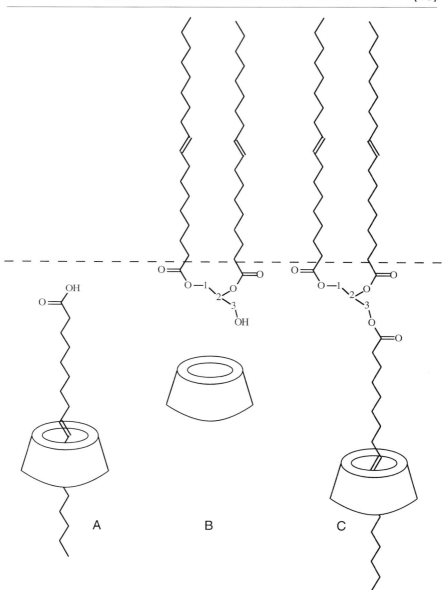

FIG. 8. Representation of β-CD action of monomolecular films. (A) Diagram of the water-soluble oleic acid/β-CD inclusion complex. (B) No complex formation between 1,2-diolein and β-CD. (C) Inclusion of a single acyl chain having the tuning fork type conformation of triolein into the β-CD cavity. (From Laurent et al.[24])

TABLE I
GLOBAL KINETIC CONSTANTS OF THE HYDROLYSIS OF MONOMOLECULAR FILMS OF DPPC, DOPC, AND DLPC AT VARIOUS SURFACE PRESSURES[a]

Global kinetic constants of hydrolysis ($cm^3\ s^{-1}\ molecule^{-1}$)

$$Q_m = \dfrac{k_{cat}}{K_M^* \dfrac{k_d}{k_p(A/V)}}$$

Surface pressure (mN/m)	DPPC		DOPC		DLPC
	No β-CD	With β-CD	No β-CD	With β-CD	No β-CD
15	9.0×10^{-17}	3.7×10^{-15}		3.8×10^{-15}	4.5×10^{-15}
20	1.8×10^{-17}	3.8×10^{-15}	7.0×10^{-17}	5.8×10^{-15}	6.5×10^{-15}
25	7.0×10^{-17}	1.9×10^{-15}	6.2×10^{-17}	5.9×10^{-15}	4.6×10^{-15}
30	$2 \times 10^{-17}\text{–}9 \times 10^{-17}$	1.2×10^{-15}		9.3×10^{-15}	1.0×10^{-15}

[a] By *Vipera berus* PLA_2 (final concentration 1.03 nM), in the absence and presence of β-CD (final concentration 0.53 mM) in the water subphase, pH 8.0, of the reaction compartment.

of the lipase hydrolysis rates of long-chain glycerides forming monomolecular films.

During hydrolysis with phospholipase A_2, the obtained apparent value of the global kinetic constant at $\pi = 20$ mN/m with dipalmitoylphosphatidylcholine (DPPC) is $Q_m = 1.8 \times 10^{-17}\ cm^3\ s^{-1}\ molecule^{-1}$. This value is 360 times smaller than the constant obtained with DLPC, which is $Q_m = 6.5 \times 10^{-15}\ cm^3\ s^{-1}\ molecule^{-1}$.

The observed difference between these global kinetic constants is probably due to physical steps involving a possible molecular reorganization, in the interface, of long-chain insoluble lipolytic products ($P^* \rightarrow P^{**}$), associated with product inhibition. β-CD has been used to check this hypothesis,[25] and leads to similar global enzymatic kinetic constant values with long-chain phosphatidylcholine [DPPC and dioleoylphosphatidylcholine (DOPC)] and medium-chain phosphatidylcholine (DLPC) (Table I).

3.2.3 Comparative Kinetics of Phospholipase A_2 Action on Liposomes and Monolayers. The literature is abundant concerning the binding of phospholipases to liposomes and the hydrolysis of various anionic or zwitterionic phospholipids.[17,26,27]

The detailed analysis of experimental data, reported in Ref. 14, leads to the conclusion that the hopping mode is adapted to describe the hydroly-

[25] M. G. Ivanova, T. Ivanova, R. Verger, and I. Panaiotov, *Colloids Surf.* **B6,** 9 (1996).
[26] M. K. Jain, J. Rogers, and G. H. de Haas, *Biochim. Biophys. Acta* **940,** 51 (1988).
[27] R. Apitz-Castro, M. K. Jain, and G. de Haas, *Biochim. Biophys. Acta* **688,** 349 (1982).

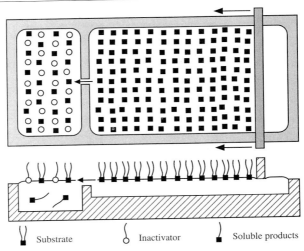

FIG. 9. Principle of the method for studying enzymatic lipolysis of mixed monomolecular films (from Piéroni and Verger.[28])

sis of DOPC liposomes with phospholipase A_2 from *V. berus* venom. The apparent global kinetic constant $Q_p = 6.4 \times 10^{-18}$ cm^3 s^{-1} molecule^{-1} was estimated. The comparison with DOPC monolayer constant ($Q_m = 7.0 \times 10^{-17}$ cm^3 s^{-1} molecule^{-1}) confirms the usual belief that liposomes are poorer substrates than monolayers.

3.3 Mixed Monolayers as Lipase Substrates

Most studies on lipolytic enzyme kinetics have been carried out *in vitro* with pure lipids as substrates. Virtually all biological interfaces are composed of complex mixtures of lipids and proteins. The monolayer technique is ideally suited for studying the mode of action of lipolytic enzymes at interfaces using controlled mixtures of lipids. Two methods exist for forming mixed lipid monolayers at the air/water interface: either by spreading a mixture of water-insoluble lipids from a volatile organic solvent, or by injecting a micellar detergent solution into the aqueous subphase covered with preformed insoluble lipid monolayers.

A new application of the zero-order trough was proposed by Piéroni and Verger[28] for studying the hydrolysis of mixed monomolecular films at constant surface density and constant lipid composition (Fig. 9). A Teflon barrier was placed transversely over the small channel of the zero-order

[28] G. Piéroni and R. Verger, *J. Biol. Chem.* **254**, 10090 (1979).

trough to block surface communications between the reservoir and the reaction compartment. The surface pressure was first determined by placing the platinum plate in the reaction compartment, where the mixed film was spread at the required pressure. Surface pressure was then measured after switching the platinum plate to the reservoir compartment, where the pure substrate film was subsequently spread. The surface pressure of the reservoir was equalized to that of the reaction compartment by moving the mobile barrier. The barrier between the two compartments was then removed to allow the surfaces to communicate. The enzyme was then injected into the reaction compartment and the kinetics was recorded as described.[20] The main purpose of this study was to describe the influence of lecithin on lipolysis of mixed monomolecular films of trioctanoylglycerol/didodecanoylphosphatidylcholine by pancreatic lipase to mimic some physiological situations. The authors used a radiolabeled pancreatic lipase (5-thio-2-nitro[^{14}C]benzoyllipase) to determine the influence of the film composition on the enzyme penetration (adsorption) and/or turnover at the interface. Of special interest is the concept of Scow et al.,[29,30] according to which triacylglycerol hydrolysis can be taken to be the first step in lateral flow lipid transport in cell membranes. Lipolytic activity was enhanced three- to fourfold in the presence of colipase, an effect that was attributed to the increase in the enzyme turnover number. When a pure triacylglycerol film was progressively diluted with lecithin, the minimum specific activity of lipase exhibited a bell-shaped curve: a mixed film containing only 20% trioctanoylglycerol was hydrolyzed at the same rate as a monolayer of pure triacylglycerol. The main conclusion drawn from this study was that considerable activation or inhibition may result as a function of the lipid composition, lipid packing, and surface defects in mixed films.

3.4 Inhibition of Lipases Acting on Mixed Substrate/Inhibitor Monomolecular Films

It is now becoming clear from the abundant lipolytic enzyme literature that any meaningful interpretation of inhibition data has to take into account the kinetics of enzyme action at the lipid/water interface. With the preceding method used, mixed substrate/inhibitor monomolecular films may be formed to study the lipase inhibition. As shown in Fig. 4, Ransac et al.[12] have devised a kinetic model, which is applicable to water-insoluble competitive inhibitors, to compare quantitatively the results obtained at several laboratories. Furthermore, with the kinetic procedure developed, it was possible to make quantitative comparisons with the same inhibitor

[29] R. O. Scow, E. J. Blanchette-Mackie, and L. C. Smith, *Circ. Res.* **39,** 149 (1976).
[30] R. O. Scow, P. Desnuelle, and R. Verger, *J. Biol. Chem.* **254,** 6456 (1979).

placed under various physicochemical situations, that is, micellar or monolayer states. Adding a potential inhibitor to the reaction medium can lead to paradoxical results. Usually, variable amounts of inhibitor are added, at a constant volumic concentration of substrate. The specific area and the interfacial concentration of substrate are continuously modified accordingly. To minimize modifications of this kind, the proposal was made to keep constant the sum of inhibitor and substrate $(I + S)$ when varying the inhibitor molar fraction $[\alpha = I/(I + S)]$, that is, to progressively substitute a molecule of substrate with a molecule of inhibitor.[31] This is the minimal change that can be made at increasing inhibitor concentrations. Of course, with this method, the classical kinetic procedure based on the Michaelis–Menten model is not valid because both inhibitor and substrate concentrations vary simultaneously and inversely. However, by measuring the inhibitory power (Z), as described by Ransac et al.[12] (see Section 2.2) or Xi(50) as used by Jain et al.,[17] it is possible to obtain a normalized estimation of the relative efficiency of various potential inhibitors. This method has been successfully used for the studies on the inhibition on phospholipase A_2 by substrate analogues.[16]

Covalent inhibition of lipases has also been performed with the mixed film method. Inhibitors used were serine reagents (phosphonates[32] or tetrahydrolipstatin[15]) or cysteine reagents [dodecyl dithio-5-(2-nitrobenzoic acid)[15]]. The results are described by Ransac et al. in Chapter [10].

3.5 Inhibition of Lipases by Proteins

Proteins can indirectly inhibit lipolysis.[7,33,34] The inhibition of lipases by various proteins was studied by lipase injection under a mixed lipid/protein monolayer, previously formed by two techniques: (1) spreading of lipid followed by injection of a protein solution into a stirred aqueous subphase and subsequent penetration of the protein into the lipid monolayer or (2) successive spreadings of the protein and the lipid. As an example, the inhibition of horse pancreatic lipase by ovalbumin and β-lactoglobulin A (β-LG) is demonstrated in Fig. 10 by the fall of surface pressure at constant area after injection of the lipase under the mixed dicaprin/protein monolayer film, formed by the first procedure. The observed inhibitor effect of β-LG is greater than that of ovalbumin. A simple explanation is that the

---

[31] G. de Haas, M. van Oort, R. Dijkman, and R. Verger, "Phospholipase A2 inhibitors: Monoacyl, monoacylamino-glucero-phosphocholines." *Biochem. Soc. Trans.* London, 1989.
[32] F. Marguet, I. Douchet, R. Verger, and G. Buono, *J. Am. Chem. Soc.* submitted (1997).
[33] M. G. Ivanova, I. Panaiotov, A. G. Bois, Y. Gargouri, and R. Verger, *J. Colloid Interface Sci.* **136**, 363 (1990).
[34] M. G. Ivanova, R. Verger, A. G. Bois, and I. Panaiotov, *Colloids Surf.* **54**, 279 (1991).

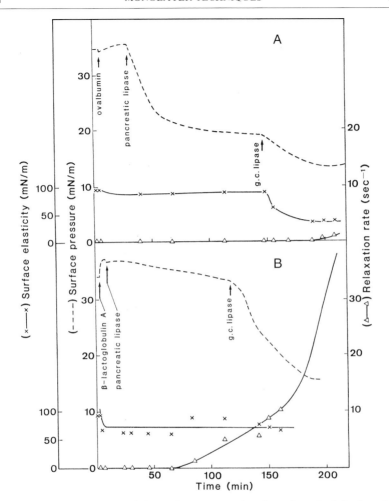

FIG. 10. Time dependence of the surface pressure (---), surface elasticity (x—x), and relaxation time (△—△) of a dicaprin film at an initial surface pressure of 35 mN/m after injection of various proteins and lipase. (a) The arrows indicate the successive injections of ovalbumin and horse lipase. (b) The arrows indicate the successive injections of β-lactoglobulin A and horse lipase; $P = 0.1$ μM; $E = 80$ pM; pH 8.0; $t = 20°$. (From Ivanova et al.[33])

protein and lipase compete for discrete adsorption sites at the interface. A comparative study of interfacial behavior (tensioactivity) of various proteins shows a good correlation with the inhibitory effect. First of all, the kinetics of protein penetration is tested. The initial rate of surface pressure increase,

TABLE II
INITIAL RATE OF SURFACE PRESSURE INCREASE[a] AND PENETRATION CONSTANT[33]

Protein	$(d\pi/dt)_{t\to 0}$ mN m^{-1} s^{-1} at π_0 = 20 mN/m	k_p cm/s at π_0 = 10 mN/m
Ovalbumin	2.5×10^{-3}	6.7×10^{-3}
β-lactoglobulin	1.1×10^{-1}	2.0×10^{-2}

[a] Surface pressure increase $(d\pi/dt)_{t\to 0}$,[7,35] is calculated following protein injection under a dicaprin monolayer; pH = 8.0.

occurring after protein injection $(\partial\pi/\partial t)_{t\to 0}$,[7,35] as well as the penetration constant k_p[33] are calculated from the kinetic curves (not shown) and presented in Table II. It is clear that the penetration capacity of β-LG is greater than that of ovalbumin. This finding is confirmed by measuring the surface rheological properties (i.e., surface elasticity E_d and relaxation rate Θ^{-1}) of the system[33] (Fig. 10). The elastic properties of the mixed dicaprin/ β-LG monolayer ($\Theta^{-1} = 0$) change into viscoelastic ones ($\Theta^{-1} > 0$) during lipolysis. As previously discussed,[36] the viscoelastic behavior corresponds to a protein-rich mixed monolayer. The experimental results confirm the assumption that lipase inhibition is due to preferential protein penetration into the monolayer film. It has been found that the amount of β-LG necessary to prevent dicaprin hydrolysis was five times smaller than that of ovalbumin.[34]

In conclusion, the inhibitory effects of proteins can be explained in terms of their kinetic competition with the lipase penetration into the monomolecular film of substrate.

3.6 Interfacial Binding and Film Recovery

Using the monomolecular film technique, several investigators have reported that an optimum occurs in the velocity-surface pressure profile. The exact value of the optimum varies considerably with the particular enzyme/substrate combination used. Qualitative interpretations have been given to explain this phenomenon. The first hypothesis, proposed by Hughes[2] and supported by later workers,[37,38] was that a packing-dependent orientation of the substrate may be one of the factors on which the regula-

[35] Y. Gargouri, G. Piéroni, C. Rivière, A. Sugihara, L. Sarda, and R. Verger, *J. Biol. Chem.* **260**, 2268 (1985).
[36] S. Taneva, I. Panaiotov, and L. Ter Minassian Saraga, *Colloids Surf.* **10**, 101 (1984).
[37] J. M. Smaby, J. M. Muderhwa, and H. L. Brockman, *Biochemistry* **33**, 1915 (1994).
[38] J. M. Muderhwa and H. L. Brockman, *J. Biol. Chem.* **267**, 24184 (1992).

tion of lipolysis depends. Of special interest is the recent approach by Peters et al.[39] on the structure and dynamics of 1,2-sn-dipalmitoylglycerol monolayers undergoing two phase transitions at 38.3 and 39.8 Å^2 per molecule. The first transition is unique for diacylglycerol molecules and is driven by a reorganization of the headgroups causing a change in the hydrophobicity of the oil/water interface. X-ray diffraction studies of different mesophases show that in the two highest pressure phases, the alkyl chains pack in an hexagonal structure relaxing to a distorted-hexagonal lattice in the lowest pressure phase with the alkyl chains tilted by ~14° in a direction close to a nearest neighbor direction.

Another interpretation was put forward by Esposito et al.,[40] who explained the surface pressure optimum in terms of changes in the lipase conformation on adsorption at the interface, resulting in an optimal conformation at intermediate values of the interfacial free energy (film pressure). It was suggested that lower and higher values of surface pressure would lead to inactive forms either because of denaturation or because the conformational changes in the enzyme structure are not sufficiently marked. This view was challenged by Verger and de Haas[3] and Pattus et al.[41] Using radiolabeled enzymes, these authors showed that the observed maxima in the velocity-surface pressure profile disappear when they are correlated with the interfacial excess of enzyme. Indeed, the main difference between the monolayer and the bulk system lies in the interfacial area to volume ratios, which differ from each other by several orders of magnitude. In the monolayer system, this ratio is usually about 1 cm^{-1}, depending on the depth of the trough, whereas in the bulk system it can be as high as 10^5 cm^{-1}, depending on the amount of lipid used and the state of lipid dispersion. Consequently, under bulk conditions, the adsorption of nearly all the enzyme occurs at the interface, whereas with a monolayer only one enzyme molecule out of hundred may be at the interface.[19] Owing to this situation, a small but unknown amount of enzyme, responsible for the observed hydrolysis rate, is adsorbed on the monolayer. To circumvent this limitation, different methods were proposed for recovering and measuring the quantity of enzymes adsorbed at the interface.[42–45]

[39] G. H. Peters, S. Toxvaerd, N. B. Larsen, T. Bjørnholm, K. Schaumburg, and K. Kjaer, *Nature Struct. Biol.* **2**, 395 (1995).
[40] S. Esposito, M. Sémériva, and P. Desnuelle, *Biochim. Biophys. Acta* **302**, 293 (1973).
[41] F. Pattus, A. J. Slotboom, and G. H. de Haas, *Biochemistry* **13**, 2691 (1979).
[42] J. Rietsch, F. Pattus, P. Desnuelle, and R. Verger, *J. Biol. Chem.* **252**, 4313 (1977).
[43] S. G. Bhat and H. L. Brockman, *J. Biol. Chem.* **256**, 3017 (1981).
[44] W. E. Momsen and H. L. Brockman, *J. Biol. Chem.* **256**, 6913 (1981).
[45] M. Aoubala, M. Ivanova, I. Douchet, A. de Caro, and R. Verger, *Biochemistry* **34**, 10786 (1995).

After performing velocity measurements, Momsen and Brockman[44] transferred the monolayer to a piece of hydrophobic paper and the adsorbed enzyme was then assayed titrimetrically. After correcting for blank rate and subphase carryover, the moles of adsorbed enzyme were calculated from the net velocity and the specific enzyme activity.

In assays performed with radioactive enzyme,[42] the film was aspirated by inserting the end of a bent glass capillary into the liquid meniscus emerging above the ridge of the Teflon compartment walls. The other end of the same capillary was dipped into a 5-ml counting vial connected to a vacuum pump. Because radioactive molecules dissolved in the subphase were unavoidably aspirated with the film constituents, the results had to be corrected by counting the radioactivity in the same volume of aspirated subphase. The difference between the two values, which actually expressed an excess radioactivity existing at the surface, was attributed to the enzyme molecules bound to the film. Because it is possible with the monolayer technique to measure the enzyme velocity expressed in μmol cm^{-2} min^{-1} and the interfacial excess of enzyme in mg/cm^2, it is easy to obtain an enzymatic specific activity value, which can be expressed as usual in μmol min^{-1} mg^{-1}.

Using radiolabeled 5,5'-dithiobis(2-nitrobenzoic acid) (DTNB), Gargouri et al.[46] investigated the interactions between covalently labeled [^{14}C]TNB-human gastric lipase ([^{14}C]TNB-HGL) and monomolecular lipid films. It is worth noting that [^{14}C]TNB-HGL is an inactive enzyme, and, moreover, to facilitate detection, that the [^{14}C]TNB-HGL concentrations used by Gargouri et al.[46] to study its binding to monomolecular films were about 40 times higher than the usual catalytic concentrations of HGL. Under these conditions, the existence of a correlation between the increase in surface pressure and the amount of protein bound to the interface was the only possible conclusion that could be drawn by the authors. They showed that in the presence of egg PC films, the total amounts of surface-bound [^{14}C]TNB-HGL decreased linearly as the initial surface pressure increased; whereas, using lipase substrates such as dicaprin films, the amounts of surface-bound inactive [^{14}C]TNB-HGL remained constant at variable surface pressures.[46]

The ELISA biotin–streptavidin system has been found to be as sensitive as the use of radiolabeled proteins.[47] In addition, biotinylation preserves the biological activities of many proteins. Using this labeling procedure, Aoubala et al.[48] (see Chapter [7] by Aoubala et al. in this volume) recently

---

[46] Y. Gargouri, H. Moreau, G. Piéroni, and R. Verger, *Eur. J. Biochem.* **180,** 367 (1989).
[47] J. L. Guesdon, T. Térnynck, and S. Avrameas, *J. Histochem. Cytochem.* **8,** 1131 (1979).
[48] M. Aoubala, I. Douchet, R. Laugier, M. Hirn, R. Verger, and A. de Caro, *Biochim. Biophys. Acta* **1169,** 183 (1993).

developed a specific double-sandwich ELISA to measure the HGL levels of duodenal contents. Another application of this method was the development of a sensitive sandwich ELISA, using biotin–streptavidin system to measure the amount of surface-bound HGL and anti-HGL monoclonal antibodies (mAbs) adsorbed to monomolecular lipid films. The detection limit was 25 and 85 pg for HGL and mAb, respectively.

By combining the sandwich ELISA technique just described with the monomolecular film technique, it was possible to measure the enzymatic activity of HGL on 1,2-didecanoyl-sn-glycerol monolayers as well as to determine the corresponding interfacial excess of the enzyme.[45] The HGL turnover number increased steadily with the lipid packing. The specific activities determined on dicaprin films spread at 35 mN/m were found to be in the range of the values measured under optimal bulk assay conditions, using tributyrin emulsion as substrate (i.e., 1000 μmol min^{-1} mg^{-1} of enzyme). At a given lipase concentration in the water subphase, the interfacial binding of HGL to the nonhydrolyzable egg yolk phosphatidylcholine monolayers was found to be 10 times lower than in the case of dicaprin monolayers.[45] However, we have to keep in mind that the surface-bound enzyme includes not only those enzyme molecules directly involved in the catalysis but also an unknown amount of protein present close to the monolayer. These enzyme molecules were not necessarily involved in the enzymatic hydrolysis of the film.

3.7 Stereoselectivity of Lipases Is Controlled by Surface Pressure

Lipases, which are lipolytic enzymes acting at the lipid/water interface, display stereoselectivity toward acylglycerols and other esters.[49–51] As stated in the introduction, biological lipids, which self-organize and orientate at interfaces, are chiral molecules, and their chirality is expected to play an important role in the molecular interactions between proteins and biomembranes. The most unusual aspect of acylglycerol hydrolysis catalyzed by pure lipases is its particular stereochemistry.[52–60] Under physiological conditions,

[49] H. Brockerhoff and R. G. Jensen, "Lipolytic Enzymes." Academic Press, New York, 1974.
[50] B. Borgström and H. L. Brockman, "Lipases." Elsevier, Amsterdam, 1984.
[51] C.-S. Chen and C. J. Sih, Angew. Chem. Int. Engl. **28**, 695 (1989).
[52] N. Morley, A. Kuksis, and D. Buchnea, Lipids **9**, 481 (1974).
[53] B. Akesson, S. Gronowitz, B. Herslöf, P. Michelsen, and T. Olivecrona, Lipids **18**, 313 (1983).
[54] R. G. Jensen, D. R. Galluzzo, and V. J. Bush, Biocatalysis **3**, 307 (1990).
[55] E. Rogalska, S. Ransac, and R. Verger, J. Biol. Chem. **265**, 20271 (1990).
[56] E. Rogalska, S. Ransac, and R. Verger, J. Biol. Chem. **268**, 792 (1993).
[57] K. Hult and T. Norin, Pure Appl. Chem. **64**, 1129 (1992).
[58] M. Holmquist, M. Martinelle, P. Berglund, I. G. Clausen, S. Patkar, A. Svendsen, and K. Hult, J. Protein Chem. **12**, 749 (1993).
[59] E. Rogalska, C. Cudrey, F. Ferrato, and R. Verger, Chirality **5**, 24 (1993).
[60] E. Cernia, M. Delfini, A. D. Magri, and C. Palocci, Cell. Mol. Biol. **40**, 193 (1994).

many other hydrolases, such as phospholipases, proteases, glycosidases, and nucleases, encounter only one optical antipode of their substrates. Lipases, on the contrary, can encounter both chiral forms of their substrates as well as molecules that are prochiral. This unique situation calls for some fundamental clarification, because the physiological consequences of lipases' stereo preferences are not known. Membrane-like lipid structures, such as monolayers, provide attractive model systems for investigating to what extent lipolytic activities depend on the chirality and other physicochemical characteristics of the lipid/water interface.

The mechanism whereby an enzyme differentiates between two antipodes of a chiral substrate may be influenced by physicochemical factors such as the temperature,[61,62] the solvent hydrophobicity,[63–66] or the hydrostatic pressure,[67] which can affect the stereoselectivity of the reaction. To achieve a measurable impact of hydrostatic pressure on a protein in a bulk solution, however, high pressures of the order of 3 kbar need to be used and monitoring the enzyme activity under these conditions is difficult. However, the surface pressure is easy to manipulate and its effects on the enzyme activity can be readily controlled. Rogalska et al.[56] investigated the assumption that the stereoselectivity, which is one of the basic factors involved in enzymatic catalysis, may be pressure dependent. When working with bulk solutions, the external pressure is not a practical variable because liquids are highly incompressible, whereas the monolayer surface pressure is easy to manipulate. To establish the effects of the surface pressure on the stereochemical course of the enzyme action, during which optical activity is generated in a racemic substrate insoluble in water, the authors developed a method with which the enantiomeric excess of the residual substrate can be measured in monomolecular films. With all four lipases tested (*Rhizomucor miehei* lipase, lipoprotein lipase, *Candida antarctica* B lipase, and HGL), low surface pressures enhanced the stereoselectivity (see Fig. 11) while decreasing the catalytic activity. This finding, which to our knowledge is unprecedented, should help to elucidate the mode of action of water-soluble enzymes on water-insoluble substrates.

In another study, Rogalska et al.[68] presented the results of an extensive comparative study on the stereoselectivity of 23 lipases of animal and

---

[61] E. Holmberg and K. Hult, *Biotechnol. Lett.* **13**, 323 (1991).
[62] L. K. Lam, R. A. H. F. Hui, and J. B. Jones, *J. Org. Chem.* **51**, (1986).
[63] S. Parida and J. S. Dordick, *J. Am. Chem. Soc.* **113**, 2253 (1991).
[64] S. H. Wu, Z. W. Guo, and C. J. Sih, *J. Am. Chem. Soc.* **112**, 1990 (1990).
[65] M. Matori, T. Asahara, and Y. Ota, *J. Ferment. Bioeng.* **72**, 413 (1991).
[66] K. Makamura, Y. Takebe, T. Kitayama, and A. Ohno, *Tetrahedron Lett.* **32**, 4941 (1991).
[67] S. V. Kamat, E. J. Beckman, and A. J. Russell, *J. Am. Chem. Soc.* **115**, 8845 (1993).
[68] E. Rogalska, S. Nury, I. Douchet, and R. Verger, *Chirality* **7**, 505 (1995).

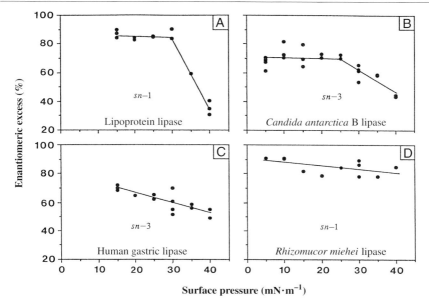

FIG. 11. Kinetic resolutions of 1,2-*rac*-dicaprin with (A) lipoprotein lipase, (B) *C. antarctica* B lipase, (C) HGL, and (D) *R. miehei* lipase. All the reactions were performed in a single compartment trough. The enantiomeric excess (ee) % was measured at 50% yield, that is, when the barrier had moved halfway across the trough.

microbial origin. Contrary to previous studies on lipase–acylglycerol chiral recognition where racemic or prochiral substrates were used, the substrates chosen here were three optically pure dicaprin isomers. The monomolecular film technique was chosen as a particularly appropriate method for use in chiral recognition studies.[69–74] To establish the effects of the surface pressure on the stereochemical course of the enzyme action, Rogalska *et al.*[68] used as lipase substrates optically pure dicaprin enantiomers 1,2-*sn*-dicaprin, 2,3-*sn*-dicaprin, and 1,3-*sn*-dicaprin, spread as monomolecular films at the air/water interface. The two former isomers are optically active antipodes (enantiomers), forming stable films up to 40 mN/m, while the latter one is

[69] N. Baba, S. Tahara, K. Yoneda, and J. Iwasa, *Chem. Express* **6,** 423 (1991).
[70] C. Böhm, M. Möhwald, L. Leiserowitz, and J. Als-Nielsen, *Biophys. J.* **64,** 553 (1993).
[71] D. Andelman and H. Orland, *J. Am. Chem. Soc.* **115,** 12322 (1993).
[72] N. G. Harvey, D. Mirajovsky, P. L. Rose, R. Verbiar, and E. M. Arnett, *J. Am. Chem. Soc.* **111,** 1115 (1989).
[73] M. Dvolaitzky and M.-A. Guedeau-Boudeville, *Langmuir* **5,** 1200 (1989).
[74] E. M. Landau, L. Levanon, L. Leiserowitz, M. Lahav, and J. Sagiv, *Nature* **318,** 353 (1985).

a prochiral compound, which collapses at a surface pressure of 32 mN/m. To our knowledge, this is the first report on the use of three diacylglycerol isomers as lipase substrates under identical, controlled physicochemical conditions. In this study, the authors showed that the regioselectivity, as well as the stereoselectivity, which are main factors involved in lipolytic catalysis of acylglycerols, are surface pressure dependent. The lipases tested displayed highly typical behavior, characteristic of each enzyme, which allowed them to classify the lipases into groups on the basis of enzyme velocity profiles as a function of the surface pressure and their preferences for a given diacylglycerol isomer, quantified using new parameters coined as stereoselectivity index, vicinity index, and surface pressure threshold. The general finding, which was true of all the enzymes tested, was that the three substrates are clearly differentiated, and the differentiation is more pronounced at high interfacial energy (low surface pressure). This finding supports the hypothesis that lipase conformational changes resulting from the enzyme/surface interaction affect the enzymes' specificities. Generally speaking, the stereo preference for either position sn-1 or sn-3 on acylglycerols is maintained in the case of both di- and triacylglycerols. We link this effect to the assumed lipase conformational changes involving the active site, occurring on the enzyme/interface interaction.

3.8 Acylglycerol Synthesis Catalyzed by Cutinase Using the Monomolecular Film Technique

When enzymatic catalysis takes place in systems with a low water content, the resulting thermodynamic equilibrium favors synthesis over hydrolysis. Hydrolytic enzymes can therefore be used to catalyze the formation of ester bonds in these systems. Among the hydrolytic enzymes, lipases have been widely used to perform esterification or transesterification reactions. Lipid monolayers have been used recently as lipase substrates with the monomolecular film technique. Melo *et al.*[75] adapted the monomolecular film technique for use in the synthesis of oleoylacylglycerols (monoolein, diolein, and triolein) as schematically shown in Fig. 12. The water subphase was replaced by glycerol and a stable film of oleic acid was initially spread on its surface. This method makes it possible to study and perform ester synthesis in a new and original system involving a specific array of self-organized lipid substrate molecules. The authors continuously recorded the surface pressure and furthermore estimated the acylglycerol synthesis after film recollection and HPLC analysis at the end of the experiment. A purified

[75] E. P. Melo, M. G. Ivanova, M. R. Aires-Barros, J. M. S. Cabral, and R. Verger, *Biochemistry* **34**, 1615 (1995).

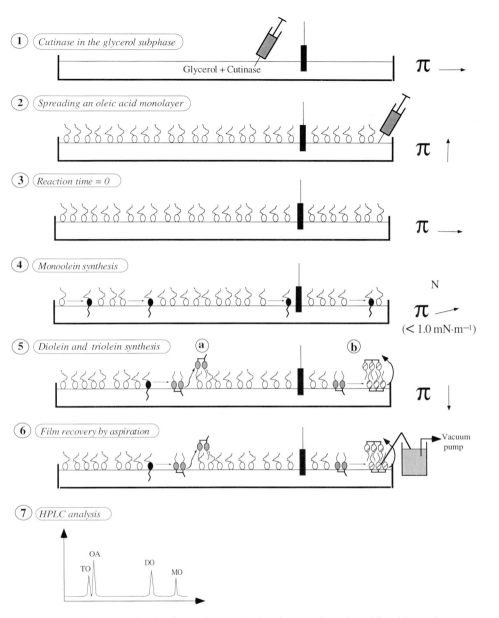

FIG. 12. Steps occurring in a first-order trough after the spreading of an oleic acid monolayer (2) over a glycerol subphase containing cutinase (1). Start of the reaction (3). Monoolein synthesis (4). Diolein and triolein synthesis (5). Film recovery (6) and HPLC analysis (7). Open ovals stand for oleic acid; solid ovals, monoolein; stippled ovals, diolein; diagonally striped ovals, triolein. Surface pressures are indicated by the slopes of the arrows on the right-hand side. The platinum plate is depicted as a thick vertical bar. Acylglycerol molecules are identified by an arrow. (From Melo et al.[75])

recombinant cutinase from *Fusarium solani*[76] was used as a biocatalyst in this study. More than 50% of the oleic acid film was acylated after 7 min of reaction. The surface pressure applied to the monomolecular film acts as a physical selectivity factor, since acylglycerol synthesis can be steered so as to produce either diolein or triolein.

Acknowledgments

Research was conducted with financial support from the BRIDGE-T-Lipase (contract BIOT-CT 91-0274 DTEE) and the BIOTECH G-lipase (contracts BIO2-CT94-3041 and BIO2-CT94-3013) programs of the European Union as well as the CNRS-IMABIO project.

[76] M. Lauwereys, P. de Geus, J. de Meutter, P. Stanssens, and G. Matthyssens, *GBF Monogr.* **16**, 243 (1991).

[14] Recovery of Monomolecular Films in Studies of Lipolysis

By WILLIAM E. MOMSEN and HOWARD L. BROCKMAN

Introduction

Understanding of the regulation of lipases has lagged behind that of other classes of enzymes as a consequence of the way in which lipases and their substrates interact. Enzymes that are water soluble and act on monomeric substrate molecules in aqueous solution have evolved a highly efficient mechanism for increasing the concentration of the correct substrate molecules, properly oriented, in the vicinity of the active site. This consists of a specific adsorption site on the protein for each of the substrates involved in the reaction. In contrast, water-soluble lipases function on water-insoluble, amphipathic substrates that reside in both the interior and the surface of a macroscopic aggregate, such as an emulsion particle. One consequence of the physical properties of the substrate molecules in the particle surface is that they are in relatively high concentrations and are oriented with respect to the surrounding aqueous phase. Lipases, though water soluble, generally have a high affinity for substrate-rich interfaces as a consequence of their ability to reversibly expose apolar groups to the aqueous milieu.[1]

[1] M.-P. Egloff, L. Sarda, R. Verger, C. Cambillau, and H. van Tilbeurgh, *Protein Sci.* **4**, 44 (1995).

This affinity for interfaces does not necessarily involve any direct interaction between the lipase and substrate and may even occur even in its absence.[2] Subsequent direct interactions between the lipase and the substrate, which culminate in catalysis, occur in the interfacial plane and are often weak, considering the effective concentration of substrate present. Thus, the reactant concentrations that immediately regulate lipase-catalyzed reactions are those in the lipid–water interface and the quantity of such interface supersedes bulk volume in importance.

The complexities of lipolytic systems present challenges for understanding the regulation of the reaction. Principally, it is necessary to determine and control the two-dimensional concentrations of enzyme and substrate in the interface as the reaction proceeds. This is because the partitioning of substrate molecules between the interfacial and bulk lipid phases and, particularly, of the proteins between the interfacial and bulk aqueous phases can be quite sensitive to small changes in lipid composition and packing density in the interface.[3] As a consequence, physiologically relevant emulsion-based assay systems, while extremely useful for monitoring lipase activity during purification, are less useful for studying the regulation of lipase adsorption and catalysis.[2] Of the many ways in which substrate can be presented to a lipase, one of the most versatile is the lipid monolayer.[2,4] The strengths of the monolayer approach are the absence of a bulk lipid phase, independent control of the lipid composition and packing density of the interface, and the amenability of the interface to physical characterization under the same conditions in which lipase adsorption and catalysis are to be measured. The traditional weakness of the approach was that it was difficult to determine how much of the lipase added to the aqueous phase was actually present in the monolayer and to determine the extent of substrate utilization.

Under certain conditions the real-time monitoring of lipase adsorption to and lipid hydrolysis in monolayers is feasible. A common strategy for monitoring substrate utilization in monolayers is to use substrates that generate water-soluble reaction products and thereby cause a drop in surface pressure as hydrolysis proceeds. This can be calibrated with the appropriate surface pressure–area isotherm or can be countered with changes in surface area to maintain constant surface pressure.[4] However, for studying the hydrolysis of more physiologically relevant, long-chain substrates

---

[2] H. L. Brockman, in "Lipases" (B. Borgström and H. L. Brockman, eds.), p. 1. Elsevier Science Publishers, Amsterdam, 1984.
[3] C. Chapus, M. Rovery, L. Sarda, and R. Verger, *Biochimie* **70**, 1223 (1988).
[4] G. Pieroni, Y. Gargouri, L. Sarda, and R. Verger, *Adv. Colloid Interface Sci.* **32**, 341 (1990).

these methods are less useful unless a sequestering agent such as albumin[5] or cyclodextrin[6] is added to the aqueous phase and the transfer of product to the aqueous phase is not rate limiting. Surface potential changes may be used to monitor generation of insoluble products generated in the absence of sequesterant. However, if the product is a minor component of the monolayer, sensitivity is low. Similar sensitivity arguments can be presented for the measurement of lipase adsorption, which, under catalytic conditions, exhibits a very low concentration in the interface.[7] Where significant quantities of protein are being adsorbed to an interface, surface radioactivity can be used to monitor that adsorption.[8]

An alternative to the continuous monitoring of lipase adsorption or extent of lipolysis is collection of the interface followed by quantitation of amount of lipase adsorbed or substrate utilized. Although this approach has the limitation of being a discontinuous assay, it allows a wider variety of techniques to be used to quantitate lipase adsorption and substrate hydrolysis. In addition, it can be carried out with very simple and inexpensive equipment. Monolayer collection has been applied in two variations. One is based on aspiration of the surface phase[9–12] and the second on the adsorption of the surface phase to a hydrophobic support (see discussion later). Although conceptually identical, there are practical differences in the data obtainable with each of these techniques. The first is sensitivity for determining net lipase adsorption. Sensitivity of the measurements is limited by the volume of the aqueous subphase, which is adventitiously collected with the monolayer.[13] When lipase adsorption is relatively low and subphase enzyme concentration is relatively high, the correction for enzyme carried over in the aqueous subphase can greatly exceed the quantity of adsorbed lipase, especially with the aspiration approach. Another is the efficiency with which the monolayer is collected. With aspiration, this can approach 100%. In contrast about 85–90% is achievable with a single surface using the hydrophobic support method. However, this latter value is relatively consistent, allowing recovery to be calibrated. In addition,

---

[5] R. O. Scow, P. Desnuelle, and R. Verger, *J. Biol. Chem.* **254**, 6456 (1979).
[6] S. Laurent, M. G. Ivanova, D. Pioch, J. Graille, and R. Verger, *Chem. Phys. Lipids* **70**, 35 (1994).
[7] J. M. Muderhwa and H. L. Brockman, *J. Biol. Chem.* **267**, 24184 (1992).
[8] R. B. Weinberg, J. A. Ibdah, and M. C. Phillips, *J. Biol. Chem.* **267**, 8977 (1992).
[9] J. Rietsch, F. Pattus, P. Desnuelle, and R. Verger, *J. Biol. Chem.* **252**, 4313 (1977).
[10] Y. Gargouri, G. Pieroni, C. Riviere, L. Sarda, and R. Verger, *Biochemistry* **25**, 1733 (1986).
[11] Y. Gargouri, H. Moreau, G. Pieroni, and R. Verger, *Eur. J. Biochem.* **180**, 367 (1989).
[12] M. Aoubala, M. Ivanova, I. Douchet, A. De Caro, and R. Verger, *Biochemistry* **34**, 10786 (1995).
[13] S. G. Bhat and H. L. Brockman, *J. Biol. Chem.* **256**, 3017 (1981).

when recovering lipid, rather than adsorbed protein, both sides of the support can be used, boosting recovery to >95%. The third, and major, difference between the two collection techniques concerns the recovery of the enzymatic activity of the lipase collected. Because aspiration generates high shear forces concomitant with the creation of air–water interface, enzymes are easily denatured. Such denaturation has been demonstrated in one study in which the two collection methods were directly compared.[13] Thus, determination of recovered lipase based on its catalytic activity is precluded.

Based on the advantages and disadvantages of the two collection methods, our laboratory has utilized the hydrophobic support method generally because of its versatility and specifically because adsorbed lipase can be determined by either its expressed catalytic activity or by protein quantitation. Measuring catalytic activity makes it unnecessary to tag the lipase with radioactivity, which may cause loss of its catalytic activity in the monolayer,[11] or with other reagents such as biotin.[12] Conversely, tagging greatly facilitates protein determination.[14,15] Moreover, using both assay procedures allows the extent of surface denaturation of lipase in the monolayer to be assessed.[14] The implementation of the hydrophobic support method is described next.

Monolayer Collection Method

Principle

Deposition of monomolecular films on a solid support is a widely used technique[16] and the use of a horizontally, as opposed to vertically, oriented planar substrate was first reported for the deposition of protein monolayers by Langmuir and Schaefer in 1938.[17] The difference between the present collection method and the horizontal deposition of monolayers is the use of a "fuzzy" hydrophobic support to facilitate the collection process with fluid lipid–protein films. The goal in normal film deposition is to maintain the collected film in a state as close as possible to that in the original monolayer. This generally requires long deposition times. For discontinuous determination of substrate hydrolysis or adsorbed lipase, however, it is important that the collection time be kept short to minimize lipase

---

[14] T. Tsujita and H. L. Brockman, *Biochemistry* **26**, 8423 (1987).
[15] G. D. Schmit, M. M. Momsen, W. G. Owen, S. Naylor, A. Tomlinson, G. Wu, R. E. Stark, and H. L. Brockman, *Biophys. J.* **71**, 3421 (1996).
[16] R. M. Swart, in "Langmuir-Blodgett Films" (G. Roberts, ed.), p. 273. Plenum Press, New York, 1990.
[17] I. Langmuir and V. J. Schaefer, *J. Am. Chem. Soc.* **60**, 1351 (1938).

adsorption/desorption or substrate hydrolysis during the collection process. The state of the collected film on the solid support is of no importance because it will be immediately subjected to additional analysis. These differences allow the use of a hydrophobic support with a larger surface area than a truly planar support would provide. The best commercially available support we have found for this purpose is filter paper that has been rendered hydrophobic by chemical modification.

Preparation of Hydrophobic Paper

Whatman 1PS phase separation filter paper is obtained from Whatman LabSales (Hillsboro, OR), or any Whatman product distributor. It is available in several sizes of disks between 9 and 27 cm in diameter (7 cm available as a special order) or as 28- × 46-cm sheets, which can be cut to fit any trough size or shape. For best results, the filter paper should be about 1 cm larger than the width or diameter of the surface to be collected. A typical circular trough used in our laboratory has a diameter of 5 cm and holds about 20 ml of aqueous phase. A 7-cm 1PS disk works well with this trough and all volumes and amounts given in this section refer to this combination. Quantities needed for other trough sizes should be proportional to paper area.

The paper contains residual surface-active reagents from its manufacture and, therefore, should be washed several times with solvent before use. Typically, thirty 7-cm disks are placed in 150 ml of solvent and gently agitated on a shaker bath for 20–30 min. The solvent is thoroughly drained and the papers briefly dried under N_2. Three washes with chloroform/methanol (2:1, v/v), two washes with chloroform, and two washes with hexane have been found to be adequate. The chloroform and methanol are reagent grade, but the hexane used is a high-purity-grade solvent used for monolayer studies (B&J Brand, Non-Spectro, from VWR Scientific Products). The 1PS disks are dried under N_2 for 10–15 min and then at ~70° for 30–60 min to remove traces of hexane. Cleaned disks are stored in tightly sealed glass jars with Teflon-lined caps. Prior to monolayer collection a clean 1PS disk is hydrated by floating it in a beaker containing high-quality water as used for monolayer studies for 10–15 min on each side. This is done because dry paper tends to pick up considerably more aqueous subphase and to curl when placed on the monolayer surface. Immediately before use, the hydrated paper is dipped into the water and excess droplets are shaken or blotted off.

Monolayer Collection

A generalized schematic for the collection of a monolayer using a piece of hydrophobic filter paper is illustrated in Fig. 1. Although the figure shows

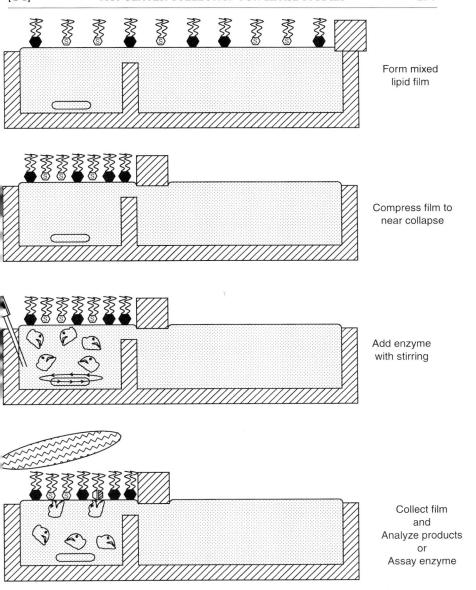

FIG. 1. Generalized scheme for formation and collection of lipid–enzyme monolayers using hydrophobic paper.

a two-compartment trough in which the monolayer is spread at large area and then compressed into the working surface, the technique is equally applicable to a single-compartment trough, which is much easier to build. The trough is usually constructed of Teflon or other fluoropolymer. The lipid monolayer, with or without protein adsorbed to it, is created at the air–water interface. The top of the aqueous phase onto which the monolayer is spread or adsorbed should be <0.1 mm above the edge of the trough in which the aqueous phase is contained. With such a small meniscus height there is a tendency for the aqueous phase to pull away from the sides of the trough, especially at high surface tension values. To minimize this, all but the top surface of the trough can be treated before use with a reagent such as Chemgrip (Berghof/America, Concord, CA) to increase its wettability. Having a small meniscus minimizes the tendency of the aqueous phase to spill over the sides of the trough when the paper is applied, especially if a plate is placed on top of it as described later.

The paper is placed on the monolayer surface and withdrawn horizontally over 1–2 sec (Fig. 2). For collection of adsorbed lipase, for which only one side of the paper is to be used, the use of a rigid plate, for example, 1/8-inch glass, to flatten the paper against the interface is recommended (Fig. 2). This maximizes monolayer–paper contact and minimizes the volume of aqueous subphase adventitiously collected with the monolayer. For the collection of lipid monolayers from large rectangular troughs, the paper can be cut to one-third to one-half of the trough length and held at a slight angle to the surface at one end of the trough. The monolayer is then compressed toward the edge of the paper, at which time the paper is lowered onto the surface to collect the compressed monolayer. This has been found useful for collection of lipid monolayers in that less paper and less solvent are needed.[18] The compression approach should not be used for measuring lipase adsorption because increasing the monolayer lipid packing density could significantly decrease lipase adsorption.[3] Another variation of the technique for the collection of monolayers of insoluble lipids is to use both sides of the paper successively. However, because no significant discrimination between insoluble lipid species has been detected when their recoveries from mixtures have been compared, this should not normally be necessary.

Recovery and Analysis of Monolayer Lipids

After removing the hydrophobic paper from the surface of the trough, it is placed in a glass or Teflon dish containing solvent, typically chloroform/methanol (2:1), containing carrier lipids. The paper is swirled and the

---

[18] W. E. Momsen and H. L. Brockman, *J. Biol. Chem.* **256**, 6913 (1981).

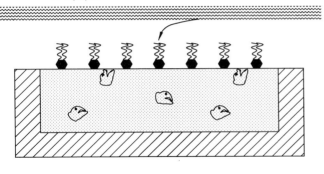

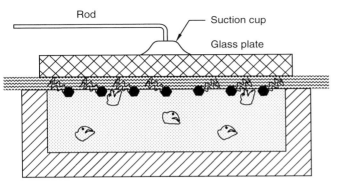

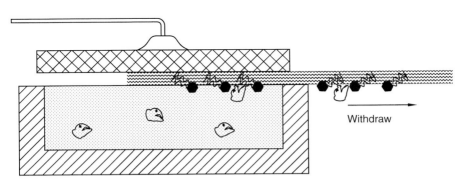

FIG. 2. Detailed scheme for lipid–enzyme monolayer collection. (a) A hydrophobic paper is carefully placed on the monolayer. (b) A glass plate is placed on top of the paper. (c) The paper is withdrawn horizontally using a continuous motion. Total elapsed time, <5 sec.

solvent is removed. This procedure is repeated one or two times depending on the importance of quantitative recovery to the particular experiment. Typical volumes used in an 80- × 40-mm (diameter × height) crystallization dish for elution of lipids from a 7-cm-diameter paper are 3, 2, and 2 ml.[19] The solvent is then evaporated from the combined extracts and the residue analyzed. Normally, a radiolabeled substrate is incorporated in the monolayer and the distribution of reactants and products is determined by thin-layer chromatography/radiochromatogram scanning.[20] Alternatively, the hydrolysis of lipids can be measured by monitoring the exchange of ^{18}O between fatty acid and the aqueous phase.[19] In our experience the presence of more than a 1000-fold molar excess of substrate and product species, that is, carrier lipids, relative to the monolayer, and the denaturing properties of the solvent have been sufficient to halt catalysis involving monolayer-derived substrate. This can be verified by repeating the experiment with a radiolabel or equivalent reporter in the substrate carrier lipid. In this case there should be no measurable hydrolysis of this lipid. The carrier lipids also provide mass for the subsequent separation of the reactants and products chromatographically.

Recovery and Analysis of Monolayer-Associated Protein

If the protein adsorbed to the collected monolayer is radiolabeled, the paper can simply be cut into pieces and deposited into a vial for scintillation counting. Alternatively, catalysis or another property of the lipase can be used for its quantitation. In this case, immediately after the monolayer is collected the paper is placed, monolayer side down, onto 4–5 ml of elution buffer contained in a crystallization dish and swirled. Aliquots of this solution are then taken for assay of lipase released from the paper. For elution of pancreatic carboxylester lipase the elution buffer contains 10 mM potassium phosphate, pH 6.6, and 0.15% Brij 35 nonionic detergent.[14] Because this buffer is that used in the catalytic assay for the enzyme, the assay can be initiated by addition of the substrate. This maximizes assay sensitivity when enzyme adsorption is low. In addition to being required to elute the lipase from the paper, the nonionic detergent helps to retain its catalytic activity. Recently, a titrametric procedure has been developed for assaying adsorbed pancreatic lipase without obligatory removal of the enzyme from the hydrophobic paper (M. Dahim and H. L. Brockman, in preparation, 1997).

The collection technique can also be used to measure lipase adsorption under conditions in which relatively high concentrations of enzyme are

[19] J. M. Muderhwa, P. C. Schmid, and H. L. Brockman, *Biochemistry* **31**, 141 (1992).
[20] B. A. Cunningham, T. Tsujita, and H. L. Brockman, *Biochemistry* **28**, 32 (1989).

added to the aqueous phase. In this case, however, a correction should be made for lipase present in the small amount of aqueous phase that is unavoidably collected with the monolayer. This correction is based on the incorporation of a marker in the aqueous phase. The choice of compound to use as a marker requires only that it not interfere with adsorption or quantitation of the lipase or other protein being studied and that it not adsorb to the monolayer. In early studies using plastic as the collection medium, simultaneous measurements of carryover using ^{32}P-phosphate and glycerol labeled with ^{3}H or ^{14}C did not reveal any significant difference in measured carryover for each sample tested. Radiolabeled acetate, ^{3}H$_2$O and similar molecules should not be used because of their volatility. For the carryover correction to be made, an aliquot of the subphase that supported the monolayer should be collected *immediately after collection of the monolayer*. Care should be taken to collect the subphase samples beneath the surface of the liquid, and the outside of the collecting syringe needle should be wiped clean of residual monolayer components that might adhere to it.

The lipases and other proteins we study are typically determined from their catalytic activity or a covalently attached radiolabel, typically ^{14}C. Particularly in the latter case, the addition of 0.02 μCi ^{32}P-phosphate/ml to the subphase to quantitate carryover is convenient because lipase and marker can be simultaneously determined by scintillation counting. Samples containing the hydrophobic paper and the two radiolabeled species are sonicated in a bath sonicator for 10 min, equilibrated $\geq$4 hr and then counted. Alternatively, if the lipase is to be eluted from the paper for subsequent assay, it may be necessary to add as much as 0.2 μCi ^{32}P-phosphate/ml to the monolayer subphase to compensate for the subsequent dilution of carried-over subphase in the lipase elution medium.

Calculations

As noted earlier, the recovery of lipid monolayers and their adherent protein is less than 100% when only one side of the paper is employed. For the determination of extent of hydrolysis or other reaction that has occurred in the interface, complete recovery is generally not necessary. This is because the calculation of extent of reaction is usually relative and we have found no evidence for discrimination with respect to the recovery of different, insoluble lipid species. Partially soluble species, like fatty acid, are recovered in lower yield, presumably because of their tendency to desorb from the monolayer before collection.[14]

For the collection of monolayers containing adsorbed lipase or other protein it is desirable to perform a simple calibration procedure in which aliquots of lipid with a known quantity of radioactivity or other quantifiable property are spread, collected, and analyzed. Figure 3 shows an example in which aliquots of radiolabeled 1-palmitoyl-2-oleoyl-sn-glycero-3-phosphocholine were spread on a single compartment trough to surface pressures within and spanning the liquid-expanded state. The slope of the line fitted through zero gives the fractional recovery, F, which in this case is 0.86, a value typically obtained. In our experience using this technique to measure the adsorption of lipases and colipase to lipid monolayers, F is essentially independent of the presence of adsorbed proteins in the interface. Thus, in subsequent determinations of lipase adsorption when the subphase concentration of the enzyme is low, the surface excess of lipase, Γ_L, is calculated by dividing the quantity of lipase associated with the paper, L_p, by F and the surface area of the trough, A:

$$\Gamma_L = L_p/AF$$

Conversely, if the subphase lipase concentration of lipase is high, a correction for subphase carryover of lipase may be necessary. To accomplish this, the quantity of marker, M_p, and lipase associated with the paper are compared with those, M_s and L_s, in known volumes, V_{sm} and V_{sL}, of aqueous

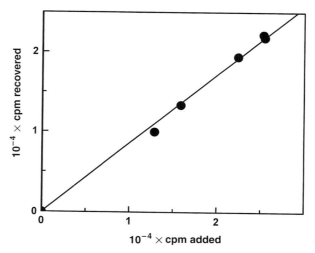

FIG. 3. Calibration of monolayer recovery using hydrophobic paper. Monolayers of 1-palmitoyl-2-oleoyl-[*oleoyl-1-^{14}C*]-sn-glycero-3-phosphocholine, were formed by spreading aliquots of the lipid dissolved in hexane/ethanol (95:5) on an argon–buffer interface. The films were equilibrated for 10 min before collection as shown in Fig. 2. Surface area, 26 cm^2; surface pressure, 0–33 mN/m; temperature, 24°; buffer, 10 mM potassium phosphate, 0.1 M NaCl, 0.01% NaN$_3$, pH 6.6.

subphase collected with a syringe or pipette immediately after collection of the monolayer. If the monolayer was eluted into buffer, V_{sL} must, of course, be corrected also for dilution into the elution buffer. From these quantities the net amount of lipase present on the paper can be calculated. This is then corrected for efficiency of recovery and the area of the monolayer to give the desired surface excess of lipase using the following equation:

$$\Gamma_L = [L_p - M_p(V_{sm}/M_s)(L_s/V_{sL})]/AF$$

Although the correction for monolayer recovery is relatively independent of monolayer composition, the actual volume of subphase collected on the paper in each experiment varies. This is illustrated in Table I for a series of experiments involving the collection of monolayers comprised of 1-stearoyl-2-oleoyl-*sn*-glycero-3-phosphocholine, either 13,16-*cis,cis*-docosadienoic acid or 1,3-dioleoylglycerol and procolipase. The collection protocol used was that described by Fig. 2 for a two-compartment trough (Fig.

TABLE I
VARIATION IN SUBPHASE CARRYOVER WITH MONOLAYER COMPOSITION

Monolayer composition								
Lipid components		Mole fraction Lipid 1	Surface concentration (pmol/cm^2)			Carryover volume		
Lipid 1	Lipid 2		Lipid	Procolipase	No. of samples	Range (μl/cm^2)	Average (μl/cm^2)	Standard deviation
SOPC[a]	None	1.0	30–250	0–20	9	2.3–5.0	3.6	0.8
DA[a]	SOPC	0.2	50–290	1–20	7	1.9–3.7	3.0	0.8
DA	SOPC	0.4	50–370	0–21	8	3.0–7.4	4.3	1.4
DA	SOPC	0.6	60–440	2–20	8	2.6–4.3	3.5	0.7
DA	SOPC	0.8	70–500	6–23	8	3.2–5.5	4.2	0.8
DA	None	1.0	80–570	9–28	14	2.3–5.2	3.5	0.9
DO[a]	SOPC	0.2	40–290	0–23	9	1.8–4.3	3.3	0.9
DO	SOPC	0.25	50–300	0–22	8	3.0–4.2	3.6	0.4
DO	SOPC	0.4	50–310	0–18	8	3.0–6.7	4.1	1.2
DO	SOPC	0.5	30–300	2–20	8	2.6–4.2	3.3	0.7
DO	SOPC	0.6	10–260	8–24	8	2.5–5.7	3.6	1.0
DO	SOPC	0.7	40–270	7–20	8	2.8–4.0	3.4	0.4
DO	SOPC	0.8	30–260	11–26	8	2.9–4.2	3.5	0.4
DO	None	1.0	30–250	14–25	12	2.3–5.2	3.8	0.8
None	None		0	18–28	16	2.6–5.5	3.9	1.0

[a] SOPC, 1-stearoyl-2-oleoyl-*sn*-glycero-3-phosphocholine; DA, 13,16-*cis,cis*-docosadienoic acid; DO, 1,3-dioleoylglycerol.

1) with a 5-cm-diameter reaction chamber. The table shows that there is considerable variation in carryover volume for any set of data. Neither the scatter in the values nor the averages show any obvious dependence on the lipid composition of the films.

Sorting the data at each lipid composition or the entire set on the basis of lipid concentration or protein surface excess in the monolayer revealed no dependency of carryover volume on these parameters. Thus, the variability of carryover volumes shown in Table I indicates that if it is necessary to correct lipase adsorption for carryover from the aqueous subphase, the correction should be experimentally determined for each lipase adsorption measurement. However, in planning experiments, the average of 3.7 ± 0.9 μl/cm^2 for all values in Table I can be compared with expected protein adsorption and its bulk concentration to initially determine if a correction may be necessary. Finally, with appropriate choice of subphase marker (e.g., ^{32}P), monolayer lipids (e.g., ^{3}H), and adsorbing species (e.g., ^{14}C), it is possible to determine carryover, monolayer recovery, and adsorbed component in the same experiment (W. E. Momsen and H. L. Brockman, unpublished, 1983).

Applications

The collection of lipid monolayers has been applied in a number of studies of lipase-catalyzed reactions. In many of these rates of substrate hydrolysis catalyzed by pancreatic carboxylester lipase or pancreatic lipase have been measured using radiolabeled substrates. Substrates used have been triacylglycerols,[21] diacylglycerols,[7,20,21] monoacylglycerols,[22] methyl oleate,[21] and cholesterol oleate.[23] In one study, the equilibrium properties of cholesterol oleate hydrolysis/synthesis were measured as a function of the presence of phospholipid.[13] Another application of the method measured the lipase-catalyzed exchange of ^{18}O from the carboxyl group of a fatty acid with ^{16}O in the aqueous medium[7,19,24] to determine mechanistic as well as catalytic rate parameters.

The quantitation of proteins adsorbed to lipid monolayers has also been achieved using hydrophobic paper. Much of this work involves carboxylester lipase[13,14,19,20,22-25] and pancreatic lipase.[7,18] In most of these, catalytic activity was used to determine lipase adsorbed. However, in two studies

[21] T. Tsujita, J. M. Muderhwa, and H. L. Brockman, *J. Biol. Chem.* **264,** 8612 (1989).
[22] J. M. Muderhwa and H. L. Brockman, *J. Biol. Chem.* **265,** 19644 (1990).
[23] S. G. Bhat and H. L. Brockman, *Biochemistry* **21,** 1547 (1982).
[24] J. M. Muderhwa and H. L. Brockman, *Biochemistry* **31,** 149 (1992).
[25] T. Tsujita, J. M. Smaby, and H. L. Brockman, *Biochemistry* **26,** 8430 (1987).

of carboxylester lipase adsorption the enzyme was inactivated with [^{3}H]diisopropylfluorophosphate.[14,25] The use of [^{3}H]prothrombin fragment 1, allowed verification of dissociation constants derived from surface pressure measurements[26] and, more recently, [^{14}C]colipase adsorption has been determined.[15] The use of radiolabeled colipase without extraction of the protein from the hydrophobic paper was necessitated by the high affinity of the cofactor for the paper. This had not be necessary with procolipase, which was readily extracted from the paper with detergent.[27] This difference emphasizes the importance of determining if the protein can be removed from the hydrophobic paper by the elution buffer when this approach is necessary for its quantitation.

Conclusion

The collection of lipid monolayers provides a simple, but effective, method for the determination of reactant distribution or adsorbed lipase in monomolecular films. In this laboratory (Hormel Institute, University of Minnesota, Austin), the technique has been applied successfully in a variety of studies of lipases and related proteins. Because its application is not specific to any particular property of lipases or their substrates, it is, in fact, a general method that is now being applied to the adsorption of other proteins to monolayer interfaces.

Acknowledgments

We express our gratitude to the many co-workers who over the years contributed to the refinement and application of the method described here. This work was supported by grants HL17371 and HL49180 from the National Institutes of Health, USPHS, and by the Hormel Foundation.

[26] L. D. Mayer, G. L. Nelsestuen, and H. L. Brockman, *Biochemistry* **22,** 316 (1983).
[27] W. E. Momsen, M. M. Momsen, and H. L. Brockman, *Biochemistry* **34,** 7271 (1995).

[15] Oil-Drop Tensiometer: Applications for Studying the Kinetics of Lipase Action

By S. LABOURDENNE, A. CAGNA, B. DELORME, G. ESPOSITO, R. VERGER, and C. RIVIÈRE

Introduction: Why Use an Oil-Drop Method for (Phospho)Lipase Kinetics?

As can be seen from the literature,[1-4] numerous techniques are available for measuring lipase activity. These can be classified into three groups on the basis of either the substrate consumption, the product formation, or the changes with time of one physical property, such as the conductivity, turbidity, or interfacial tension.

Among the interfacial methods, monomolecular film technology at the air/water interface has been extensively developed and used at our laboratory.[5-8] With this technique, it is possible to measure and control some important interfacial parameters, such as the surface pressure (interfacial free energy), the molecular area of the substrate, and the surface excess of the water-soluble lipases.[6,9,10] One prerequisite of the monolayer technique, however, is that the insoluble monomolecular film of substrate should generate water-soluble products during the reaction process. This is why synthetic medium-acyl chain lipids are mainly used as substrates for lipolytic enzymes.[11-13]

We recently developed a new method using a nontensioactive agent such as β-cyclodextrin present in the aqueous subphase to trap the single long-chain lipolytic products generated during the hydrolysis of monomo-

[1] H. Brockerhoff and R. G. Jensen, "Lipolytic Enzymes." Academic Press, New York, 1974.
[2] H. L. Brockman, *Methods Enzymol.* **71**, 619 (1981).
[3] R. G. Jensen, *Lipids* **18**, 650 (1983).
[4] H. S. Hendrickson, *Anal. Biochem.* **219**, 1 (1994).
[5] R. Verger, and G. H. de Haas, *Annu. Rev. Biophys. Bioeng.* **5**, 77 (1976).
[6] R. Verger, *Methods Enzymol.* **64**, 340 (1980).
[7] R. Verger and C. Rivière, *Revue Française des Corps Gras* **1**, 7 (1987).
[8] S. Ransac, H. Moreau, C. Rivière, and R. Verger, *Methods Enzymol.* **197**, 49 (1991).
[9] S. Ransac, C. Rivière, C. Gancet, R. Verger, and G. H. de Haas, *Biochim. Biophys. Acta* **1043**, 57 (1990).
[10] M. Aoubala, M. Ivanova, I. Douchet, A. De Caro, and R. Verger, *Biochemistry* **34**, 10786 (1995).
[11] G. Zografi, R. Verger, and G. H. de Haas, *Chem. Phys. Lipids* **7**, 185 (1971).
[12] D. G. Dervichian, *Biochimie* **53**, 25 (1971).
[13] R. Verger and G. H. de Haas, *Chem. Phys. Lipids* **10**, 127 (1973).

lecular films of long-chain glycerides[14] or phospholipids.[15] Nevertheless, the question remains as to whether the behavior of the lipid film at the air/water interface is actually representative of what is occurring either at an oil/water interface or in a complex biomembrane. In 1987, Nury et al.[16] established at our laboratory that one can gain unique information by measuring the variations in the oil/water interfacial tension ($\gamma_{o/w}$) as a function of time during lipase hydrolysis. These authors adapted the well-known *oil-drop method* for use in studying the rate of lipase hydrolysis of natural long-chain triacylglycerols. The accumulation of insoluble hydrolysis products at the surface of the drop is responsible for the $\gamma_{o/w}$ decrease, which in turn is correlated with changes with time in the oil-drop profile.[16] The theoretical basis of the calculation of $\gamma_{o/w}$ from a hanging drop profile, using the Laplace equation, has been extensively described in the physics literature since 1938.[17-25]

Nury et al.[16] demonstrated that this method constitutes a reliable, sensitive, and convenient means of investigating lipase kinetics by taking oil-drop pictures to determine the interfacial tension from the accurately measured diameters of the drops. The main drawbacks of this technique are the lengthy film processing and profile analysis and the fact that it does not yield real-time measurements. Nury et al.,[26] Grimaldi et al.,[27] and Cagna et al.[28] developed a new setup whereby the oil-drop profile was automatically

---

[14] S. Laurent, M. G. Ivanova, D. Pioch, J. Graille, and R. Verger, *Chem. Phys. Lipids* **70**, 35 (1994).
[15] M. G. Ivanova, T. Ivanova, R. Verger, and I. Panaiotov, *Colloids Surf.* **B6**, 9 (1996).
[16] S. Nury, G. Piéroni, C. Rivière, Y. Gargouri, A. Bois, and R. Verger, *Chem. Phys. Lipids* **45**, 27 (1987).
[17] J. M. Andreas, E. A. Hauser, and W. B. Tucker, *J. Phys. Chem.* **42**, 1001 (1938).
[18] E. Stauffer, *J. Phys. Chem.* **69**, 1933 (1965).
[19] M. Lin, "Transition de phase d'alcools aliphatiques à l'interface liquide/liquide sous différentes pressions hydrostatiques." Thèse de Doctorat d'État, Université de Provence, France (1981).
[20] H. H. J. Girault, D. J. Schiffrin, and B. D. V. Smith, *J. Electroanal. Chem.* **137**, 207 (1982).
[21] H. H. J. Girault, D. J. Schiffrin, and B. D. V. Smith, *J. Coll. Interface Sci.* **101**, 257 (1984).
[22] S. H. Anastasiadis, J. K. Chen, J. T. Koberstein, A. F. Siegel, J. E. Sohn, and J. A. Emerson, *J. Coll. Interface. Sci.* **119**, 55 (1987).
[23] P. Cheng, D. Li, L. Boruvka, Y. Rotenberg, and A. W. Neumann, *Colloids Surf.* **43**, 151 (1990).
[24] N. R. Pallas and Y. Harisson, *Colloids Surf.* **43**, 169 (1990).
[25] J. Satherley, H. H. J. Girault, and D. J. Schiffrin, *J. Coll. Interface Sci.* **136**, 574 (1989).
[26] S. Nury, N. Gaudry-Rolland, C. Rivière, Y. Gargouri, A. Bois, M. Lin, M. Grimaldi, J. Richou, and R. Verger, *GBF Monogr.* **16**, 123 (1991).
[27] M. Grimaldi, A. Bois, S. Nury, C. Rivière, R. Verger, and J. Richou, *Opto.* **91**, 104 (1991).
[28] A. Cagna, G. Esposito, C. Rivière, S. Housset, and R. Verger, Paper presented at 33rd International Conference on the Biochemistry of Lipids, Lyon, France, 1992.

digitized and analyzed by image processing; the interfacial tension was calculated in real time using the Laplace equation. The oil-drop methodology can be used to monitor lipase kinetics by following with time the decrease of interfacial tension due to the lipase action. It is possible to monitor accurately the lipase kinetics by increasing the drop volume to keep $\gamma_{o/w}$ constant.

Oil-Drop Technology

Automatic Oil-Drop Tensiometer

Figure 1 shows a diagram of the experimental setup. An integrating sphere light source (2), a thermostatted cuvette (3) containing the oil drop within a water phase, and a charge coupling device (CCD) camera attached to a telecentric lens (6) are aligned on an optical bench (1). A drop of liquid A is delivered from a microsyringe filled with liquid A (4), controlled by a dc motor drive (5), into a 25° thermostatted optical glass cuvette (1 × 2 × 4.3 cm, Hellma, France) containing liquid B (3). The microsyringe is attached, through a Luer-lock device, to a U-shaped stainless steel laboratory pipetting canula (14 G × 4-inch, Becton-Dickinson, Franklin Lakes) with a flat-cut tip having an external and internal diameter of 2 and 1 mm, respectively.

Depending on the volumetric masses of the two fluids, the drop is either mounting (e.g., oil drop in water) or hanging (e.g., water drop in oil). The equatorial drop diameters range from 4 to 6 mm. After formation of the drop, its profile is digitized and analyzed through the CCD camera coupled to a video image profile digitizer board (Imaging Technology, model PCVision Plus) connected to an IBM-PC compatible microcomputer with S-VGA screen (7). To retain permanent visual control, the drop image is continuously visualized on a video monitor (8). The drop profile is processed according to the fundamental Laplace equation applied to the oil-drop profile.[29] The computer calculates three characteristic parameters of the drop, namely, the area, volume, and interfacial tension ($\gamma_{o/w}$). The average standard accuracy of the interfacial tension measurements is roughly 0.1 mN × m^{-1}.

The frequency of the measurements can be set up from one measurement every 5 sec when high precision is required (precise mode) up to five measurements per second (fast mode). For highly specific requirements, interfacial tension, drop volume, and area can even be recorded up to a

---

[29] S. Labourdenne, N. Gaudry-Rolland, S. Letellier, M. Lin, A. Cagna, G. Esposito, R. Verger, and C. Rivière, *Chem. Phys. Lipids.* **71,** 163 (1994).

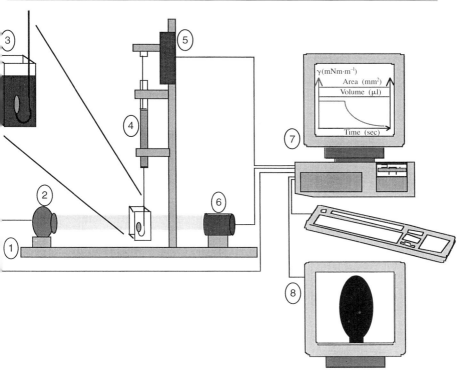

FIG. 1. Diagram of the experimental setup. (1) Optical bench; (2) integrated sphere light source (exit port aperture: 1 inch, with a 10-W halogen lamp, Labsphere, North Sutton, N.H.); (3) drop formation device with thermostatted cuvette; (4) syringe (Exmire Microsyringe, GLL 250, ITO Corporation, Fuji, Japan) containing liquid A; (5) dc motor with a 100 counts per revolution optical encoder (Maxon Motor, Switzerland) driving the piston of the syringe; (6) telecentric gagging lens (59LGF 410 and 59LGA 450 with a magnification of 0.5, Melles Griot, Rochester, New York) and CCD camera (VCM 6453 Philips, Netherlands); (7) personal computer (i486); (8) video monitor.

frequency of 25 times per second for about 5 sec, depending on the size of the drop.

The control unit can work according to one of several modes:

1. Record and plot the changes in $\gamma_{o/w}$ as a function of time (nonregulated mode).
2. Maintain $\gamma_{o/w}$ at a constant endpoint value (barostatic regulation mode) by exerting a feedback regulation on the dc motor that automatically increases the drop volume with time. This possibility is of

interest for investigating the interfacial tension-dependent phenomena occurring at the interface during kinetic experiments.[29]
3. An endpoint value of volume or area can be defined by the operator and the control is automatically performed by the control unit. This possibility is of particular interest in following the mass transfer phenomena through the interface.

A set of analytical tools built into the unit is provided to analyze the results to avoid external data processing.

1. Cut off some parts of the experimental results (extraction of the significant results).
2. Smooth experimental data by the least-squares or Fourrier transformation methods.
3. Fit data to models (linear, exponential, and hyperbolic, or operator defined) and superimpose curves on each other.
4. Display the variations of interfacial tension versus drop area or volume, or the variations of area versus volume.

For measurements that need to be carried out at high temperatures, which cannot be easily reached with a standard thermostat, the syringe and the thermostatted jacket of the cell can be heated (up to 100° and even higher). The heating of these components is separately provided as electrical heating cartridges having an electric power of 160 W for the syringe and 320 W for the cell under a voltage of 25 V. The temperature is controlled by a temperature probe fitted within the thermostatted jacket of the cell and an electronic regulator. The temperature endpoint is selected as needed.

Preparation of Materials

Cleanliness of materials is a strict requirement for the oil-drop methodology to obtain reliable results. As for the monolayer technique,[8] the water used for washing and buffer preparation was distilled and filtered on a Super Q Millipore unit made up of four columns: 0.22-μm filter, active charcoal, Ionex, and Organex. Glassware was cleaned with a sulfochromic mixture, then rinsed with specially purified water to wipe out any trace of tensioactive agents that could affect interfacial tension measurements. Before each experiment, the glass cuvette designed to contain the oil drop within a water phase was carefully washed with alcohol then rinsed with tap water and purified water. The syringe and the U-shaped stainless needle were sequentially washed with chloroform, methanol, purified water, and once again with methanol and chloroform. Both syringe and needle were dried using filtered compressed air.

Buffers

Buffers usually used for kinetics experiments were (1) 10 mM Tris-HCl (pH 8, 150 mM NaCl, 21 mM CaCl$_2$, 1 mM EDTA) for pancreatic lipase or (2) 10 mM acetic acid-acetate (pH 5, 100 mM NaCl, 20 mM CaCl$_2$, 1 mM EDTA) for gastric lipase. After dissolution of salts and pH adjustment, buffers were refiltered through a 0.22-μm membrane.

Reliability and Reproducibility

With a mounting soybean oil drop in water, we first compared the results obtained with the automatic oil-drop tensiometer experiments with those calculated using the previously described two-diameter method.[16] Both methods gave similar results (28.7 and 27.1 mN · m^{-1}, respectively).[29] It is noteworthy that the stability of the characteristic values of $\gamma_{o/w}$ obtained can be used as criteria to assess the absence of tensioactive impurities. The consistency and accuracy of the $\gamma_{o/w}$ values were therefore checked throughout this study before injecting the enzyme in each kinetic experiment. From the results of a series of five measurements carried out with mounting drops of various sizes of purified n-decane in water we derived a mean $\gamma_{o/w}$ value of 50.8 ± 0.3 mN · m^{-1}, independent of the drop volume,[29] which was quite similar to the value (50.1 mN · m^{-1}) published in the literature.[30] Generally speaking, the relative error ranged between 0.4 and 1.4% as calculated from the statistical distribution of more than 600 measurements performed on a standardized drop. We can therefore state that the oil-drop tensiometer gives reliable $\gamma_{o/w}$ values, provided that perfect conditions of cleanliness and optical alignment are fulfilled. The negligible variations with time (which averaged less than 1.5 × 10^{-3} mN · m^{-1} · min^{-1}) in the n-decane/water interfacial tension measured over a period of 200 min were taken to indicate the absence of any significant contamination by tensioactive agents in the materials used, as well as the reliability of the overall experimental procedure described earlier.[29] Table I shows a comparison of our values with those from the literature at air/water and n-decane/water interfaces.[14,24,30–32]

Principle of Studying Lipase Kinetics with the Oil-Drop Tensiometer

Edible soybean oil from Lesieur Company, containing linoleic acid (>50%), linolenic acid (>9%), and less than 0.1% free fatty acids was further

---

[30] R. C. Weast, "Handbook of Chemistry and Physics." CRC Press, Boca Raton, Florida, 1988–1989.
[31] N. R. Pallas and B. A. Pethica, *Colloids Surf.* **6,** 221 (1983).
[32] S. Labourdenne, M. G. Ivanova, O. Brass, A. Cagna, and R. Verger, *Colloids Surf.* **6,** 173 (1996).

TABLE I
SUPERFICIAL AND INTERFACIAL TENSION VALUES OBTAINED WITH THE OIL-DROP
TENSIOMETER AND COMPARISON WITH THOSE FROM THE LITERATURE[a]

Interfaces	Interfacial tension measurements (mN · m^{-1})	Interfacial tension data from literature (mN · m^{-1})
Air/water at 25°	71.739 ± 0.148	72.040 ± 0.100
	72.194 ± 0.133	71.989 ± 0.035
Air/water at 37°	70.101 ± 0.303	70.0
n-Decane/water at 25°	50.814 ± 0.337	50.1

[a] From Weast.[30]

purified through a column of silicic acid (Merck) equilibrated in hexane-ethylether as described previously.[16] The oil-drop methodology requires the oil to be carefully freed from any natural tensioactive compounds such as free fatty acids, diglycerides, and monoglycerides. As a matter of fact, those contaminants would lower the interfacial tension values because of their amphipathic character (see Fig. 2).

A drop of 20–100 µl of purified soybean oil is delivered from the

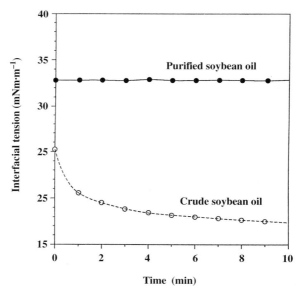

FIG. 2. Variations with time of the oil/water interfacial tension of purified or crude soybean oil.

microsyringe through the U-shaped stainless steel needle into the thermostatted cuvette containing 5 ml of an aqueous buffer and continuously stirred with a small magnetic bar. The lipase samples are injected into the aqueous phase (using Hamilton syringes) after or before the formation of the oil drop. Soybean oil mainly contains unsaturated long-chain fatty acids, which are released by the enzyme's action at the oil/water interface. The interfacial tension decreases on the accumulation of lipolytic products at the oil/water interface because of their amphipatic character.

As shown in Fig. 3, a good linear relationship was found to exist between the initial rate of the interfacial tension decrease and the porcine pancreatic lipase concentration up to 1.3 mg · liter^{-1}. This relationship is independent of the type of lipase. These data strongly suggest that the rate-limiting catalytic step is directly correlated with the enzyme action. If the product desorption would have been rate limiting, we would have observed a nonlinear dependency of the velocity on the lipase concentration, since the product desorption only depends on their physicochemical properties at the oil/water interface and is not enzyme catalyzed.

In the limiting situation, when the whole interface is covered by multilayers of lipolytic products, the enzymatic reaction is drastically slowed down.

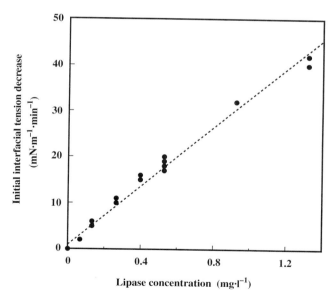

FIG. 3. Dependence of the initial rate of interfacial tension decrease on the porcine pancreatic lipase concentration. Specific activity is 7300 IU × mg^{-1} in the presence of a fivefold molar excess of colipase.

To overcome these intrinsic limitations, we determined, on the one hand, the initial velocities (initial interfacial tension decrease), that is, the conditions for which the proportion of the interface occupied by lipolytic products can be said to be negligible. On the other hand, we have developed a method for keeping $\gamma_{o/w}$ constant, that is, by maintaining a constant surface density of lipolytic products. Using the equipment described earlier and keeping the interfacial tension at a fixed endpoint value, it is possible to monitor the enzyme kinetics using the increase with time of the size (area) of the drop. Figure 4 shows the kinetics of the lipolysis of a mounting soybean oil drop formed in 5 ml of an aqueous phase (10 mM Tris-HCl buffer, pH 8) where the pancreatic lipase sample is injected. Interfacial tension ($\gamma_{o/w}$), volume (V), and area (A) are recorded as a function of time.

It can be observed that (1) the interfacial tension decreases following the lipase injection (first arrow) and (2) the barostatic regulation (second arrow) results in a linear increase of the drop volume and area. One can therefore monitor lipase kinetics by measuring the increase in area (dA) as a function of time to keep $\gamma_{o/w}$ constant. Throughout the kinetic process, the lipase therefore acts on triglyceride substrate molecules present at a

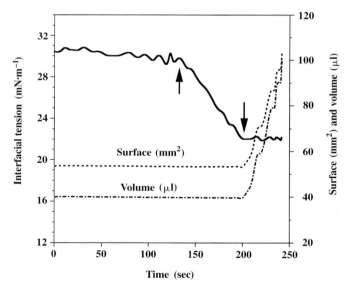

FIG. 4. Kinetic recordings of interfacial tension, drop volume, and area during the lipolysis of a mounting soybean oil drop (initial volume 50 μl) formed within an aqueous phase (10 mM Tris-HCl buffer, pH 8, 150 mM NaCl, 21 mM CaCl$_2$, 1 mM EDTA). First arrow marks the injection of the pancreatic lipase solution (final concentration 2 IU · ml^{-1}). Second arrow indicates the beginning of barostatic regulation.

preset and constant interfacial composition. The surface density (Γ) of the lipolytic products (diglycerides and free fatty acids in a 1:1 molar ratio) was established from independently determined surface compression isotherms.[32] The total amount of lipolytic products appearing per time unit (dn/dt) at the surface of a single oil drop is equal to the oil-drop surface increase (dA) multiplied by the surface density of the lipolytic products (Γ), as shown in Eq. (1):

$$\frac{dn}{dt} = \frac{dA}{dt} \times \Gamma \qquad (1)$$

Schematic Procedure for Studying Phospholipase A_2 Kinetics

A mounting drop of soybean oil was formed at zero time within a buffered solution containing an emulsion of Intralipid 20%, diluted 1000-fold. The egg phosphatidylcholine molecules, serving as an emulsifying agent for the soybean oil of the Intralipid emulsion, were slowly exchanged and adsorbed onto the oil drop, building up a monomolecular film (Fig. 5A). This process was followed by a spontaneous decrease with time of the oil/water interfacial tension. After a 15-min period of incubation of the soybean oil drop in the diluted Intralipid emulsion, a phospholipase A_2 solution (final concentration 33 nM) was injected into the diluted emulsion surrounding the oil drop (Fig. 5B). A faster decay in $\gamma_{o/w}$ was then observed.

This experiment has some drawbacks, however, since phospholipase A_2 seems to act at the same time on lecithin molecules coating the oil drop and those covering the Intralipid emulsion particles. Consequently, the interfacial tension decrease could result from the hydrolysis products (lysolecithin and free fatty acid) generated either from the oil drop or from the Intralipid emulsion. To overcome this drawback, we designed a new protocol in which the diluted Intralipid emulsion was replaced after an incubation period of 15 min by a fresh buffer solution (Fig. 5C) before the enzyme injection (Fig. 5D). The interfacial tension remained constant after the buffer perifusion and decreased only after the phospholipase A_2 injection. Under these conditions, the observed interfacial tension decrease could be attributable only to the hydrolysis of the phospholipid film adsorbed onto the drop surface. This new approach may therefore be the first step toward developing original phospholipase activity assays at an oil/water interface.

Hydrolysis of Soybean Oil by Human Pancreatic Lipase

Human pancreatic lipase (HPL) and porcine pancreatic colipase were purified in the laboratory by Y. Gargouri and J. de Caro, using methods

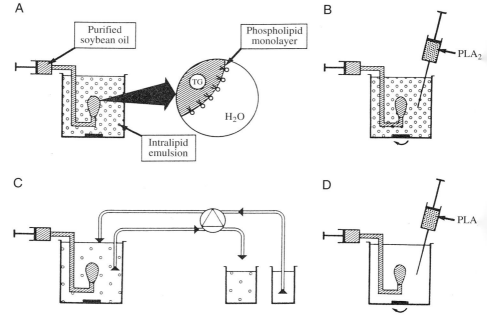

FIG. 5. Diagram of the procedure used to study phospholipase A_2 kinetics. (A) Formation of a phospholipid monomolecular film, on a mounting drop of soybean oil, by adsorption from an Intralipid emulsion 20%, diluted 1000-fold. (B) Injection of phospholipase A_2 (final concentration 33 nM) within the diluted Intralipid emulsion surrounding the drop formed in a 24-ml glass cuvette. (C) Perifusion of the Intralipid emulsion diluted with fresh buffer (as described) using two synchronous peristaltic pumps with equal flow rates (0.2 ml · min^{-1}). (D) Injection of phospholipase A_2 (final concentration 33 nM) within the fresh buffer used to replace the diluted Intralipid emulsion.

described previously.[33-35] Polyclonal antibodies against native HPL were prepared in the laboratory by Aoubala *et al.*[36] Kinetics studies of lipolysis with biotinylated HPL were carried out with the oil-drop tensiometer with the aim of estimating the fraction of enzyme bound to the oil/water interface. Biotinylated HPL (HPL*) was prepared by the procedure previously set up in our laboratory by Aoubala *et al.*[36] for studying the interfacial

[33] A. De Caro, C. Figarella, J. Amic, R. Michel, and O. Guy, *Biochim. Biophys. Acta* **490**, 411 (1977).
[34] M. F. Maylié, M. Charles, M. Astier, and P. Desnuelle, *Biochem. Biophys. Res. Comm.* **52**, 291 (1973).
[35] C. Chapus, P. Desnuelle, and E. Fogglizzo, *Eur. J. Biochem.* **115**, 99 (1981).
[36] M. Aoubala, L. de la Fournière, I. Douchet, A. Abousalham, C. Daniel, M. Hirn, Y. Gargouri, R. Verger, and A. de Caro, *J. Biol. Chem.* **270**, 3932 (1995).

binding of human gastric lipase to lipid monolayers. Biotin labeling does not result in any appreciable loss of catalytic activity.

Kinetics of HPL Hydrolysis in the Absence or Presence of Colipase. The lipolysis at the oil/water interface was monitored by recording the decrease with time of the interfacial tension following the lipase injection into the aqueous phase. Figure 6A shows the effects of colipase or biotinylated HPL injections on the decrease of interfacial tension. Colipase, at the final

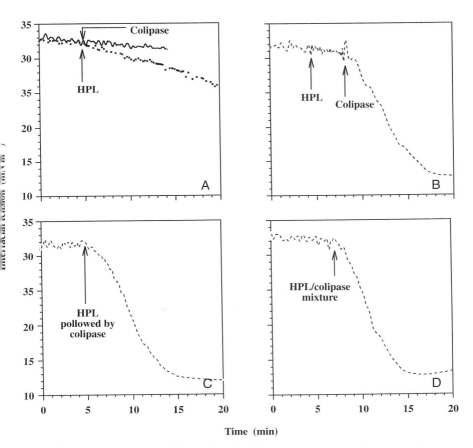

FIG. 6. Variation as a function of time of interfacial tension at a soybean oil/water interface: (A) Separate colipase (——) and pure HPL (----) injections. (B) HPL was injected first and colipase second. (C) HPL and colipase were injected simultaneously. (D) The preformed HPL/colipase mixture (1:5 molar ratio) was injected. The bulk phase (volume of 5 ml) was composed of 10 mM Tris-HCl buffer, pH 8, 150 mM NaCl, 21 mM CaCl$_2$, and 1 mM EDTA and was stirred continuously. The final concentrations of colipase and HPL were 0.2 μg · ml^{-1}.

concentration of 0.2 μg · ml^{-1}, did not exhibit any detectable tensioactive effect at the surface of the soybean oil drop. These findings are in agreement with the results of previous studies on the effects of colipase on the hydrolysis of monomolecular films by lipase.[37] Furthermore, pure HPL* displayed very low enzymatic activity levels on soybean oil substrates, and the rate of the decrease of interfacial tension was 0.4 ± 0.1 mN · m^{-1} · min^{-1} (see Fig. 6A).

Figure 6B shows a rapid decrease of interfacial tension (6 ± 2 mN · m^{-1} · min^{-1}) as a function of time after an HPL* injection followed 4 min later by a colipase injection. Figure 6C shows the effects of an injection of HPL*, immediately followed by a colipase injection. Figure 6D shows the decrease of the interfacial tension occurring after the injection of an HPL*/colipase preformed mixture (1/5 molar ratio). Note that in all cases where an efficient lipolysis occurred, the same final interfacial tension value (12 ± 1 mN · m^{-1}) was reached. Colipase strongly increased the activity of HPL (approximately 10- to 15-fold). The stimulatory effect of colipase does not depend on the mode of the cofactor injection. However, the lag time required to obtain a maximal decrease of interfacial tension is shorter with a preformed lipase/colipase mixture than with separate injections of HPL and colipase.

Oil-Drop Recovery and Estimation of the Fraction of Bound Enzyme. At the end of a kinetic experiment, the oil drop and a part of the aqueous phase surrounding it were recovered using a Pasteur pipette adapted to the tip of a pipetting device (Pipetman P200) and transferred into a glass tube containing 300 μl of a gum arabic mixture used to ensure an efficient emulsification. The emulsification of the recovered oil drop containing adsorbed HPL* was performed for 30 sec at 50° with an Ultra-Turax T25 device (Janke and Klunkle, Staufen, Germany) in a 10% w/v gum arabic solution in TBS, which also contains 0.05% v/v Tween 20 and 0.05% BSA, previously filtered through a cheese cloth having a pore size of 50 μm. As a control, in parallel with the oil-drop recovery procedure, an equal volume of the aqueous phase was sampled from the cuvette then emulsified with the gum arabic mixture after addition of 50 μl of pure soybean oil.

The enzyme-linked immunosorbent assay (ELISA) test procedure was adapted from the method developed in our laboratory by Aoubala *et al.*[10] for studying the interfacial binding of human gastric lipase to lipid monolayers. All the ELISA tests were performed in the 96 wells of polyvinylchloride microplates (Maxisorb, Nunc, Roskilde, Denmark). Wells were coated with 250 ng of a specific polyclonal anti-HPL antibody solubilized in 50 μl of 150 mM Tris-HCl buffer containing 140 mM NaCl and 3 mM KCl (TBS)

[37] R. Verger, J. Rietsch, and P. Desnuelle, *J. Biol. Chem.* **252**, 4319 (1977).

and incubated overnight at 4°. Subsequent steps of the procedure were as described previously[10] except for the preparation of HPL* samples. The wells, previously coated with polyclonal anti-HPL antibody, were incubated for 2 hr at 37° with 50 μl of various dilutions of the emulsion, as described earlier for the recovered biotinylated lipase (HPL*). A reference curve was drawn using pure HPL* samples of known concentrations, emulsified as described earlier in the gum arabic mixture after adding of 50 μl of pure soybean oil. Each assay was carried out in duplicate.

The reliability of the ELISA sandwich was tested with HPL*. The recovery levels of HPL* injected into the glass cuvette were determined at each oil-drop kinetic experiment and found to be good (90% ± 10). This means that the amount of HPL* lost by adsorption to the walls of the glass cuvette or during the emulsification procedure is to be regarded as negligible. To calculate the amount of HPL* adsorbed to the oil-drop surface, Eq. (2) was used:

$$\Gamma_E = \frac{[O,B]V_a - [B]V_b}{S_d} \quad (2)$$

where Γ_E is the surface excess of HPL* adsorbed to the oil surface, expressed as $ng \cdot cm^{-2}$, V_b the volume of the aspirated bulk phase, S_d the oil-drop surface, $[B]$ the HPL* concentration in the aspirated bulk phase, V_a the total volume of the oil drop plus the simultaneously aspirated bulk phase and $[O, B]$ the average of HPL* concentration in the aspirate (oil-drop and bulk phase).

Figure 7 gives the plots of the maximum rates of decrease in the interfacial tension and the corresponding interfacial HPL* excess resulting from a series of kinetic experiments interrupted after a predetermined period of time. The amount of adsorbed HPL* at the drop surface was time dependent and the maximum amount of adsorbed HPL* was equal to 1% of the total HPL* injected.

We clearly established that colipase induces a highly significant increase in the turnover of HPL, whereas the adsorption of this enzyme to the triacylglycerol surface does not require the assistance of the cofactor (Fig. 7). Colipase probably contributes to stabilizing the open HPL conformation preexisting at the interface, giving the active site increased accessibility and efficiency.[38] A nonexclusive alternative hypothesis is that colipase could have "substrate channeling" effects on the interfacial activity of HPL, as recently suggested by Momsen et al.[39]

[38] H. van Tilbeurgh, M.-P. Egloff, C. Martinez, N. Rugani, R. Verger, and C. Cambillau, *Nature* **362**, 814 (1993).
[39] W. E. Momsen, M. M. Momsen, and H. L. Brockmann, *Biochemistry* **34**, 7271 (1995).

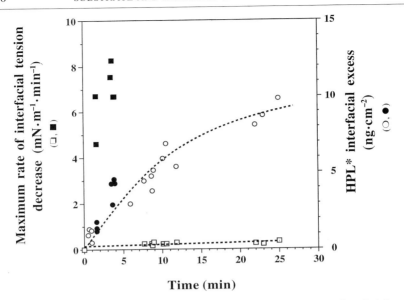

FIG. 7. Maximal rate of interfacial tension decrease (□, ■) and the HPL* interfacial excess (○, ●) as a function of time. At each point, the oil drop is recovered and the interfacial HPL* excess determined. The measurements were performed with (●, ■) and without (○, □) colipase (0.2 $\mu g \cdot ml^{-1}$, final concentration). The bulk phase is described in Fig. 6.

Kinetics of HPL Hydrolysis in the Presence of Bile Salts. To evaluate the effects of bile salts on HPL* hydrolysis, we injected them after the formation of the drop, before the injection of the HPL/colipase mixture (see Fig. 8). Following the bile salt injection, a rapid decrease of the interfacial tension was observed and a new equilibrium value was reached. With the oil-drop technique, it is unfortunately impossible to study the lipase hydrolysis at NaTDC concentrations above their initial micellar concentration (CMC), because the drop surface is saturated by bile salt molecules and consequently the interfacial tension does not vary when further tensioactive agents (lipolytic products) are generated (see Fig. 8B).

After the HPL/colipase injection and after a given period of lipolysis, the oil drop was recovered with a view toward determining the HPL* interfacial excess. Figures 9A and 9B illustrate the kinetics of an oil-drop hydrolysis by HPL*, in the presence of colipase, as a function of the NaTDC concentration. Throughout the NaTDC concentration range investigated, the rate of hydrolysis was found to be roughly proportional to the lipase concentration for 0.1 and 0.2 $\mu g \cdot ml^{-1}$ (see Fig. 9A). More important, the presence of NaTDC, up to 100 μM, was not found to affect the hydrolysis

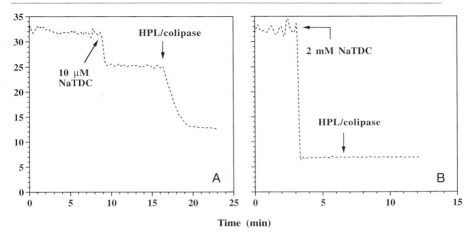

FIG. 8. Effects of bile salts and of an injection of a preformed HPL/colipase mixture (1:5 molar ratio) on the oil/water interface. HPL final concentration was 0.2 μg · ml^{-1}.

rates significantly. However, note that for various initial $\gamma_{o/w}$ values, various final $\gamma_{o/w}$ values were reached. For instance, in the absence of NaTDC, $\gamma_{o/w}$ decreased from 32 to 12 mN · m^{-1}, whereas in the presence of 400 μM NaTDC, $\gamma_{o/w}$ decreased from 14 to 5 mN · m^{-1}. In the absence of colipase, the presence of bile salts enhanced the hydrolysis rate from 0.4 ± 0.1 mN · m^{-1} · min^{-1} without NaTDC to 2 mN · m^{-1} · min^{-1} at 10 and 100 μM NaTDC (data not shown).

Figure 9B illustrates the fact that the presence of NaTDC (up to 100 μM) had no significant effects on the interfacial HPL* excess. Note that the lag time is drastically reduced in the presence of bile salts at concentrations lower than CMC. Assuming that lag time is corresponding to a rate-limiting step associated with lipase adsorption ($E \leftrightarrow E^*$) as previously suggested,[40] we can conclude from our data that the presence of negative charges at the drop surface facilitates the binding of the HPL/colipase to the oil surface. This interpretation is in agreement with the shortening of the lag phase previously observed in the presence of ionised fatty acids.[41]

Figures 9A and 9B show that the presence of bile salts below the CMC is not affecting either the HPL binding or the hydrolysis rates at the oil/water interface. This observation means that the HPL/colipase complex has

---

[40] L. de la Fournière, M. G. Ivanova, J.-P. Blond, F. Carrière, and R. Verger, *Colloids Surf.* **B2**, 585 (1994).

[41] T. Wieloch, B. Borgström, G. Piéroni, F. Pattus, and R. Verger, *J. Biol. Chem.* **257**, 11523 (1982).

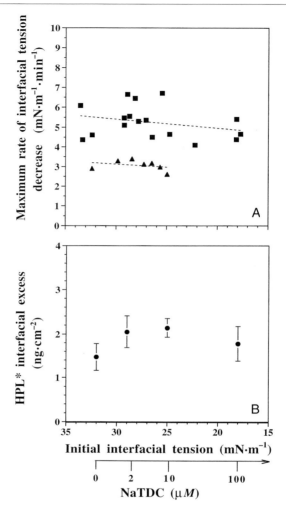

FIG. 9. The maximal rates of interfacial tension decrease at a soybean oil/water interface induced by HPL*/colipase mixture (1:5 molar ratio) as a function of the NaTDC concentration. (A) (▲) 0.1 µg · ml^{-1}, (■) 0.2 µg · ml^{-1} HPL*. (B) (●) Interfacial excess of HPL* (0.2 µg · ml^{-1}) in the presence of colipase (1:5 molar ratio). The interfacial excess of HPL* was measured 100 sec after its injection into the aqueous phase. The bulk phase is described in Fig. 6.

the same binding and catalytic efficiency at electrically neutral or negatively charged oil/water interfaces. Our data agree with findings made some years ago by Momsen and Brockmann,[42] using the silicon-coated glass bead system. These authors showed that pure pancreatic lipase readily adsorbs to a substrate surface covered by bile salts below the CMC. At the oil/water interface, in the presence or absence of bile salts below their CMC, colipase behaves like a true lipase cofactor, that is, it increases the enzyme turnover and does not affect the interfacial adsorption of the enzyme.

Comparison between the Monomolecular Film Technique and the Oil-Drop Method. Lipid/water interfaces are obviously not all equivalent, and they do not act as "inert mirrors" for lipolytic enzymes, whereas the "interfacial quality" plays an important regulatory role in enzymatic processes. To look more closely at the "interfacial quality," Ivanova *et al.* studied lipid hydrolysis by HPL on various monomolecular films of dicaprin, dilaurin, diolein, and purified soybean oil. With long-chain lipids (diolein and soybean oil), we use β-cyclodextrin (β-CD) to solubilize the lipolytic products as described previously.[15]

The adapted version of the Michaelis–Menten kinetic scheme for use at the air/water interface has been described and used by previous authors.[5,6,8,43] In the case of the hydrolysis of monolayers of medium-chain lipids, it was experimentally checked that the lipolytic products of the reaction desorbed rapidly into the water subphase.[44] Raneva *et al.*[45] recently adapted this original kinetic model to study phospholipase A_2 action on liposomes of long-chain phosphatidylcholine spread at the air/water interface. This kinetic scheme may also be used to describe the enzymatic hydrolysis at the oil/water interface.[46]

Based on these kinetic data, we calculated the global catalytic constant (Q), which takes into account the substrate surface concentration and the HPL concentration in the bulk phase. From Table II, we can see first that HPL does not require the presence of the cofactor to hydrolyze lipid monolayers; this is in agreement with previous data on the hydrolysis of dicaprin by HPL.[40]

Table II also shows that HPL hydrolyzes monomolecular films of di- and triglycerides with various chain lengths at about the same rate. For the sake of comparison, the global kinetic constants (Q) of HPL at the oil-drop surface are given in Table II. With the oil-drop system, in the absence

---

[42] W. E. Momsen and H. L. Brockmann, *J. Biol. Chem.* **251**, 384 (1976).
[43] M. K. Jain and O. G. Berg, *Biochim. Biophys. Acta* **1002**, 127 (1989).
[44] R. Verger, M. C. E. Mieras, and G. H. de Haas, *J. Biol. Chem.* **248**, 4023 (1973).
[45] V. Raneva, T. Ivanova, R. Verger, and I. Panaiotov, *Colloids Surf.* **B3**, 357 (1995).
[46] S. Labourdenne, O. Brass, M. Ivanova, A. Cagna, and R. Verger, *Biochemistry* **36**, 3423 (1996).

TABLE II
COMPARISON BETWEEN THE GLOBAL KINETICS CONSTANTS $(Q)^a$

	Oil-drop method				Monolayer technique	
	Soybean oil				Diolein	Soybean oil
	−NaTDC	+NaTDC	Dicaprin	Dilaurin	+β-CD	+β-CD
HPL	0.31	0.95	4.7	3.2	3.2	3.1
HPL + CO	1.40	1.70	5.4	4.1	3.5	2.9

^{a} Q is expressed in (ml × s^{-1} × molecule^{-1}) × 10^{14}, as measured with the oil-drop method (in the presence or absence of 10 μM NaTDC) using various di- and triglycerides as substrates and with the monolayer technique at a surface pressure π of 12 mN · m^{-1} (in the presence of 0.77 mM β-cyclodextrin) using long-chain acylglycerol monomolecular films.

of colipase, the global kinetic constant was found to be 10-fold lower than the value obtained using monomolecular films of soybean oil in the presence of β-CD. In the presence of colipase, the global kinetic constant was found using the oil-drop method to be only two times lower. Once again, we can conclude that the type of interface used, that is, oil/water versus air/water, strongly influences the catalytic activity of HPL as well as the colipase

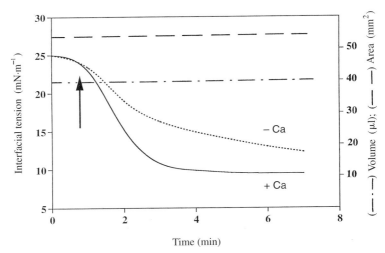

FIG. 10. Kinetics of Lipolase (Lipase from Novo Nordisk, cloned from the fungus *H. lanuginosa* and expressed in *A. oryzae*), showing the interfacial tension decrease with or without calcium ions (3 mM). Arrow indicates lipase injection. Notice the faster decrease in the presence of 3 mM Ca^{2+} [99% pure CaCl$_2$, Merck (Merck, Darmstadt, Germany)] in 0.1 M TAPS buffer, pH 9, filtered through a 0.45-μm filter. Triolein [C18:1, [*cis*]-9 from Sigma (Sigma, St. Louis)]. (Reprinted with permission from Ladefoged *et al.*[47])

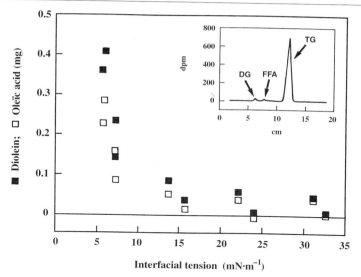

Fig. 11. Amount of free fatty acids (□) and diglycerides (■) produced by the action of cutinase on a 40-μl drop of ^{14}C-labeled triolein as a function of the interfacial tension. These lipolytic products were detected using thin-layer chromatography (TLC) and scintillating counting. Insert: TLC pattern from a kinetic experiment at $\gamma_{o/w} = 7$ mN · m^{-1} with 10 mM Tris buffer, pH 9. Temperature 30°. (Reprinted with permission from Flipsen et al.[48])

effect. It is worth noting that the effect of colipase is more pronounced at the oil/water interface, contrary to what occurs with soybean oil monomolecular films spread at the air/water interface.

Results Obtained with the Oil-Drop Methodology at Other Laboratories

The oil-drop methodology has been found useful for investigating differences in lipase activity toward triolein at various interfacial tensions. Ladefoged et al.[47] have investigated the effect of calcium ions in the buffer when performing hydrolysis of triolein at pH 9 by a lipase cloned from the fungus *Humicola lanuginosa*, expressed in *Aspergillus oryzae* and commercially available as Lipolase (Novo Nordisk, Bagsvaerd, Denmark). As shown in Fig. 10, they observed that the decrease of the interfacial tension with time

[47] C. Ladefoged, A. Cagna, and E. Gormsen, "Lipase activity as a function of interfacial tension using the rising drop method on a new oil drop tensiometer," Paper presented at Enzyme Engineering XII, Deauville, France, 1995.

is much slower without calcium present. This could be explained by the lack of formation of calcium soaps. These workers have also used the principle of the barostat at the oil/water interface with which it is possible to monitor the enzyme kinetics at a fixed interfacial tension endpoint value by increasing the drop surface with time as recently described.[32] Plotting the various dA/dt values as a function of the interfacial tension shows a continuous decrease in rate with decreasing interfacial tension and can be considered to be a unique representation of each lipase at the oil/water interface.

Flipsen et al.[48] recently showed that hydrolysis of triolein by the lipolytic enzyme cutinase appeared to be linear with time. To correlate the rate of hydrolysis with the effect of the lipolytic products on the interfacial tension, these workers analyzed these products using ^{14}C-labeled triolein (Fig. 11). They observed, first, that only diolein and oleic acids are produced (see insert, Fig. 11). Although less than 2% of the triglycerides were hydrolyzed, the surface area that can be occupied by the fatty acids formed exceeded the drop surface area by a factor of 5000. The amount of lipolytic products formed possibly results in multilayers of products at the interface. In this particular case, the presence of 1 mM CaCl$_2$ does not affect the rate of hydrolysis but it has great impact on the absolute value of the interfacial tension. Flipsen et al. also studied the influence of oleic acid on the hydrolysis and the adsorption phenomena using mixtures of oleic acid (0-3%) and triolein. It was observed that the initial rate of interfacial tension decrease on lipase injection is lowered in the presence of oleic acid.

In conclusion, when evaluating new lipases, surface activity at the oil/water interface is considered to be a critical parameter, in addition to the catalytic turnover.

Acknowledgments

This work was supported by the BRIDGE-T-Lipase Programme of the European Communities under contract BIOT-CT910274 (DTEE) and by BIOTECH G-Program BIO 2-CT94-3041.

Thanks are due to Claus Ladefoged (Novo Nordisk A/S, Denmark) and Jacco Flipsen (Unilever, Netherlands) for permission to use material presented in Figures 10 and 11, respectively.

[48] J. Flipsen, H. van der Hijden, A. Slotboom, and B. Verheij, "Cutinase kinetics at the triolein-water interface. An oil drop tensiometer study," Poster presented at NATO-ASI on Engineering of/with Lipases, Póvoa de Varzim, Portugal, 1995.

[16] A Critical Reevaluation of the Phenomenon of Interfacial Activation

By FRANCINE FERRATO, FRÉDÉRIC CARRIERE, LOUIS SARDA, and ROBERT VERGER

1. Introduction

The four main classes of biological substances are carbohydrates, proteins, nucleic acids, and lipids. The first three of these substances have been clearly defined on the basis of their structural features, whereas the property that is common to all lipids is a physicochemical one. Lipids are in fact a group of structurally heterogeneous molecules. The molecules are all soluble in apolar and slightly polar solvents such as benzene, ether, and chloroform, and insoluble or partly soluble in water. Lipids have been classified by Small[1] depending on how they behave in the presence of water. This makes it possible to distinguish between polar and apolar lipids (e.g., hydrocarbon, carotene).

The polar lipids can be further subdivided into three classes. Class I consists of those lipids that do not swell in contact with water and form stable monomolecular films (these include long-chain triacylglycerols, long-chain diacylglycerols, phytols, retinols, vitamins A, K, and E, waxes, and numerous sterols). The class II lipids (which include phospholipids, monoacylglycerols, and fatty acids) spread evenly on the surface of water, but since they become hydrated, they swell up and form well-defined lyotropic (liquid crystalline) phases such as liposomes. The class III lipids (such as lysophospholipids, bile salts, short-chain triacylglycerols, and medium-chain phospholipids) are partly soluble in water and form unstable monomolecular films, and, beyond the critical micellar concentration (CMC) level, micellar solutions.

In vitro experiments have been carried out for a long time and by many investigators using synthetic short- and medium-chain esters as putative lipase substrates, such as methyl butyrate, triacetin, tripropionin, and tributyrin. In aqueous media, these esters can exist either as an isotropic solution in the form of monomers, micelles, and adsorbed monomolecular films or can exist as a turbid emulsion, which appears above the solubility limit. These various physicochemical forms of ester molecules are depicted schematically in Fig. 1. They usually coexist as a complex equilibrium between

---

[1] D. M. Small, *J. Am. Oil Chemists Soc.* **45,** 108 (1968).

328 SUBSTRATE AND INHIBITOR CHARACTERIZATION [16]

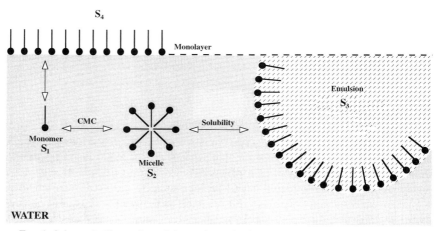

FIG. 1. Schematic illustration of the various physicochemical states of ester molecules in aqueous media. Monomer (S_1), micelle (S_2), emulsion (S_3), and adsorbed monolayer (S_4) coexist in equilibrium and represent potential lipase monomeric substrate (S_1) or supersubstrates (S_2, S_3, and S_4). The dotted area represents the isotropic solution and the dashed area represents the turbid emulsion.

potential lipase *monomeric substrate* (S_1) or super substrates (S_2, S_3, and S_4, respectively).

Lipases are carboxylic ester hydrolases and have been termed glycerolester-hydrolase (EC 3.1.1.3) in the international system of classification. They greatly differ with respect to both of their origins (which can be bacterial, fungal, plant, mammalian, etc.) and their kinetic properties. They can catalyze *in vitro* the hydrolysis, or synthesis, of a wide range of different carboxylic esters; however, they all show a higher specific activity toward glyceridic substrates. Under physiological conditions, because natural triacylglycerols are water insoluble, lipases, which are generally soluble in water, catalyze the hydrolysis of carboxylic ester bonds at lipid/water interfaces.

2. Definition of Lipases and Interfacial Activation

What exactly is a lipase? Is it enough to say that it is a carboxylic esterase that specifically hydrolyzes triacylglycerols? In 1958, Sarda and Desnuelle[2] defined lipases in kinetic terms, based on the *interfacial activa-*

[2] L. Sarda and P. Desnuelle, *Biochim. Biophys. Acta* **30**, 513 (1958).

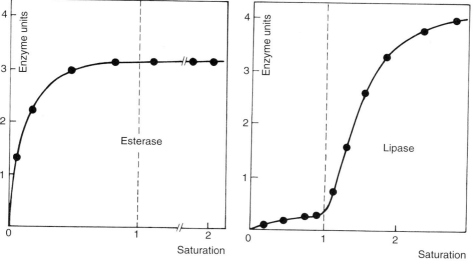

FIG. 2. Hydrolysis rate as a function of the amount of a partly water-soluble ester. Dashed vertical lines represent the limit of solubility of the ester used. Such kinetic behaviors have been commonly used to discriminate between esterases (left panel) and lipases (right panel). (Adapted from Sarda and Desnuelle.[2])

tion phenomenon, as shown schematically in Fig. 2. This property was not to be found, for example, among the enzymes that have been classified as esterases, that is, those acting only on carboxylic ester molecules that are soluble in water. The interfacial activation phenomenon was in fact first recognized as far back as 1936 by Holwerda et al.[3] and then in 1945 by Schønheyder and Volqvartz.[4] It was known that the activity of lipases is enhanced on insoluble substrates (such as emulsions) than on the same substrates below the solubility limit. It therefore emerged from the studies mentioned that lipases might constitute a special category of esterases that is highly efficient at hydrolyzing molecules having a carboxylic ester group and aggregated in water. This property was used for a long time to distinguish between lipases and esterases. In 1960, Desnuelle et al.[5] postulated that a conformational change of lipases might be associated with the phenomenon of interfacial activation, by analogy with the proteolytic activation of pancreatic zymogens.

[3] K. Holwerda, P. E. Verkade, and A. H. A. de Willigen, *Rec. Trav. Chim. Pays-Bas* **55**, 43 (1936).
[4] F. Schønheyder and K. Volqvartz, *Acta Physiol. Scand.* **9**, 57 (1945).
[5] P. Desnuelle, L. Sarda, and G. Ailhaud, *Biochim. Biophys. Acta* **37**, 570 (1960).

The preceeding hypothesis gradually drifted and was then progressively transformed to cover an idealized concept, far away from real experimental facts and artifacts. Before the elucidation of the first 3D structures of lipases, "interfacial activation" had been taken to mean a hypothetical conformational change occurring as the result of interfacial adsorption. Worse still, *interfacial activation* has sometimes been wrongly taken to refer to the increase in the catalytic activity of an enzyme on a triacylglycerol substrate occurring either on mechanical emulsification or in the presence of a tensioactive agent.

The results of recent lipase research have nevertheless shown how careful one has to be about extrapolating any kinetic and/or structural characteristics observed in all lipases in general. The catalytic activity of many lipolytic enzymes has been measured using carboxylic esters, which are partly soluble in water, and many differences have been found to exist between the resulting profiles.

As shown in Fig. 3, the same partly water-soluble substrate (TC_2: triacetin or triethanoylglycerol) and the same purified porcine pancreas lipase (PPL) gave very different profiles, depending on the presence of sodium chloride[6] (lower panel) or gum arabic $(GA)^2$ (upper panel) in the assay medium.

These differences can be understood by taking into account the fact that high ionic strength will in general reduce the value of the critical micellar concentration, with a corresponding increase in the micellar size. Consequently, in the case of experiments carried out by Entressangles and Desnuelle[6], at 0.1 M NaCl, the presence of isotropic aggregates (micelles) below the solubility limit could explain the relatively high catalytic activity found by these authors, in contrast to Sarda and Desnuelle,[2] who performed their kinetic experiments in the absence of ionic strength and in the presence of GA (see Fig. 3).

Furthermore, Sarda and Desnuelle[2] reported that purified fractions of PPL were relatively less active than crude fractions (probably contaminated by pancreatic esterases) when tested on an isotropic solution of triacetin (see Fig. 3, upper panel). Thus these authors proposed a radical point of view: Lipases, contrary to esterases, are enzymes completely unable to act on a true monomeric solution of ester. To shed more light on this fundamental question, Entressangles and Desnuelle[6] measured, by differential optical absorbance, the appearance of molecular aggregates (micelles?) using the micellar solubility properties of dyes such as benzopurpurin 4B or iodine or Sudan III (see Fig. 4, left panel).

---

[6] B. Entressangles and P. Desnuelle, *Biochim. Biophys. Acta* **159,** 285 (1968).

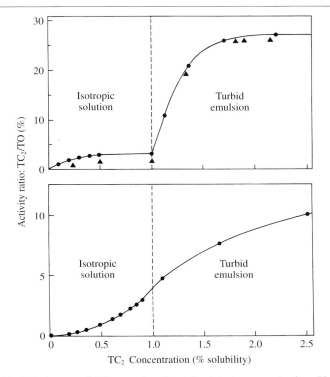

FIG. 3. Hydrolysis rate of TC_2 as a function of substrate concentration. Upper panel: Experiment from Sarda and Desnuelle,[2] using TC_2 emulsified in GA. The ordinates are expressed in percent of the maximal rate observed on triolein or trioleoylglycerol (TO). ●, results obtained with a crude lipase preparation (specific activity of 400 μmol · min^{-1} · mg^{-1}); ▲, results obtained with a purified lipase preparation (specific activity of 4230 μmol · min^{-1} · mg^{-1}). Lower panel: Experiment from Entressangles and Desnuelle.[6] Initial hydrolysis rates were measured titrimetrically with a radiometer pH-stat at 25° and pH 7.0 in a thermostatted vessel containing 15 ml of a TC_2 solution or 15 ml of an emulsion, both prepared in 0.1 M NaCl. The 0.1 M NaCl solution was placed in the glass pH-stat vessel and a known amount of glyceride was added. The mixture was agitated for 30 min at 1400 rpm and agitation was maintained after addition of lipase. Lipase was purified by chromatography on DEAE-cellulose and its specific activity was about 4000 μmol · min^{-1} · mg^{-1} (using TO as substrate). The rates are expressed in percent of the maximal rate observed with the same amount of lipase and an excess of a TO emulsion. The total concentration of the substrate (solution + emulsion) is expressed in multiples of the saturation. The vertical dashed line indicates the saturation point.

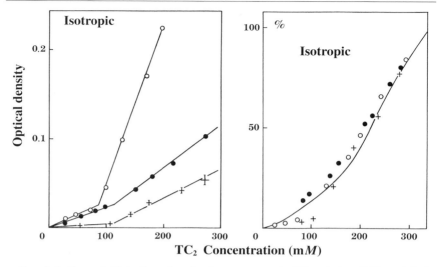

FIG. 4. Left panel: Detection of molecular aggregates in TC_2 solution. Differential absorbance read at 510 (○) and 355 (●) nm for benzopurpurin 4B and iodine, respectively. Crosses (+) indicate the solubilization of Sudan III, measured by the absorbance of the solutions at 505 nm. For this latter determination, the standard errors are indicated by the vertical bar of the crosses. All assays were carried out at 25°. Right panel: Correlation between lipase activity and the formation of aggregates in TC_2 solution. The continuous curve indicates lipase activity versus TC_2 concentration. The other symbols indicate the responses of the techniques used for the detection of aggregates. The symbols are the same as in the left panel, namely, open circles for benzopurpurin, closed circles for iodine and crosses for Sudan III. The ordinates are expressed in percent of their maximal values at the saturation point. (From Entressangles and Desnuelle.[6])

As shown in the right panel of Fig. 4, these authors correlated lipase activity with the formation of aggregates in triacetin solution. Furthermore, identical V_{max} values were obtained by extrapolation from an isotropic solution or a turbid emulsion of triacetin,[6] or tripropionin or tripropanoyl-glycerol (TC_3) (see Fig. 5). It appeared therefore, within the relatively high experimental error associated with this kind of study, that the maximal rate of lipolysis of the same substrate in an aggregated or an emulsified form was similar, if not identical.

3. Interfacial Activation: Facts and Artifacts

The greatest caution must be exercised both when performing and interpreting kinetic measurements with lipids. First, it is essential to check that the initial lipase velocity is proportional to the amount of enzyme

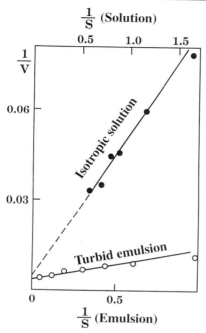

FIG. 5. Lineweaver–Burk representation for the action of PPL on aggregates and emulsified particles of TC_3. The abscissa are the reciprocal values of the total TC_3 concentration minus the critical concentration for the solutions or minus the saturation for the emulsions. The ordinates are the reciprocal values of the total activities measured in the solutions and the emulsions, respectively. (From Entressangles and Desnuelle.[6])

used, both below and above the solubility limit. Substrate depletion, in the monomeric range of substrate concentration, is sometimes a major experimental limitation. Second, it is also essential to check that the same lipase active site, and not other unspecific sites, is responsible for the measured catalytic activity on monomeric substrates. Control experiments with nonenzymatic proteins or inhibited lipase should be performed. Third, since the medium is heterogeneous, adding any amphiphilic compound to the system is liable to modify both quantitatively and qualitatively the physicochemical properties of the interface. The presence of a phospholipid substrate in the micellar state will lead, for example, to the aggregation of pancreatic phospholipase A_2 into multimolecular lipoprotein complexes.[7]

[7] P. Soares de Araujo, M. Y. Rosseneu, J. M. H. Kremer, E. J. J. van Zoelen, and G. H. de Haas, *Biochemistry* **18**, 580 (1979).

Some lipases, such as gastric lipases, rapidly become denatured at an interface with a pure tributyrin or tributanoylglycerol (TC_4) emulsion. Consequently, it is impossible to assess experimentally what interfacial activation may have occurred. In addition, some esters that are partly soluble in water sometimes form monomolecular adsorption films on the surface of the air bubbles produced by stirring the reaction mixture. This artifact is responsible for great disparity between initial velocity measurements, depending on whether or not efficient mechanical stirring methods are used. Brockman et al.[8] studied the enhancement of hydrolysis of TC_3 by pancreatic lipase in the presence of siliconized glass beads. This is a very interesting interfacial system that can be considered a spherical monolayer of TC_3 at a liquid/solid interface. The binding of the enzyme to the surface was shown to be reversible and diffusion controlled. In addition, the hydrolytic reaction on the surface appeared to be first order with respect to the amount of adsorbed enzyme and first order with respect to the concentration of TC_3 at the solid/liquid interface. Brockman et al.[8] reported an enhancement of the velocity on the surface of the siliconized glass beads of 3 orders of magnitude as compared to the homogeneous reaction, and they ascribed this activating effect to the increased local concentration of the substrate at the interface. We should mention, however, that although the authors measured directly the fraction of enzyme adsorbed on the interface, this determination was done in the absence of TC_3. From their results we can calculate that the ratio of TC_3 molecules present as monomer in solution to TC_3 molecules adsorbed to the interface is about 300. We see that the large amount of substrate in the bulk phase could act as a "competitive" inhibitor. In that case, the enzyme concentration present at the interface is unknown. Nevertheless, this method was used by other authors as evidence of the increase of the hydrolysis rate of a TC_4 solution by dromedary pancreatic lipase after addition to the assay system of siliconized glass beads.[9] It has in fact been established that the lipid/water *interfacial quality,* in terms of the tension at the interface, is one of the most decisive parameters when working with lipolytic enzymes.[10-12] This unfortunately means that valid comparisons can be made only between data obtained under strictly identical conditions, preferably in the same laboratory. This is the reason why we decided to reevaluate the phenomenon of interfacial activation using synthetic short-chain triacylglycerols as substrates such as triacetin, tripropionin, and tribu-

---

[8] H. L. Brockman, J. H. Law, and F. J. Kézdy, *J. Biol. Chem.* **248,** 4965 (1973).
[9] H. Mejdoub, J. Reinbolt, and Y. Gargouri, *Biochim. Biophys. Acta* **1213,** 119 (1994).
[10] R. Verger and G. H. de Haas, *Annu. Rev. Biophys. Bioeng.* **5,** 77 (1976).
[11] R. Verger and G. H. de Haas, *Chem. Phys. Lipids* **10,** 127 (1973).
[12] S. Ransac, H. Moreau, C. Rivière, and R. Verger, *Methods Enzymol.* **197,** 49 (1991).

tyrin. It appears worthwhile, however, to measure the aqueous solubility of these short-chain triacylglycerols by two independent and complementary methods, as shown in Fig. 6.

The maximal decrease in the surface tension indicates the point where a closely packed adsorbed monolayer is formed at the air/water interface, in equilibrium with the monomeric solution (see Fig. 1). This point corresponds as well to the appearance of aggregates.[1] The turbidimetric method reveals the appearance of large-size droplets, indicating that the solubility limit is reached.

As a result, both the surface tension and the turbidity measurements gave comparable values for the solubility of TC_2 (318 mM) and TC_3 (12 mM), which are very close to those reported previously by Entressangles and Desnuelle.[6] In the case of TC_4, we found a solubility limit value of close to 0.410 mM (see Fig. 6). We then decided to repeat the experiments carried out originally by Sarda and Desnuelle,[2] in 1958, using TC_2 as substrate. However, in addition to PPL we used two other recombinant pancreatic lipases: the recombinant guinea pig pancreatic lipase related protein 2 (r-GPL-RP2) and a chimeric r-GPL-RP2 mutant in which the C-terminal domain was substituted by the HPL C-terminal domain, as described recently by Carrière et al.[13] The kinetic data are illustrated in Fig. 7.

As expected, PPL in the presence of its cofactor colipase displays a clear interfacial activation in the absence or presence of GA in sharp contrast to r-GPL-RP2 and its chimeric mutant. These latter enzymes possess a large deletion within the lid domain giving free access to the lipase active site. The lack of interfacial activation in r-GPL-RP2 is probably a consequence of a maximal enzymatic activity being displayed on soluble substrates.[14] In fact interfacial activation of classical lipases may be viewed as a depressed action on soluble esters rather than an increased interfacial activity on insoluble substrates. In that sense, r-GPL-RP2 is a true lipase distinguished by the absence of interfacial activation. The phenomenon of interfacial activation in HPL is probably linked to the ability of the lid domain to undergo a conformational change from an open to a closed conformation in the presence of a lipid/water interface. These conformations are probably stabilized by structural elements that are only present in the homologous protein core of HPL.[13,15] To check if the phenomenon of interfacial activation was also observable under different conditions of

---

[13] F. Carrière, K. Thirstrup, S. Hjorth, F. Ferrato, P. F. Nielsen, C. Withers-Martinez, C. Cambillau, E. Boel, L. Thim, and R. Verger, *Biochemistry* **36**, 239 (1997).

[14] A. Hjorth, F. Carrière, C. Cudrey, H. Wöldike, E. Boel, D. M. Lawson, F. Ferrato, C. Cambillau, G. G. Dodson, L. Thim, and R. Verger, *Biochemistry* **32**, 4702 (1993).

[15] C. Withers-Martinez, F. Carrière, R. Verger, D. Bourgeois, and C. Cambillau, *Structure* **4**, 1363 (1996).

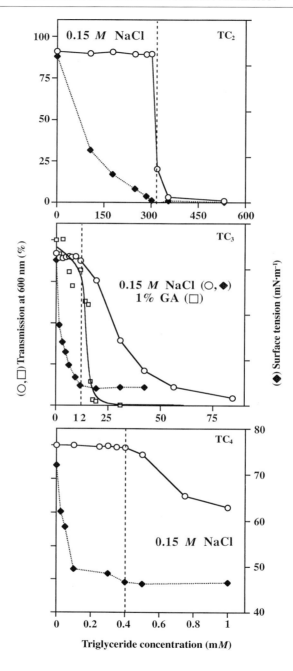

emulsification, we performed experiments at various GA concentrations, ranging from 0 to 6.6% (see Fig. 8). As reported previously by Entressangles and Desnuelle[6] (Fig. 3, lower panel), we confirmed (Fig. 8) that the maximal enzymatic activity expressed by PPL on TC_2 was around 10% of the rate measured on a triolein emulsion. We then turned our attention to a partly water-soluble (0.410 mM) synthetic short-chain triacylglycerol, TC_4, known to be a very good lipase substrate, to determine if the interfacial activation phenomenon was also observable under conditions where lipases display very high turnover rates. Such experiments are illustrated in Fig. 9, using PPL, cutinase (*Fusarium solani* cutinase), and native (nHPL) as well as recombinant HPL (rHPL).

As shown in Fig. 9, when using a very good substrate such as TC_4 for revealing the interfacial activation phenomenon, the situation becomes much less clear. Except in the case of rHPL (Fig. 9, panel E), the other

---

FIG. 6. Aqueous solubility of short-chain triacylglycerols used as lipase substrates. 1°/ Measurement of the decrease in the surface tension.[11,12] We used a Beckman LM 600 electromicrobalance with a platinum plate attached to the beam of the balance. The surface tension was monitored and the equilibrium value was obtained after approximately 1 min. Each TC_3 and TC_4 solution was prepared by injecting, with a Hamilton syringe, a given volume of the above pure triacylglycerols into a 250-ml volumetric flask containing a 0.15 M NaCl solution at pH 7.0 or pH 8.0 for TC_3 or TC_4, respectively. Deionized water, further purified with a Millipore Super Q system was used throughout all the experiments. The volumetric flasks had been previously cleaned with concentrated sulfochromic acid. Each flask was then hand shaken and placed into a thermostatted waterbath at 37°. Each flask was shaken again before pouring its contents into the reaction compartment (100 ml) of a zero-order Teflon trough, stirred by two magnetic bars. Before each utilization, the Teflon trough was cleaned with tap water, then gently brushed in the presence of distilled ethanol, before being washed again with tape water and finally rinsed with double distilled water. 2°/Turbidity measurements. The triacylglycerol solutions were poured into a 3-ml quartz cuvette placed in a thermostatted jacket (37°); the optical density was measured at 600 nm using a Kontron Uvikon 810 spectrophotometer. In the case of TC_4, the solutions used were the same as those described earlier for the surface tension measurements. Each 250-ml volumetric flask was hand shaken before each turbidity measurement. In the case of TC_2 and TC_3, a given volume of the pure triacylglycerol was injected, using a Hamilton syringe, into a stirred glass vessel (35 ml) thermostatted at 37° (TTT 80 titration assembly) containing 15 ml of a 0.15 M NaCl solution at pH 7.0. Between the experiments, the glass vessel was cleaned with distilled water and then with pure ethanol. After stirring the solution for 5 min, a sample was taken from the thermostatted glass vessel and measured spectrophotometrically as described earlier. The solubility measurements were also performed in 1% (final concentration) GA solution. In this latter case 1 g of GA powder (Sigma) was dissolved, under stirring, in 100 ml of boiling water and then filtered through glass wool. The TC_3 solutions were prepared by emulsifying, three times for 30 sec in a Waring blender, a given amount of the triacylglycerol in 15 ml of 1% (final concentration) GA placed in a 50-ml flask. A sample was taken and measured spectrophotometrically as described earlier. [1°/ corresponds to experiments labeled (♦) and 2°/ to experiments labeled (○, □).]

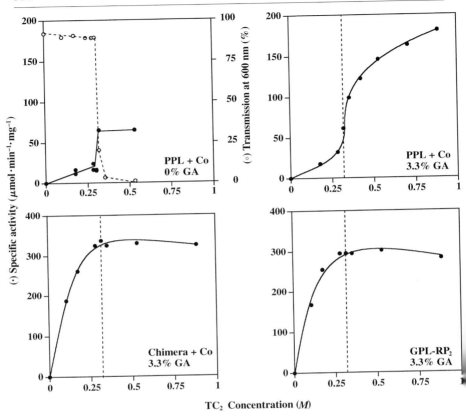

FIG. 7. TC_2 as substrate for measuring the interfacial activation of PPL, r-GPL-RP2, and the chimera. TC_2 solutions were systematically prepared by mixing, three times for 30 sec in a Waring blender, a given amount of TC_2 in 15 ml of 10% GA. Before each assay, 5 ml of the solutions were added to 10 ml of purified water (Millipore Super Q system) in a thermostatted (25°) pH-stat vessel. The pH-stat (TTT 80 radiometer) was equipped with a 250-μl syringe containing 0.1 N NaOH. The lipase activity was measured potentiometrically at pH 7.0, because at pH 8.0 the spontaneous hydrolysis of TC_2 reaches relatively high levels. The assay was carried out on a mechanically stirred solution of substrate in the reaction vessel. Spontaneous hydrolysis was recorded in the pH-stat mode for 2 min before lipase injection, and this background value was subtracted from the activity measurement. One international lipase unit is the amount of enzyme catalyzing the release of 1 μmol fatty acid per minute. The PPL (18 μg) and chimera (4 μg) samples were saturated with a fivefold molar excess of colipase. The r-GPL-RP2 (7.4 μg) was found to be insensitive to the presence of colipase (data not shown). The specific activity was expressed in units per milligram of protein. The protein concentrations were estimated on the basis of spectrophotometric measurements, using the absorption coefficient of 13.3 at 280 nm for pure PPL and an amino acid analysis for the chimera and the r-GPL-RP2. In the left upper panel, the open symbols give the solubility of TC_2 measured by turbidimetry as described. The solubility limit (318 mM) is indicated by vertical dashed lines.

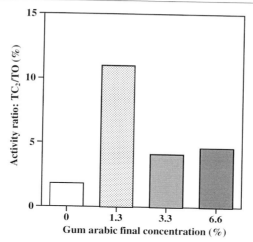

FIG. 8. Effects of GA concentration on the maximal rate of hydrolysis of TC_2 by PPL. TC_2 emulsions were systematically prepared by mixing, three times 30 sec in a Waring blender, a given amount of TC_2 in 2, 5, or 10% GA solutions, respectively. Before each assay, 5 ml of pure water was added to 10 ml of the above TC_2 emulsions to obtain a final concentration of 886 mM in TC_2. The PPL (17.2 μg) sample was saturated with a fivefold molar excess of colipase. For further experimental details, see the legend of TC_2 assays (Fig. 7). The reference test (15 ml, final volume), using emulsified TO (3.3 mM TO in 3.3% GA, final concentrations) was performed in a TTT 80 pH-stat radiometer at pH 8.0 and 37° in 100 mM NaCl, 2 mM $CaCl_2$, 0.33 mM Tris-HCl, and 4 mM sodium taurodeoxycholate, final concentrations.

enzymes tested (PPL and cutinase) either in the presence of 0.15 M NaCl or 0.33% GA, display no obvious interfacial activation. These puzzling results illustrate the fact that not all short-chain triacylglycerols are optimally suited for measuring interfacial activation. To find an acceptable compromise between lipase turnover rates and the water solubility of substrates, we used TC_3, as illustrated in Fig. 10.

From results presented in Fig. 10 it is clear that TC_3, emulsified in GA, is by far the best system to evidence the interfacial activation phenomenon in the case of PPL. It is also clear that in the absence of GA, (i.e., in pure water; see Fig. 10, panel A), the lipase activity declined with time. As a result of the nonlinear kinetics observed, probably due to an irreversible interfacial denaturation, the measurements of initial velocities are inaccurate and irreproducible. The presence of 0.15 M NaCl improves the assay conditions, allowing one to evidence the interfacial activation phenomenon (Fig. 10, panel B). It is worth noting from Fig. 10 that increasing concentrations of GA gradually shift the takeoff point of lipase activity, up to TC_3 concentrations above its solubility limit (12 mM). We do not have any

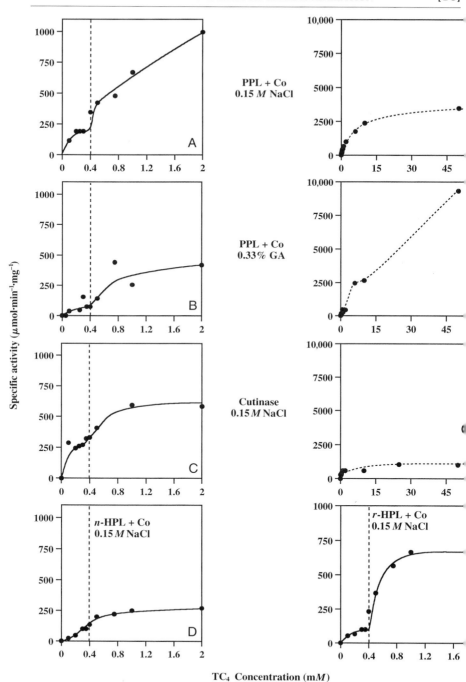

plausible explanation for such an odd observation. However, one has to recall that GA is basically a natural polysaccharide (from the acacia tree) used since antiquity in pharmaceutical preparations. As a conclusion, interfacial activation can be best evidenced in a reproducible manner using TC_3 as substrate, if one can avoid the many possible experimental artifacts discussed.

4. Do Any Structural Elements Responsible for Interfacial Activation Exist?

The first three-dimensional structures of lipases to be elucidated[16–19] suggested that the interfacial activation phenomenon might be due to the presence of an amphiphilic surface loop covering the active site of the enzyme in solution, just like a lid. When contact occurs with a lipid/water interface, this lid might undergo a conformational rearrangement as the result of which the active site becomes accessible. It is worth noting, however, that the hydrolysis of a substrate having the form of a truly monomeric

[16] L. Brady, A. M. Brzozowski, Z. S. Derewenda, E. Dodson, G. Dodson, S. Tolley, J. P. Turkenburg, L. Christiansen, B. Huge-Jensen, L. Norskov, L. Thim, and U. Menge, *Nature* **343,** 767 (1990).

[17] F. K. Winkler, A. d'Arcy, and W. Hunziker, *Nature* **343,** 771 (1990).

[18] H. van Tilbeurgh, L. Sarda, R. Verger, and C. Cambillau, *Nature* **359,** 159 (1992).

[19] H. van Tilbeurgh, M.-P. Egloff, C. Martinez, N. Rugani, R. Verger, and C. Cambillau, *Nature* **362,** 814 (1993).

FIG. 9. TC_4 as substrate for measuring the interfacial activation of PPL, cutinase, nHPL, and rHPL. Since the solubility limit of TC_4 is very low, it is advisable to prepare the millimolar solutions of TC_4 in 0.15 M NaCl in 250-ml volumetric flasks in order to obtain the desired concentration (for instance, 7.3 μl of pure TC_4 in 250 ml gives a 0.1 mM TC_4 solution). These flasks were kept in a thermostated waterbath (37°). Before each assay, 15 ml of the TC_4 solution was placed in the thermostated pH-stat vessel. The spontaneous nonenzymatic hydrolysis at pH 8.0 was recorded before lipase injection and then subtracted from the activity measured at pH 8.0 and 37°, using a pH-stat TTT 80 radiometer, as described previously in the case of TC_2 assays (Fig. 7). The PPL (2.5 μg, panels A, A'), nHPL (7 μg, panel D), and rHPL (1.5 μg, panel E) were saturated with a fivefold molar excess of colipase. Cutinase (2.87 μg, panels C, C') was injected without any cofactor. Preparatory to the measurements on emulsified TC_4 in 0.33% GA (final concentration), the TC_4 emulsions were systematically prepared by mixing, three times 30 sec in a Waring blender, a given amount of TC_4 in about 100 ml of 1% GA. Before use, 5 ml of the emulsions was added to 10 ml of pure water in the thermostated (37°) pH-stat vessel. The PPL (2.5 μg, panels B and B') was saturated with a fivefold molar excess of colipase. Deionized water, purified with a Millipore Super Q system, was used throughout the experiments. The TC_4 concentration in panels A, B, C, D, and E ranged from 0 to 2 mM and the specific activity from 0 to 1000, whereas in panels A', B', and C' the TC_4 concentration ranged from 0 to 60 mM and the specific activity from 0 to 10,000.

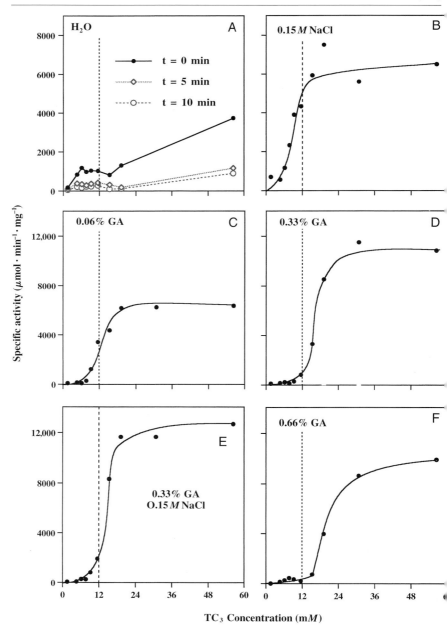

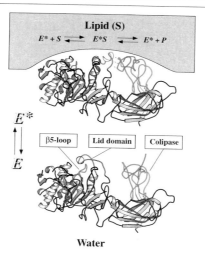

Fig. 11. Structure of the HPL–procolipase complex in the closed conformation (E), and structure of the HPL–procolipase complex in the open conformation ($E*S$). These two figures show the conformational changes in the lid, the $\beta 5$ loop, and the colipase during interfacial activation. (Adapted from van Tilbeurgh et al.[18,19])

solution might well also require the lid to be open without any interfacial activation being involved. In fact, interfacial activation seems to have to do with the respective lifetimes of the open and closed forms of lipases, as schematically illustrated in Fig. 11.

Fig. 10. TC_3 as substrate for measuring the interfacial activation of PPL. The TC_3 solutions were systematically prepared by mixing, three times 30 sec in a Waring blender, a given amount of TC_3 in 15 ml of either pure water or 0.15 M NaCl or GA. Before each assay, 5 ml of the solutions was added to 10 ml of pure water in the thermostatted (37°) pH-stat vessel. Deionized water, purified with a Millipore Super Q system, was used throughout all the experiments. The spontaneous nonenzymatic hydrolysis was recorded before lipase injection and then subtracted from the activity measured at pH 7.0 and 37°, using a pH-stat TTT 80 radiometer, as described previously in the case of the TC_2 assays (Fig. 7). The PPL samples used were saturated with a fivefold molar excess of colipase and 7.6 μg (panel A), 1.52 μg (panels B, D, and E), 0.86 μg (panel C), or 2.17 μg (panel F) was injected. The solubility limit of TC_3 (12 mM) is indicated by vertical dashed lines. The PPL activity was proportional to the amount of enzyme injected: we measured the same specific activity (4700 μmol · min^{-1} · mg^{-1}) on TC_3 (11.5 mM) in 0.06% GA (final concentration) by injecting either 0.429 μg, or 0.858 μg or 1.287 μg of PPL (data not shown). We checked that bovine serum albumin (1%, final concentration) had no detectable catalytic activity on a TC_3 solution (7.7 mM) or on a TC_3 emulsion (15.33 mM). In experiments performed in pure water (A), nonlinear kinetics were recorded. The rate of hydrolysis decreased with time and the instantaneous velocities were measured at 0 (●), 5 (◇), and 10 (O) min after lipase injection.

In the framework of the European Bridge-T-Lipase project (1990 to 1994), several new three-dimensional structures as well as biochemical and kinetic data provided new insights into lipases at the molecular and atomic levels.[20] It emerged from these studies that lipases do not all subscribe to the phenomenon of interfacial activation. The main exceptions noted so far are the lipases from *Pseudomonas glumae*,[21,22] *Pseudomonas aeruginosa*,[23,24] and *Candida antarctica* B,[25] all of which nevertheless have an amphiphilic lid covering the active site. In view of our reinvestigation of the interfacial activation of pancreatic lipases, one cannot exclude that the optimal experimental conditions have not yet been found to observe this phenomenon with other lipases.

However, some new pancreatic lipases have been recently identified.[26] Comparisons between their primary amino acid sequences have shown that they have a fairly high degree of homology, but they can nevertheless be divided into three subgroups: (1) the "classical" pancreatic lipases; (2) the pancreatic lipase-related proteins of type 1 (PLRP1), and (3) the pancreatic lipase-related proteins of type 2 (PLRP2). Although the kinetic properties of the classical pancreatic lipases, particularly with regard to interfacial activation, have been quite fully documented, it was only recently that the PLRP2 lipases of the coypu[27] and the guinea pig[14] were found to show no interfacial activation. Surprisingly, the coypu lipase has a 23 amino acid lid, which is homologous to that of the classical pancreatic lipases, whereas the guinea pig lipase has a mini-lid consisting of only five amino acid residues. Other lipases, such as *Staphylococcus hyicus*, show interfacial activation only with some substrates.[28]

Last, the pancreatic phospholipase A_2 (PLA$_2$) has no identified amphiphilic lid, and yet shows a high degree of interfacial activation using dihepta-

---

[20] L. Thim, M. R. Egmond, O. Misset, R. Verger, and R. D. Schmid, "Biotechnology research for innovation, development and growth in Europe (1990–1994)," Paper presented at European Commission D.G. XII, Brussels, Belgium, 1995.

[21] A. Deveer, "Mechanism of activation of lipolytic enzymes." Thesis, University of Utrecht, Utrecht, The Netherlands (1992).

[22] M. E. M. Noble, A. Cleasby, L. N. Johnson, M. R. Egmond, and L. G. J. Frenken, *FEBS Lett.* **331,** 123 (1993).

[23] K.-E. Jaeger, S. Ransac, H. B. Koch, F. Ferrato, and B. W. Dijkstra, *FEBS Lett.* **332,** 143 (1993).

[24] K.-E. Jaeger, S. Ransac, B. W. Dijkstra, C. Colson, M. Vanheuvel, and O. Misset, *FEMS Microbiol. Rev.* **15,** 29 (1994).

[25] J. Uppenberg, M. T. Hansen, S. Patkar, and T. A. Jones, *Structure* **2,** 293 (1994).

[26] T. Giller, P. Buchwald, D. Blum-Kaelin, and W. Hunziker, *J. Biol. Chem.* **267,** 16509 (1992).

[27] K. Thirstrup, R. Verger, and F. Carrière, *Biochemistry* **33,** 2748 (1994).

[28] M. G. van Oort, A. M. T. J. Deveer, R. Dijkman, M. Leuveling Tjeenk, H. M. Verheij, G. H. de Haas, E. Wenzig, and F. Götz, *Biochemistry* **28,** 9278 (1989).

noylphosphatidylcholine as substrate.[29] It is worth noting, however, that the interfacial activation of PLA_2 occurs close to the CMC and that medium-chain phospholipids do not form emulsions (class III lipids; see Section 1). Another structurally very similar PLA_2 (*Naja melanoleuca*), however, shows no interfacial activation with medium-chain phospholipid substrates.[30] The three-dimensional structure of porcine pancreatic phospholipase A_2, present in a 40-kDa ternary complex with micelles and competitive inhibitor, has been determined recently by van den Berg *et al.*[31] using multidimensional heteronuclear nuclear magnetic resonance (NMR) spectroscopy. Whereas free in solution, Ala 1, Leu 2 and Trp 3 are disordered, with the alpha amino group of Ala 1 pointing out into the solvent. In the ternary complex, these residues have an α-helical conformation with the alpha amino group buried inside the protein. As a consequence, the important conserved hydrogen bonding network, which is also seen in the crystal structures, is present only in the ternary complex, but not in free PLA_2. Comparison of the NMR structures of the free enzyme and the enzyme in the ternary complex indicates that conformational changes play a role in the interfacial activation of PLA_2.

5. Interfacial Activation of Lipase in Nonaqueous Media

The question of whether lipases can be activated by adsorption onto an interface in organic solvents was addressed by Louwrier and colleagues,[32] using *Rhizomucor miehei* lipase as a model.[33] In aqueous solution, this enzyme was shown to undergo a marked interfacial activation. However, lipase (either lyophilized or precipitated from water with acetone) suspended in ethanol or 2-(2-ethoxyethoxy)ethanol containing triolein exhibited no jump in catalytic activity when the concentration of triolein exceeded its solubility in these solvents, thereby resulting in formation of an interface. To test whether the lack of interfacial activation was due to the insolubility of the enzyme in organic media, lipase was covalently modified with poly (ethylene glycol). The modified lipase, although soluble in nonaqueous media, was still unable to undergo interfacial activation, regardless of the hydrophobicity of the interface. This inability was found to be caused by

---

[29] W. A. Pieterson, J. C. Vidal, J. J. Volwerk, and G. H. de Haas, *Biochemistry* **13**, 1455 (1974).
[30] J. H. van Eijk, H. M. Verheij, R. Dijkman, and G. H. de Haas, *Eur. J. Biochem.* **132**, 183 (1983).
[31] B. van den Berg, M. Tessari, R. Boelens, R. Dijkman, R. Kaptein, G. H. de Haas, and H. M. Verheij, *J. Biomol. NMR* **5**, 110 (1995).
[32] A. Louwrier, G. J. Drtina, and A. M. Klibanov, *Biotech. Bioeng.* **55**, 1 (1996).
[33] J. S. Okkels, A. Svendsen, S. A. Patkar, and K. Borch, *in* "Engineering of/with Lipases" (F. Xavier Malcata, ed.), p. 203. Kluwer Academic Publishers. Dordrecht, 1996.

the absence of adsorption of lipase onto interfaces in organic solvents, presumably because of the absence of the hydrophobic effect (the driving force of lipase adsorption onto hydrophobic interfaces in water) in such media. The uncovered lack of interfacial adsorption and activation suggests that the "lid" covering the active center of the lipase remains predominantly closed in nonaqueous media, thus contributing to diminished enzymatic activity.[32] Schrag *et al.*[34] came recently to the opposite conclusion by comparing the crystallization conditions of various microbial lipases and their observed conformations. These authors suggested that the open conformation of *Candida* as well as those of three *Pseudomonas* lipases are determined by the solution conditions (i.e., the presence of n-propanol), perhaps the dielectric constant.

6. Is Interfacial Activation a Criterion for Defining Lipases?

In conclusion, one can suggest that the molecular explanations for the "interfacial activation" phenomenon had to be investigated not only at the level of the lipase 3D structure, but also in the dynamics of organized multimolecular structures as well as in the interfacial conformations (interfacial quality) of lipids used as lipase substrates. Among other physicochemical parameters, one can imagine that the radius of curvature of the super substrates (S_2, S_3, and S_4 depicted in Fig. 1) could play an important role during the interfacial docking of lipases and hence could modulate the catalytic activity, through subtle conformational changes.

Interfacial activation as well as the presence of a lid domain are therefore not in the least appropriate criteria on the basis of which to determine whether such an esterase belongs to the lipase subfamily. Interfacial activation was sometimes wrongly taken as a criterion for predicting the existence of a lid domain in lipases with an unknown 3D structure!

Because naturally occurring triacylglycerols are totally insoluble in water, in contrast to short-chain triacylglycerols, interfacial activation can be said, in light of this overview, to be little more than an artifact that has stimulated the imaginations of many biochemists, but which has not turned out to be of any very great physiological significance. Lipases might therefore be quite pragmatically redefined as carboxyl esterases that catalyze the hydrolysis of long-chain acylglycerols. In fact, they are simply fat splitting "ferments."

---

[34] J. D. Schrag, Y. Li, M. Cygler, D. Lang, T. Burgdorf, H.-J. Hecht, R. Schmid, D. Schomburg, T. J. Rydel, J. D. Oliver, L. C. Strickland, C. M. Dunaway, S. B. Larson, J. Day, and A. McPherson, *Structure* **5,** 187 (1997).

Acknowledgments

Research was carried out with financial support from the BRIDGE-T-Lipase (contract BIOT-CT 91-0274 DTEE) and the BIOTECH G-lipase (contracts BIO2-CT94-3041 and BIO2-CT94-3013) programs of the European Union as well as the CNRS-IMABIO project.

Section III
Biocatalytic Utility

[17] Screening Techniques for Lipase Catalyst Selection

By U. ADER, P. ANDERSCH, M. BERGER, U. GOERGENS, B. HAASE,
J. HERMANN, K. LAUMEN, R. SEEMAYER, C. WALDINGER,
and M. P. SCHNEIDER

1. Introduction

The use of lipases from microbial and mammalian sources allows the preparation of a wide variety of enantiomerically pure or selectively functionalized molecules such as carboxylic acids, esters, derivatives of dicarboxylic acids, primary and secondary alcohols, diols, triols (i.e., glycerol and polyols, including carbohydrates), nucleosides, inositols, and many more. Clearly, these nearly unlimited possibilities have made these enzymes very attractive for applications in organic syntheses. Their convenient commercial availability, high stability toward temperature and solvents, their general ease of handling, broad substrate tolerance, and high enantioselectivities have all added to their widespread use among organic chemists.

Faced with the necessity of finding the most suitable lipase for the solution of a synthetic problem, organic chemists usually have to address one of the following (most likely in the order shown): adopting an "established" lipase with known selectivities and substrate tolerance toward the transformation of a given substrate; finding the best lipase among commercially available enzymes for the conversion of a given target molecule; finding a new lipase for the conversion of a defined target molecule; most likely of interest mainly to enzyme producers and in an attempt to make new lipases with interesting properties available to the scientific community, screening lipases—or microorganisms producing such enzymes—for their substrate tolerance, enantioselectivity, solvent dependence, and other properties.

2. Lipases as Biocatalysts

It has been well established in the literature since the mid-1980s[1] that lipases (EC 3.1.1) catalyze both the hydrolysis and synthesis of esters of

---

[1] L. Poppe and L. Novak, "Selective Biocatalysis." VCH Verlag, Weinheim, 1992; S. Servi, ed., "Microbial Reagents in Organic Synthesis," NATO ASI Series, Vol. 381. Kluwer Academic, Dordrecht, 1992; C.-H. Wong and G. M. Whitesides, "Enzymes in Synthetic Organic Chemistry." Pergamon, Elsevier, Oxford, 1994; K. Drauz and H. Waldmann, "Enzyme Catalysis in Organic Synthesis." VCH Verlag, Weinheim, 1995.

widely different structures and molecular backbones. Frequently, these transformations are highly chemo-, regio-, and enantioselective.

To be successful in the practical application of lipases, a number of important lipase properties toward the components of carboxylic esters must be determined such as substrate tolerance regarding the carboxylic acid moiety; substrate tolerance regarding the alcohol component; the fatty acid profile, for example, toward triglycerides; enantioselectivities by enantiomer differentiation and differentiation of enantiotopic groups; solvent tolerance; and the regioselectivities in multifunctional systems such as di- and tricarboxylic acids and derivatives of polyols such as glycerol, monosaccharides, and nucleosides.

Although the three-dimensional structures of several lipases, determined by X-ray crystallography of these enzymes, have been reported[2] and the molecular mechanism of catalysis is becoming quite well understood, most of the parameters mentioned still have to be determined experimentally using screening experiments. From these, however, working models can be deduced that frequently prove to be extremely valuable for the adaptation of new molecules (see Section 5.1.6).

Probably the most important current applications of lipases is their use in the preparation of enantiomerically pure carboxylic acids and alcohols, both having variable structures and numbers of substituents.

Using substrates of this nature, in this chapter we address a number of the problems outlined, largely by using examples from our laboratory.

3. Analytical Tools Essential for the Success of Screening Experiments

The most essential requirements for successful screening experiments clearly are—next to the proper choice of substrate—fast and reliable analytical methods that allow the convenient monitoring of the studied biotransformations.

For qualitative or semiquantitative information or a simple "yes" or "no" answer, visual or optical methods can be employed; alternatively reactions can be followed by thin-layer chromatography (TLC). Preferably, however, separations by gas chromatography (GC) or high-performance liquid chromatography (HPLC) are used in which case both substrates and resulting products should be baseline separated.

If the ultimate goal of the studied biotransformation constitutes the preparation of single enantiomers (enantiomerically pure compounds) by kinetic resolution of a racemic mixture or by enantiotopic differentiation,

---

[2] B. Rubin, *Structural Biol.* **1,** 568 (1994).

all of the corresponding substrates and possible products must be separable in one single experiment if such a screening method is to provide reliable answers within a reasonable timescale. Because so-called "chiral columns" with enantiomerically pure stationary phases are commercially available[3] for a wide range of separations, conditions can normally be found for the solution of such an analytical problem. Sometimes structural modifications of the substrates may be required to meet the previously stated requirements.

4. Qualitative Screening

4.1. A First Look: Qualitative Testing for Enzyme Activity

Numerous simple and inexpensive test systems have been developed by individual research groups that allow a rapid first look at the quality of the enzyme toward a variety of different substrates. Among other systems we found that a commercially available test system, marketed by Bio Merieux of France (API ZYM)[4] provides the first qualitative information of a given enzymatic preparation without the need for extensive analytical work or elaborate spectrophotometric procedures. The API ZYM test is a semiquantitative micromethod that allows the screening of various enzyme preparations from a multitude of sources (tissues, cells, biological fluids, microorganisms, soil samples, etc.) using very small sample quantities.

Experimental. Test trays with 10 or 20 microwells containing a series of 2-naphthyl esters of known concentration derived from fatty acids of various chain lengths are provided in the microwells of the test kit. The enzyme preparations, dissolved or suspended in a small amount of water, are added to these microwells, which are then covered by a plastic lid. After an incubation time of generally 4 hr at 37°, a drop of a solution of Fast Blue BB is added to all of the microwells. After a development period of 5 minutes, excess Fast Blue BB is destroyed by exposing the microwells to a powerful visible light source for just a few seconds. The remaining color of the test samples is derived solely from the azo dye produced by

---

[3] *Chiral HPLC column for optical resolution:* Supplier: Daicel (U.S.A.) Inc., 611 West 6th Street, Room 2152, Los Angeles, California 90017 USA: chiralcel OB (application, secondary alcohols); chiralcel OD (application, secondary alcohols); and crownpak CR + (application, dl-aminoacids). Supplier: Merck, E. Merck 64271 Darmstadt: welk-*o*-1 (chiral selector with π-donor and π-acceptor groups for efficient differentiation of enantiomers with either π-acidic or π-basic groups). *chiral GC column for optical resolution:* Supplier: CS Chromatographie Service GmbH, Postfach 1207, 52374 Langerwehe, Germany: FS-Cyclodex beta-I/P (application, secondary alcohols) and FS-Cyclodex gamma-I/P (application, lactones).
[4] API SYSTEM, La Balme les Grottes, F-38390 Montalieu, Vercieu, France.

the coupling reaction of 2-naphthol with Fast Blue BB. A color chart calibrated in intervals of 5 nmol (5 → 40 nmol) is provided with the test kit and serves as a measure of the enzyme activity toward the test substrates in the microwells.

For better comparison of results, the concentrations of enzyme solutions should be standardized, for example, by using enzyme preparations with identical protein contents (determined using the method of Lowry or others). The principle of the method described above in chemical terms is outlined in Fig. 1.

Using the previously described method, relative and semiquantitative substrate specificities of a given enzyme toward fatty acids ranging from butyric acid (C_4) to stearic acid (C_{18}) can be determined simultaneously in one single experiment. For an alternative method, also leading to the fatty acid profile of a given lipase, see Section 4.4. Clearly, provided some effort is made to synthesize the required substrates, this principle can also be extended to any new carboxylic acid such as the individual target molecules to be studied. The method is by no means limited to the determination of lipase activities but can also be applied to the determination of side activities such as protease, glycosidase, and phosphatase activities. Examples of the results obtained, chosen randomly from numerous available data, are shown in Fig. 2. Clearly, with this method, although allowing a first glimpse of the activities of the enzymes toward certain substrates, the enantioselectivities of such enzymes regarding the target substrates cannot be determined this way.

4.2. A Second Look: More Analytical Work

In another variation of this theme, racemic esters—with the carboxylic acid or the alcohol constituting the chiral part of the molecule—are used (Figs. 3 and 4).

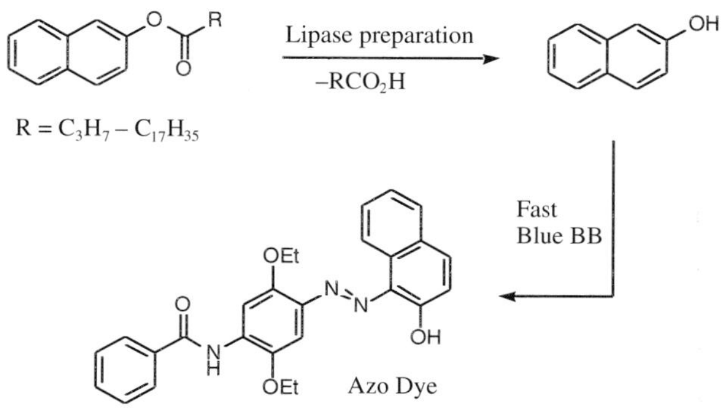

FIG. 1. The API ZYM test: chemical reactions involved.

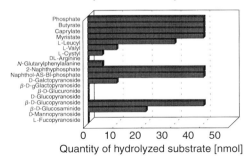

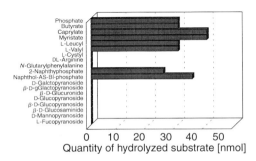

Fig. 2. Lipase screening using the API ZYM test.

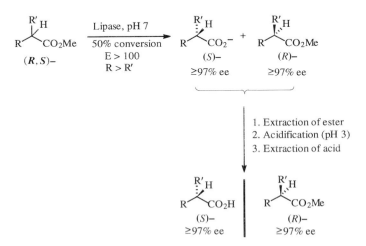

Fig. 3. Enantioselective hydrolysis of racemic esters: schematic procedure.

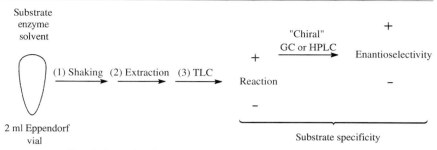

FIG. 4. Screening for substrate specificity and enantioselectivity.

Experimental. One millimol of the target molecule in question is suspended in 1–1.5 ml of 0.1 M phosphate buffer, pH 7, contained in a 2-ml Eppendorf vial to which variable amounts of the enzyme preparations are added. After shaking the vials at room temperature for 24–72 hr, unreacted ester and released alcohols are extracted with diethylether while the produced carboxylates remain in solution. After adjusting the pH of the remaining aqueous solutions to pH 2–3, the released carboxylic acids are extracted. All fractions are checked simply by TLC for initial information regarding the extent of transformation and are then analyzed in detail via "chiral" GC or HPLC for more information regarding the enantioselectivity of the resolution in question. Clearly this method is rather tedious and is preferably applied to determine the substrate tolerances of new lipases using a defined set of substrate molecules based on structural and functional variety. Results obtained this way will lead to new lipases with new substrate specificities.

If a single target molecule or a group of closely related building blocks has to be transformed, one would try to apply a more elegant method, making use of the modern methods of enantiomer separation as shown later in two typical examples (see Sections 5.1.2 and 5.1.4).

4.3. A Third Look: Qualitative Testing for Enantioselectivity

This can be achieved in limited cases using another simple procedure, especially where lipase-catalyzed resolution of racemic esters for the production of optically pure carboxylic acids or alcohols is the desired transformation.

For this, suspensions of the enantiomeric esters in an agar-type medium are filled in simple petri dishes. Solutions of the enzymes are added and the clearing of the treated areas is observed visually. If the pure enantiomeric ester is hydrolyzed, the products of hydrolysis are soluble in the

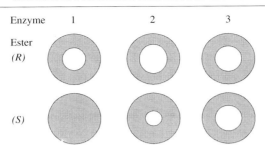

FIG. 5. Screening for lipase-catalyzed resolutions using pure enantiomers as test substrates.

medium—a fatty acid and an alcohol (compare Fig. 3)—causing the treated area to clear. If no reaction takes place, the area stays translucent. In the ideal case, only one enantiomer is transformed, in this case by enzyme 1[5] (Fig. 5).

Obviously this method is not applicable to the production of enantiomers derived from dicarboxylic acids or certain diols or where the differentiation of enantiotopic ester groups is the desired transformation.

4.4. A Fourth Look: The Fatty Acid Profile of Lipases

It is well known that lipases from different sources show different substrate selectivities toward both carboxylic esters and triglycerides. Consequently, the specificities of numerous microbial lipases have been studied in aqueous emulsions.[6] The major drawback of this method is the long time required to obtain a full specificity spectrum of a lipase because in a normal assay the reaction rate of only one triglyceride is determined. Another disadvantage is the physical emulsion effects, which cannot be neglected especially with long-chain triglycerides. Another possible approach is the determination of relative reaction rates for the hydrolysis or alcoholysis of fatty acid alkyl esters with different chain lengths in organic solvents.[7] However, due to the structural differences between fatty acid alkyl esters and triglycerides, it is not clear whether these results can be applied to both

---

[5] Method used by Chiroscience, described in S. C. Stinson, *Chiral Drugs*, C&EN **9**, October 1995.

[6] M. Iwai, S. Okumura, and Y. Tsujisaka, *Agric. Biol. Chem.* **39**, 1063 (1975); M. Sugiara, M. Isobe, *Chem. Pharm. Bull.* **23**, 1226 (1975); H. Borgström and H. L. Brockman, *in* "Lipases." Elsevier, Amsterdam, 1984.

[7] J. Baratti, M. S. Rangheard, G. Langrand, and C. Triantaphylides, *in* "Proceedings of the World Conference on Biotechnology for the Fats and Oils Industry," (T. Applewhite, ed.) p. 164, *J. Am. Oil Chem. Soc., 1987.*

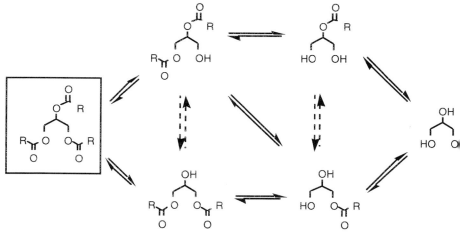

FIG. 6. Lipase-catalyzed glycerolysis of triglycerides in an organic solvent: reaction equilibria.

classes of compounds. Using triglycerides as substrates we have developed a rapid method that allows the complete determination of the fatty acid chain length profile of lipases in one single experiment and in very short time[8] (compare also Section 4.1).

The principle can be explained in simple terms. If an equimolar mixture of triglycerides with different acyl residues is incubated in an organic solvent in the presence of glycerol and a lipase, a very complex reaction system is established in equilibrium state (Fig. 6). However, in the beginning of the incubation the major reaction is the consumption of triglycerides. The reverse reaction and secondary reactions of the primary products can largely be neglected.

The situation can therefore be described conveniently by Eq. (1):

$$\begin{bmatrix} \text{triglyceride}_1 + \text{glycerol} \rightarrow \text{diglyceride}_1 + \text{monoglyceride}_1 \\ \text{triglyceride}_2 + \text{glycerol} \rightarrow \text{diglyceride}_2 + \text{monoglyceride}_2 \\ \hline \text{triglyceride}_N + \text{glycerol} \rightarrow \text{diglyceride}_N + \text{monoglyceride}_N \end{bmatrix} \quad (1)$$

When acting on insoluble substrates, the kinetics of lipase-catalyzed reactions is rather complex. By contrast, Michaelis–Menten kinetics have been obtained for lipase-catalyzed reactions in organic media. The solution of

[8] M. Berger and M. P. Schneider, *Biotechnol. Lett.* **13**, 333 (1991); M. Berger and M. P. Schneider, *Biotechnol. Lett.* **13**, 641 (1991).

Eq. (1) using the Michaelis–Menten equation, followed by integration, yields Eq. (2):

$$\alpha_1 * \log(S_1)/(S_1)_0 = \alpha_2 * \log(S_2)/(S_2)_0 = \ldots = \alpha_N * \log(S_N)/(S_N)_0 \quad (2)$$

where

$S_{1,2,\ldots,N}$ = concentration of substrate 1, 2, ..., N
$(S_{1,2,\ldots,N})_0$ = start concentration of substrate 1, 2, ..., N
$\alpha_N = (V/K)_1/(V/K)_N$, where V = maximum velocity and K = Michaelis constant
S_1 is reference with $\alpha_1 = 1$

The competition factor α defined as the ratio of the catalytic powers (V/K) for each substrate was found to be very useful for the prediction of reaction rates.[9] In this case α describes the relative reaction rates for triglycerides with different chain lengths and therefore gives the fatty acid chain length profile for any lipase toward triglycerides in organic solvents.

Experimental. Five milliliters of a 0.1% solution of triglycerides from tributyrin to triolein in *tert*-butyl methyl ether (TBME), 100 mg of glycerol, immobilized on silica gel, and 10 mg of the lipase preparation were incubated. The mixture was stirred at room temperature, and every 30 min aliquots of 0.5 ml were taken. The solvent was evaporated under a stream of dry nitrogen, 0.2 ml of a 10:1 mixture of bis(trimethylsilyl)acetamide (BSTFA) and trimethylchlorosilane (TMCS) was added and the mixture heated at 70° for 3 hr. Then 0.2 μl of the resulting mixture was analyzed by gas chromatography. A typical GC diagram is shown in Fig. 7.

Analysis of the GC diagrams revealed that the reaction profiles are indeed linear in the first 2 hr of incubation up to a conversion of about 30%. From these data the competition factors are calculated using Eq. (2). At higher degrees of conversion, the reverse reaction and secondary reactions cannot be neglected and Eqs. (1) and (2) are no longer valid. The results for five lipases are shown in Table I. In all cases, trilaurin was used as a reference ($\alpha = 1$). Graphic representations of the results for three lipases of clearly different fatty acid profiles are shown in Figs. 8–10.

The method described allows the determination of a specificity profile of lipases toward triglycerides in organic media. The short analysis time and the fair accuracy (± 5–10%) make it a practicable tool for the screening of lipases useful for the selective transformations of triglycerides in organic solvents. In a qualitative sense, the data seem to be transferable also to the selectivities of lipases toward other ester substrates.

---

[9] J. Baratti and H. Deleuze, *Biochim. Biophys. Acta* **911**, 117 (1987).

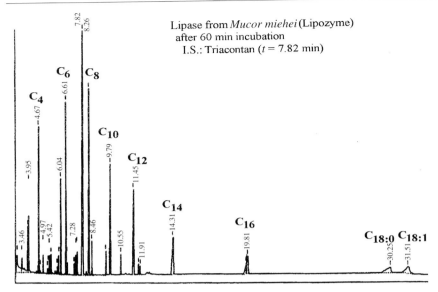

Fig. 7. Fatty acid profile of lipases: GC diagram.

5. Practical Applications

5.1. Enantioselective Transformations

5.1.1. Time Dependence of Conversions and E Values. Lipases catalyze basically three types of reactions: hydrolysis, esterification (reverse hydrolysis), and acyl transfer (alcoholysis). Both for the preparation of carboxylic acids and alcohols the hydrolytic reaction mode is frequently used. Representative for many examples in the literature is the lipase-catalyzed hydrolysis of a racemic carboxylic ester, as shown in Fig. 11.

TABLE I
ALPHA VALUES OF LIPASE PREPARATIONS

Lipase	C_4	C_6	C_8	C_{10}	C_{12}	C_{14}	C_{16}	$C_{18:0}$	$C_{18:1}$
Porcine pancreatic lipase (PPL)	2.33	1.30	1.37	1.72	1	0.39	0.77	1.33	1.58
Lipozym	0.66	0.74	1.91	1.12	1	1.07	1.29	1.23	0.07
Candida rugosa	6.41	1.35	2.00	1.60	1	1.15	1.87	2.28	2.07
Humicola lanuginosa	0.32	0.56	0.77	1.04	1	0.92	0.35	0.26	0.13
Geotrichum candidum	—	0.96	0.83	1.50	1	1.58	5.56	4.44	19.2

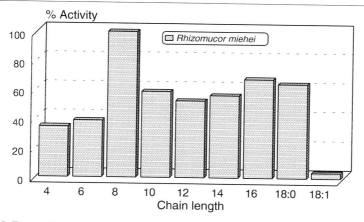

Fig. 8. Fatty acid chain length profile for lipase from *Rhizomucor miehei* (Lipozyme).

Typically these reactions are carried out under pH-stat conditions (e.g., at pH 7), the carboxylic acid produced being neutralized by continuous addition of NaOH from an autoburette. The amount of added NaOH is therefore a measure of the substrate conversion and thus allows convenient monitoring of these reactions. Typical time/conversion curves are shown in Fig. 12. If the enantioselectivity of the lipase is high (i.e., the lipase can differentiate well between the two enantiomers), the reaction will terminate after 50% conversion (Fig. 12a) and both the product (RCO_2^-) and unreacted substrate (RCO_2R') (Fig. 11) obtained will be essentially optically pure.

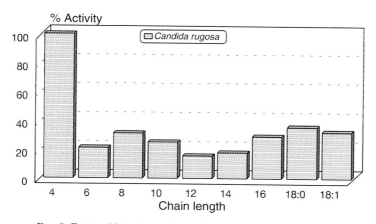

Fig. 9. Fatty acid chain length profile for lipase from *C. rugosa*.

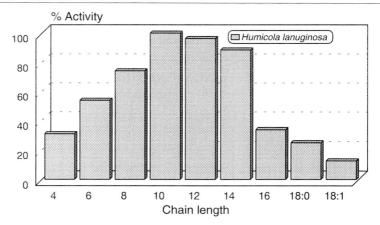

FIG. 10. Fatty acid chain length profile for lipase from *H. lanuginosa*.

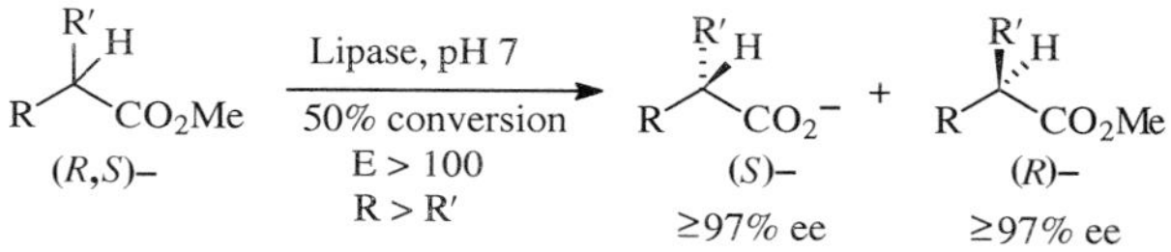

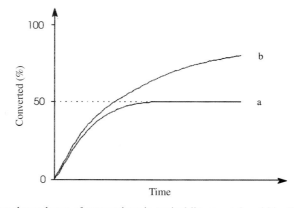

FIG. 11. Lipase-catalyzed hydrolysis of a racemic ester: enantiomer differentiation.

FIG. 12. Time dependence of conversions in typical lipase-catalyzed kinetic resolutions.

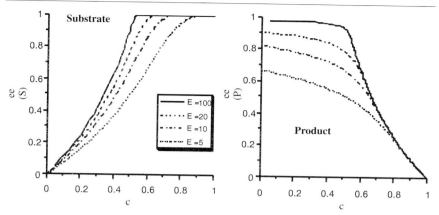

FIG. 13. Kinetic resolutions: enantiomeric purities of substrate and product.

The quality of such resolution depends both on the enantioselectivity of the lipase E and the achievable conversions. Based on the kinetics of all resolutions in the presence of a chiral auxiliary,[10] Chen and Sih have correlated the enantioselectivity of a lipase with the enantiomeric purities of product and substrate at any given conversion[11]:

$$\text{E value } E = \frac{\ln[(1-c)(1-ee(S))]}{\ln[(1-c)(1+ee(S))]} = \frac{\ln[1-c(1+ee(P))]}{\ln[1-c(1-ee(P))]} \quad (3)$$

where c is conversion. For different E values the dependence of enantiomeric purities in relation to the achieved conversion is plotted in Fig. 13. From Fig. 13 it becomes obvious (and this is true for every such kinetic resolution—both for carboxylic acids and alcohols and regardless of the substrate structure) that only with very high enantioselectivities can both the product and the unreacted substrate be obtained in nearly optically pure form. In case of lower E values the lesson to remember is simple: Only the remaining, unreacted substrate can be obtained optically pure via a conversion of >50%. For this, however, one has to pay with a concominant loss of chemical yield and a rather low optical yield of the product. In such a case the titration curve in Fig. 12 would ressemble curve (b).

With this principle, initial information regarding both the substrate tolerance of a given lipase and their enantioselectivities can be obtained

[10] V. S. Martin, S. S. Woodward, T. Katsuki, Y. Yamada, M. Ikeda, and K. B. Sharpless, *J. Am. Chem. Soc.* **103,** 6237 (1981).
[11] C.-S. Chen and C. J. Sih, *Angew. Chem.* **101,** 711 (1989).

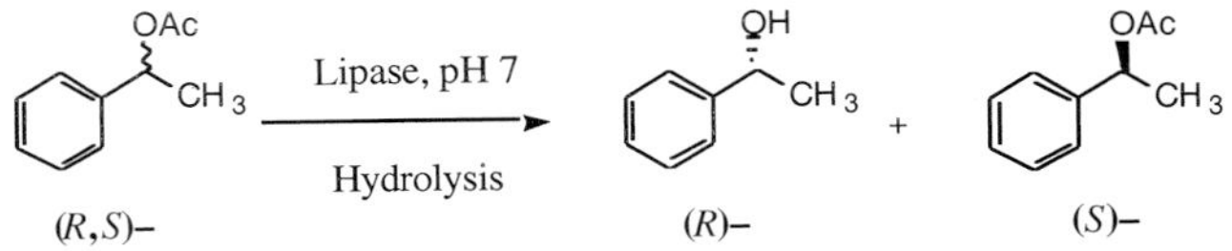

FIG. 14. Resolution of secondary alcohols via lipase-catalyzed hydrolysis.

by simply recording titration curves like those shown in Fig. 12. Note, however, that such an approach will only lead to semiquantitative information regarding the suitability of a given lipase for the purpose in question. Especially in cases of high enantioselectivities, the exact determination of the enantiomeric purities using reliable analytical tools—preferably so-called "chiral columns"—will be inevitable. A facile screening system of this type is described in Section 5.1.2. However, the method described here is probably the most widely used for the initial determination of resolution qualities following simple incubation tests, such as the ones described earlier. The method can also be used for the corresponding resolution of numerous secondary alcohols as shown in Fig. 14.

5.1.2. Screening for the Enantioselectivities of Lipases. If a single target molecule or a group of closely related building blocks has to be prepared in enantiomerically pure form by lipase-catalyzed resolution, one should try to apply a more elegant method, making use of the modern methods of enantiomer separation as shown below.[12] It is well established that the biological activities of β-adrenergic blockers of the general formula **3** (Fig. 15) reside largely in the (S)-enantiomers of these molecules. Consequently, numerous attempts have been made to prepare these compounds in enantiomerically pure form. (S)-**3** can be correlated retrosynthetically with a number of potential precursers, for example, the corresponding α-chloro-, α-bromo-, or α-azido-derivatives [(R-) **3–5**; Fig. 15], which in principle could serve as starting materials for the title compounds.

[12] U. Ader and M. P. Schneider, *Tetrahedron: Asymmetry* **3**, 205 (1992).

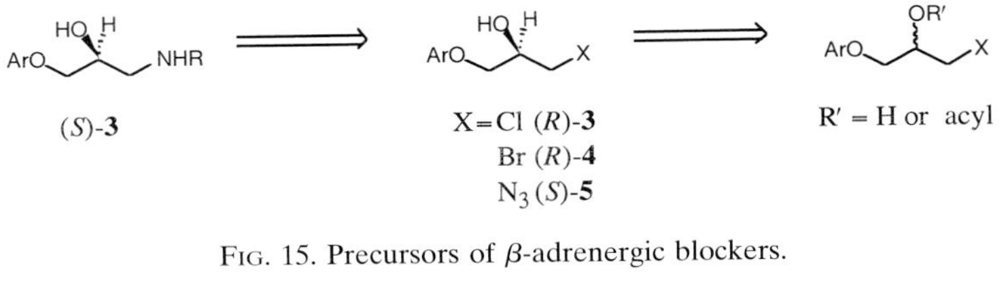

FIG. 15. Precursors of β-adrenergic blockers.

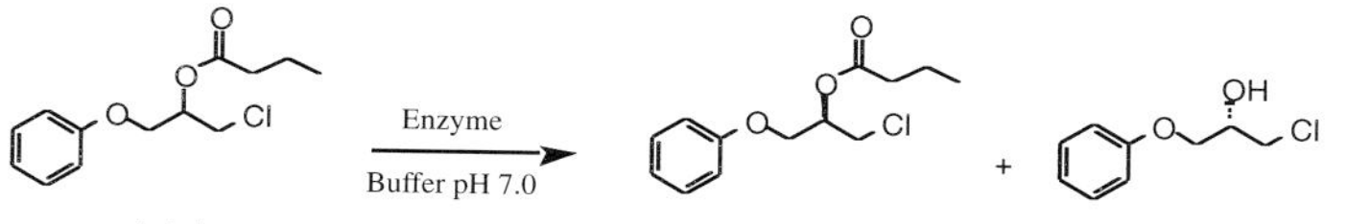

FIG. 16. Enantioselective hydrolysis of *rac* **5**.

Lipases are well known for their ability to differentiate enantiomers. In view of our previous experience[13] in this area and based on literature data, we felt that the enantioselective hydrolysis of the corresponding esters could well provide a facile route to the desired intermediates in optically pure form (Fig. 16). For this purpose, however, the most suitable biocatalyst and intermediate had to be found.

Target-oriented screening for suitable enzymes is usually a very time-consuming and tedious enterprise. We have therefore developed a facile screening system for this purpose based on the enantioselective separation of the reaction products by HPLC on chiral supports. Using a commercially available column (Chiralcel OB)[14] all products resulting from the enantioselective, enzymatic hydrolyses can be separated simultaneously. A direct determination of the achieved conversions, as well as configurations and enantiomeric purities of both substrates and products, can be achieved in one single experiment. The obtained data allow a rapid determination of the enantioselectivities[15] displayed by the employed biocatalysts and thus a rapid evaluation of their synthetic usefulness. Using Fig. 17, a simple visual comparison already reveals that the lipase from *Pseudomonas* sp. leads to much better results as compared to the lipase from *Candida* sp.

The results obtained in the hydrolyses of 2-butyloxy-1-chloro-3-phenoxypropan [(±)-**5a**] with a series of biocatalysts are summarized in Table II and shown in Fig. 17 for two lipases of widely different enantioselectivities. From Table II (compare also Fig. 17) it is obvious that the best suited biocatalysts are the lipases from *Mucor miehei* and *Pseudomonas* species. In view of the much higher specific activity the lipase from *Pseudomonas* sp. was chosen for further experiments.

---

[13] U. Goergens and M. P. Schneider, *J. Chem. Soc. Chem. Commun.* 1065 (1991); R. Seemayer and M. P. Schneider, *Recl. Trav. Chim Pays-Bas* **110**, 171 (1991); K. Laumen and M. P. Schneider, *J. Chem. Soc. Chem. Commun.* 598 (1988); K. Laumen, D. Breitgoff, and M. P. Schneider, *J. Chem. Soc. Chem. Commun.* 1459 (1988).
[14] T. Shibata, K. Mori, and Y. Okamoto, in "Chiral Separations by HPLC" (Krstulovic M. P., ed.), pp. 336–398. Ellis Horwood Ltd., United Kingdom, 1989.
[15] For calculation of conversion and enantiomeric ratio, see C.-S. Chen and C. J. Sih, *Angew. Chem.* **101**, 711 (1989).

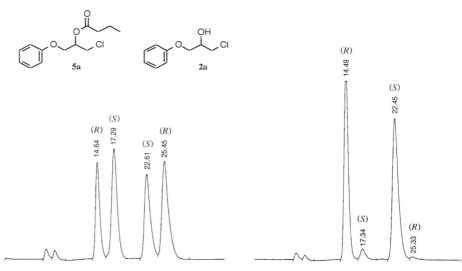

FIG. 17. Enzymatic hydrolyses of (±)-**5a** in the presence of different lipases. HPLC (Daicel OD) tracings after ~50% conversion.

5.1.3. Solvent Dependence of Lipase-Catalyzed Transformations. As already mentioned earlier, enantiomerically pure secondary alcohols of the type described in the preceding section can be prepared either by enantioselective enzymatic hydrolysis (Fig. 14) or by the corresponding esterification of a racemic alcohol. The stereochemical outcome is often complementary (Fig. 18).

Enantioselective esterifications of that type are usually carried out under the conditions of irreversible acyl transfer using vinyl acetate as acyl donor and organic solvents of generally low water content as reaction media. A detailed account of this method with several examples can be found in Chapter [19] of this book.

As is evident from the available literature and our own experience, both the rates of transformations and the enantioselectivities are often strongly dependent on the type of solvent used. The choice of the best solvent can therefore be crucial for the success of the desired esterification. To determine the influence of the type of solvent on the esterification of certain secondary alcohols this reaction was studied choosing racemic 2-phenylethanol as substrate and vinyl acetate as acyl donor. From previous

TABLE II
LIPASE-CATALYZED HYDROLYSES OF (±)-5a IN PRESENCE OF VARIOUS LIPASES

Enzyme	t (25%) (hr)	Products	% e.e.	Conv.	E
Lipase from *Aspergillus oryzae*	—	(±)-(2a)	—	—	—
Lipase from *Aspergillus niger*	—	(±)-(2a)	—	—	—
Lipase from *Aspergillus sojae*	12	(R)-(2a)	19.6	0.44	2
		(S)-(5a)	24.7		
Lipase from *Candida lipolytica*	74	(R)-(2a)	70.7	0.44	42
		(S)-(5a)	90.5		
Lipase from *Candida* species	2	(S)-(2a)	20.7	0.62	<2
		(R)-(5a)	12.7		
Lipase from *Chromobacterium viscosum*	<2	(R)-(2a)	66.9	0.47	14
		(S)-(5a)	75.6		
Lipase from *H. lanuginosa*	28	(R)-(2a)	61.0	0.39	60
		(S)-(5a)	94.0		
Lipase from *G. candidum*	12	(R)-(2a)	48.2	0.44	6
		(S)-(5a)	60.2		
Lipase from *Mucor miehei*	53	(R)-(2a)	44.4	0.31	>100
		(S)-(5a)	98		
Lipase from *Penicillium roquefortii*	—	(±)-(2a)	—	—	—
Lipase from *Rhizopus delemar*	52	(S)-(2a)	20.9	0.62	2
		(R)-(5a)	12.7		
Lipase from PPL	19	(R)-(2a)	43.7	0.35	15
		(S)-(5a)	82		
Cholesterinesterase from *Pseudomonas fluorescens*	<1	(R)-(2a)	24.6	0.47	2
		(S)-(5a)	27.6		
Lipase from *Pseudomonas* species	<2	(R)-(2a)	86.2	0.47	>100
		(S)-(5a)	96.3		
Porcine liver esterase (PLE)	6	(S)-(2a)	14.0	0.90	1
		(R)-(5a)	1.5		

work,[16] we know that this substrate was converted with very high enantioselectivity, leading to enantiomerically pure products in the presence of a lipase from *Pseudomonas* sp. (SAM-II), both using hydrolysis and esterification (see Fig. 18). Therefore, it was chosen for a comprehensive study regarding the solvent dependence of its esterification (see Fig. 18) in the presence of 15 commonly used solvents (Table III, Fig. 19).[17]

EXPERIMENTAL. Ten millimoles of racemic 2-phenylethanol were dissolved in 20 ml of the corresponding solvent. To this solution 30 mmol of vinylacetate and 200 mg of the lipase were added. The mixtures were stirred

[16] K. Laumen and M. P. Schneider, *J. Chem. Soc. Chem. Commun.* 598 (1988); K. Laumen, D. Breitgoff, and M. P. Schneider, *J. Chem. Soc. Chem. Commun.* 1459 (1988).
[17] R. Seemayer, unpublished Ph.D. thesis, Wuppertal, 1991.

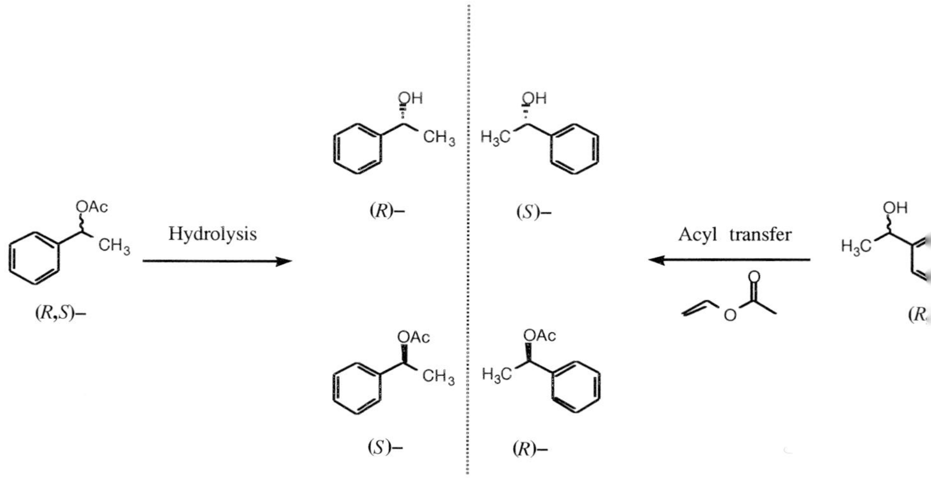

FIG. 18. Lipase stereochemistry: hydrolysis versus esterification.

at room temperature on a multipoint stirrer plate. In regular intervals, samples were drawn and analyzed by GC.

In Fig. 20, the amount of (*R*)-acetate being produced after a reaction time of 24 hr is plotted against the solvent used. In view of the high

TABLE III
ESTERIFICATION OF 2-PHENYLETHANOL IN VARIOUS
SOLVENTS: TIMES REQUIRED FOR 50% CONVERSION

Solvent	t [days]
n-Hexane	1
Toluene	1, 5
Diethylether	2
t-Butyl methyl ether	2
Vinyl acetate	3
Methylisobutylketone	3
t-Butanol	5
Ethyl acetate	5
Dichloromethane	6
Acetone	6
Tetrahydrofurane	7
Acetonitrile	13
DMA	After 14 days conversion <5%
DMF	After 14 days conversion <5%
DMSO	After 14 days conversion <5%

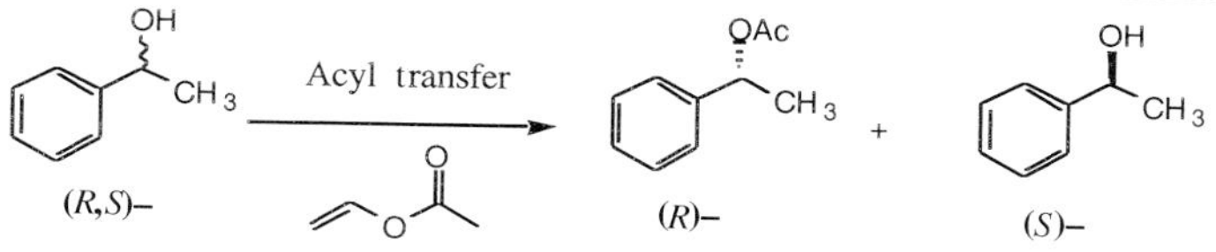

FIG. 19. Enantioselective esterification of 2-phenylethanol.

enantioselectivity of the reaction, the maximum achievable yield is of course 50%. In Table III, the times required to reach 50% conversion are listed for all solvents studied.

From Table III and Fig. 20, it is obvious that the highest rates of transformations are observed using solvents of low polarity whereas solvents such as DMSO, DMA, and DMF are totally unsuitable for this transformation.

Note, however, that—although with somewhat varying rates—a wide variety of solvents can be employed. This is very useful indeed because for an efficient esterification the substrates should preferably be completely dissolved. Clearly not all substrates are soluble in solvents of low polarity

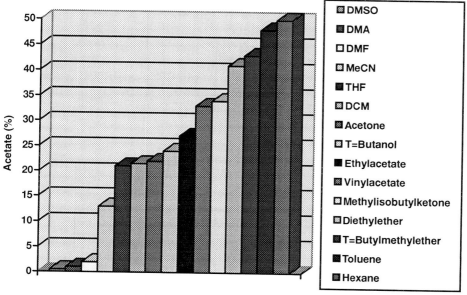

FIG. 20. Esterification of 2-phenylethanol. Solvent dependence (DMSO to n-hexane from left to right).

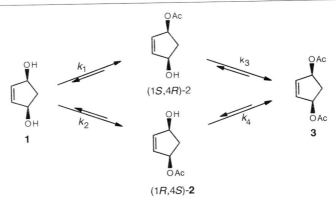

FIG. 21. Lipase-catalyzed esterification of a *meso*-diol **1**: kinetic scheme.

and being able to use acetone instead of *n*-hexane may make all the difference (see Section 5.1.5).

Correlation between the observed experimental results, that is, the relative rates of transformation and the solvent polarity, expressed as so-called log *P* values was attempted and was reasonably good. However, in other cases studied in our laboratory no correlation whatsoever with either log *P* values or other solvent parameters was observed. It seems that better correlations can be found if the water activities of the employed solvents are taken into account. For this aspect please see Chapter [21] in this book.

As pointed out earlier, considerable differences are also frequently observed regarding the enantioselectivities of lipase-catalyzed esterifica-

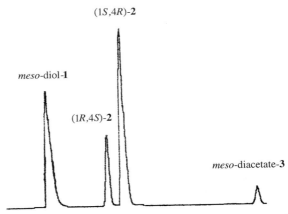

FIG. 22. Esterification of the *meso*-diol **1**: GC diagram after 1 hr.

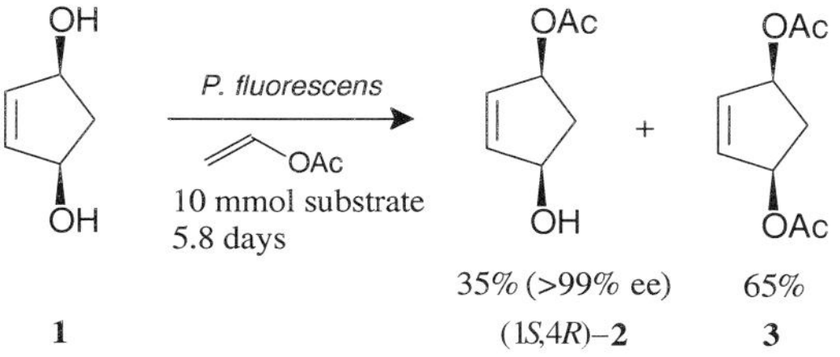

FIG. 23. Esterification of *meso*-diol **1**.

tions depending on the employed solvent. To determine these factors, however, the enantiomeric purities of the resulting products must be determined in a way similar to the examples described in Sections 5.1.2 and 5.1.4.

5.1.4. Lipase-Catalyzed Enantiotopic Differentiations. While lipase-catalyzed resolutions of racemates usually provide single enantiomers or mixtures thereof, the situation becomes much more complicated in cases where lipases are being used for the differentiation of enantiotopic ester groups. Thus the lipase-catalyzed esterification of the *meso*-diol **1** under the conditions of irreversible acyl transfer (see Chapter [19] in this book) using vinyl acetate as acyl donor produces—prior to completion—a mixture of four different molecules: starting material **1**, the two enantiomers (1*S*,4*R*)-**2**, (1*R*,4*S*)-**2** and the *meso*-diacetate **3** (Figs. 21 and 22).

To evaluate the enantioselectivities of lipases in this transformation, it is essential to monitor the absolute and relative amounts of all components in this mixture at any given time of the reaction. This can indeed conveniently be achieved using a chiral GC column (Cyclodex beta I/P, Fig. 22).[18]

The time/conversion data (a detailed analysis is beyond the scope of this chapter) obtained with several lipases clearly revealed that the transformations are not only dependent on the corresponding enzyme, but also on the solvent employed as well as the reaction time. Using the kinetic parameters ($k_1 > k_2$, $k_3 = 0$, $k_4 > 0$), the reaction was optimized to produce the desired product (1*S*,4*R*)-**2** with high optical purity (Figs. 22 and 23).

5.1.5. Combined Screening Efforts—A Case Study. The target was clear and well defined. For the synthesis of enantiomerically pure *myo*-inositol polyphosphates, starting essentially from *myo*-inositol (**1**, Fig. 24) itself, a central building block was needed that was believed could be prepared by enantioselective lipase-catalyzed esterification of a suitably protected derivative such as **2** (Fig. 24). The indicated derivation was required (1) to obtain a *meso* derivative with a reduced number of hydroxy groups, and (2) to increase the solubility in organic solvents.

[18] Supplier: CS Chromatographie Service GmbH, Postfach 1207, 52374 Langerwehe, Germany.

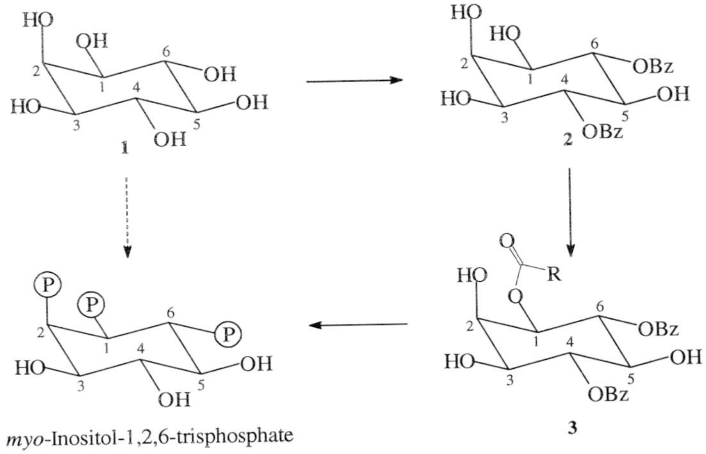

FIG. 24. From *myo*-inositol to enantiomerically pure polyphosphates.

Required now was a lipase that could differentiate between the enantiotopic hydroxy groups in positions 1 and 3 of **2** by selectively esterifying one of them. Since **2** is a *meso* compound this would in one step lead to only one enantiomer **3** without the need for further separations. Typically, and in complete agreement with the preceding examples in this chapter, the following problems had to be addressed: finding (1) a suitable enzyme, (2) the best solvent, and (3) the most suitable acyl donor.

5.1.5.1. SUCCESS BY PATIENCE—SCREENING FOR THE LIPASE. Since **2**—in spite of the two aryl substituents—is only soluble in more polar solvents such as THF, acetone, and acetonitrile, one of these had to be used for the screening experiments. To cover the most likely fatty acid profile of most lipases, a mixture of equal molar amounts of different vinyl esters (derived from acetic, butyric, and caproic acid) was used as acyl donors.

Experimental. We dissolved 388 mg (1 mmol) of **2** in 10 ml THF to which 20 mg of the corresponding lipase and 92 μl (1 mmol) vinyl acetate, 125 μl (1 mmol) vinyl butyrate, and 160 μl (1 mmol) vinyl caproate were added. The mixtures were stirred at room temperature and analyzed at regular intervals by TLC and HPLC for new products.

It was like looking for a needle in a haystack. No conversions were observed with all commercially available lipases, and new enzymes had to be found. After intensive contacts with enzyme producers and the screening of more than 60 enzymes, finally a new lipase was found in the portfolio of Boehringer Mannheim, which caused the breakthrough. A lipoprotein lipase from a *Pseudomonas* strain led to the desired product with high

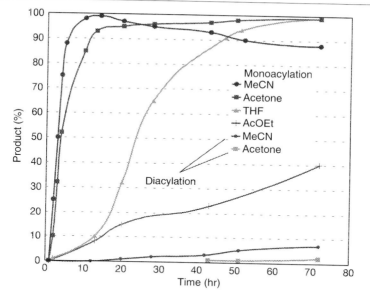

Fig. 25. Esterification of **2**: solvent influence.

specific activity and extremely high enantioselectivity.[19] For best performance the lipase was used immobilized on SiO_2.

5.1.5.2. SOLVENT SCREENING. As already mentioned, defined substrate **2** was only soluble in more polar solvents. The best solvent for the transformation should lead to the required, enantiomerically pure monoester **3** with high selectivity and as rapidly as possible. There should be as little diacylation as possible.

Using vinyl butyrate as acyl donor the solvent influence was studied in detail.

Experimental. We dissolved 388 mg (1 mmol) of **2** in the corresponding solvent (acetonitrile, acetone, tetrahydrofurane, ethyl acetate) to which 40 mg lipoprotein lipase from *Pseudomonas* sp., 500 mg of silica gel, and 375 µl (3 mmol) vinyl butyrate were added. The mixtures were stirred at 35° and the conversions followed by HPLC.

From the results obtained (Fig. 25), it is obvious that acetone is the best solvent for this transformation with the formation of the desired product of **3** in high purity in near quantitative yield and within a reasonable reaction time.

[19] P. Andersch and M. P. Schneider, *Tetrahedron: Asymmetry* **4**, 2135 (1993).

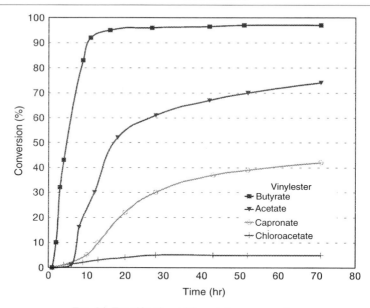

Fig. 26. Esterification of **2**: acyl donor screening.

5.1.5.3. SCREENING FOR THE ACYL DONOR. For preparative scale preparations the most suitable acyl donor had to be found.

Experimental. We dissolved 388 mg (1 mmol) of **2** in 10 ml acetone to which 40 mg lipase (LPL), 500 mg silica gel, and 3 mmol of the corresponding vinyl esters were added. The reactions were again monitored by HPLC. From the results shown in Fig. 26 it is absolutely clear that vinyl butyrate is the most suitable acyl donor for the desired transformation.

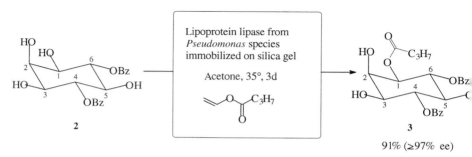

Fig. 27. Enantioselective esterification of **2**.

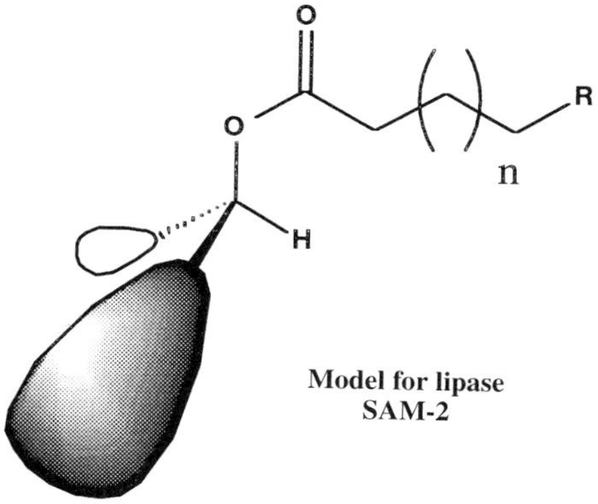

FIG. 28. Preferred absolute configurations of converted substrates (hydrolysis and esterification) in the presence of lipases from *Pseudomonas* species.

In conclusion, by using screening techniques of the type described in the preceding sections, the optimal conditions for the desired transformation were determined as summarized in Fig. 27. With these conditions the reactions could now be carried out routinely on a 10–20 g scale and several hundred grams of the chiral building block **3** were prepared this way. The immobilized enzyme is recovered and reused for at least 10 conversions. Enantiomerically pure building block **3** has since been used for the synthesis of various enantiomerically pure *myo*-inositol polyphosphates (Fig. 24).[20]

5.1.6. Success by Prediction: Expanding the Synthetic Scope of Lipases. With increasing structural information regarding the accepted substrates, the observed rates of transformation, absolute configurations, and enantiomeric purities, the development of models with predictive power becomes possible. Thus from more than 100 molecules converted in our laboratory with lipases from *Pseudomonas* species we were able to deduce a (cartoon) model for this enzyme from which the suitability of substrates and the preferred absolute configuration of the resulting products could be predicted[21,22] (Fig. 28). Numerous secondary alcohols of that general structure were converted in a multitude of laboratories around the world supporting this simple model.

[20] P. Andersch and M. P. Schneider, *Tetrahedron: Asymmetry* **7**, 349 (1996).
[21] U. Ader, P. Andersch, M. Berger, U. Goergens, R. Seemayer, and M. P. Schneider, *Pure Appl. Chem.* **64**, 1165 (1992).
[22] For similar attempts see R. J. Kazlauskas, *J. Org. Chem.* **56**, 2656 (1991); K. Sakai, *Tetrahedron: Asymmetry* **1**, 395 (1990); *ibid.* **2**, 733 (1991).

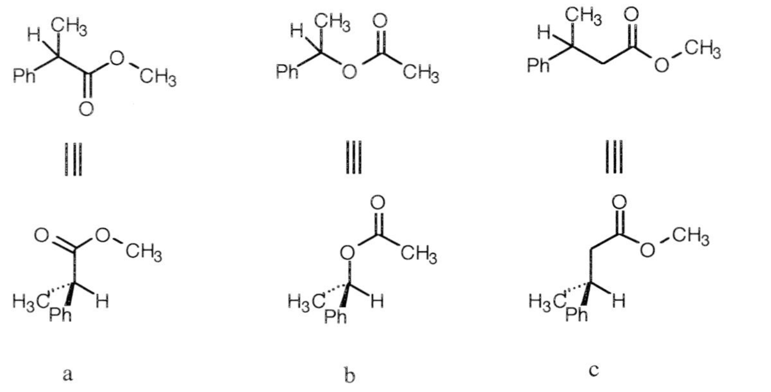

Prediction: **no substrate**   Known: **good substrate**   Prediction: **good substrate**

FIG. 29. Prediction of substrate tolerance based on a working model (see Fig. 28).

In an attempt to increase the predictive power of this model and under the assumption that the exact orientation of the ester carbonyl group at the serine moiety of the lipase plays an important role, we were able to predict correctly[23,24]—and prior to the actual experiments—the results to be expected from the transformations of two different classes of substrates that are sterically but not structurally related, the 2-phenylpropionic ester (a) and the 3-phenylbutyric ester (c) (Fig. 29).

Based on the preceding model, we know that the lipase from *Pseudomonas* species transforms the acylated alcohol (Fig. 29b) with high enantioselectivity and with the absolute configuration shown. The model consideration would then clearly predict that the carboxylic ester (Fig. 29a) would be no substrate for the enzyme, while the carboxylic ester (Fig. 29c) would be converted with the absolute configuration shown. This was confirmed by the experiment. This material was resolved into the two enantiomeric products with the expected absolute configuration and high enantiomeric purity (93% e.e.)[24] (Fig. 30).

Similarly, although—based on the model of Fig. 28—it was to be expected that the acetate of racemic lactic acid esters would be hydrolyzed with little or no enantioselectivity, we were able to predict that the corresponding *tert*-butyl esters would be excellent substrates leading to the corresponding products with the predicted absolute configuration (Fig. 31). This

[23] K. Laumen and M. P. Schneider, unpublished, 1986.
[24] K. Laumen, Ph.D. thesis, Wuppertal, 1987.

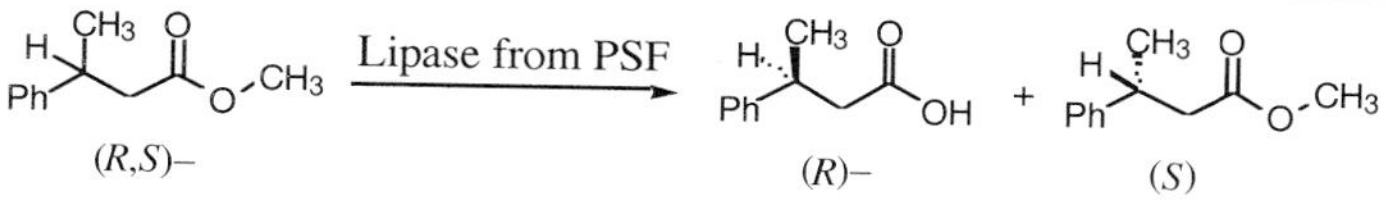

FIG. 30. Based on prediction: enantioselective hydrolysis of (*R,S*) ester.

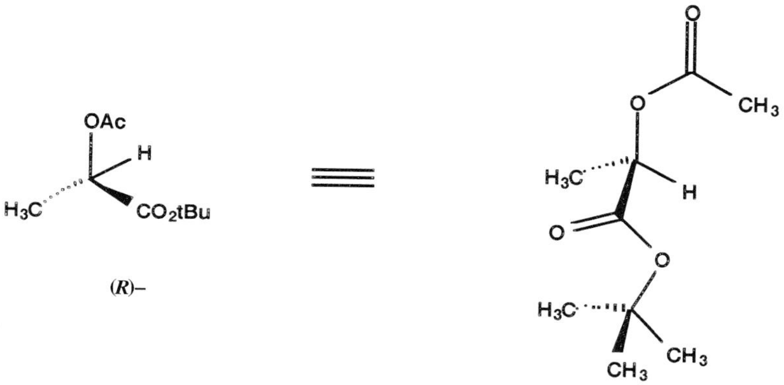

FIG. 31. Correlation of substrate and model (see Fig. 28).

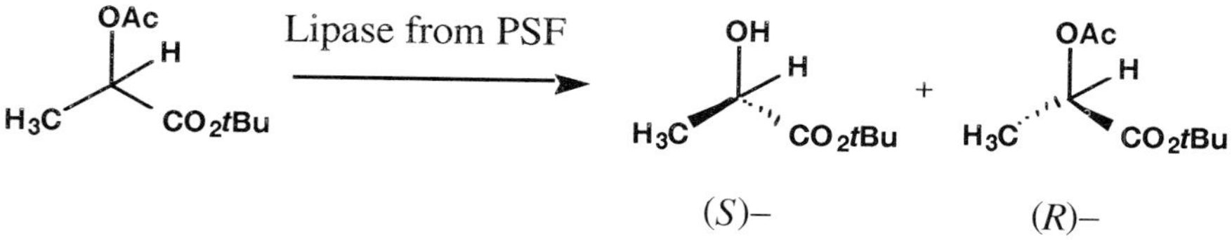

FIG. 32. Based on prediction: resolution of lactic acid esters.

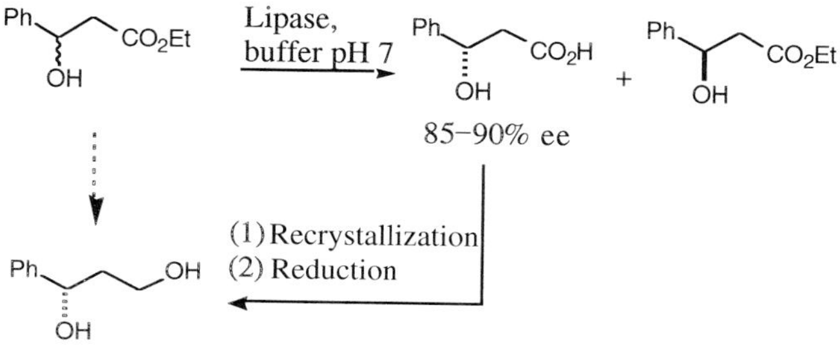

FIG. 33. Following the model: resolution leading to a building block for fluoxetine.

was confirmed experimentally by us[25] (Fig. 32) and independently by Scilimati et al.[26] As expected the enantioselectivities are extremely high (E $\gg$ 100, calculated E = 450).

The same model consideration led to the successful resolution of yet another carboxylic acid (Fig. 33), a useful building block for the production of fluoxetine in optically pure form.[27] It can be predicted that with the increasing number of X-ray structures of lipases becoming available,[2] together with methods for molecular modeling, a further refinement of our understanding regarding the prediction of lipase-catalyzed transformation will result.

5.1.7. Success by Experience: The Trojan Horse Approach. From numerous experiments with lipases from *Pseudomonas* species in our laboratory, the structural requirements for an enantioselective conversion of substrates were well known and established. A working model was developed (Fig. 28) that allowed reliable predictions regarding substrate tolerance of new substrates of that type, probable enantioselectivities, and the obtained absolute configurations. However, regardless of all knowledge, the success of these lipase-catalyzed transformations will always depend on the substrate structure. The question to be asked was simple: if the substrate to be converted was unsuitable for the enzyme, could we modify its structure in order to accommodate the requirements of the enzyme? The answer was yes! The principle is outlined in Fig. 34.

Although simple alkyl-substituted secondary alcohols are resolved with little or no enantioselectivity by this enzyme, we believe that by the introduction of a bulky substituent we could fool the enzyme by providing the essential structural requirement in accordance with the working model (Fig. 28). If this bulky substituent was to be chosen in a way that it could be removed or manipulated in a variety of ways this would open up a completely new approach to numerous target molecules previously inaccessible by this lipase.

After a few exploratory experiments the (*t*-butylthio) *t*-BuS-group turned out to be the functional group of choice both regarding its ease of introduction, removal, manipulation, and last, but not least, its ability to induce extremely high enantiomeric purities.

The method was employed successfully for the highly enantioselective production of β-hydroxythioethers and oxiranes[28] (Fig. 35), as well as a wide variety of δ-lactones[29] derived therefrom.

---

[25] A. Almsick, Ph.D. thesis, Wuppertal, 1989.
[26] A. Scilimati, T. K. Ngooi, and C. J. Sih, *Tetrahedron Lett.* **29**, 4927 (1988).
[27] U.S. patent 4,921,798.
[28] U. Goergens and M. P. Schneider, *Tetrahedron: Asymmetry* **3**, 1149 (1992).
[29] B. Haase and M. P. Schneider, *Tetrahedron: Asymmetry* **4**, 1017 (1993).

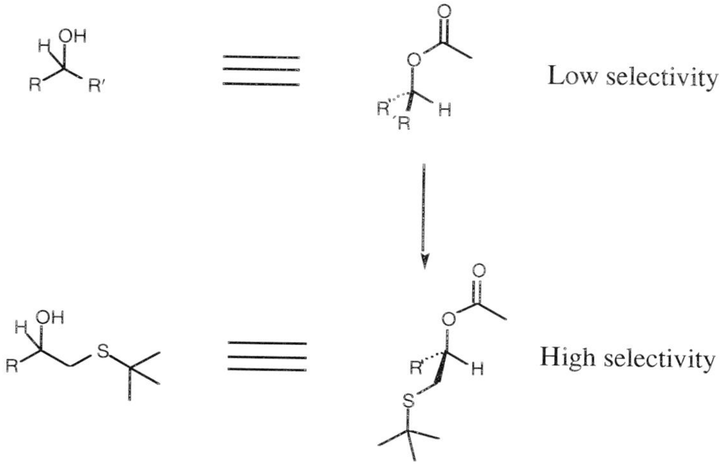

FIG. 34. Modifying the substrate to accommodate the enzyme, producing a Trojan horse.

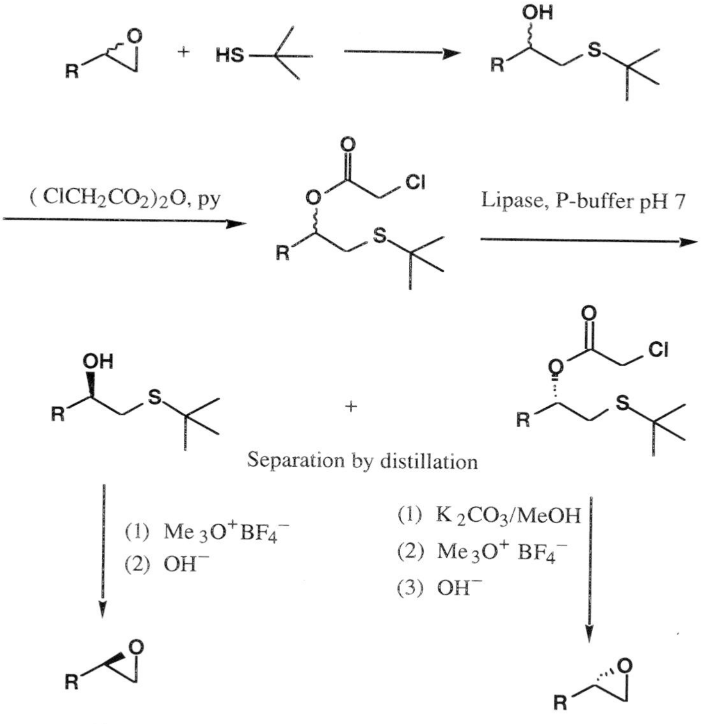

FIG. 35. Enantioselective hydrolysis of a Trojan horse.

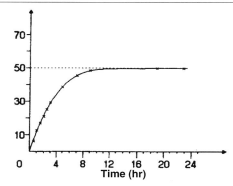

Fig. 36. Enantioselective hydrolysis of chloroacetates derived from β-hydroxythioethers: time/conversion curves.

The obtained enantiomeric purities both documented by the time/conversion curves (Fig. 36, compare Fig. 12) and their analysis by chiral GC (Fig. 37) were among the highest ever observed in our laboratory. Interestingly, the influence of the t-butyl group dominates all other substituents and n-alkyl groups up to C_{12} are considered small by the enzyme in reference to the working model (Fig. 28).

In conclusion, we can state that, also, substrate modifications can be the solution of a synthetic problem involving lipase-catalyzed transformations

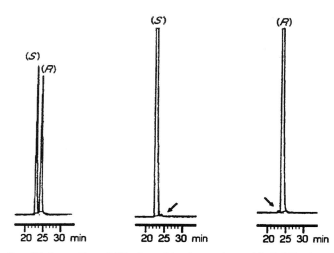

Fig. 37. Separation of β-hydroxythioethers using a chiral GC column.

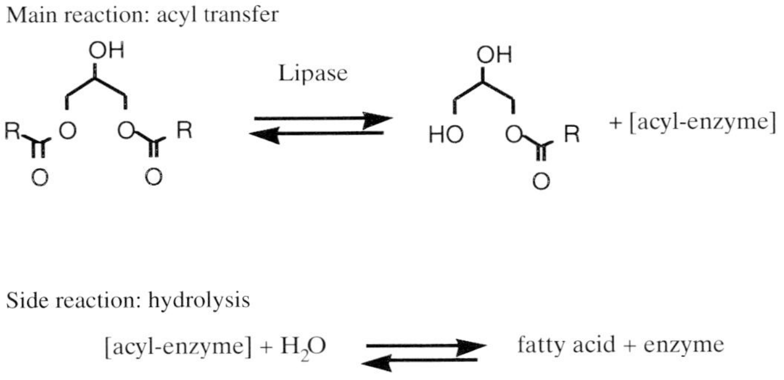

Main reaction: acyl transfer

$$\text{R-C(=O)-O-CH}_2\text{-CH(OH)-CH}_2\text{-O-C(=O)-R} \xrightleftharpoons{\text{Lipase}} \text{HO-CH}_2\text{-CH(OH)-CH}_2\text{-O-C(=O)-R} + [\text{acyl-enzyme}]$$

Side reaction: hydrolysis

$$[\text{acyl-enzyme}] + H_2O \rightleftharpoons \text{fatty acid} + \text{enzyme}$$

FIG. 38. Equilibria displayed by 1,3-specific lipases in aprotic organic media.

using an "established" enzyme for which a working model has been developed.

5.2. Regioselective Transformations

5.2.1. The Regioselectivities of Lipases.

Regioselective transformations of glycerol and glycerides, including natural fats and oils by lipase-catalyzed transformations, are of considerable synthetic and practical interest. For this a detailed understanding of the substrate tolerances and regioselectivities is essential. Literature data regarding the regioselectivities of lipases—obtained from studies in aqueous or biphasic media—are inherently unreliable due to the notorious problem of acyl group migrations associated with these systems. Based on the observation that mono- and diglycerides are stable toward acyl group migrations in aprotic, organic solvents we were able to develop a facile method for a reliable determination of regioselectivities displayed by microbial lipases toward glycerides.[30] The principle for such a determination can be simply explained as follows.

If a 1,3-sn-diglyceride is incubated with a lipase in an aprotic, organic solvent, an equilibrium between 1,3-sn-diglyceride, 1(3)-monoglyceride, and an acyl enzyme is established. If the lipase is strictly 1,3-specific, only this equilibrium will exist (Fig. 38). If, however, the lipase is only 1,3-selective or even unspecific, other equilibria will arise that yield additional products such as 1,2-diglyceride, 2-monoglyceride, and triglyceride, respectively (Fig. 39). In aprotic organic solvents with low water content (<2%) the hydrolysis of the acyl enzyme, leading to the free fatty acid, remains a side reaction in both cases.

---

[30] M. Berger and M. P. Schneider, *Biotechnol. Lett.* **13**, 641 (1991).

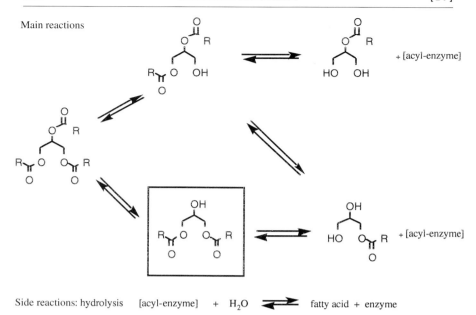

FIG. 39. Equilibria displayed by unspecific lipases.

We now found that for a given lipase the ratio of 1,3- and 1,2-diglyceride approaches a constant value during the incubation, which is independent from all other parameters of the system such as water content and solvent. It can therefore be taken as a measurement of the regioselectivity of lipases in organic solvents.

In analogy to the %e.e. or %d.e. values used for the description of enantiomeric or diastereomeric excess, we would like to suggest the definition of a similar value for the regioisomeric excess as [Eq. (4)]

$$RE = \%r.e. = \%(1,3\text{-diglyceride}) - \%(1,2\text{-diglyceride}) \qquad (4)$$

and call this the RE value of the lipase.

As substrate for these investigations 1,3-sn-dilaurin was used. It is easily available in very high regioisomeric purity (>99.5%) and the medium-chain length of the fatty acid residue allows the use of GC for rapid analysis of the resulting equilibrium constituents.

EXPERIMENTAL. One hundred milligrams of 1,3-sn-dilaurin and 10 mg of the lipase preparation were incubated in 5 ml of *tert*-butyl methyl ether (*t*-BuOMe). The mixture was stirred at room temperature and every 24 hr aliquots of 0.5 ml were taken. The solvent was evaporated under a stream of dry nitrogen, 0.2 ml of a 10:1 mixture of BSTFA and TMCS was added and the mixture heated at 70° for 3 hr. Then 0.2 ml of the resulting mixture was analyzed by gas chromatography.

In all experiments we observed that the ratio between 1,3- and 1,2-dilaurin is only time dependent in the first hours of incubation and therefore unsuitable for the determination of comparable RE values. After reaction times of >72 hr, however, the ratio between 1,3- and 1,2-dilaurin remains constant and is characteristic for each enzyme. The time required to reach this constant ratio is of course dependent on the enzyme activity and is proportional to the amount of enzyme employed. The RE values can be determined according to Eq. (4) from the measured ratio of 1,3- and 1,2-dilaurin (Table IV).

Based on these values and by subjective definition, lipases can be classified as 1,3-specific (RE > 90), 1,3-selective (90 > RE > 70), and unspecific (RE < 70). As already pointed out, the ratio of 1,3- and 1,2-dilaurin resulting from an incubation with all studied lipases is independent of all other parameters such as water content and type of solvent. In the presence of larger amounts of water, hydrolysis of the products is no longer a side reaction and the total amount of 1,3- and 1,2-dilaurin is decreasing. However, regardless of the water content in the system the ratio of 1,3- and 1,2-dilaurin remains unaffected and can be used for the determination of the RE value. For reasons of practicability the total water content of the system should not exceed 2%. A variety of different solvents was tested and although the total amounts of 1,2- and 1,3-dilaurin showed differences,

TABLE IV
RE VALUES OF LIPASE PREPARATIONS

Source of lipase	RE value (%r.e.)	Property
P. fluorescens	79	1,3-selective
Mucor javanicus	93	1,3-specific
Rhizomucor miehei (Lipozyme)	84	1,3-selective
Pseudomonas sp.	22	Unspecific
C. rugosa	80	1,3-selective
C. viscosum	93	1,3-specific
G. candidum	59	Unspecific
Rhizopus arrhizus	88	1,3-selective
R. delemar	98	1,3-specific
Rhizopus javanicus	96	1,3-specific
Rhizopus niveus	98	1,3-specific
Penicillium cambertii	16	Unspecific
Porcine pancreas	96	1,3-specific

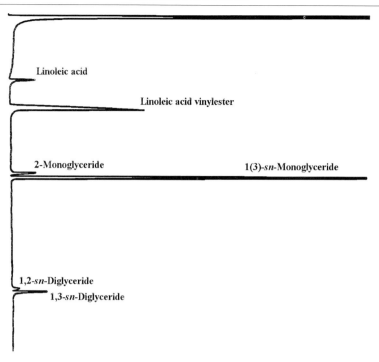

Fig. 40. Irreversible acyl transfer of linoleic acid vinyl ester onto glycerol in the presence of a lipase from *R. niveus*: GC diagram after 24 hr.

the ratio of the two products reached a constant value, allowing sufficiently long incubation times.

In agreement with the general understanding that lipases are more active in apolar organic solvents than in more polar systems (see Section 5.1.3) the required constant ratio of 1,3- and 1,2-dilaurin is reached faster in a nonpolar solvent after a reasonably short incubation time. In other words, the determination of RE values should preferably be carried out in apolar solvents such as *n*-hexane or TBME.

The RE values determined are valid in aprotic solvents having a total water content between 0.05 and 1.5%. In view of the shorter times required for analysis, unpolar solvents like *n*-hexane or TBME are preferred. Thus, the RE values obtained can be used efficiently as a screening parameter for lipase preparations in relation to the synthesis of regioisomerically pure mono- and diglycerides.[31]

[31] M. Berger and M. P. Schneider, *J. Am. Oil Chem. Soc.* **69**, 961 (1992); M. Berger and M. P. Schneider, *J. Am. Oil Chem. Soc.* **69**, 955 (1992).

5.2.2. *Another Look at Regioselectivities.* Alternatively, the regioselectivities displayed by lipases toward glycerol can also be determined directly by analyzing the reaction products of its esterification. Using the vinylester of linoleic acid as acyl donor, the obtained reaction products are a clear reflection of the lipase selectivities.

EXPERIMENTAL. One millimole of linoleic acid vinyl ester and 5 mmol of glycerol, immobilized onto 1.2 g of silica gel, are suspended in 2 ml of *t*-BuOMe. Ten milligrams of the corresponding crude lipase preparation is added to the reaction mixture. At regular intervals (24, 48, and 72 hr), samples are drawn, converted into the corresponding trimethylsilyl ethers (where applicable) and analyzed by GC. Essential for successful screening are again baseline separations of all molecules involved (Fig. 40).

Using this method all essential information such as conversion, regioisomeric purity, and product ratios can be determined in one single experiment. Using Eq. (5), the regioselectivities can be calculated, leading to the

TABLE V
RE VALUES OF DIFFERENT LIPASES[a]

No.	Lipase from:	RE value[b]	RE value[c]	Property
1	*Rhizopus oryzae*		91	1,3-specific
2	*R. delemar*	98	83	1,3-selective
3	*R. arrhizus*	88	84	1,3-selective
4	*R. niveus*	98	92	1,3-specific
5	*A. niger*		91	1,3-specific
6	*Pseudomonas* species		96	1,3-specific
7	*Pseudomonas* species (SAM II)	22	74	1,3-selective
8	*P. fluorescens*	79	68	Unspecific
9	*M. javanicus*	93	89	1,3-selective
10	*Rhizomucor miehei* (Lipozyme)	84	88	1,3-selective
11	*H. lanuginosa*		90	1,3-specific
12	*P. roquefortii*		94	1,3-specific
13	*Pseudomonas* species		86	1,3-selective
14	*M. javanicus* (Amano)		90	1,3-specific
15	Lipase AK		95	1,3-specific
16	*C. viscosum*	93	87	1,3-selective
17	*Candida cylindracea*		84	1,3-selective
18	*C. lipolytica*		64	Unspecific
19	*G. candidum*	59	65	Unspecific
20	*P. cambertii*		68	Unspecific
21	*C. rugosa*	80	78	1,3-selective
22	*Pseudomonas cepacia* (SAM I)		84	1,3-selective

[a] RE value = % 1(3)-monoglyceride − % 2-monoglyceride, from three screening experiments RE > 90 = 1,3-specific, 90 > RE > 70 = 1,3-selective, RE < 70 = unspecific.
[b] Saturated fatty acids.
[c] Unsaturated fatty acids.

results summarized in Table V for the lipase-catalyzed acyl transfer of linoleic acid onto glycerol. The results are compared with data obtained from using the 1,3-dilaurin approach described earlier. Although the numerical values are (not surprisingly) somewhat different, the general picture is comparable. The obtained RE values are very valuable indeed for providing reliable information regarding the best choice of lipase for the synthesis of isomerically pure glycerides.[32]

$$\text{RE value} = \%\ 1(3)\text{-monoglyceride} - \%\ 2\text{-monoglyceride} \quad (5)$$

6. Conclusion

In this chapter, we tried to summarize—quite subjectively—what we believe are among the most important types of screening experiments commonly to be executed in laboratories engaged in the synthetic applications of lipases. Clearly, this is not a comprehensive review covering all possible screening methods but rather is intended to inform the interested reader about the most important issues related to the application of these enzymes. It is also intended to encourage all of those involved in this area to use their knowledge and imagination for the discovery of more facile screening methods as a basis for more efficient lipase-catalyzed transformations. In this sense, it is hoped that this chapter will be outdated as rapidly as possible.

[32] C. Waldinger and M. P. Schneider, *J. Am. Oil Chem. Soc.* **73,** 1513 (1996).

[18] Kinetics, Molecular Modeling, and Synthetic Applications with Microbial Lipases

By KARL HULT and MATS HOLMQUIST

Introduction

The investigations of lipases in recent years have resulted in tremendous progress in our understanding of their structure and function. Despite this fact, the knowledge required for a rational approach to be taken in the selection or engineering of lipases for particular synthetic applications remains incomplete. Still, a rigorous understanding of lipase function in terms of the kinetics, thermodynamics, and substrate recognition processes associated with the catalysis is needed to be able to predict correctly and control the outcome of a lipase-catalyzed reaction.

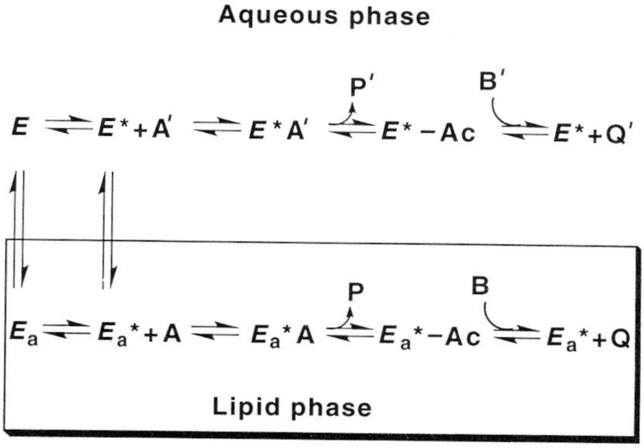

Aqueous phase

$$E \rightleftharpoons E^* + A' \rightleftharpoons E^*A' \xrightarrow{P'} E^* - Ac \xrightarrow{B'} E^* + Q'$$

$$E_a \rightleftharpoons E_a^* + A \rightleftharpoons E_a^*A \xrightarrow{P} E_a^* - Ac \xrightarrow{B} E_a^* + Q$$

Lipid phase

FIG. 1. Kinetic model for lipase catalysis. E, enzyme in solution, closed inactive form; E^*, enzyme in solution, open active form; E_a, enzyme adsorbed to the interface, closed inactive form; E_a^*, enzyme adsorbed to the interface, open active form; E^*A', E_a^*A, Michaelis–Menten complexes; E^*-Ac, E_a^*-Ac, acyl-enzyme intermediates; A, A', B, B', substrates (B is water in hydrolysis); P, P', Q, Q', products.

In this chapter, we describe methods for the investigation of lipase catalysis by means of enzyme kinetics and molecular modeling. These procedures give information about different aspects of the catalytic process and provide insight, in a complementary manner, into the details and molecular mechanisms involved in lipase action and substrate recognition. We also describe procedures to perform convenient and efficient lipase-catalyzed resolutions.

Kinetics

A model of interfacial kinetics of lipolysis was proposed by Verger and de Haas.[1] The model includes an adsorption equilibrium of the water-soluble lipase to an interface. The adsorption step is followed by the binding of the substrate to the active site of the enzyme and a subsequent catalytic turnover. In this model the lipase changes conformation concomitantly with, or after, adsorption to the water/lipid interface. In models suggested by Martinelle and Hult[2] (Fig. 1) and Rubin,[3] the lipase exists in two conformations (open active and closed inactive) in solution. The two states are in free equilibrium favoring the closed form in solution. When an interface

[1] R. Verger and G. H. de Haas, *Ann. Rev. Biophys. Bioeng.* **5**, 77 (1976).
[2] M. Martinelle and K. Hult, *in* "Lipases: Their Structure, Biochemistry and Application" (P. Wooley and S. Petersen, eds.), p. 159. Cambridge University Press, Cambridge, 1994.
[3] B. Rubin, *Nature Struct. Biol.* **1**, 568 (1994).

is introduced, the adsorption of the lipase changes the overall equilibrium in favor of the open conformation because that is stabilized by the hydrophobic surrounding (low dielectric constant) at the interface.

The studies of lipase kinetics should be performed in such a way that one can distinguish between the following: (1) the activity of the lipase in solution, (2) the adsorption of the lipase to the interface, and (3) the activity of the lipase adsorbed to the interface. Only then can a correct evaluation of the contributions to the overall activity be made.

Lipase Activity with Soluble Substrates

The activity of lipases toward water-soluble substrates can be measured if great care is taken to compensate for the activity of lipase adsorbed to the cuvette walls. The activity of the adsorbed lipase is the main contribution to the overall activity of lipases that are highly interfacially activated (*Humicola lanuginosa* lipase and *Rhizomucor miehei* lipase), while it is a minor contribution to others (*Candida antarctica* lipase B).[4]

Procedure. The hydrolysis of *p*-nitrophenyl acetate (PNPA) is followed at 400 nm using a spectrophotometer equipped with a magnetic stirring device in the cuvette. Incubate the lipase in 50 mM MOPS-KOH buffer, pH 7.5, for 30 min in a new polystyrene cuvette before the total activity is measured. The reaction is started by the addition of PNPA dissolved in acetonitrile to a final concentration of 1 mM, keeping the acetonitrile concentration at 1% (v/v). Measure the activity of adsorbed lipase after the lipase solution has been poured out of the cuvette and the cuvette has been washed twice with buffer (50 mM MOPS-KOH, pH 7.5), by the addition of 1 mM PNPA in buffer. The activity of the free lipase is calculated as the difference between the total activity and the activity of the adsorbed lipase. In this assay the contribution to the total activity of dissolved lipase is high for *C. antarctica* lipase A and *C. antarctica* lipase B, but below the detection limit for *H. lanuginosa* lipase.[4]

Lipase Binding to a Solid Interface

The amount of lipase adsorbed to, for example, a cuvette wall can be measured using radioactively labeled protein. The protein can be labeled using the method of Langone.[5]

Procedure. Incubate a solution of the labeled lipase, 0–0.02 mM lipase in 1 ml 50 mM MOPS-KOH, pH 7.5, for 30 min in a new polystyrene cuvette. Empty the cuvette and wash twice with the buffer. The amount of lipase adsorbed on the cuvette can be measured directly with a Packard

---

[4] M. Martinelle, M. Holmquist, and K. Hult, *Biochim. Biophys. Acta* **1258**, 272 (1995).
[5] J. J. Langone, *Methods Enzymol.* **70**, 221 (1980).

Tri-Carb 1500 liquid scintillation analyzer (Downers Grove, IL) after placing the cuvette in a scintillator counting vial in the presence of Ready Safe (Beckman Instruments, Fullerton, CA). Alternatively, the lipase is removed from the surface by washing with 10% w/w sodium dodecyl sulfate (SDS) and measurement of the radioactivity in the collected washing solution. *Candida antarctica* lipase B formed a monolayer at the polystyrene surface at saturation, while *H. lanuginosa* lipase formed a more loosely packed layer.[4]

Lipase Affinity to a Substrate Interface

The affinity of a lipase toward a substrate interface can be measured by monitoring the reaction rate at a constant lipase concentration but varying the interfacial area of the substrate. The observed rate takes a measure of the enzyme amount bound to the interface. This is an efficient method for following the effect of point mutations on interfacial binding.[6]

Procedure. Add 0.5 mmol of tributyrin to 5 ml water solution of 5% (w/v) gum arabic and 0.2 M $CaCl_2$. Prepare an emulsion by sonicating the mixture for 1 min. Dissolve the enzyme in 10 mM MOPS-KOH buffer, pH 7.5. Start the reaction by adding the enzyme to a thermostatted (25°) substrate solution (1.5 ml). Measure the activity using a Radiometer pH-stat equipped with an ABU91 autoburette (1 ml) connected to a VIT90 videotitrator. Run the reaction for 5 min under nitrogen. Vary the available substrate area by diluting the original tributyrin emulsion directly to the desired concentration.

Lipase Activity at an Interface

All lipase activity originates from interfacial catalysis when emulsions of substrates with low water solubility are used, because there is no monomeric substrate in the solution. If the surface area of the substrate emulsion is in excess, all lipase will be bound to the water/lipid interface of the emulsion. The reaction rate observed will then reflect a substrate specificity, which does not involve the binding of the lipase to the interface. The combination of enzyme and substrate concentrations used must meet the following criteria: a change in enzyme concentration must cause a proportional change in reaction rate, whereas the rate must be unchanged if the substrate concentration is altered. This method can be used to compare substrate specificities between lipases and mutated lipases.[7]

[6] M. Holmquist, M. Martinelle, I. G. Clausen, S. Patkar, A. Svendsen, and K. Hult, *Lipids* **29**, 599 (1994).
[7] M. Martinelle, M. Holmquist, I. G. Clausen, S. Patkar, A. Svendsen, and K. Hult, *Protein Eng.* **9**, 519 (1996).

Procedure. Prepare the substrate emulsion by sonicating for 1 min 1.0 mmol of substrate in 5 ml of 5% (w/v) gum arabic and 0.2 M $CaCl_2$. Dissolve the lipase in 10 mM MOPS-KOH, pH 7.5. Start the reaction with the addition of lipase (5–10 μl) to 1.5 ml stirred thermostatted substrate emulsion. Follow the reaction by maintaining the pH with NaOH (10–100 mM), using a Radiometer pH-stat as above. Check the reaction rate with a range of substrate and enzyme concentrations to prove the validity of the method.

Lipase Kinetics in Organic Solvent

One of the difficulties in lipase kinetics is to determine values for k_{cat} and K_m. This is due to the binding of the lipase to the interface and the difficulties in varying the substrate concentration at the interface. The substrate emulsion can be diluted with another hydrophobic substance,[6] but the properties of the interface may change and the concentration of the substrate in the interface will be difficult to define. The method of choice is instead to use a homogenous organic phase consisting of an organic solvent and dissolved substrate. The lipase is used as a dry powder or better immobilized at a low load at the surface of a carrier to avoid diffusion limitations. The method affords apparent K_m values, which depend on the choice of solvent and reflect the solubility of the substrate in the solvent as well as the affinity of the substrate to the enzyme. Thus K_m values measured in different solvents cannot be compared without further thermodynamic analysis.[8,9]

Measurements of lipase kinetics in organic media clearly show that lipases (and many other enzymes) do not need a continuous water phase to be active. Even if no free water is needed, the thermodynamic water activity in the system is extremely important. Thus all ingredients must have a controlled water activity before use. This can be achieved by equilibration against salt solutions with known water activities before use.[10]

The activity of lipases in organic media compared to water emulsions varies a lot. Some lipases, such as *C. antarctica* lipase B, express the same activity in hydrophobic solvents as in water emulsions,[9] whereas *H. lanuginosa* lipase and *R. miehei* lipase are much less active in organic media.[11,12]

---

[8] K. Ruy and J. S. Dordick, *Biochemistry* **31**, 2588 (1992).
[9] M. Martinelle and K. Hult, *Biochim. Biophys. Acta* **1251**, 191 (1995).
[10] P. Halling, *Biotechnol. Techniques* **6**, 271 (1992).
[11] M. Holmquist, I. G. Clausen, S. Patkar, A. Svendsen, and K. Hult, *J. Protein Chem.* **14**, 217 (1995).
[12] M. Holmquist, M. Martinelle, P. Berglund, I. G. Clausen, S. Patkar, A. Svendsen, and K. Hult, *J. Protein Chem.* **12**, 749 (1993).

Procedure. The initial rates in the esterification (octanol and octanoic acid) or transesterification (octanol and ethyl octanoate) catalyzed by a lipase can be measured using the following procedure. All components are preequilibrated against salt solutions with defined water activities.

Immobilized lipase (typically 2 mg/ml, containing 2 mg protein/g carrier) is weighed in a 3-ml vial. Solvent (heptane or acetonitrile) and acyl acceptor substrate (octanol) are added and the vial is sealed with a Teflon membrane. Allow the mixture to equilibrate 10 min at 25° under top-stirred agitation. Start the reaction by adding the acyl donor (acid or ester) and solvent to a total volume of 2 ml. The vial should be well filled to minimize exchange between the liquid and gas phase in the reaction vessel. Samples of 5 μl each are withdrawn at appropriate times. It is essential to draw samples before the consumption of the acyl donor forms products (water or ethanol), which will be competitive substrates with the acyl acceptor. With low efficient acceptors, less than 0.01% of the acyl donor should be consumed.

The samples are analyzed by gas chromatography. Use, for example, a Perkin Elmer 8500 GC (Norwalk, CT) equipped with a CP-sil 19CB (Chrompack, Middelburg, The Netherlands; 25 m, 0.32 mm I.D., d_f = 0.2 μm) and N_2 as carrier gas. Use an oven temperature gradient between 70 and 250° and an injector and detector temperature of 220 and 350°, respectively. This method proved to be useful in the determination of the kinetic constants for *C. antarctica* lipase B.[9]

Molecular Modeling

Lipase Structure in Different Media

To determine how the surrounding solvent might affect the molecular structure of a lipase, molecular dynamics simulations of the protein can be performed with different solvent environments. The solvent medium can be modeled either as a continuum without a molecular representation with the dielectric constant set to that of the solvent, or as discrete solvent molecules.

When parameters are sought to describe a discrete solvent molecule, the calculated physical properties of the model such as density and heat of vaporization should be in agreement with experimentally obtained data to ensure the quality of the model. Simulations of *R. miehei* lipase in different environments showed that vacuum is a reasonably good model for a hydrophobic organic solvent.[13] Therefore, unless discrete solvent molecules are for some reason believed to play a significant role in the process that is to be modeled, the calculation may instead be done with the lipase

[13] M. Norin, F. Hæffner, K. Hult, and O. Edholm, *Biophys. J.* **67,** 548 (1994).

surrounded by vacuum. This saves a significant amount of computer time that may be better used to perform longer simulations resulting in a more thorough search of the conformation space.

Procedure. The coordinates for the three-dimensional structure of the lipase molecule can be downloaded from the Brookhaven Protein Data Bank.[14,15] The protein is centered in a box of solvent and all solvent molecules within a distance of 3 Å from any protein atom are discarded from the system. Discrete solvent water molecules may be described by the simple point charge (SPC) model.[16] By using standard Lennard–Jones parameters from the GROMOS force field,[17] an ester solvent such as methyl hexanoate can be modeled with four fractional charges.[13] As for the protein, nonpolar hydrogen atoms are included in the carbons as united hydrocarbon groups.

The calculations can be performed on a workstation using the GROMOS program.[17] The manual inspection of molecular structures can be done using the molecular modeling software SYBYL (Tripos Associates, 1994) running on a Silicon Graphics workstation. Both the C (charged) and D (uncharged) version of GROMOS force field may be used in the simulations. In the D version, all amino acids residues that may carry net charges are neutral and treated as dipoles. Periodic boundaries are applied to all simulations including solvent. Nonbonded interactions are truncated at 10 Å. Bond lengths are constrained using the SHAKE algorithm.[18] Before the dynamics simulation the system is energy minimized using 1000 steps of the steepest descent method. The dynamics simulations are performed at 300 K with a time step of 2 fsec. The simulated molecules are equilibrated for 50 psec and the data collected for analyses are obtained in the subsequent 150 psec.

Conformational States

To estimate energies and identify intramolecular interactions during the lid opening process in a lipase and to be able to rationalize how the open active state of the enzyme is stabilized in a hydrophobic environment, a computational method combining molecular mechanics and dynamics can be used. If the structures of the closed and open form of the lipase are

[14] F. C. Bernstein, T. F. Koetzle, G. J. B. Williams, E. F. Meyer, M. D. J. Brice, J. R. Rodgers, O. Kennard, T. Shimanouchi, and M. Tasumi, *J. Mol. Biol.* **112**, 535 (1977).
[15] E. Abola, F. C. Bernstein, S. H. Bryant, T. F. Koetzle, and J. Weng, Data Commission of the International Union of Crystallography. Bonn/Cambridge/Chester, pp. 107–132 (1987).
[16] H. J. C. Berendsen, J. P. M. Postma, W. F. van Gunsteren, and J. Hermans, *in* "Intermolecular Forces" (B. Pullman, ed.). Riedel, Dordrecht, Holland, 1981.
[17] W. F. van Gunsteren and H. J. C. Berendsen, GROMOS Software Manual, GROMOS Molecular Simulation. Biomos B.V., Gröningen, The Netherlands, 1987.
[18] J.-P. Ryckaert, G. Ciccotti, and H. J. C. Berendsen, *J. Comp. Phys.* **23**, 327 (1977).

known, the lid opening event can be modeled by moving the lid between its two conformations by restraining the pseudotorsional angles of the Cα carbons within the lid in small steps.[19] After each incremental change of the pseudotorsional angles, molecular dynamics is used to relax the system. This method allows for the calculation of the energy difference between the two conformational states of the enzyme. By performing the simulation on mutated lipases and in environments with different dielectric constants, it is possible to identify amino acid residues responsible for the stabilization of the two conformations of the lipase and to clarify the nature of these interactions.

Electrostatic forces almost solely stabilize the active open-lid conformation of a lipase (*R. miehei* lipase) in a hydrophobic environment.[19–21] The quality of the model can be tested by simulating both the lid opening and closing events. The modeled process should be reversible in terms of energy and structure of the two extreme conformational states.

Procedure. The simulations are performed on a Silicon Graphics workstation running the program SYBYL (Tripos Associates, 1995). In the dynamics simulations the SYBYL implementation of the AMBER united atom force field can be used.[22] The coordinates for the three-dimensional structure of the lipase molecule are downloaded from the Brookhaven Protein Data Bank.[14,15] Polar hydrogens are added to the structure using the BIOPOLYMER module in the SYBYL program.

All water molecules are removed in the structure. The molecule is energy minimized for 300 steps using the steepest descent method followed by 1000 steps using the conjugate gradient method with a distant-dependent dielectric constant ($\varepsilon = r$). Both the closed and open state of the enzyme are refined in this manner. The resulting structures are used as the two extreme states when the conformational change of the enzyme is modeled.

The pseudotorsional angles of the residues in the lid are restrained in steps (*R. miehei* lipase, 21 steps). The pseudotorsional angles are defined as the torsional angles formed by the Cα trace of the protein. The restraints are formed by adding an extra harmonic penalty term to the force field.

$$E_{tot} = E_{ff} + 1/2C^*\Sigma_i\{[\varphi_i - \varphi_{ic} - \lambda/20^*(\varphi_{io} - \varphi_{ic})]^2\}$$

with λ being the step number, E_{tot} the total energy of the system, E_{ff} the energy given by the original force field, φ_i the current pseudotorsional angle, φ_{ic} and φ_{io} the pseudotorsional angles of the closed and the open

[19] M. Norin, O. Olsen, A. Svendsen, O. Edholm, and K. Hult, *Protein Eng.* **6**, 855 (1993).
[20] M. Holmquist, M. Norin, and K. Hult, *Lipids* **28**, 721 (1993).
[21] G. H. Peters, O. H. Olsen, A. Svendsen, and R. C. Wade, *Biophys. J.* **71**, 119 (1996).
[22] S. J. Weiner, P. A. Kollman, D. A. Case, U. C. Singh, C. Ghio, G. Alagona, S. Profeta, Jr., and P. Weiner, *J. Am. Chem. Soc.* **106**, 765 (1984).

lipases, respectively. Constant C set to 8.3 kJ/mol deg^2. For the conversion from the closed to the open lipase, λ goes from 0 to 20, while it goes from 20 to 0 in the opposite direction.

For each of the 21 steps, the structure is minimized for 200 steps with the appropriate restraining potential using the Powell method in SYBYL. After that restrained molecular dynamics is performed, in which the structure is heated for 1.2 psec with successively increasing temperatures up to 300 K followed by equilibration at that temperature for 4.8 psec. During the dynamic simulations all atoms within the lid of the lipase are allowed to move together with side chains of residues within a 6-Å distance from any atom in the lid in its closed position.

The simulation is continued at 300 K using a temperature coupling constant of 50 fsec. The nonbonded list is updated every 25 fsec and the SHAKE algorithm[18] is applied to all bonds allowing a time step of 2 fsec. Data are sampled every 0.5 psec during the simulation. Finally, the structure is energy minimized, keeping the restraining potential, in 500 steps or until the rms gradient is <0.4 kJ/Å using the Powell method. The final structure is then used for further analysis of energy and structure, and it is taken as the input structure for the next step.

Substrate Binding

To locate specific binding sites that may determine the substrate recognition ability of an enzyme, the active site can be mapped by means of a computer program called GRID.[23] This allows for the identification of regions within the active site that show high affinity for probes that mimic functional groups of the substrate. The affinities of such probes are calculated by means of a molecular force field. By dissecting a substrate into a set of different probes and subsequently searching for favorable interactions between the probes and the protein, it is possible to determine how the substrate can coordinate with the active site of a lipase.[24] Binding sites for water molecules having a role in the catalytic mechanism can also be identified.

Procedure. Download the coordinates for the three-dimensional structure of the lipase molecule from the Brookhaven Protein Data Bank.[14,15] Remove inhibitors, if present, and water molecules from the structure. Process the structure with the GRIN module of the GRID program to prepare the input file for the GRID calculation. The GRID calculation is subsequently performed with one of the standard probes supplied with the program. If the calculations are intended to model the lipase when acting

[23] P. J. Goodford, *J. Med. Chem.* **28,** 849 (1985).
[24] M. Norin, F. Hæffner, A. Achour, T. Norin, and K. Hult, *Protein Sci.* **3,** 1493 (1994).

TABLE I
RELATIONSHIP BETWEEN THE ENANTIOMERIC RATIO (E) OF THE
ENZYME AND THE ENANTIOMERIC EXCESS (ee) OF THE PRODUCT
IN AN ENZYME-CATALYZED RESOLUTION[a]

E	$\Delta\Delta G\ddagger$ (kJ/mol)	$\Delta\Delta G\ddagger$ (kcal/mol)	ee of product (%)
1	0	0	0
10	5.8	1.4	68.3
100	12	2.8	94.4
1000	23	4.1	>99.9

[a] The data were calculated with $T = 300$ K and 48% conversion of the substrate.

at a hydrophobic interface or in an organic solvent, a dielectric constant of 4 for the environment can be chosen. The LEAU=2 option in the GRID program is used to model water bridges between a water probe and the protein. The active site of the lipase is centered in a box with the dimensions $20 \times 20 \times 30$ Å

Interactions between the probe and the protein are calculated at grid points that are 0.5 Å apart. The results from the GRID calculations can be displayed as contour plots at desired interaction energy levels on the graphics display, together with the structure of the lipase molecule.

Calculation of Enantioselectivity

The ability of an enzyme to distinguish between two competing enantiomeric substrates is defined by the enantiomeric ratio E. This is the ratio between the specificity constants (k_{cat}/K_m) of the enzyme for the two competing enantiomers denoted S and R.[25]

$$E_S = (k_{cat}/K_m)_S/(k_{cat}/K_m)_R$$

The enantiomeric ratio E of an enzyme is related to the difference in free energy of activation ($\Delta\Delta G\ddagger$) between the reactions of the two enantiomers, according to the expression: $\Delta\Delta G\ddagger = -RT \ln E$ in which R is the gas constant and T the temperature.[26] The relationship between the enantiomeric ratio and the optical purity (enantiomeric excess: $ee_S = ([S] - [R])/([R] + [S])$ or $ee_R = ([R] - [S])/([R] + [S]))$[27] of the reaction product at $T = 300$ K and 48% conversion is shown in Table I. As seen, a $\Delta\Delta G\ddagger$ for

[25] C.-S. Chen, Y. Fujimoto, G. Girdaukas, and C. J. Sih, *J. Am. Chem. Soc.* **104**, 7294 (1982).
[26] M. Norin, K. Hult, A. Mattson, and T. Norin, *Biocatalysis* **7**, 131 (1993).
[27] A. Bassindale, "The Third Dimension in Organic Chemistry." Wiley, New York, 1984.

the two competing reactions, corresponding approximately to the energy of one hydrogen bond, is sufficient to bring about a very high optical purity of the product in an enzyme-catalyzed resolution. The energy contribution of a hydrogen bond is usually considered to be about 20 kJ/mol.[28]

With currently available molecular modeling techniques, it is possible to calculate the enantioselectivity of enzymes showing an E value higher than 10. The geometries and energies of the three-dimensional structures, representing the energy minima of the diastereomeric enzyme–substrate transition state complexes, can be calculated using a method that combines molecular mechanics and dynamics.[24,26,29,30] Because the free energy of the ground state is the same for the two enantiomeric substrates, the difference in free energy between the diastereomeric transition state complexes is a direct measure of the enantioselectivity, $E_S = (k_{cat}/K_m)_S/(k_{cat}/K_m)_R$ of the enzyme.

Procedure. All calculations can be performed with a Silicon Graphics workstation equipped with the molecular modeling software SYBYL (Tripos Associates, 1994). Energy minimizations are performed by means of the Powell minimizer in SYBYL. Download the coordinates for the three-dimensional structure of the lipase from the Brookhaven Protein Data Bank.[14,15] Water molecules tightly bound to the lipase should be included in the calculations. To identify these water molecules, the GRID program[23] can be used. In a first energy minimization for 150 steps, only the water molecules are allowed to move. Subsequently, with all atoms allowed to move, the structure is minimized until the rms gradient is less than 0.04 kJ/Å. It is important that the structure is energy minimized before using it as input in the following molecular dynamics protocol.

The model of the tetrahedral oxyanion transition state, in which the O_γ atom of the active serine residue is covalently bound to the carbonyl carbon atom of the substrate, can be built using SYBYL. The charges of the substrate atoms may be calculated by a semiempirical method[31] using the software MOPAC 6.0 ESP. The transition state model is subsequently docked into the active site of the lipase and oriented so as to form all catalytically essential hydrogen bonds to the active site.

The negatively charged oxygen atom of the transition state model is coordinated with the oxyanion hole. The histidine residue in the catalytic triad is to be protonated[32] and allowed to form a bifurcated hydrogen bond

---

[28] A. Fersht, "Enzyme Structure and Mechanism." W. H. Freeman, New York, 1985.
[29] M. Holmquist, F. Hæffner, T. Norin, and K. Hult, *Protein Sci.* **5**, 83 (1996).
[30] C. Orrenius, F. Hæffner, D. Rotticci, N. Öhrner, T. Norin, and K. Hult, *Biocatalysis and Biotransformations* (in press) (1997).
[31] B. H. Besler, K. M. Merz, Jr., and P. A. Kollman, *J. Comput. Chem.* **11**, 431 (1990).
[32] A. A. Kossiakoff and S. A. Spencer, *Biochemistry* **20**, 6462 (1981).

to O_γ of the catalytic serine residue and to the oxygen atom of the alcohol moiety of the substrate, in analogy with the active site function of serine proteases. For reviews of active site function in serine proteases see Refs. 33 and 34.

The positions of the hydrogen atoms in the lipase are calculated by means of the BIOPOLYMER module of SYBYL. Amino acid residues within a 6-Å distance from any atom of the transition state model are treated by the all-atom AMBER force field,[35] whereas the united atom AMBER force field[22] is applied to the rest of the amino acid residues in the lipase. Water molecules can be represented by means of the TIP3P water model.[36] The simulations are performed in vacuum with a distance-dependent dielectric function ($\varepsilon = r$). Nonbonded interactions are truncated at $r > 8$ Å.

Initially, the transition state of the substrate is subjected to 200 steps of energy minimization with the lipase kept in a locked conformation. In the subsequent calculations amino acid residues and water molecules within a 6-Å distance from the substrate are allowed to move freely during the simulations. This cutoff distance should be chosen so that it defines a volume well including the active site and the bound substrate. This allows for all interactions between the substrate and the protein to be treated in as much detail as possible to obtain the best model of the complex. In a region between 6 and 8 Å away from the substrate, only the amino acid side chains and water molecules are allowed to move. The rest of the lipase is kept rigid during the simulations.

The lipase–transition state complex is energy minimized for 500 steps. These initial energy minimization steps, described earlier, are done to obtain a relaxed structure for the complex before the subsequent molecular dynamics simulations. The simulations are equilibrated by raising the temperature to 300 K during 1.2 psec, followed by a simulation for 4 psec. A time step of 2 fsec is used in all simulations. By periodically lowering and raising the temperature between 300 and 1 K, during molecular dynamics simulation, the conformational space is searched for local energy minima of the complex. The temperature is kept at 300 K for 6 psec and then slowly decreased to 1 K with temperature decrements of 10 K and an interval length of 0.05 psec.

[33] J. Kraut, *Ann. Rev. Biochem.* **46**, 331 (1977).
[34] A. Warshel, G. Naray-Szabo, F. Sussman, and J. K. Hwang, *Biochemistry* **28**, 3629 (1980).
[35] S. J. Weiner, P. A. Kollman, D. T. Nguyen, and D. A. Case, *J. Comput. Chem.* **7**, 230 (1986).
[36] W. L. Jorgensen, J. Chandrasekhar, J. D. Madura, R. W. Impey, and M. L. Klein, *J. Chem. Phys.* **79**, 926 (1983).

Subsequently the temperature is stepwise raised from 1 to 300 K with increments of 50 K with an interval length of 0.2 psec. The total simulation time comprising 10 cycles is 107 psec. Ten different conformers corresponding to energy minima of the lipase–transition state complex are collected at 1 K and subsequently refined through energy minimization for 5000 steps or until the rms gradient is less than 0.04 kJ/Å. The conformer with the lowest potential energy is selected and considered to represent the global energy minimum of the complex.

Applications

Lipases are versatile catalysts for the resolution of alcohols. Acids may also be resolved using lipases, but the broad substrate specificity seen for alcohols does not have its counterpart with acids. The reason might be that lipases have evolved to bind the substrate by the acyl part, which will form the acyl enzyme, and not by the alcohol part, which should freely be able to leave the enzyme to avoid product inhibition. It is therefore natural that substitutions or other variations of the acyl moiety will lead to poor fit in the active site.

Resolution of Alcohols

The resolution of alcohols catalyzed by serine hydrolases will follow the general scheme seen in Fig. 2, in which the enzyme (HO-Enz) and the

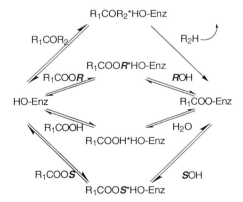

FIG. 2. Schematic representation of the resolution of a racemic alcohol catalyzed by a serine hydrolase. HO-Enz, enzyme with the hydroxy group of the catalytically active serine; R_1COO-Enz, acyl enzyme; R_1COOR, R_1COOS, the two enantiomers of the racemic ester; ROH, SOH, the two enantiomers of the racemic alcohol; R_1COR_2, acyl donor for esterification; R_1COOH, carboxylic acid from the acyl donor; R_2H, leaving group from the acyl donor; *, denotes a complex between ligand and enzyme.

acyl enzyme (R_1COO-Enz) are emphasized.[37] There are two competitive situations in the scheme, one around the enzyme and another around the acyl enzyme. Both these points are important for the enantiomeric outcome of a resolution. An intrinsic high enantioselectivity of the lipase can be ruined if the wrong reaction system is chosen because most reaction steps in the scheme are reversible.

Hydrolysis. The most simple situation is the hydrolysis of a racemic ester (R_1COO***R*** and R_1COO***S***). In that case the acyl donor, R_1COR$_2$, is not present and the scheme in Fig. 2 gets reduced. The equilibrium can be shifted toward hydrolysis by keeping the pH high and titrate away the formed acid (R_1COOH). Soap formation by the addition of Ca^{2+} shifts the equilibrium further. A problem, which increases during the conversion of the substrate, is that the formed alcohol (e.g., ***R***OH being the fast-reacting enantiomer) may accumulate at the interface and start to compete with water as nucleophile to react with the acyl enzyme. The effect is a decrease in the enantioselectivity at high conversions as the fast-reacting ester is reformed and has to compete with the slow-reacting ester once more for the free enzyme.

Esterification. The kinetic scheme for the esterification of an alcohol with a free acid is the same as that of hydrolysis, but this time the water activity of the system should be kept low.

The complete scheme in Fig. 2 describes the esterification of a racemic alcohol (***R***OH and ***S***OH) using an activated acyl donor (R_1COR$_2$). Water is an important part the system, both as substrate and protein ligand, even if this type of reaction is run under low water activity conditions. The reaction is driven toward esterification of the alcohol by removing the leaving group of the acyl donor (R_2H). The use of vinyl esters relies on tautomerization of the formed vinyl alcohol to acetaldehyde, which makes that part of the scheme virtually irreversible. Other possibilities are to use ethyl or thioethyl esters and evaporate the formed ethanol and ethanethiol. These acyl donors usually afford higher initial rates than if the free acid is used. The different acyl donors have different advantages and disadvantages.[37] Vinyl esters are usually good acyl donors, but they are too reactive to be used with amines. In addition, the formed acetaldehyde will react with free amino groups of the substrate or the enzyme. The lipase might get inactivated or change enantioselectivity as a consequence of such a reaction.[38]

For some lipases ethyl esters are good substrates. These esters are inert, but have the drawback that the formed ethanol is a good nucleophile and

---

[37] N. Öhrner, M. Martinelle, A. Mattson, T. Norin, and K. Hult, *Biocatalysis* **9**, 105 (1994).
[38] H. K. Weber, H. Stecher, and K. Faber, *Biotechnol. Lett.* **17**, 803 (1995).

will act as a potent product inhibitor by competing for the acyl enzyme with the racemic alcohol. Ethanol can be removed by running the reaction under reduced pressure, which can be combined with a control of the water activity.

Thioesters are versatile acyl donors because the formed thiols have low boiling points and can easily be evaporated. The formed thiols act as very poor nucleophiles for the acyl enzyme. Thioesters can be used to advantage in transesterification of volatile alcohols, but are too reactive to be used with amines. The only drawback with thioesters is that their K_m values are high, which has two consequences: it is hard to saturate the enzyme and they do not compete as well with the formed ester products for the free enzyme. An excellent mathematical treatment of lipase-catalyzed resolutions has been published by Straathof et al.[39]

Procedures for the resolution of chiral alcohols are given later, with rac-2-octanol as an example. The procedures are general and should be useful for the resolution of a number of chiral alcohols with any lipase.

Resolution of Racemic Alcohols by Hydrolysis of Racemic Esters

Emulsify by sonification 1 mmol of rac-2-octyl octanoate in 10 ml 0.2 M CaCl$_2$ containing gum arabic (5%, w/v). Adjust pH to 7.5 and add 50 units (0.1 mg) of C. antarctica lipase B dissolved in 0.1 ml 10 mM TES-HCl buffer, pH 7.5. Run the reaction of 25°. Maintain pH at 7.5 and follow the rate of hydrolysis of the ester by automatic addition of 1 M NaOH using a pH-stat (Radiometer pH-stat equipped with an ABU91 autoburette and a VIT90 videotitrator). Stop the reaction close to 50% conversion by the addition of 1 M HCl to pH 1. Extract the produced alcohol and remaining ester with diethyl ether and separate them with flush chromatography on silica gel.

Resolution of Racemic Alcohols by Esterification

In esterification reactions the acyl moiety can be transferred from a number of acyl donors to the chiral alcohol. Here four examples are given, which differ in the leaving group of the acyl donor and how this group is removed from the reaction vessel.

General Procedure. Add to a reaction vessel rac-2-octanol (4 mmol), acyl donor (4 mmol), and 50 mg immobilized C. antarctica lipase B (Novozyme 435, Novo Nordisk A/S, Bagsvaerd, Denmark). Incubate at 40° and follow the esterification of the alcohol by gas chromatography. Solvent is not used if the alcohol is soluble enough in the acyl donor.[37]

---

[39] A. J. J. Straathof, J. L. L. Rakels, and J. J. Heijnen, *Biocatalysis* **7**, 13 (1992).

Octanoic Acid. The free acid affords a low initial reaction rate (10–50% of that of other acyl donors). The low K_m for the acid leads to a high specificity constant, which is beneficial at high conversions because the acyl donor competes well with the products for the free lipase. The result is an only slightly longer reaction time than with the other acyl donors. The water activity can be controlled and the produced water evaporated by applying reduced pressure to the reaction mixture.

Ethyl Octanoate. Ethyl octanoate is a good and cheap acyl donor, but the product ethanol is also a very good nucleophile, which competes with the substrate alcohol for the acyl enzyme. To achieve high conversions reduced pressure (30–100 mm Hg) must be used to evaporate ethanol. This method has been shown to be useful for the resolution of many alcohols.[37,40]

S-Ethyl Thiooctanoate. S-ethyl thiooctanoate has the benefit of having a leaving group, ethanethiol, which evaporates at room temperature and atmospheric pressure. In addition, ethanethiol is an extremely bad nucleophile with *C. antarctica* lipase B. The only drawback is the high K_m, which makes it difficult to saturate the enzyme with the acyl donor. The reaction may well be run in an open vessel to allow the evaporation of the thiol. S-ethyl thiooctanoate usually affords higher optical purities than ethyl octanoate.[41] Another useful thioester is S-methyl thiobutyrate. It is commercially available and has a very volatile leaving group.

Vinyl Octanoate. Vinyl esters have been used for many years as acyl donors.[42] They afford a good displacement of the equilibrium toward product formation. The selectivity is usually good with vinyl esters,[37] but there might be a decrease in selectivity if the vinyl ester is used as the solvent.[43]

Resolution of Acids

Lipase-catalyzed resolution of racemic acids (Fig. 3) is kinetically more complicated than the resolution of racemic alcohols (Fig. 2). In the resolution of racemic acids, two different acyl enzymes will be formed (**S**COO-Enz and **R**COO-Enz). In addition, both the formation and the breakdown of the acyl enzymes will pass through transition states that might be involved in the discrimination between the two enantiomers. The substrate competitions are centered around three points, the free enzyme and the two acyl enzymes, and not around two points as in the resolution of alcohols. The

---

[40] N. Öhrner, M. Martinelle, A. Mattson, T. Norin, and K. Hult, *Biotechnol. Lett.* **14,** 263 (1992).
[41] H. Frykman, N. Öhrner, T. Norin, and K. Hult, *Tetrahedron Lett.* **34,** 1367 (1993).
[42] M. Degueil-Castaing, B. de Jeso, S. Drouillard, and B. Maillard, *Tetrahedron Lett.* **28,** 935 (1987).
[43] C. Orrenius, T. Norin, K. Hult, and G. Carrea, *Tetrahedron: Asymmetry* **6,** 3023 (1995).

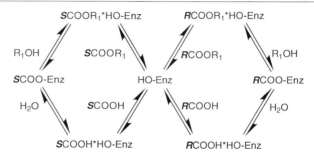

FIG. 3. Schematic representation of the resolution of a racemic acid catalyzed by a serine hydrolase. HO-Enz, enzyme with the hydroxy group of the catalytically active serine; RCOO-Enz, SCOO-Enz, diasteriomeric acyl enzymes; RCOOR$_1$, SCOOR$_1$, the two enantiomers of the racemic ester or acid if R_1 = H; R_1OH, alcohol; *, denotes a complex between ligand and enzyme.

reactions at all these points can influence the enantioselectivity of the catalyzed transformation.

The scheme outlined in Fig. 3 is valid both for the esterification and hydrolysis of racemic acids and esters, respectively, as all reactions are reversible. The scheme is also valid for acyl transfer from one ester to another if the free acids (SCOOH and RCOOH) are changed to esters (SCOOR$_2$ and RCOOR$_2$) and water to an alcohol (R_2OH). The enantioselectivity might be altered by the selection of esters and alcohols.

If, for example, the substrate is a vinyl ester, the acylation of the free enzyme will be essentially irreversible and the attacking alcohol (R_1OH) will not contribute to the enantioselectivity because it will be governed during the formation of the acyl enzyme. The competition around the free enzyme will in this case dominate the enantioselectivity.

If, however, a slow reacting alcohol is combined with a reversible acyl donor (an acid or an ester) so that the two acyl enzymes have time to reach equilibrium over the free enzyme, deacylation becomes important for the selectivity.[44]

Hydrolysis. A lipase-catalyzed hydrolysis reaction can be performed in batch mode in a buffered aqueous solution or in a reaction vessel where the pH is held constant at a preset value by means of a pH-stat instrument. If the enantioselectivity of the enzyme employed in the resolution is very high, the reaction may simply be carried out in a buffered aqueous solution

[44] K. Hult, *in* "Microbial Reagents in Organic Synthesis" (S. Servi, ed.), pp. 289–298. Kluwer Academic Publishers, Dordrecht, The Netherlands, 1992.

because the reaction will virtually stop at a conversion close to 50%. Thus there is no risk of running the reaction past 50% conversion, which will reduce the optical purity of the product. However, if the enantioselectivity of the lipase is lower but still high enough for a particular application, the conversion must be carefully monitored to be able to stop the reaction at the conversion that will give the optimal yield and enantiomeric purity of the product. In this situation a pH-stat device may be instrumental as the conversion can determined in real time from the amount of sodium hydroxide consumed.

In principal, an identical methodology can be used for the resolution of racemic carboxylic acids as described in the previous section. However, from a practical and applied point of view, performing the reaction in organic media may be particularly advantageous when an acid substrate is to be resolved. Due to the amphiphilic nature of alkanoic acids, they are particularly challenging to isolate in high yield from an emulsion. Their surface active properties often promote extensive formation of foam and poor phase separation during extractive workup. Therefore, we concentrate here on procedures for the resolution of chiral acids through esterification in organic media.

Esterification. Here we describe a method that can be used for the resolutions of 2-methylalkanoic acids, examples of useful building blocks for the synthesis of chiral pharmaceuticals, pesticides, fragrances, and pheromones. This method has been successfully used to produce both enantiomers of 2-methyldecanoic[45] and 2-methyloctanoic acid[46] in close to absolute enantiomeric purity in multigram preparative scale. The procedure represents a simple and straightforward alternative to lengthy asymmetric synthesis of these synthons. The method relies on three *Candida rugosa* lipase-catalyzed resolution steps. Note, however, that the steps including the repeated lipase-catalyzed esterification are only necessary if both enantiomers are desired in close to 100% ee. The (*R*)-enantiomer can be obtained in an optically pure state by performing only the first lipase-catalyzed step and running the reaction past 50% conversion.

Procedure. Prepare a cyclohexane solution containing 2-methylalkanoic (2-methyldecanoic or 2-methyloctanoic) acid (0.15 M), 1-alkanol (C18 or C20, respectively; 0.11 M), internal standard (eicosane or tetracosane 6.1 mg/ml solution), and Na_2SO_4/Na_2SO_4; $10H_2O$ (for a_w = 0.8; 0.2 and 0.1

---

[45] P. Berglund, M. Holmquist, E. Hedenström, K. Hult, and H.-E. Högberg, *Tetrahedron: Asymmetry* **4**, 1869 (1993).
[46] H. Edlund, P. Berglund, M. Jensen, E. Hedenström, and H.-E. Högberg, *Acta Chem. Scand.* **50**, 666 (1996).

mmol/ml solution). Stir the mixture for 15 min in a sealed flask at room temperature. Add 20 mg particles of immobilized *C. rugosa* lipase (2 mg crude lipase/mg particles) and stir the reaction mixture by means of a magnetic stirring device. Follow the esterification reaction by gas chromatography. Use, for instance, a Perkin Elmer 5000 instrument equipped with a FFAP-CB column (Chrompack, 25 m, 0.32 mm i.d., $d_f = 0.32$ μm), an on-column injector, and a FID detector. Stop the reaction close to 50% conversion by filtering off the enzyme. The unreacted 2-methylalkanoic acid is extracted from the solution (1 vol) into aqueous sodium carbonate (10%, 5 × 0.2 vol) and saturated sodium chloride solution (0.2 vol). The combined water phase (1 vol) containing the carboxylate salt and some alcohol is acidified to pH 1 with hydrochloric acid (6 M) and extracted with diethyl ether (5 × 0.5 vol). The combined ether phase (1 vol) is washed with saturated sodium chloride solution (0.2 vol), dried (MgSO$_4$) and evaporated to dryness yielding the (*R*)-2-methylalkanoic acid (about 75–90% ee).

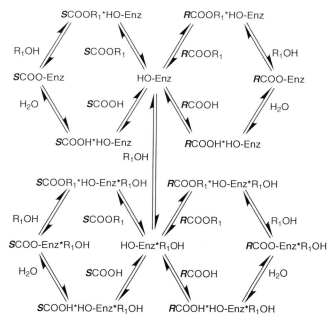

FIG. 4. Schematic representation of the resolution of a racemic acid with enantioselective alcohol inhibition catalyzed by a serine hydrolase. HO-Enz, enzyme with the hydroxy group of the catalytically active serine; *R*COO-Enz, *S*COO-Enz, diastereomeric acyl enzymes; *R*COOR$_1$, *S*COOR$_1$, the two enantiomers of the racemic ester or acid if R$_1$ = H; R$_1$OH, alcohol; *, denotes a complex between ligand and enzyme.

This crude (*R*)-acid should be reesterified, in the second lipase-catalyzed step with a 6 fold excess of the 1-alcohol (1-decanol and 1-dodecanol for 2-methyldecanoic acid and 2-methyloctanoic acid, respectively) to a predetermined conversion (25–45%, depending on the starting ee). This should give the enatiopure (*R*)-2-methylalkanoic acid.

The remaining (*S*)-ester/alcohol mixture from the first lipase-catalyzed step is hydrolyzed in aqueous methanolic potassium hydroxide solution. This step may lead to 1–2% ee loss. The (*S*)-ester is dissolved in 10% KOH in ethanol (0.33 mg ester/ml). The mixture is stirred at room temperature for 3 hr. The solution is diluted with water, acidified with hydrochloric acid, and partitioned between diethyl ether and water. The water phase (1 vol) is extracted with diethyl ether (5 × 1 vol). The combined organic phase (1 vol) is extracted with aqueous sodium carbonate solution (10%, 5 × 1 vol). The pooled carbonate solution (1 vol) is acidified to pH 1 with hydrochloric acid (6 M) and extracted with diethyl ether (5 × 1 vol). The ether phase is dried (MgSO$_4$) and concentrated to give (*S*)-methylalkanoic acid.

The crude (*S*)-acid obtained is used directly in an additional lipase-catalyzed esterification reaction performed in the same manner as that described earlier. Run the reaction to 50–70% conversion, which should yield an enantiopure (*S*)-ester. This ester can be transformed to (*S*)-2-methyl-1-alcohol in unchanged ee using lithium aluminum hydride. If needed this alcohol can be converted to the (*S*)-2-methylalkanoic acid in high yield and unchanged ee via oxidation with Jones reagent.[45]

Enantioselective Inhibition. Alcohol substrate inhibition is common in lipase kinetics.[9] In the general case such an inhibition will just decrease the reaction rate, but if the alcohol binds so that one enantiomer is inhibited more than the other, the inhibition becomes enantioselective. The CRL-catalyzed esterification of 2-methyl alkanoic acids was shown to follow the scheme in Fig. 4. The substrate alcohol can bind to the free enzyme to form an enzyme alcohol complex (HO-Enz*R$_1$OH), which is active but with another enantioselectivity than the free enzyme.[47] Molecular modeling was used to give a molecular understanding of the process.[29]

---

[47] P. Berglund, M. Holmquist, K. Hult, and H.-E. Högberg, *Biotechnol. Lett.* **17,** 55 (1995).

[19] Ester Synthesis via Acyl Transfer (Transesterification)

By P. ANDERSCH, M. BERGER, J. HERMANN, K. LAUMEN, M. LOBELL, R. SEEMAYER, C. WALDINGER, and M. P. SCHNEIDER

1. Introduction

In the literature of the last decade, ester hydrolases and, in particular, microbial lipases have found widespread application as biocatalysts for the solution of synthetic organic problems.[1,2] Due to their role in nature as catalysts for the hydrolysis of triglycerides (fats and oils), lipases are ubiquitous in the world of mammals, plants, and microorganisms such as bacteria and fungi. Many microbial lipases are used for various industrial applications and are therefore produced in considerable quantities. Synthetic organic chemists have greatly profited from this fact, and numerous lipases are commercially available from a variety of sources. Although lipases have been designed by nature for the hydrolysis of triglycerides, they are also capable of catalyzing the selective hydrolysis and synthesis of unnatural esters with widely differing structures. This was also the basis for their widespread application in organic synthesis. As a further advantage for synthetic applications, many microbial lipases have proved to be highly stable in a variety of media ranging from water and various buffer systems to numerous organic solvents. Their application in aprotic organic solvents allows the catalysis of esterifications—the topic of this chapter—thereby greatly enhancing the synthetic scope of these enzymes.

Lipases derived from thermophilic microorganisms are becoming increasingly available. They show high thermostability, which allows them to be used at rather elevated temperatures. Last, but not least, lipases frequently display in their reactions high enantioselectivities and this, in combination with the synthetically very desirable high substrate tolerance, makes them particularly useful for the preparation of enantiomercially pure compounds such as carboxylic acids, esters, derivatives of dicarboxylic acids, primary and secondary alcohols, and diols and triols of various structures. Frequently their explicit regioselectivities allow the preparation of selec-

---

[1] L. Poppe and L. Novak, *Chap. 3 in* "Selective Biocatalysis." VCH, Weinheim, Germany, New York, 1992.

[2] K. Drauz and H. Waldmann, "Enzyme Catalysis in Organic Synthesis: A Comprehensive Handbook," Vols. I and II. VCH, 1995; D. H. G. Crout and M. Christen, "Modern Synthetic Methods," Vol. 5. Springer Verlag, Germany, 1989.

tively functionalized, multifunctional molecules such as glycerol and polyols including carbohydrates, nucleosides, and inositols. In summary, microbial lipases have become versatile and well-established synthetic tools in the hands of organic chemists. Their convenient commercial availability, high stability toward temperature and solvents, their general ease of handling, broad substrate tolerance, and high enantio- and regioselectivities have all added to their widespread use in organic synthesis.

2. Sources of Lipases

Due to their biological role in the hydrolytic degradation of natural fats and oils, lipases are ubiquitous in practically all types of organisms, such as plants, mammals, and microorganisms. In view of well-established fermentation processes, microbial lipases have developed—possibly with the exception of porcine pancreatic lipase—into the most widely used biocatalysts for synthetic organic purposes. Lipases are extracellular enzymes and are therefore excreted into the surrounding fermentation medium from which they can be isolated after removal of the biomass (whole cells) by simple precipitation processes. Due to their widespread industrial application—ranging from food processing to additives of detergents—commercially available lipases are usually available as crude mixtures in which the desired biocatalyst represents only a small fraction. Experience has shown, however, that these impurities—proteins or others—pose no serious problems for most synthetic applications. Due to the commercial availability of certain lipases and based on the observation that synthetic organic chemists showed only little interest in the isolation of their "own" lipases, the use of these biocatalysts was concentrated on a rather limited number of enzymes available from established enzyme producers such as Amano, Boehringer Mannheim, Fluka, Gist Brocades, Novo, and Sigma. Recently, several producers of "enzyme kits" have entered the marketplace, clearly indicating the increased interest in the application of these biocatalysts.

3. Modes of Application

Due to their high stability and ease of handling, most lipases can be used in a variety of forms for example, as native enzymes, using batch or membrane reactors; immobilized covalently, adsorptive on ion exchange resins, hollow fibers; as cross-linked enzyme crystals; chemically modified, for example, with PEG; and in various reaction media: aqueous, biphasic, and organic solvents.

Advantageously and thus greatly facilitating their widespread application in the synthetic organic laboratory, the usually available crude enzyme preparations can be handled just as any other chemical reagent and added directly to the corresponding reaction mixture. Simple batch reactors, often just flasks made from standard laboratory glassware, are frequently used for this purpose. The biocatalyst is usually discarded during workup. Clearly this is not a very economical way of operation and only tolerable in small-scale laboratory experiments and/or with less valuable biocatalysts. For larger scale operations or in applications with expensive biocatalysts, their reuse must be achieved in one of the following ways: (1) immobilization, (2) entrapment in hollow fibers, or (3) entrapment in a membrane reactor.

In contrast with many other enzymes where immobilizations using covalent bond-forming reactions (e.g., using Eupergit, Röhm GmbH, Darmstadt, Germany) have been applied successfully, these methods seem to be only moderately successful using microbial lipases. This is most likely due to the highly hydrophobic nature of the accessible surface of these enzymes, providing insufficient functionalities for chemical transformations. Thus most immobilized enzyme preparations are prepared using adsorptive techniques, with diatomaceous earths or ion-exchange resins (e.g., Duolite, Röhm GmbH, Darmstadt, Germany) being among the more frequently used carriers. Although this latter method is completely sufficient for the application of immobilized lipases in organic solvents, their use in aqueous media is often accompanied with considerable loss of enzyme activity due to washout.

Recently Amano Pharmaceuticals (Nagoya, Japan) introduced a new lipase preparation derived from *Pseudomonas cepacia* immobilized on chemically modified ceramics (Toyonite 200). This preparation is stable both in aqueous and organic media. Columns, and especially loop reactors using these immobilized enzymes, have proved to be highly effective mainly for esterification reactions (see Section 5.2.1.2).

Another interesting, recent approach is the preparation of so-called "cross-linked enzyme crystals" using microcrystalline enzymes and glutardialdehyde as linking agent.

From an industrial point of view it is advantageous to combine lipase-catalyzed reactions with facile methods for the product separation, thereby retaining the biocatalyst. Entrapment of the lipase in hollow fibers or the use of membrane reactors have been the methods of choice.

4. Lipases: Mechanism of Catalysis

Lipases catalyze both the hydrolysis and esterification of a wide variety of structurally different esters, and both reaction modes can, in principle, be used in the above sense. In this chapter, lipase-catalyzed esterifications are the main topic, and the key for an understanding of the issues described therein resides in the mechanism of lipase-catalyzed transformations in general.

From the available structural data,[3] we know that the active site of lipases consists of a catalytic triad formed by the amino acids Glu/Asp–His–Ser. Exemplified here by *Rhizomucor* lipase (RmL) the catalytic triad consists of aspartic acid (Asp 203), histidine (His 257), and serine (Ser 144). This is in complete agreement with the general situation found in so-called serine proteases, and thus the general mechanism of lipase catalysis can be formulated as outlined in Fig. 1.

The combined action of the Asp–His moiety causes an increase of electron density on the primary hydroxyl group of serine, thus allowing a nucleophilic attack on the carbonyl groups in functions such as CO_2H, CO_2R, and CONH. There is now sufficient evidence that the stabilization of the primarily formed tetrahedral intermediate is essential for this initiating step of the catalytic cycle. In case of RmL, this stabilization is provided in the so-called oxyanion hole by hydrogen bonding to the amino acid residues Ser-82 and Leu-145 (Fig. 2).[4]

The importance of the oxyanion hole is further underlined by the fact that partial exchange of amino acids (Asp–Glu) or even the removal of amino acids in the active site by point mutations is tolerated to a certain degree (with loss of specific activity) so long as the oxyanion hole remains intact.[5]

Collapse of the tetrahedral intermediate by loss of the corresponding leaving group (ROH, RNH) leads to the so-called acyl enzyme, the central intermediate of the catalytic cycle (see Fig. 1). Here the acyl group is covalently attached to the serine function of the enzyme molecule. It can be considered an *activated ester*, which can be subject to nucleophilic attack by various nucleophiles. Next to H_2O, alcohols (ROH), various other nucleophiles such as hydrogen peroxide (H_2O_2),[6] ammonia (NH_3),[7] amines,[8] and oximes[9] can be employed (Fig. 1). In organic solvents, that is, in the

---

[3] B. Rubin, *Struct. Biol.* **1**, 568 (1994).
[4] A. M. Brzozowski, U. Derewenda, Z. S. Derewenda, G. G. Dodson, D. M. Lawson, J. P. Turkenburg, F. Björkling, B. Huge-Jensen, S. A. Patkar, and L. Thim, *Nature* **351**, 491 (1991).
[5] P. Carter and J. A. Well, *Proteins: Struc. Func. Genet.* **7**, 335 (1990); D. R. Corey and C. S. Craik, *J. Am. Chem. Soc.* **114**, 1784 (1992).
[6] F. Björkling, H. Frykman, S. E. Godtfredsen, and O. Kirk, *Tetrahedron* **48**, 4587 (1992).
[7] R. A. Sheldon, M. C. de Zoete, A. C. Kock-van Dalen, and F. van Rantwijk, *Biocatalysis* **10**, 307 (1994); R. A. Sheldon, M. C. de Zoete, A. C. Kock-van Dalen, and F. van Rantwijk, *J. Chem. Soc., Chem. Commun.* 1831 (1993).
[8] V. Gotor, R. Brieva, C. González, and F. Rebolledo, *Tetrahedron* **47**, 9207 (1991); V. Gotor, E. Menéndez, Z. Mouloungui, and A. Gaset, *J. Chem. Soc., Perkin Trans. I* 2453 (1993).
[9] V. Gotor and E. Menéndez, *Synlett.* 699 (1990).

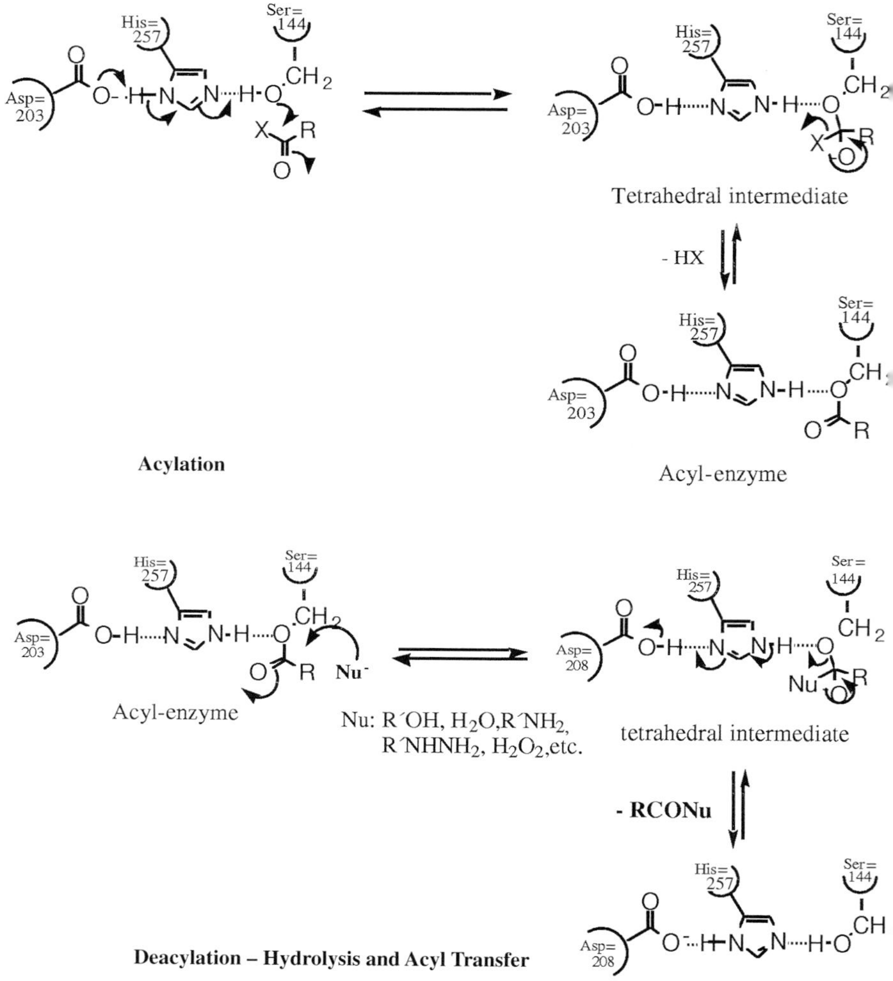

Acylation

Deacylation – Hydrolysis and Acyl Transfer

FIG. 1. Lipases: mechanism of catalysis.

absence of appreciable amounts of water, various acyl transfer reactions can be carried out, including lactone formations,[10–12] peptide syntheses,[13] and polycondensations.[14]

[10] A. L. Gutman, K. Zuobi, and T. Bravdo, *J. Org. Chem.* **55,** 3546 (1990).
[11] H. Yamada, S. Ohsawa, T. Sugai, and H. Ohta, *Chem. Lett.* 1775 (1989).
[12] M. Lobell and M. P. Schneider, *Tetrahedron: Asymmetry* **4,** 1027 (1993).
[13] A. J. Margolin and A. M. Klibanov, *J. Am. Chem. Soc.* **109,** 3802 (1987).
[14] J. S. Wallace and C. J. Morrow, *J. Polym. Sci. Part A: Polym. Chem.* **27,** 2253 (1989).

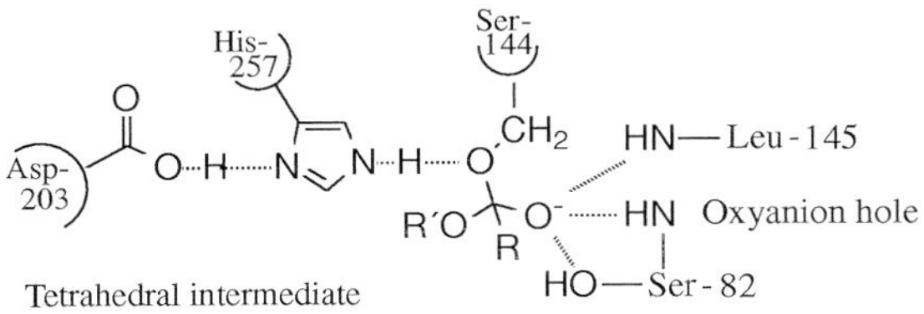

FIG. 2. Lipase catalysis: stabilization of the tetrahedral intermediate, the oxyanion hole (see Fig. 1).

4.1. Lipase-Catalyzed Reactions

From the above-described mechanistic considerations, four types of reactions can be derived that cover the majority of lipase-catalyzed transformations useful for organic synthetic applications (Fig. 3). Figure 3a shows ester hydrolysis and ester synthesis by direct esterification; 3b shows ester

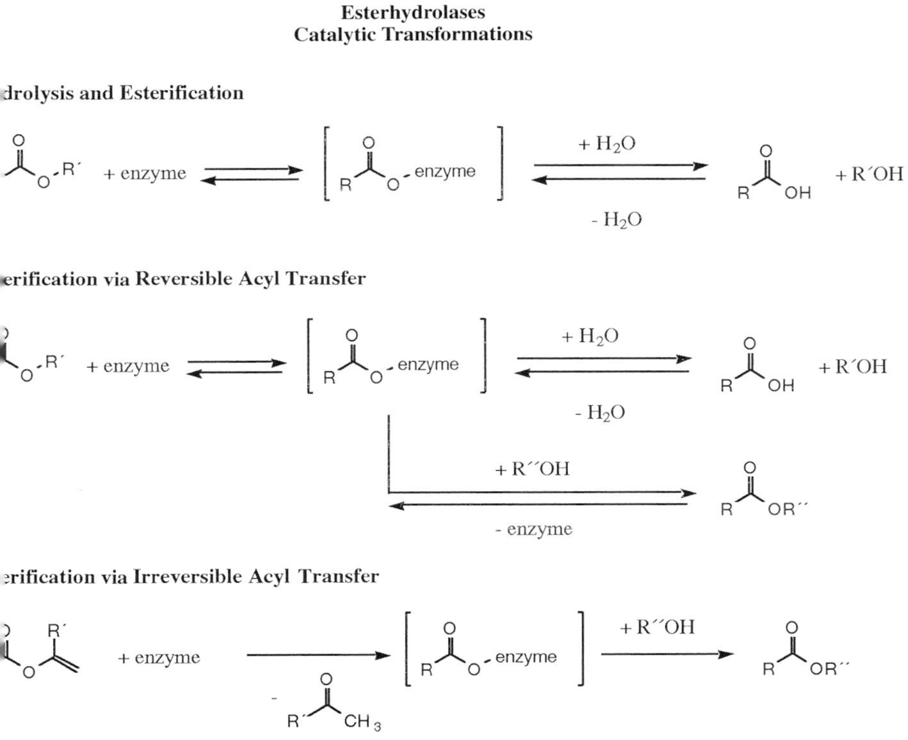

FIG. 3. Lipase-catalyzed reactions.

synthesis via reversible acyl transfer; and 3c shows ester synthesis via irreversible acyltransfer.

4.1.1. Hydrolysis and Esterification. In aqueous media, with a large excess of water present, ester hydrolysis is clearly the dominating reaction. However, under low water conditions such as in nearly anhydrous aprotic solvents, the reverse reaction, that is, esterification, can be achieved. If the water content of the medium can be controlled effectively and produced water is removed by either applying vacuum or drying agents (e.g., molecular sieves) relatively good product yields can be obtained.

4.1.2. Reversible Acyl Transfer. Esterifications by enzymatic acyl transfer are particularly attractive for synthetic applications in organic syntheses since no water is involved. If a suitable donor ester (see discussion of acyl donors later), which often also serves as the solvent, is used in excess, the acyl groups from the acyl enzyme intermediates can be successfully transferred onto various hydroxy compounds such as primary and secondary alcohols, diols, triols, carbohydrates, and nucleosides. The success of such acyl transfer reactions depends largely on the nucleophilicity of the acceptor alcohol R″OH (Fig. 3b). On inspection of Fig. 3b it is evident that the desired transfer can be achieved efficiently only if the initially eliminated alcohol R′OH is less nucleophilic than the substrate alcohol R″OH or at least of comparable nucleophilicity. Thus ester syntheses of primary alcohols can indeed be achieved using simple acyl donors such as EtOAc, MeOAc, or tributyrin, albeit at relatively low reaction rates. However, little or no esterification takes place with secondary alcohols or other substrates of low nucleophilicity. In other words, enzymatic esterifications by acyl transfer only proceed with acceptable rates when the reverse reaction of the acyl enzyme with R′OH can be minimized. This problem can be overcome in one of the following ways: (1) by continous removal of the liberated alcohol R′OH (e.g., MeOH) by applying vacuum or using molecular sieves or (2) by using esters as acyl donors in which the alcohol moiety R′OH is of extremely low nucleophilicity, such as trichloroethanol.[15]

4.1.3. Irreversible Acyl Transfer.[16] Significantly higher rates of esterification are achieved using the irreversible reaction mode (Fig. 3c) in which the acyl enzyme is produced without the chance of a back reaction. The use of enol esters for this purpose was first described in 1987 for a lipase-catalyzed esterification although without an enantioselective application.[17] Here, the liberated (nucleophilic) enol is tautomerizing immediately into

---

[15] G. Kirchner, M. P. Scollar, and A. M. Klibanov, *J. Am. Chem. Soc.* **107,** 7072 (1985).
[16] K. Faber and S. Riva, *Synthesis* 895 (1992).
[17] M. Degueil-Castaing, D. De Jeso, S. Drouillard, and B. Maillard, *Tetrahedron Lett.* **28,** 953 (1987).

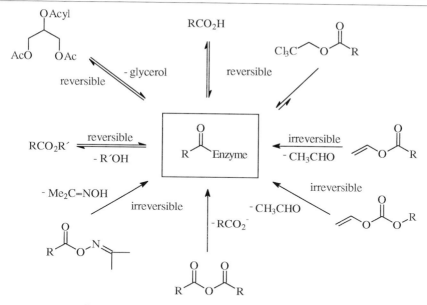

FIG. 4. Acyl donors for lipase-catalyzed esterifications.

the corresponding (electrophilic) carbonyl compound, for example, acetaldehyde or acetone. The thus irreversibly produced acyl enzyme has, in the absence of water, only the possibility of reacting with the sole nucleophile present—the alcohol to be acylated. By using this reaction mode, reaction rates are increased by a factor of 10 to 1000 as compared to reversible acyl transfer reactions.[18] As an alternative to enol (vinyl, isopropenyl) esters, anhydrides,[19] oxime esters,[9] and vinyl carbonates[20] have been used for the same purpose. Examples for frequently used acyl donors, both for reversible and irreversible acyl transfer, are summarized in Fig. 4.

4.1.3.1. SYNTHESES OF VINYL ESTERS. Enol esters and vinyl esters are probably the most widely employed acyl donors for lipase-catalyzed enantio- and regioselective transformations. In this section, we describe the synthesis of these molecules in detail.

The most economical and convenient method for the preparation of vinyl esters proved to be the transvinylation of free carboxylic acids in the

---

[18] J.-M. Fang and C.-H. Wong, *Synlett.* 393 (1994).
[19] S. Ozaki and L. Ling, *Tetrahedron Lett.* **34**, 2501 (1993).
[20] M. Pozo, R. Pulido, and V. Gotor, *Tetrahedron* **48**, 6477 (1992).

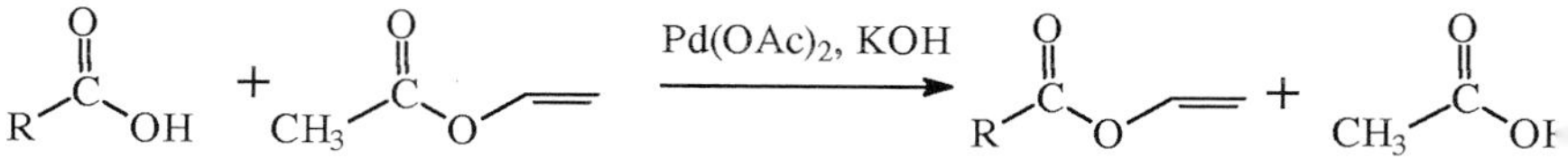

$$\underset{R}{\overset{O}{\|}}\underset{OH}{\overset{}{C}} + \underset{CH_3}{\overset{O}{\|}}\underset{O}{\overset{}{C}}\diagup\!=\xrightarrow{Pd(OAc)_2, KOH} \underset{R}{\overset{O}{\|}}\underset{O}{\overset{}{C}}\diagup\!= + \underset{CH_3}{\overset{O}{\|}}\underset{OH}{\overset{}{C}}$$

FIG. 5. Synthesis of fatty acid vinyl esters.

presence of vinyl acetate catalyzed by Pd(OAc)$_2$ (Fig. 5).[21] The method is illustrated for three different types of substrates: saturated, unsaturated, and hydroxylated carboxylic acids.

4.1.3.1.1. Synthesis of vinyl myristate: experiment. A mixture of 20.0 g (88 mmol) of myristic acid, 200 ml (2.16 M) of vinyl acetate, 200 mg of Pd(OAc)$_2$, and 1.0 g (1.8 mmol) of KOH was stirred at room temperature while the reaction progress was monitored via thin layer chromatography (TLC) (hexane : ether: 8 : 2; R$_F$, 0.73). After 24-hr the reaction was complete. The mixture was filtered and the residue washed with 20 ml vinyl acetate. After evaporation of the vinyl acetate *in vacuo* the crude reaction product was purified by flash chromatography (silica gel; n-heptane). Obtained were 16.82 g (75%) of vinyl myristate (m.p. 23°).[22]

4.1.3.1.2. Synthesis of vinyl oleate: experiment. Oleic acid [2.82 g (10 mmol)] was dried over 0.3 g 3 Å molecular sieves and then dissolved in 50 ml vinyl acetate. Ten mg Pd(OAc)$_2$ and 5.6 mg (0.1 mmol) KOH were added. The reaction mixture was stirred at room temperature in the dark under an inert gas atmosphere (Ar). The conversion was followed by TLC. After 24 hr the reaction mixture was filtered and the unreacted vinyl acetate evaporated *in vacuo*. The crude product (95% yield) was filtered over a short silica gel column and further purified by flash chromatography on silica gel (eluent, n-heptane : t-BuOMe, 9 : 1), resulting in 2.87 g (93%) of vinyl oleate with a chemical purity of 99%.[23]

4.1.3.1.3. Synthesis of vinyl-12-hydroxydodecanoate: experiment. A mixture of 2.16 g (10 mmol) of 12-hydroxydodecanoic acid, 100 ml (1.08 M) of vinyl acetate, 350 mg of Pd(OAc)$_2$, and 56 mg (1 mmol) of KOH was stirred at room temperature while the reaction progress was monitored via TLC. After 24 hr the reaction was complete. The mixture was filtered and the residue washed with 20 ml vinyl acetate. After evaporation of the vinyl acetate *in vacuo* the crude reaction product was purified by flash chromatography (silica gel; n-hexane : t-BuOMe, 6 : 1). Obtained were 2.06 g (85%) of 12-hydroxydodecanoic acid vinyl ester.[21]

[21] M. Lobell and M. P. Schneider, *Synthesis* 375 (1994).
[22] G. Machmüller, Diploma thesis, Wuppertal (1996).
[23] C. Waldinger and M. P. Schneider, *J. Am. Oil Chem. Soc.* **73**, 1513 (1996).

4.1.4. Synthetic Applications. Microbial lipases are the "reagents" of choice for the selective transformation of hydroxy compounds. This is clearly reflected in the large number of publications describing reactions of this kind. Although all of the above-described reactions (Fig. 3) are potentially useful for organic synthetic applications, in practice only two of them, ester hydrolysis and irreversible acyl transfer using enol esters, are employed routinely. In the following sections, examples of selective esterifications of numerous compounds are described with the main emphasis placed on enantio- and regioselectivity.

From Fig. 3 and the discussions thereafter, it is obvious that lipase-catalyzed esterifications have to be carried out in organic solvents, either aprotic or of low nucleophilicity. With the exception of highly polar solvents like DMF or DMSO, which have been shown to be unsuitable for lipase-catalyzed esterifications (they can, however be used in acyl transfer reactions catalyzed by certain proteases) a wide variety of organic solvents can be employed, ranging from aprotic hydrocarbons of low polarity (*n*-hexane), ethers (*t*-BuOMe, THF), and ketones (acetone), via sterically hindered alcohols (2-methyl-2-butanol), to chlorinated hydrocarbons (CH_2Cl_2, $CHCl_3$) and acetonitrile. Thus, depending on the solubility of the employed substrate, the most suitable solvent can be selected. Note, however, that the rates of esterification can vary widely depending on the solvent used. This aspect is discussed in Chapter [17] in this volume. Another important issue is the control of the water content in such media. This aspect is discussed in Chapter [21] in this volume.

Although, as pointed out earlier, esterifications using enol esters are by far the most widely applied methods, examples of other acyl donors are included below. For a more systematic arrangement of the material, the reactions have been organized according to the substrate structure and the type of selectivity.

4.2. Selectivities of Microbial Lipases

Both the enantioselectivities and regioselectivities of lipases have been widely used for the selective preparation of enantiomerically pure compounds and selectively functionalized or protected molecules.

4.2.1. Enantioselectivities. Microbial lipases, like other esterhydrolases, display essentially three different modes of substrate recognition (schematically represented in Fig. 6): (1) enantiomer differentiation in racemic precursors; (2) enantiotope differentiation in achiral precursors carrying a prochiral center; and (3) enantiotope differentiation in achiral *meso*-compounds.

Enantiomer differentiation

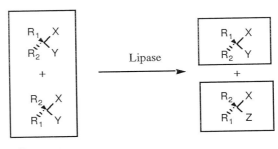

Racemate

Differentiation of enantiotopic groups

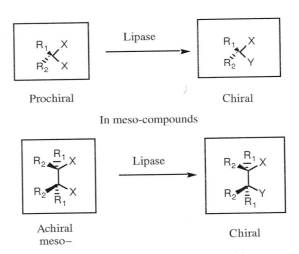

FIG. 6. Lipases: modes of substrate recognition.

4.2.2. Lipase-Catalyzed Resolutions. Starting with racemic substrates, lipase-catalyzed kinetic resolutions can lead to enantiomerically pure diastereomers, which can be separated by classical techniques such as extraction, distillation, and chromatography. Clearly, in these cases, a maximum of 50% of one particular enantiomer can be obtained, and economic ways for recycling of the undesired enantiomer must be developed. The quality of such resolutions depends both on the enantioselectivity of the lipase E *and* the achieved (achievable) conversions. Based on the kinetics of all resolutions in the presence of a chiral auxiliary,[24] Sih has correlated the enantioselectivity of a lipase with the enantiomeric purities of product and substrate at any given conversion.[25] This aspect is discussed extensively in Chapter [17] in this volume.

Only with very high enantioselectivities can both the product and the unreacted substrate be obtained in nearly optically pure form. In case of lower E values, one should remember that only the remaining, unreacted substrate can be obtained optically pure via a conversion of >50%. For this, however, one has to pay with a concomitant loss of chemical yield and a rather low optical yield of the product.

4.2.3. Lipase-Catalyzed Enantiotope Differentiation. Starting with achiral substrates (i.e., molecules without any inherent chirality), enantiotope differentiation of the functional ester groups leads to only one enantiomer with—in theory—100% optical and chemical yield.

The same is true for the conversion of achiral *meso*-compounds in which the differentiation of the enantiotopic functions turns them into one enantiomer, again—in theory—with 100% optical and chemical yield.

4.2.4. Regioselectivities. Lipases also display distinct regioselectivities, which are frequently employed for the selective functionalization of polyols including glycerol, monosachamides, disaccharides, and nucleosides and other natural products. Note that in the case of pure carbohydrates, nucleosides or most natural products enantioselectivity plays no role. The substrates to be functionalized are enantiomerically pure and only selected functions are derivatized.

5. Synthetic Applications

5.1. Esterification of Primary Alcohols

For the selective esterification of primary hydroxy functions, practically all available acyl donors can be employed, albeit with widely differing rates of transformation. As already pointed out, reasonable yields of products can also be obtained using the mode of reversible acyl transfer, however, higher yields are usually obtained using the mode of irreversible acyl transfer with vinyl esters.

Frequently, in organic synthetic protocols, a simple and mild method is required by which hydroxy functions can be esterified under neutral conditions at ambient temperatures and with high efficiency. This can in

---

[24] V. S. Martin, S. S. Woodward, T. Katsuki, Y. Yamada, M. Ikeda, and K. B. Sharpless, *J. Am. Chem. Soc.* **103**, 6237 (1981).

[25] C. J. Sih, C.-S. Chen, Y. Fujimoto, and G. Girdaukas, *J. Am. Chem. Soc.* **104**, 7294 (1982).

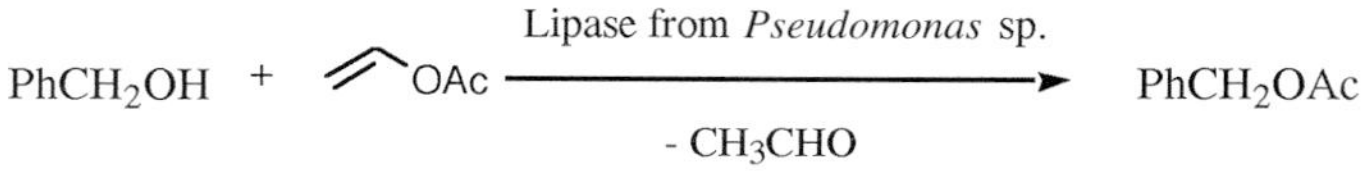

PhCH$_2$OH + $\diagup\!\!\diagdown$OAc $\xrightarrow[-\text{CH}_3\text{CHO}]{\text{Lipase from }Pseudomonas\text{ sp.}}$ PhCH$_2$OAc

FIG. 7. Esterification of benzyl alcohol.

fact be easily achieved using lipase-catalyzed esterifications in the presence of vinyl esters as acyl donors. The method is exemplified here by the esterification of benzyl alcohol (Fig. 7). Under extremely mild conditions, this esterification leads to the product ester in quantitative yield and excellent purity.

5.1.1. Synthesis of Benzylacetate: Experiment. Benzyl alcohol (10.3 g, 0.1 mol) and vinyl acetate (21.5 g, 0.25 mol) are dissolved in 150 ml of *n*-hexane contained in a 250-ml round, bottomed flask. After the addition of 200 mg of a lipase from *Pseudomonas* sp. (SAM-II; 1600 U, measured with a tributyrin standard), the flask is sealed with a polyethylene (PE) stopper and the mixture is stirred at room temperature. The course of the reaction is followed by TLC (SiO$_2$; Et$_2$O:*n*-hexane at 1:4, R_f of benzyl alcohol, 0.12; R_f of benzyl acetate, 0.41). The reaction is complete after 20 hr (benzyl alcohol is no longer detectable by TLC). The enzyme is filtered, washed with *n*-hexane, and stored in a refrigerator for further use. The filtrate is concentrated on a rotary evaporator and the residue is distilled *in vacuo*. The yield of benzyl acetate is 14.7 g (98%); b.p.$_{18}$ is 100°, and the purity is 99.2% (GC).

5.1.2. Primary Alcohols by Kinetic Resolution. It can be stated, in general, that the enantioselectivities displayed by microbial lipases toward primary alcohols and their esters—in which the functionality is removed by one additional methylene group from the center of chirality—are considerably lower when compared to the corresponding transformations of secondary alcohols. Consequently, kinetic resolutions in the classical sense, that is, leading to 50% of each enantiomer, are impossible in most cases.[26] In agreement with the kinetics of such resolutions, only the unconverted substrate can be obtained with high enantiomeric purity. Clearly, if only 20% of product can be obtained, such a transformation can hardly be termed a resolution or useful.[27]

[26] High enantiometric purities were claimed for molecules of this kind with certain substitution patterns: O. Nordin, E. Hedenström, and H.-E. Högberg, *Tetrahedron: Asymmetry* **5**, 785 (1994); D. L. Dellinck and A. L. Margolin, *Tetrahedron Lett.* **31**, 6797 (1990); P. Ferraboschi, P. Grisenti, A. Manzocchi, and E. Santaniello, *J. Chem. Soc., Perkin Trans. I* 1159 (1992).

[27] F. Effenberger and S. Barth, *Tetrahedron: Asymmetry* **4**, 823 (1993).

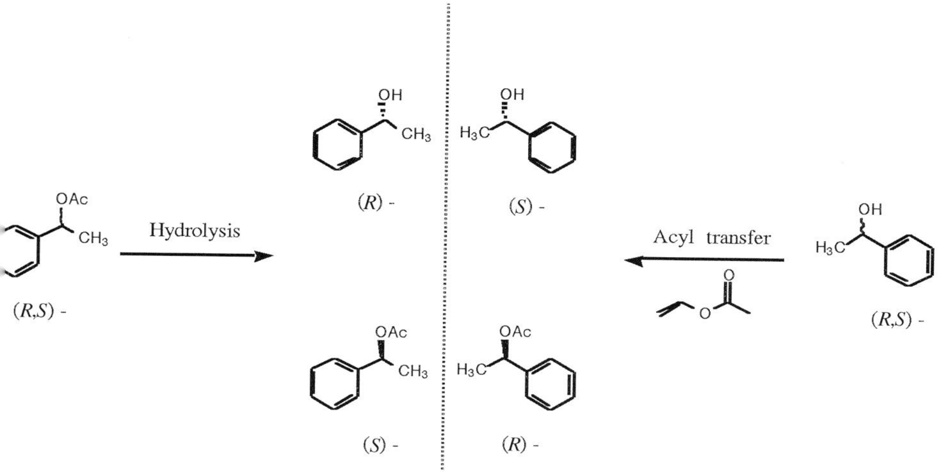

FIG. 8. Lipase-catalyzed hydrolyses and esterifications: stereochemical outcome.

5.2. Secondary Alcohols by Kinetic Resolution

Next to enantioselective reductions of the corresponding ketones by alcohol dehydrogenases, the lipase-catalyzed resolutions of secondary alcohols are the methods of choice for their preparation in enantiomerically pure or enriched form. Probably thousands of such resolutions have been carried out in recent years, and it is beyond the scope of this chapter to give a comprehensive summary of all the structures involved. However, the principle is nearly always the same with certain variations regarding the reaction conditions.

Secondary alcohols can be resolved either by enantioselective (1) hydrolysis of their esters or (2) esterification (Fig. 8). As mentioned earlier, for esterifications the method of irreversible acyl transfer using vinyl or isopropenyl esters has become the method of choice.[17,28,29] The two reactions are complementary regarding the stereochemical outcome: in both reaction modes the lipase recognizes the same absolute configuration in the offered substrate leading to the formation of the corresponding enantiomeric products.

At first, both methods seem equally suitable for the purpose of resolution. However, there are two limiting factors: (1) the widely differing rates

---

[28] C.-H. Wong and H. M. Sweers, *J. Am. Chem. Soc.* **108,** 6421 (1986).
[29] K. Laumen, D. Breitgoff, and M. P. Schneider, *J. Chem. Soc. Chem. Commun.* 1459 (1988); Hoechst AG, Patent EP 032 1918 (20.12.1988).

of transformation in both reaction modes and (2) the concomitant dependence of enantiomeric purities (E values; see Section 4.2.2 and Chapter [17] in this volume). While in the hydrolytic mode, the reactions are rapidly driven to completion both by the large excess of water and the irreversible removal of the produced carboxylic acids under pH-stat conditions; this is not the case in the corresponding esterifications. Earlier studies in our laboratory have shown[30] that beyond a threefold molar excess of vinyl ester generally no increase of reaction rate can be achieved. Even using the acyl donor, for example, vinyl acetate as solvent, does not cause a distinct improvement. To obtain both product and substrate with high optical purity, it is absolutely essential to reach 50% conversion. Since this occasionally requires extensively long reaction times, the method of esterification often only allows (provided the enzyme displays very high enantioselectivity $E \geq 100$) the production of the product (!) ester with high optical purity. Some improvement can be achieved by using immobilized lipases in batch or column[31] operations.

Although numerous molecules with secondary alcohol substructures have been resolved this way in recent years, the number of microbial lipases so far employed is remarkably small. A great advantage for synthetic organic applications, as mentioned earlier, is that many of these biocatalysts show an extremely broad substrate tolerance toward molecules of this type with widely different substitution pattern. With a mere handful of lipases the majority of known transformations was carried out; thus they are ideal catalysts for the described purposes.

5.2.1. Enantioselective Resolutions of Secondary Alcohols by Esterification: Irreversible Acyl Transfer. Enantiomerically pure secondary alcohols with variable molecular structure are useful chiral auxiliaries and building blocks for various analytical and synthetic applications.[31,32] Lipase-catalyzed resolutions of these materials have been found to be highly attractive alternatives to conventional methods due to the extremely high enantioselectivities displayed by these enzymes. In the examples that follow a number of such resolutions are described using esterifications via acyl transfer.[29]

5.2.1.1. (R)- AND (S)-PHENYLETHANOL: EXPERIMENT. Dissolve 1.22 g (10 mmol) of (R,S)-1-phenylethanol (freshly distilled and free of acetophenone) and 2.6 g (30 mmol) of vinyl acetate in 15 ml of t-BuOMe (Fig. 9). After the addition of 200 mg lipase from *Pseudomonas* sp. (SAM-II; 1600 units, standard: tributyrin) this mixture is stirred at room temperature while the

---

[30] K. Laumen and M. Schneider, unpublished (1987).
[31] R. Seemayer, PhD thesis, Wuppertal, 1992; K. Laumen, R. Seemayer, and M. P. Schneider, *J. Chem. Soc. Chem. Commun.* 49 (1990).
[32] K. Laumen and M. P. Schneider, *J. Chem. Soc. Chem. Commun.* 598 (1988).

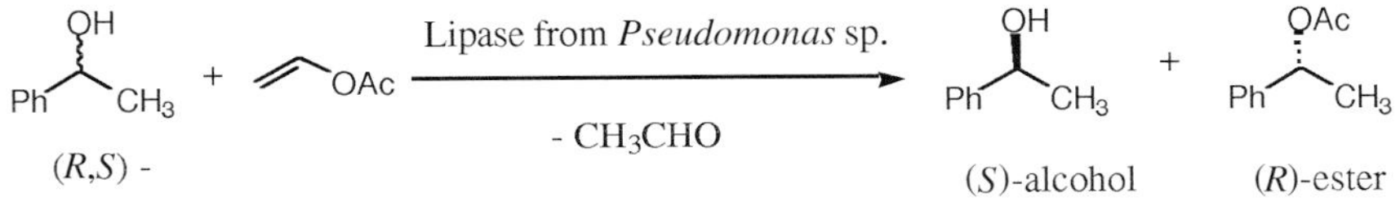

FIG. 9. Lipase-catalyzed resolution of (R,S)-1-phenylethanol.[29]

reaction is monitored by GC. The desired conversion of 50% is usually reached after 48 hr. The enzyme is filtered off (#3 fritted-glass filter) and stored in a refrigerator for further use. After the solvent and excess vinyl acetate are removed, the remaining mixture consists of (S)-1-phenylethanol and (R)-1-acetoxy-1-phenylethane. These two products are separated by column chromatography (SiO$_2$; Et$_2$O:hexane at 1:4). Recovered are 0.58 g (4.8 mmol, 48%) of (S)-1-phenylethanol {R_f 0.08; $[\alpha]_D^{20} = -45°$, $c = 5$, MeOH, 99% enantiomeric excess} and 0.78 g (4.7 mmol, 47%) of (R)-1-acetoxy-1-phenylethane {R_f 0.48; $[\alpha]_D^{20} = 114°$, $c = 2$, MeOH, 99% ee}. The (R)-1-acetoxy-1-phenylethane was hydrolyzed chemically (0.3 g K$_2$CO$_3$ in 10 ml of MeOH for 1 hr) leading, after conventional work, to 0.5 g (R)-1-phenylethanol {$[\alpha]_D^{20} = 45.5°$, $c = 5$, MeOH, 99% enantiomeric excess}.

The preceding experimental procedure has been checked independently[33] and found reproducible in principle with the exception that the remaining alcohol of the above resolution was only obtained with relatively low enantiomeric purity (85–92% ee). That, in turn, can only mean that the enzymatic esterification was not carried to the required conversion of 50%. An inspection of the kinetics clearly reveals that at low conversions only the product (here the R-acetate) is optically pure; the remaining (S)-1-phenylethanol is of lower optical purity. To achieve high optical purities for both product and substrate, the reactions should really be followed by GC to determine the achieved conversion as exactly as possible. We found that improved rates of esterification are achieved if a 3 M excess of vinyl acetate (based on the racemate) is employed.[34]

Secondary alcohols of this type have been the target for similar resolutions by numerous research groups, including ours, whereby the substitution pattern can be varied widely. Thus the aryl group can be replaced by the corresponding pyridyl moiety without any loss of enantioselectivity,[35] nonaromatic substituents can be used so long as they are sufficiently bulky.

---

[33] S. Roberts, ed., in "Preparative Biotransformations." J. Willey & Son, New York, 1993.
[34] R. Seemayer, unpublished; compare K. Laumen, D. Breitgoff, R. Seemayer, and M. P. Schneider, J. Chem. Soc. Chem. Commun. 148 (1989).
[35] R. Seemayer and M. P. Schneider, Tetrahedron: Asymmetry 3, 827 (1992).

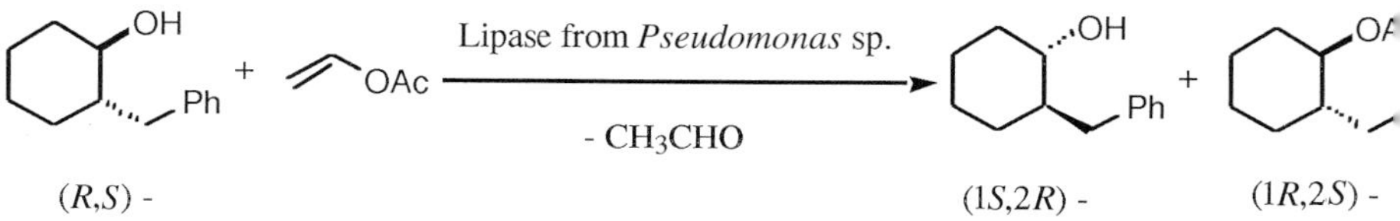

FIG. 10. Lipase-catalyzed resolution of *trans*-2-benzylcyclohexanol.[34]

Last, but not least, the secondary alcohol moiety can even be part of a cyanohydrin group.[36]

5.2.1.2. LIPASE-CATALYZED RESOLUTION OF *trans*-2-BENZYLCYCLOHEX-ANOL. Also alicyclic secondary alcohols can be resolved using in principle the same general method, here exemplified by *trans*-2-benzylcyclohexanol. Secondary alcohols of that general substructure are attractive chiral auxiliaries, and enantiomerically pure phenylcyclohexanol has been used effectively as a substitute for chiral inductions in replacement of 8-phenylmenthol.[37]

Experimental. Dissolve 1.9 g (10 mmol) of (*R,S*)-*trans*-2-benzylcyclohexanol and 2.6 g (30 mmol) vinyl acetate in 15 ml of *t*-BuOMe (Fig. 10). After addition of 400 mg lipase from *Pseudomonas* sp. (SAM-II; 3200 U, standard: tributyrin) the mixture is stirred for 48 hr at room temperature. The enzyme is filtered off, the solvent and excess vinyl acetate are removed *in vacuo*, and the residue [consisting of (1*S*,2*R*)-2-benzylcyclohexanol and (1*R*,2*S*)-1-acetoxy-2-benzylcyclohexanol] is separated by column chromatography (SiO$_2$; Et$_2$O:*n*-hexane at 1:2). Recovered are 0.76 g (40%) of (1*S*,2*R*)-2-benzylcyclohexanol {$[\alpha]_D^{20}$ = 55.5°, c = 1, MeOH, >95% ee} and 0.95 g (41%) of (1*R*,2*S*)-1-acetoxy-2-benzylcyclohexane {$[\alpha]_D^{20}$ = −19.7°, c = 1.2, MeOH, >95% ee}. The desired 50% conversion can only be achieved within reasonable reaction times by strongly increasing the amount of employed enzyme. Clearly, the most convenient way to achieve this is with a continuous column reactor. Consequently, with the same lipase (*Pseudomonas* sp. immobilized on kieselguhr, 5.0 g equal to 900 mg native enzyme), a solution of the reactants [racemic alcohol (30 mmol), *t*-BuOMe (80 ml), and vinyl acetate (90 mmol)] is continuously circulated over the column (loop reactor). Greatly enhanced rates of transformations are achievable this way and the desired 50% conversion is reached three to five times faster as compared to the batch procedure.[31]

Clearly, and as pointed out earlier, the method is not limited to these simple examples. It proved nearly unlimited in its application to numerous

---

[36] Y.-F. Wang, S. T. Chen, K. K-C. Liu, C.-H. Wong, *Tetrahedron Lett.* **30**, 1917 (1989); M. Inagaki, J. Hiratake, T. Nishioka, and J. Oda, *J. Am. Chem. Soc.* **113**, 9360 (1991).

[37] J. K. Whitesell, D. James, and J. F. Carpenter, *J. Chem. Soc. Chem. Commun.* 1449 (1985).

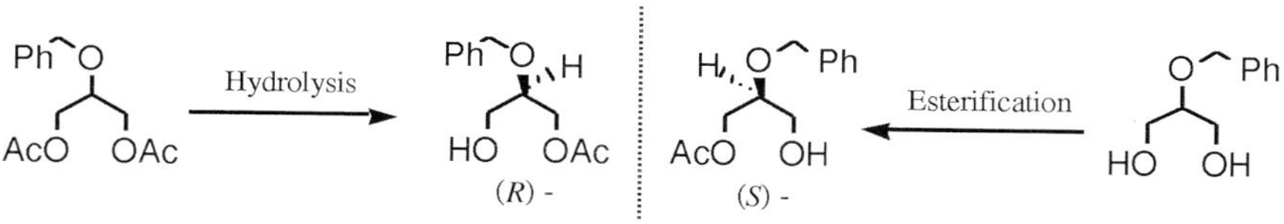

FIG. 11. Enantiotope differentiations in achiral 1,3-diols: stereochemical outcome.

other classes of compounds carrying secondary hydroxy groups on their carbon backbone. Thus numerous other cycloalkanols such as cyclopentanols,[38] tetrahydrofurans,[39] and 1,2-cycloalkanediols[40] were resolved in our laboratory both by enantioselective hydrolyses and esterifications using essentially the same procedure. Also molecules with axial chirality such as 1,1'-binaphthyl-2,2'-diol have been resolved this way.[41]

Given the broad substrate tolerance of many microbial lipases in combination with their frequently high enantioselectivities, together with the fact that chiral or prochiral centers can reside in both components of a given ester, it becomes clear that products of widely different structures are accessible via these routes.

5.3. Monoesters of 1,3-diols: Enantiotope Differentiation

In contrast to the difficulties encountered in the lipase-catalyzed resolution of monofunctionalized primary alcohols, there are numerous examples in the literature for the enantiotope differentiation of such functions in achiral diols carrying a prochiral center. As outlined earlier, and provided the reactions terminate after the (selective) conversion of only one functional group, enantiomerically pure or enriched monoesters of opposite absolute configurations result from the corresponding hydrolyses and esterifications, exemplified in Fig. 11 by the production of enantiomers with glycerol substructures.

Enantiomerically pure glycerol derivatives are of considerable interest as chiral building blocks for the synthesis of a wide variety of biologically active molecules. As an alternative to routes starting from the chiral pool (e.g., D-mannitol, L-serine) the preparation of optically active monoesters derived from 2-O-benzylglycerol was investigated by several research

---

[38] R. Seemayer and M. P. Schneider, *Recl. Trav. Chim. Pays-Bas* **110**, 171 (1991); Y. F. Wang and C. H. Wong, *J. Org. Chem.* **53**, 3127 (1988).
[39] R. Seemayer and M. P. Schneider, *J. Chem. Soc., Perkin Trans I* 2359 (1990).
[40] R. Seemayer and M. P. Schneider, *J. Chem. Soc., Chem. Commun.* 49 (1991).
[41] M. Inagaki, J. Hiratake, T. Nishioka, and J. Oda, *J. Agric. Biol. Chem.* **53**, 1879 (1989).

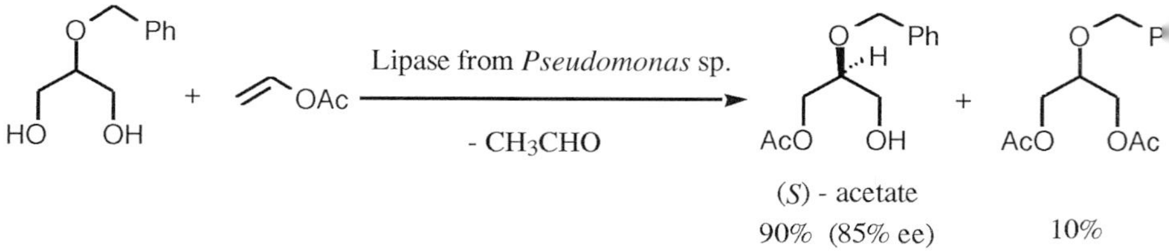

FIG. 12. Enantioselective esterification of 2-O-benzylglycerol.[45]

groups, both using the method of hydrolysis[42] and esterification.[43] In the esterification mode, using the condition of irreversible acyl transfer, optical purities ranging from 88 to 96% ee were reported depending on the reaction conditions. It seems that the selectivity increases with decreasing reaction temperature.[44]

5.3.1. *Enantioselective Esterification of an Achiral 1,3-Diol: 2-O-Benzylglycerol: Experiment.* Dissolve 52 g (0.285 mol) of 2-O-benzylglycerol and 30.7 g (0.357 mol) of vinyl acetate in 250 ml of *t*-BuOMe (Fig. 12). After the addition of 1.5 g lipase from *Pseudomonas* sp. (SAM-II; 12,000 units, standard: tributyrin) the mixture is stirred at room temperature while the reaction progress is monitored by TLC (SiO$_2$; Et$_2$O:*n*-hexane at 1:2; R_f of the diol, 0.05; R_f of the monoacetate, 0.28; R_f of the diacetate, 0.54). After the 2-O-benzylglycerol is completely esterified (62 hr), the enzyme is filtered off (recovery: 1.4 g) and the filtrate is concentrated on a rotary evaporator. Recovered was 64 g of a mixture of the mono- and diacetate in a 9:1 ratio as determined by gas chromatography using a 20-m OV-17 column at 180°. The retention time of the monoacetate under these conditions is 9.3 min, and the R_T of the diacetate is 12.4 min. After fractional distillation 57 g (90%) of (S)-1-acetoxy-2-O-benzylglycerol was obtained {[α]$_D^{20}$ = −16.3°, *c* = 1.7, CHCl$_3$ stabilized with amylene; 85% ee}.

5.3.2. *Esterification of Achiral meso-Diols: Enantiotope Differentiation.* Achiral diols with the *meso* configuration can be converted into enantiomerically pure or enriched monoesters either by lipase-catalyzed hydrolysis or esterification. Using the same lipase, both approaches are complementary regarding the stereochemical outcome and lead to opposite enantiomers. Thus achiral 1,2-cycloalkanedimethanols can be converted in one step into

---

[42] D. Breitgoff, K. Laumen, and M. P. Schneider, *J. Chem. Soc. Chem. Commun.* 1523 (1986); V. Kerscher and W. Kreiser, *Tetrahedron Lett.* **28**, 531 (1987).
[43] Y.-F. Wang, J. J. Lalonde, M. Momongan, D. E. Bergbreiter, and C.-H. Wong, *J. Am. Chem. Soc.* **110**, 7200 (1988); N. Baba, K. Yoneda, S. Tahara, J. Iwase, T. Kaneko, and M. Matsuo, *J. Chem. Soc. Chem. Commun.* 1281 (1990).
[44] Y. Terao, M. Murata, and K. Achiwa, *Tetrahedron Lett.* **29**, 5173 (1988).
[45] Hoechst AG, Patent DE 3,624,703 (22.7.1986).

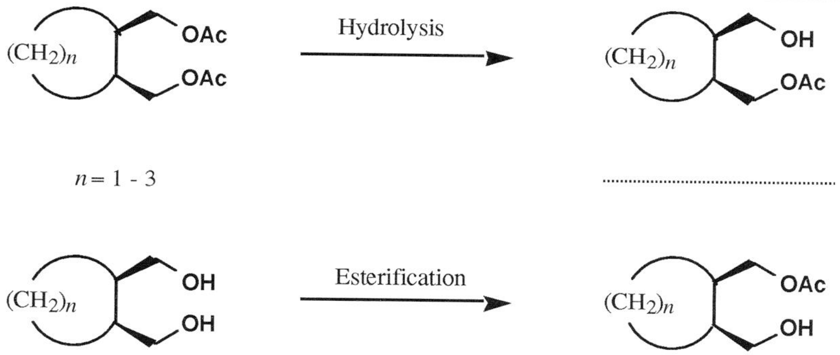

FIG. 13. Enantiotope differentiations in *meso*-diols.

the single enantiomers of opposite absolute configuration and comparable optical purity (Fig. 13).

While such esterifications would now be carried out normally via the method of irreversible acyl transfer using vinyl esters, also in this case the method of reversible acyl transfer can be employed. As already outlined, this is only successful if the nucleophilicity of the hydroxy groups is sufficiently high, which is true for the primary hydroxy groups in such substrates. Nevertheless, as is clear from the time dependence of the reactions—using both EtOAc and MeOAc as acyl donors (Fig. 14)—such transformations require considerably extended reaction times as compared with the method of irreversible acyl transfer, and cannot be driven to completion within reasonable reaction times.

5.3.2.1. SYNTHESIS OF (1S,2R)-1-ACETOXYMETHYL-2-HYDROXYMETHYLCYCLOBUTANE: EXPERIMENT.[46] Dissolve 0.581 g (5 mmol) of 1,2-dihydroxymethylcyclobutane in 20 g of EtOAc (MeOAc) to which 200 mg of lipase from *Pseudomonas* sp. (SAM-II) has been added. The mixture is stirred at room temperature, while the reaction progress is monitored by GC (see Fig. 14). After the reaction comes to a near standstill (6 days) the resulting suspensions are filtered and the residue washed with EtOAc. The solvent is removed and the resulting residue chromatographed on silica gel (Et$_2$O, R_f = 0.43). Obtained were 0.65 g (82%) and 0.63 g (80%) respectively, of the title compound, with an enantiomeric purity of >95% ee. The same transformation can also be carried out using the method of irreversible acyl transfer with vinyl acetate as acyl donor. In this case, the reactions are complete within a few hours as compared to days.[47]

---

[46] K. Laumen, Ph.D. thesis, Wuppertal, 1987.
[47] U. Ader, D. Breitgoff, P. Klein, K. E. Laumen, and M. P. Schneider, *Tetrahedron Lett.* **30**, 1793 (1989).

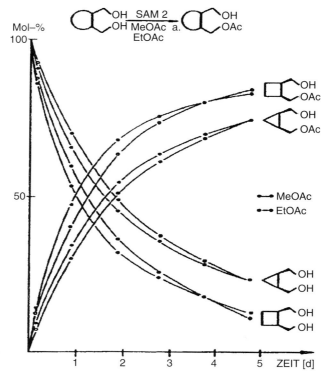

FIG. 14. Esterification of *meso*-1,2-cycloalkanedimethanols.

Cyclopentanoid natural products (e.g., prostaglandins, prostacyclins, thromboxanes) and structural analogues thereof display a great variety of biological activities and have been major synthetic targets for many years. Consequently, facile routes to chiral building blocks for the synthesis of these materials were of considerable interest. In this connection the enantioselective preparation of monoesters derived from 1,4-dihydroxycyclopent-2-ene (Fig. 15) was studied both by lipase-catalyzed hydrolysis[48] and esterification.[49,50]

5.3.2.2. SYNTHESIS OF (1S,4R)-1-ACETOXY-4-HYDROXYCYCLOPENT-2-ENE (2): EXPERIMENT. Dissolve 1 g (10 mmol) of 1,4-dihydroxycyclopent-2-ene (Fig. 15, **1**) and 5.44 ml (60 mmol) vinyl acetate in 20 ml of *t*-BuOMe to which 50 mg of lipase from *Pseudomonas* sp. (SAM-II) has been added.

[48] K. Laumen and M. P. Schneider, *J. Chem. Soc. Chem. Commun.* 1298 (1986).
[49] F. Theil, H. Schick, G. Winter, and G. Reck, *Tetrahedron* **47**, 7569 (1991).
[50] J. Hermann, Diploma thesis, Wuppertal, 1992.

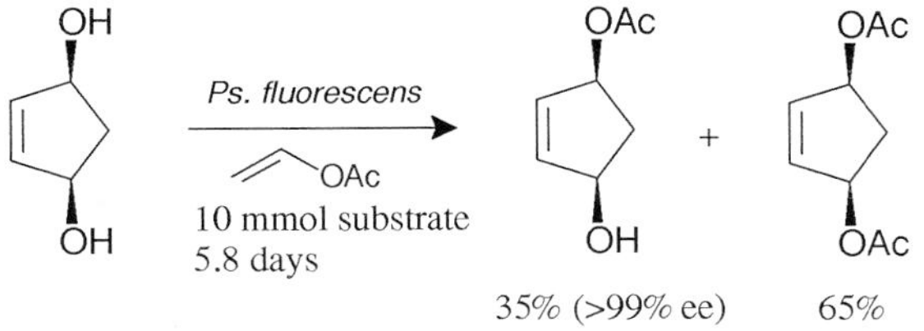

FIG. 15. Enantioselective esterification of 1,4-dihydroxycyclopent-2-ene.

The mixture is stirred at room temperature while the reaction progress is monitored by GC. After the reaction comes to a near standstill (5.8 days) the resulting suspension is filtered and the residue is washed with diethyl ether. The solvent is removed and the resulting products separated by chromatography (silica gel, n-hexane:EtOAc at 9:1, $R_f = 0.4$). Obtained were 0.5 g (35%) of (1S,4R)-1-acetoxy-4-hydroxycyclopent-2-ene (Fig. 15, **2**) with $[\alpha]_D^{20} = -65.6°$, $c = 2.325$, $CHCl_3$, and an enantiomeric purity of >99% ee and 1.2 g (65%) of meso-1,4-diacetoxycyclopent-2-ene (Fig. 15, **3**). The enantiomeric excess of **2** was determined by gas chromatography using a 50-m Cyclodex β-IP column. Similar results are obtained using pancreatine.[49]

5.3.2.3. ENANTIOTOPE DIFFERENTIATIONS IN *meso*-SUBSTRATES: INOSITOLS. The high synthetic potential of lipase-catalyzed esterifications is further underlined in the following example where the differentiation of four different hydroxy groups was achieved in one step. For the synthesis of enantiomerically pure *myo*-inositol phosphates—an important class of second messengers—an enantiomerically pure central building block was required. To avoid classical resolutions or an application of the so-called *meso* trick, both of which involve separations of diastereomers, we decided to explore the capabilities of lipases to differentiate between enantiotopic hydroxy groups in suitably protected derivatives of *myo*-inositol. We were indeed able to demonstrate that 4,6-protected *myo*-inositols (see Fig. 16),

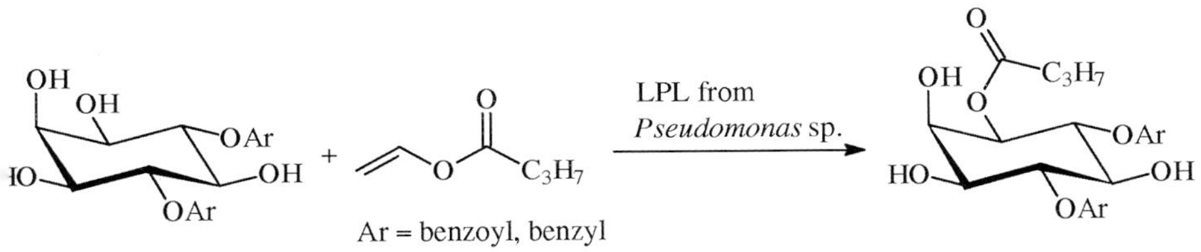

FIG. 16. Differentiation of enantiotopic hydroxy groups in *myo*-inositol derivatives.

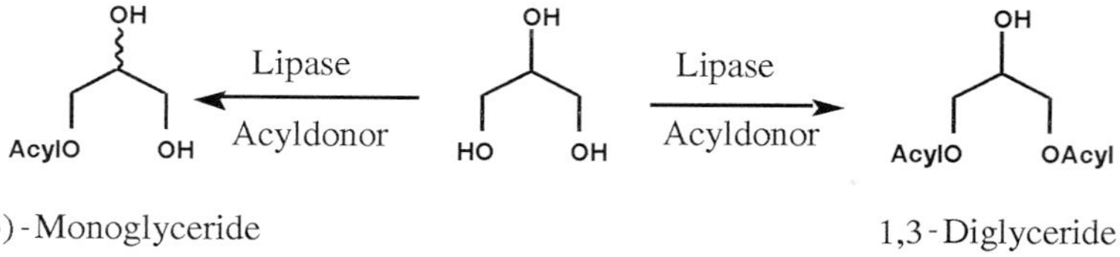

FIG. 17. Regioselective transformations of glycerol.

molecules carrying four different hydroxy groups, can be converted this way in one step into the required, enantiomerically pure product.[51]

5.3.2.4. SYNTHESIS OF 1D-1-O-BUTYRYL-4,6-O-DIBENZOYL-myo-INOSITOL: EXPERIMENT. Suspend 10 g of silica gel (70–230 mesh) and 1.25 g of lipoprotein lipase from *Pseudomonas* sp. (lyophilized, Boehringer Mannheim) in 150 ml of dry acetone. To this suspension add 11.65 g (30 mmol) of 4,6-di-O-benzoyl-myo-inositol and 12.4 ml (100 mmol) of vinylbutyrate. The mixture is stirred at 35° while the reaction progress is monitored via TLC and HPLC. After complete conversion of the starting material, the reaction mixture is filtered. Wash the mixture of lipase and silica gel two times with 50 ml acetone. From the filtrates the solvent is removed and the residue recrystallized from CH_2Cl_2:n-hexane to yield 10.72 g (78%) of the pure product. An additional 1.85 g (13%) of the material is isolated from the mother liquids thus resulting in a near quantitative conversion of the achiral starting material into the required enantiomerically and chemically pure product. It has since been used for the synthesis of several *myo*-inositol phosphates.[52]

5.4. Regioselective Esterifications of Glycerol

The regioselective esterification of glycerol can be used, depending on the stoichiometry of the transformation, for the production of isomerically pure 1(3)-*rac*-monoglycerides and or 1,3-diglycerides (Fig. 17).[53]

Products of this kind are not only interesting as nonionic surfactants for cosmetic and pharmaceutical applications, but also as attractive synthetic building blocks for the preparation of a wide variety of molecules such as defined triacylglycerols, phospholipids, drug conjugates, and reagents for the lipid modification of proteins.[54]

---

[51] P. Andersch and M. P. Schneider, *Tetrahedron: Asymmetry* **4**, 2135 (1993); K. Laumen and O. Ghisalba, *Biosci. Biotech. Biochem.* **54**, 2046 (1994).
[52] P. Andersch and M. P. Schneider, *Tetrahedron: Asymmetry* **7**, 349 (1996).
[53] M. Berger and M. P. Schneider, *Biotechnol. Lett.* **13**, 333 (1991).
[54] M. Berger and M. P. Schneider, *Fat. Sci. Technol.* **95**, 169 (1993).

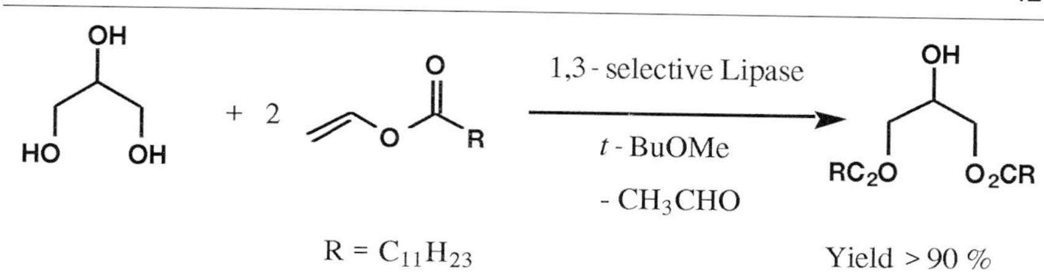

ESTER SYNTHESIS

$$\underset{\text{HO}\quad\text{OH}}{\overset{\text{OH}}{\bigwedge}} + 2 \underset{\text{O}}{\overset{\text{O}}{\bigvee}}\text{R} \xrightarrow[-\text{CH}_3\text{CHO}]{\substack{1,3\text{-selective Lipase}\\ t\text{-BuOMe}}} \underset{\text{RC}_2\text{O}\quad\text{O}_2\text{CR}}{\overset{\text{OH}}{\bigwedge}}$$

R = C₁₁H₂₃   Yield > 90 %

FIG. 18. Regioselective esterification of glycerol: irreversible acyl transfer.

For the regioselective esterification of glycerol in aprotic organic solvents two major problems had to be solved: (1) solubility—glycerol is insoluble in all nonpolar solvents; and (2) selectivity—the synthesis of pure regioisomers.

After it was discovered that partial glycerides are stable toward otherwise notorious acyl group migrations in aprotic organic solvents and given the fact that lipases are particularly active in these media (see Chapter [17] in this volume) the conditions for lipase-catalyzed esterifications were well defined. The problem of glycerol being insoluble in aprotic, nonpolar organic solvents remained. It was solved by serendipity. We simply discovered that glycerol, if adsorbed onto a solid support such as silica and suspended in such solvents, behaves as if homogeneously dissolved in such media. While trial experiments showed practically no conversion even after 7 days if glycerol itself was employed, the adsorbed material, suspended in such solvents together with an acyl donor and a suitable lipase, was quantitatively converted within 24 hr. This observation now revealed the possibilities for producing regioisomerically pure mono- and diglycerides in large quantities—up to 1 kg using laboratory equipment. Using so-called 1,3-specific lipases with high RE values[53] (see also Chapter [17] in this volume) a wide variety of regioisomerically pure 1,3-diglycerides was prepared using both the methods of reversible and irreversible acyl transfer.

5.4.1. Synthesis of Solid 1,3-sn-Diacylglycerols via Irreversible Acyl Transfer

5.4.1.1. 1,3-DILAURIN: EXPERIMENT.[55] Mechanically mix 46 g (0.5 mol) of glycerol with 46 g silica gel (70–230 mesh) until the liquid is completely adsorbed and a free-flowing "dry" powder is obtained (Fig. 18). This material is suspended in 1 liter of t-BuOMe to which 226 g (1 mol) of vinyl laurate and 2 g of the immobilized lipase from *R. miehei* (Lipozyme) has been added. The mixture is stirred at room temperature for 24 hr while

---

[55] M. Berger, K. Laumen, and M. P. Schneider, *J. Am. Oil Chem. Soc.* **69**, 955 (1992).

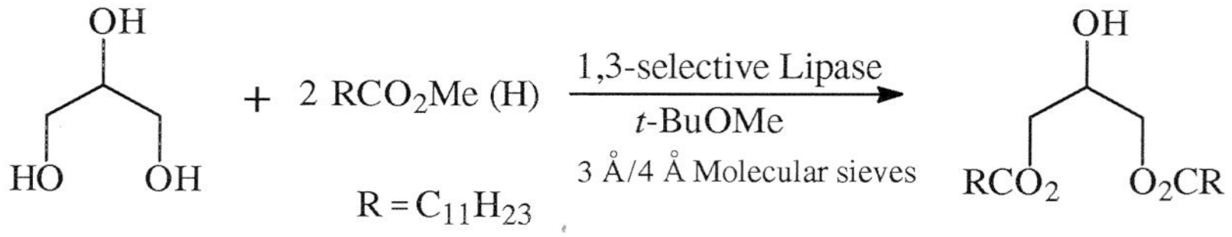

FIG. 19. Synthesis of 1,3-dilaurin via reversible acyl transfer or direct esterification.

the reaction progress is monitored by TLC. After removal of the solid components (immobilized biocatalyst and silica gel) and evaporation of the solvent, a crude reaction mixture is obtained containing about 85% of 1,3-dilaurin (GC). The solid mixture is recrystallized from dry methanol to yield 190 g (84%) regioisomerically pure (>99.5%) 1,3-sn-dilaurin as a white powder (m.p., 56.5°).

As mentioned earlier, the method of irreversible acyl transfer is frequently preferred because of the high yields of products obtainable in a short time or in cases in which the methods of reversible acyl transfer are not successful due to the low reactivity of the acceptor alcohol such as in the esterification of secondary hydroxy groups. Since in the esterification of glycerol primary hydroxy groups are to be esterified the method of reversible acyl transfer can indeed be employed successfully as shown next.

5.4.2. Synthesis of Solid 1,3-sn-Diglycerides via Reversible Acyl Transfer or Direct Esterification

5.4.2.1. SYNTHESIS OF 1,3-DILAURIN: EXPERIMENT.[55] Adsorb 4.6 g (50 mmol) of glycerol onto 4.6 g of silica gel using the procedure described above. This material is suspended in 100 ml of t-BuOMe (Fig. 19). Add 22 g (100 mmol) of methyl laurate [alternatively 20 g (100 mmol) of lauric acid] together with 2 g of the lipase from *R. miehei* (Lipozyme) and 5 g of molecular sieves (4 Å; alternatively, 3 Å). The mixture is stirred at room temperature while the reaction progress is monitored by TLC. After completion (48 hr) the solid materials are removed by filtration, the solvent evaporated, and the solid residue recrystallized from dry methanol to yield 16.7 g (70%) of regioisomerically pure 1,3-dilaurin (>99.5%) [18.2 g (80%)].

5.4.3. Synthesis of Liquid 1,3-sn-Diacylglycerols via Irreversible Acyl Transfer.[55] Because liquid derivatives of that kind cannot be purified by recrystallization, the highest possible yield of 1,3-diglycerides had to be achieved in the enzymatic reaction step. To achieve this goal, a highly regioselective lipase had to be used such as the lipase from *Rhizopus delemar* (see Chapter [17] in this volume).

5.4.3.1. SYNTHESIS OF 1,3-sn-DICAPRYLIN: EXPERIMENT. Adsorb 0.92 g (10 mmol) of glycerol onto 1 g of silica gel (70–230 mesh). Suspend in 20

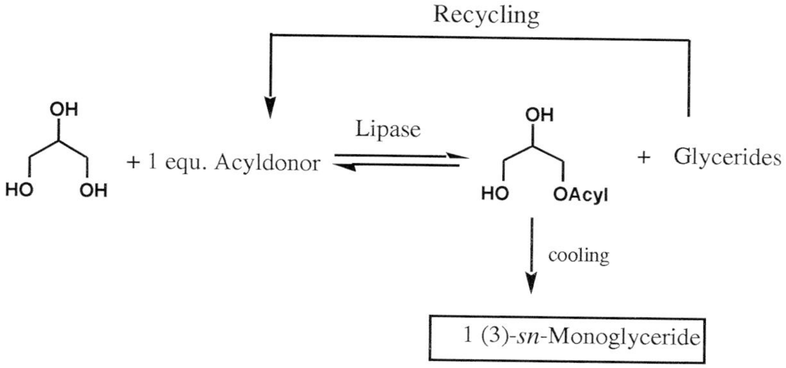

FIG. 20. Synthesis of regioisomerically pure 1(3)-*rac*-monoglycerols: isolation and recycling of by-products.

ml of *t*-BuOMe. Add 3.4 g (20 mmol) of vinyl caprylate and 50 mg of lipase from *R. delemar* to the suspension. The mixture is stirred at room temperature and the reaction progress is monitored by TLC. After the reaction is complete (96 hr) the solid components are removed by filtration, and the evaporated solvent obtained is a crude reaction mixture containing 91% of the desired 1,3-*sn*-dicaprylin. This material is further purified by column chromatography on silica gel (*n*-hexane : diethylether at 1 : 1) yielding 3.0 g (80%) of regioisomerically pure (>98%) material.

5.4.4. Synthesis of 1(3)-rac-Monoacylglycerols of High Regioisomeric Purity.[56] The synthesis of regioisomerically pure 1(3)-*rac*-monoacylglycerols proved to be less straightforward, especially under stoichiometric reaction conditions and without the need for tedious purification steps. From exploratory experiments it became quite clear that the primarily produced monoglycerides proved to also be good substrates for the lipase employed, resulting in the production of large quantities of diglycerides. Even with a tenfold excess of glycerol, the yield of the desired monoglycerides thus never exceeded 60%. A simple solution for all problems was found in the compartmental separation of the two steps of the process—synthesis and isolation (Fig. 20).

The enzymatic esterification is carried out in the reactor vessel using stoichiometric quantities of glycerol and the corresponding acyl donor. The obtained reaction mixture is circulated into the second vessel—termed a *separator*—in which the desired monoglycerides are separated out at lower temperatures. All other products—including unreacted acyl donor—

[56] M. Berger and M. P. Schneider, *J. Am. Oil Chem. Soc.* **69**, 961 (1992).

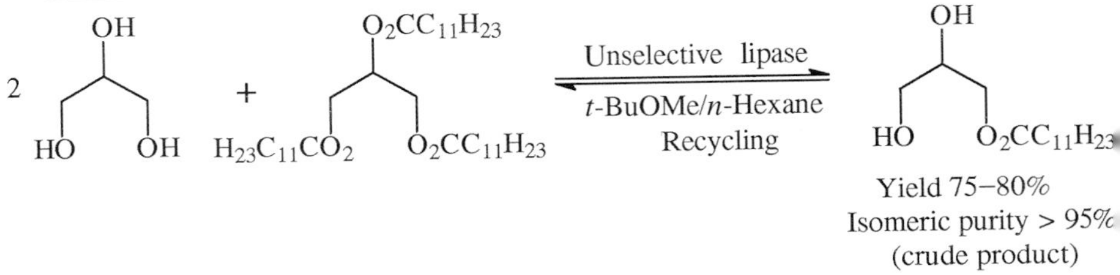

$$2 \;\; \underset{HO \quad OH}{\overset{OH}{\triangle}} + \underset{H_{23}C_{11}CO_2 \quad O_2CC_{11}H_{23}}{\overset{O_2CC_{11}H_{23}}{\triangle}} \underset{\substack{t\text{-BuOMe}/n\text{-Hexane} \\ \text{Recycling}}}{\overset{\text{Unselective lipase}}{\rightleftharpoons}} \underset{HO \quad O_2CC_{11}H_{23}}{\overset{OH}{\triangle}}$$

Yield 75–80%
Isomeric purity > 95%
(crude product)

FIG. 21. Lipase-catalyzed glycerolysis of triglycerides.

remain in solution and are fed back to the reactor, which contains both the enzyme and the glycerol on the solid support. Thus the whole transformation can be carried out under stoichiometric conditions and does not require any further purification step. The reactions can be carried out under the conditions of irreversible and reversible acyl transfer. Thus all acyl donors described above, such as free fatty acids, alkylesters (with the addition of molecular sieves to the reactor, see earlier discusssion) and vinyl esters, can be used. This includes triglycerides and thus also natural fats and oils. Interestingly, in these cases, unspecific lipases are best suited. They can make use of all acyl groups including those in the 2-position of glycerides (Fig. 21).

5.4.5. *1(3)-rac-Mono-Acylglycerols via Reversible Acyl Transfer Using Triglycerides as Acyl Donors*

5.4.5.1. SYNTHESIS OF 1(3)-*rac*-MONOLAURIN: EXPERIMENT. A 250-ml three-necked flask maintained at 25° and equipped with an efficient mechanical stirrer (the reactor) is connected via a filter disk and membrane pump with another 250-ml round-bottom flask (the separator), which was placed in a cooling bath maintained at 2°. The separator was connected with the reactor in the same manner to provide a solvent cycle.

Adsorb 4.6 g (50 mmol) of glycerol onto 10 g of silica gel in the previously described way and the resulting powder is suspended in 100 ml of *t*-BuOMe. Add 10.5 g (17 mmol) of trilaurin together with 300 mg lipase from *P. fluorescens* and the resulting mixture is stirred and circulated through the system for 48 hr. During the course of the reaction 12.2 g (90%) of 1(3)-*rac*-monolaurin precipitated in the separator as a colorless solid, which was isolated by filtration under reduced pressure. The isomeric purity of this "crude" material was determined by GC to be 98%.

Also unsaturated fatty acids or triglycerides derived thereof can be employed. This approach is illustrated in the examples below using vinyl oleate and triolein as acyl donors.[23]

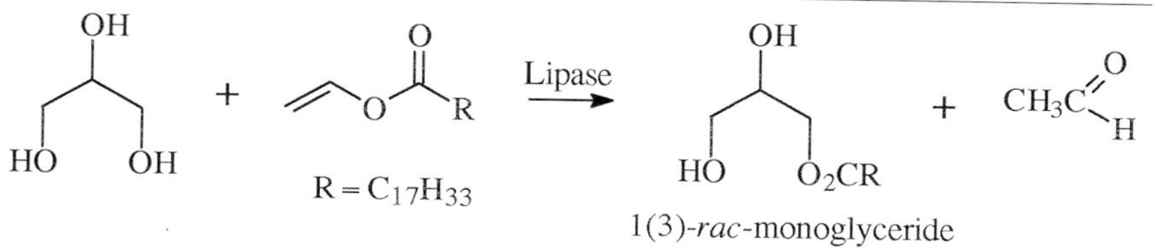

$$\text{glycerol} + \text{vinyl oleate} \xrightarrow{\text{Lipase}} \text{1(3)-rac-monoglyceride} + CH_3CHO$$

$R = C_{17}H_{33}$

FIG. 22. Synthesis of 1(3)-*rac*-monoolein: irreversible acyl transfer.

5.4.6. Synthesis of 1(3)-rac-Monoolein via Irreversible Acyl Transfer: Experiment.
Dissolve 24.68 g (80 mmol) of vinyl oleate in 1 liter of *t*-BuOMe to which 46 g (50 mmol) glycerol (immobilized onto 92 g of silica gel) and 1 g lipase from *Penicillium roquefortii* has been added (Fig. 22). The reaction mixture is stirred at room temperature and the reaction progress is monitored by TLC. After removal of the solid components (immobilized biocatalyst, silica gel), evaporation of solvent and unreacted vinyl ester, a crude mixture was obtained that contained about 85% of the desired 1(3)-*rac*-monoolein. It was purified by flash chromatography on silica gel (diethyl ether:*n*-hexane at 3:2) leading to 24.2 g (85%) of the chemically (99%) and regioisomerically pure (97% regioisomerical excess) product, m.p., 34°.

5.4.6.1. SYNTHESIS OF 1(3)-*rac*-MONOOLEIN VIA REVERSIBLE ACYL TRANSFER USING TRIOLEIN: EXPERIMENT.[23] A 2-liter three-necked flask maintained at 50° and equipped with an efficient mechanical stirrer (the reactor) is connected via a filter disk and membrane pump with another 2-liter round-bottom flask (the separator), which is placed in a cooling bath maintained at −30°. The separator is connected with the reactor in the same manner to provide a solvent cycle. Adsorb 328 g (3.5 mol) of glycerol onto 700 g of silica gel in the previously described way with the resulting powder suspended in 3.5 liters of *t*-BuOMe. Add 100 g (113 mmol) of triolein (containing 85% oleic acid) to 10 g lipase from *Candida antarctica* (Novozym 435). Stir the resulting mixture and circulate through the system for 21 hr.

During the course of the reaction 87.4% of 1(3)-*rac*-monoolein and 12.6% of 1,3-diolein precipitates in the separator as a colorless solid, which is isolated by filtration under reduced pressure (determined by GC from the reaction mixture). Recrystallization from *n*-hexane leads to 84.9 g (69%) of 1(3)-*rac*-monoolein with an isomeric purity of 88% (GC), m.p., 31°.

5.5. Regioselective Transformations: Carbohydrates

The direct esterification of carbohydrates is of importance both to food technologists and organic chemists. Sugar esters of long-chain fatty acids constitute an important group of nonionic surfactants. Due to their composi-

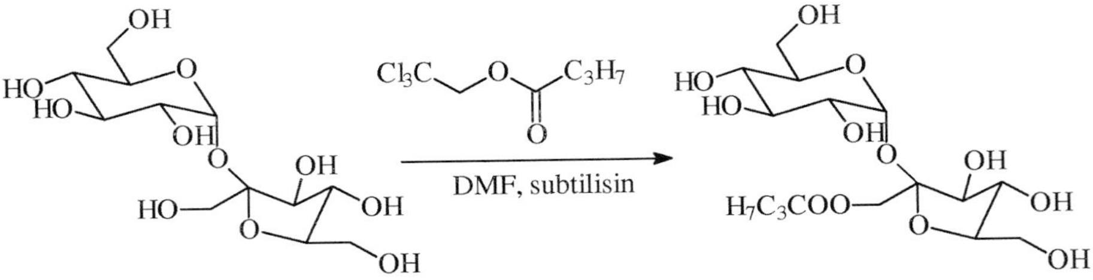

FIG. 23. Protease-catalyzed esterification of sucrose.

tion they are completely biodegradable. They are attractive emulsifiers for the processing of foods and many other applications. The selective functionalization of mono- and disaccharides leads to molecules that can be used as synthetic building blocks for further transformations.

A major requirement for the lipase-catalyzed esterification of carbohydrates and/or derivatives thereof is their at least partial solubility in aprotic or nonpolar solvents in which these enzymes usually display their highest specific activities. This is no problem, if derivatives are used that carry only a small number of hydroxy groups (Section 5.5.6) such as protected systems, glycals,[57] or anhydrosugars.[58] With increasing numbers of hydroxy groups such as those found in most hexoses and disaccharides the situation becomes much more difficult and either (1) polar solvents like pyridine, DMSO, DMF, or DMA must be used or, alternatively, (2) temporary or permanent derivatizations of the insoluble starting materials must be carried out. Examples for both approaches are given next.

In earlier experiments pyridine[59] was successfully employed for the esterification of several monosaccharides such as glucose, galactose, mannose, and fructose using a large excess of activated esters as acyl donors. Reasonable regioselectivities were achieved with moderate conversions. Using two proteases from *Bacillus subtilis* (subtilisin), trichloroethylbutyrate as acyl donor, and anhydrous DMF as solvent, the butyrates of glucose and several disaccharides were prepared (Fig. 23).[60]

After the introduction of an acyl group in the sugar moiety, thus derived molecules frequently become soluble in classical organic solvents such as methylene chloride or tetrahydrofuran and can be further esterified using

---

[57] E. W. Holla, *Angew. Chem.* **101**, 222 (1989); E. W. Holla, *Angew. Chem. Int. Ed. Eng.* **101**, 200 (1989).
[58] R. Seemayer, N. Bar, and M. P. Schneider, *Tetrahedron: Asymmetry* **3**, 1123 (1992).
[59] M. Therisod and A. M. Klibanov, *J. Am. Chem. Soc.* **108**, 5638 (1986).
[60] S. Riva, J. Chopineau, A. P. G. Kieboom, and A. M. Klibanov, *J. Am. Chem. Soc.* **110**, 584 (1980).

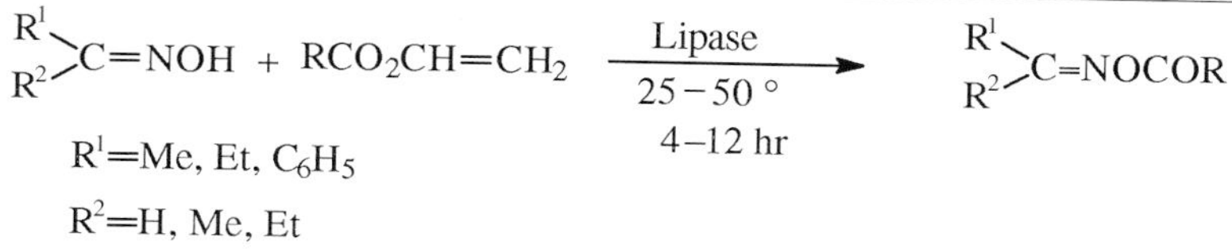

$$\underset{R^2}{\overset{R^1}{>}}C=NOH + RCO_2CH=CH_2 \xrightarrow[25-50°]{\text{Lipase}} \underset{R^2}{\overset{R^1}{>}}C=NOCOR$$

R^1=Me, Et, C_6H_5

R^2=H, Me, Et

4–12 hr

FIG. 24. Lipase-catalyzed synthesis of oxime esters.

simple lipases.[61] Using pyridine again as a solvent and oxime esters as acyl donors for the method of irreversible acyl transfer (see earlier discussion), a wide variety of pentoses and hexoses were esterified regioselectively.[62] The primary hydroxy groups of these sugars (glucose, galactose, mannose) were esterified preferentially. The required oxime esters can be synthesized either chemically[63] or enzymatically by lipase-catalyzed oximolysis of the corresponding vinyl esters[9] (Fig. 24).

5.5.1. Synthesis of Hexose Monoesters: General Procedure. The following experimental procedure was transcribed from the original publication by the authors and not checked in our laboratory. It seems that these transformations are rather slow under the given conditions and thus require large amounts of enzyme.

Experiment. Dissolve 0.45 g (2.5 mmol) of α-D-glucopyranose in 10 ml of dry pyridine to which 0.65 g (3.25 mmol) of O-octanoyl acetoxime and 2 g of lipase from *P. cepacia* have been added. The reaction mixture is stirred at room temperature using a magnetic stirrer while the reaction progress is monitored by TLC. After 3 days the reaction is terminated, the enzyme filtered off, the solvents evaporated under reduced pressure, and the resulting syrup further purified by column chromatography leading to 0.66 g (86%) 6-O-octanoyl-α-D-glucopyranose.

Note, however, that for applications in the food sector, solvents such as pyridine, DMSO, and DMF are of low acceptance due to their toxicity and for many years scientists have searched for a practical solution to overcome this problem.

Recently, 2-methyl-2-butanol has been employed for the esterification of various sugars and sugar alcohols such as sorbitol. Again using the lipase from *C. antarctica B* (Novozym 435) in the mode of direct esterification with oleic acid as acyl donor the produced water was removed in a soxhlet

---

[61] M. Therisod and A. M. Klibanov, *J. Am. Chem. Soc.* **109**, 3977 (1987).
[62] R. Pulido, F. Lopez Ortiz, and V. Gotor, *J. Chem. Soc., Perkin Trans. I* 2891 (1992); V. Gotor and R. Pulido, *J. Chem. Soc., Perkin Trans. I* 491 (1991).
[63] H. Metzger, in Houben-Weyl, Methoden der organischen Chemie, 4th Ed., Vol. X/4 Stickstoff verbindungen I, p. 180. Thieme, Stuttgart, 1968.

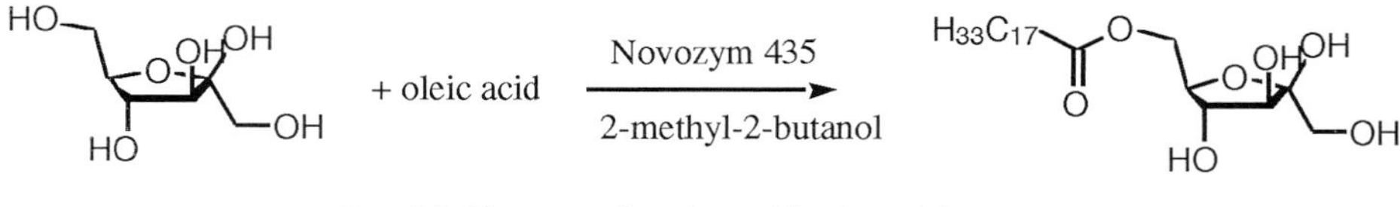

FIG. 25. Lipase-catalyzed esterification of fructose.

extractor using molecular sieves. Thus the equilibrium shifted toward the ester products and high yields of products were achieved[64] (Fig. 25).

High product yields are claimed (93%), however the prospect of scale-up remains unclear as does the composition of the products (mono-, diesters, ratio of pyranose and furanose esters, etc.). In a technical scale synthesis of sugar esters, the problem of solubility was overcome by prior acetalization of glucose using ethanol in the presence of an acidic ion-exchange resin followed by a solvent-free, lipase-catalyzed esterification using free fatty acids and the lipase from *C. antarctica* (Fig. 26).[65]

5.5.1.1. SYNTHESIS OF ETHYL-D-GLUCOPYRANOSIDE: EXPERIMENT.[66] Suspend glucose (500 g, 2.78 mol) and a strong acid cation resin (100 g Amberlyst 15, BDH Chemicals, Palmerston North, New Zealand) in ethanol (2000 ml, 34.3 mol). The mixture is stirred at 80° for 16 hr. The progress of the reaction is followed by HPLC. The ion-exchange resin is removed by filtration and the solution is treated with activated carbon (10 g). After filtration the ethanol is removed *in vacuo* giving ethyl-D-glucopyranoside (1:1 mixture of α and β anomers) as a syrup (578 g, quantitative yield).

5.5.1.2. SYNTHESIS OF 6-O-DODECANOYL-D-GLUCOPYRANOSIDE: EXPERIMENT.[65] To a mixture of crude ethyl D-glucopyranoside (528 g, 2.78 mol, prepared as described earlier) and dodecanoic acid (751 g, 3.75 mol) in a stirred batch reactor at 70° add immobilized lipase derived from *C. antarctica* (29 g, prepared as described in Danish Patent Application No. 3250/88). Stirring continues under reduced pressure (0.05 bar), and the progress of the ester synthesis is monitored by HPLC. After 23 hr the enzyme is removed by filtration (at 70°). Excess fatty acid is removed by repeated molecular distillation (105°, 4×10^{-2} mbar) yielding 96% (1050 g) of crude product along with 2% ethyl D-glucoside and 2% of of a mixture of diesters (HPLC analysis). The crude product is purified by chromatography. Identification by ^{1}H NMR analysis showed a 1:1 mixture of α and β anomers. High yields and high regioselectivities were achieved and the

---

[64] A. Ducret, A. Giroux, M. Trani, and R. Lortie, *Biotechnol. Bioeng.* **48**, 214 (1995).
[65] F. Björkling, S. E. Godtfredsen, and O. Kirk, *J. Chem. Soc. Chem. Commun.* 934 (1989); K. Adelhorst, F. Björkling, S. E. Godtfredsen, and O. Kirk, *Synthesis* 112 (1990).
[66] Novo Nordisk A/S, European Patent 0 394 280 B1 (1988).

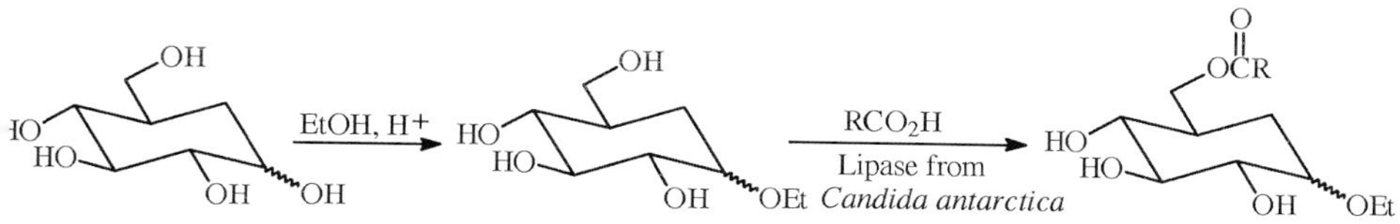

FIG. 26. Enzyme-assisted synthesis of 6-O-acylated glucose.

reaction was scaled up into the kilogram range. It can be assumed that pilot plant experiments were done up to a ton scale.

5.5.2. Esterifications of Carbohydrates Using Solubizing Agents. To achieve the required minimal solubility of carbohydrates in nontoxic organic solvents, the regioselective lipase-catalyzed esterifications of mono- and disaccharides in the presence of phenylboronic acid as complexation/solubilizing agent[67] were recently described. By simply mixing phenylboronic acid with the corresponding carbohydrate, an *in situ* formation of complexes takes place, leading to materials that are soluble in the solvent system employed and that can be simply destroyed after the enzymatic esterification by addition of water (Fig. 27).

5.5.3. Esterification of Glucose Using Phenylboronic Acid as Complexation Agent.[68] Although all reactions have so far probably only been carried out on a relatively small scale and no detailed experimental data regarding the absolute amounts or a prospect for scale-up have been given, the following experimental procedure can be derived from this publication.

EXPERIMENT. Using anhydrous t-butanol (t-BuOH) as solvent, suspensions of 0.1 M glucose and 0.2 M phenylboronic acid are prepared. The suspensions are shaken at 45° and 250 rpm for about 8 hr after which all previously undissolved glucose disappeared, indicating that complex formation had occurred. Then 0.3 M vinyl acrylate and 100 mg/ml lipoprotein lipase from *Pseudomonas* sp. are added and the resulting suspensions shaken at 250 rpm for 1–4 days until all of the starting material disappears. The enzyme is removed by filtration, water added, and after a reaction time of 2 hr the solvent t-BuOH is removed under reduced pressure. The precipitated phenylboronic acid is removed from the aqueous phase by filtration and the remaining solution lyophilized. The solid residue is washed with diethylether. From 0.35 g of the complex, 0.18 g (77%) of 6-O-acroylglucose was obtained.

5.5.4. Monoacylation of Fructose. Using the same principle, the corresponding esterification of fructose was described. Interestingly, in this case

---

[67] H.-O. Park, D.-S. Lee, and S. C. Shim, *Biotechnol. Lett.* **14**, 111 (1992).
[68] I. Ikeda and A. M. Klibanov, *Biotechnol. Bioeng.* **42**, 788 (1993).

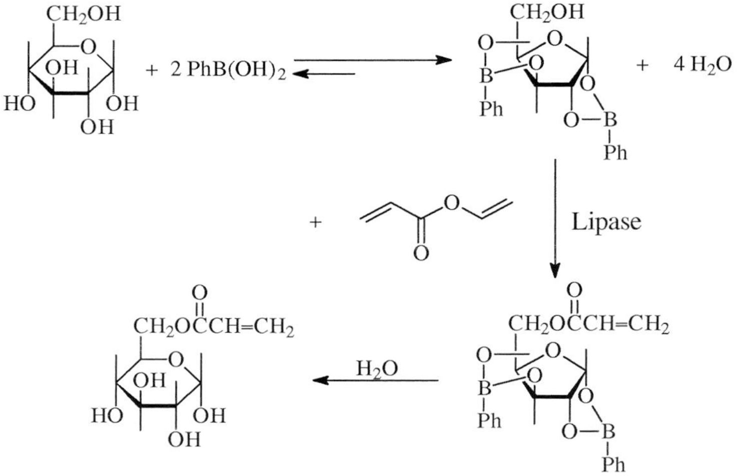

FIG. 27. Esterification of D-glucose in the presence of phenylboronic acid.

the transformation can be carried out even in *n*-hexane using the commercially available, immobilized lipase from *R. miehei* (Lipozyme IM 60).[69] Again this transformation has so far been carried out only with minute quantities of material and the possibilities of upscaling them to a technically useful level remains speculative at this time.

5.5.5. Regioselective Esterifications of 2-Deoxy-D-Ribose and D-Ribose. While in all of the examples previously described the required solubility for enzymatic esterifications was either achieved with the use of highly polar solvents like pyridine, DMSO, DMF, or DMA, it was recently demonstrated that 2-deoxy-D-ribose and D-ribose can be selectively esterified simply in THF and without the need for prior derivatizations.

For an effective route toward 2′-deoxynucleosides and the synthesis of a suitable triacylated intermediate the selective acylation of the 5-position in 2-deoxy-D-ribose in its furanoid form was required. This was very conveniently achieved by direct acylation in the presence of a commercially available lipase from *C. antarctica* (Novozym 435), which was known to be highly selective for primary hydroxy groups in monosaccharides (see earlier examples) and under conditions of irreversible acyl transfer using propionic anhydride as acyl donor[70] (Fig. 28).

5.5.5.1. SYNTHESIS OF 2-DEOXY-5-*O*-PROPIONYL-D-RIBOFURANOSE: EXPERIMENT. Dissolve 1.07 g (8 mmol) of 2-deoxy-D-ribose in 60 ml of dry

[69] A. Schlotterbeck, S. Lang, V. Wray, and F. Wagner, *Biotechnol. Lett.* **15**, 61 (1993).
[70] A. K. Prasad, M. D. Sorensen, V. S. Parmar, and J. Wengel, *Tetrahedron Lett.* **36**, 6163 (1995).

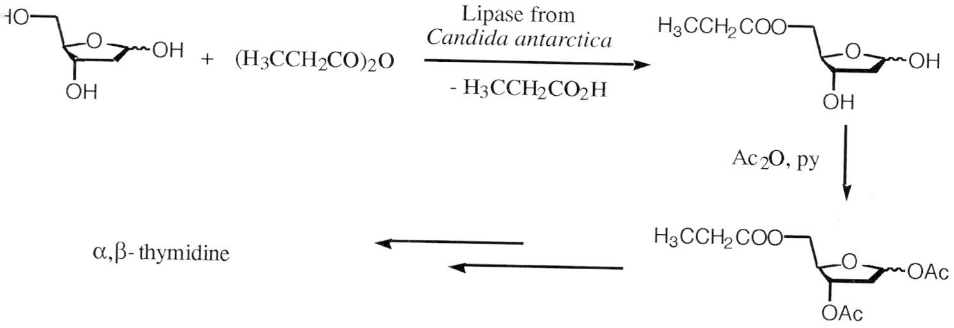

FIG. 28. Regioselective esterification of 2-deoxy-D-ribose.

THF to which 3.12 g (24 mmol) of propionic anhydride and 1.6 g lipase from *C. antarctica* (Novozym 435) have been added. Stir the mixture at 50° for 1.5 hr until the TLC shows complete conversion of the starting material. The obtained 2-deoxy-5-*O*-propionyl-D-ribofuranose is not isolated but transformed directly into the corresponding 1,3-diacetate, the required building block for 2′-deoxynucleosides. For this, the mixture is filtered and the solvent volume reduced to 10 ml under reduced pressure. Acetic anhydride (20 ml, 212 mmol) and pyridine (4 ml, 49 mmol) are added and stirring continues for 12 hr at room temperature. The reaction mixture is poured onto ice/H_2O (70 ml), stirred for 30 min, and extracted with dichloromethane (2 × 100 ml). The separated organic phase is washed with water (2 × 100 ml) and dried. After removal of the solvent under reduced pressure, 2.08 g (95%) of 2-deoxy-1,3-di-*O*-acetyl-5-*O*-propionyl-D-ribofuranose is obtained as a colorless oil.

The high product yield and the provided spectroscopic data clearly indicate that the enzymatic esterification was achieved with very high regioselectivity indeed. The method was further extended to the esterification of D-ribose with an additional secondary hydroxyl group in the 2-position.

5.5.6. *Regioselective Differentiation of Secondary Hydroxy Groups in Carbohydrates.* The preceding examples all deal with the regioselective esterification of primary hydroxy groups in carbohydrates. However, lipases are also capable of differentiating between secondary hydroxy groups in a variety of ways. Again, as in all examples using esterification reactions, the molecules to be esterified have to be soluble in the reaction medium. This, in turn, is closely related to the molecular structures of the substrates. Molecules with two or three hydroxy groups are frequently soluble in solvents of low to medium polarity, whereas molecules like glucose with five hydroxy groups will have to be modified—at least temporarily—by suitable protection groups.

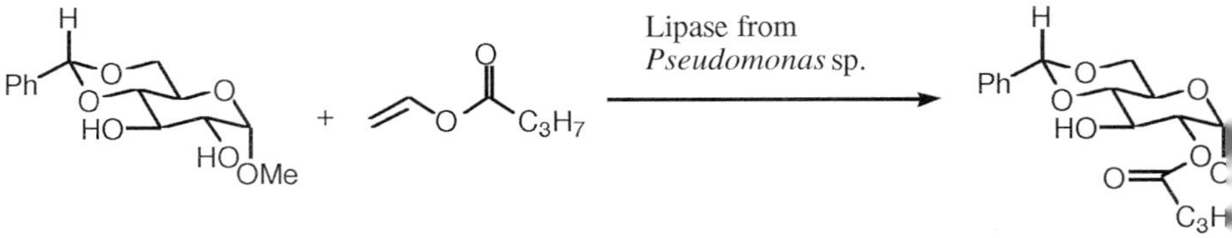

FIG. 29. Regioselective esterification of glucose derivatives.

If, in addition to the anomeric center, the highly polar (and more reactive) primary hydroxy group of glucose is temporarily protected, for example, by formation of the corresponding benzylidene derivative, the resulting molecules are soluble in practically every organic solvent and are thus ideal substrates for lipase-catalyzed esterifications. The remaining hydroxy groups in the 2- and 3-positions of glucose can be differentiated with high regioselectivity using the lipase from *Pseudomonas* sp. (SAM-II)[71] and also from *P. cepacia*[72] with the 2-position being esterified exclusively (Fig. 29). The resulting, isomerically pure glucose derivatives are obtained in near quantitative yield.

5.5.6.1. SYNTHESIS OF 2-O-BUTYRYL-4,6-O-BENZYLIDENE-α-METHYL-D-GLUCOPYRANOSIDE: EXPERIMENT. Dissolve 1.41 g (5 mmol) of 4,6-O-benzylidene-α-methyl-D-glucopyranoside in 20 ml *t*-BuOMe to which 3.42 g (30 mmol) of vinyl butyrate and 200 mg of the lipase from *Pseudomonas* sp. (SAM-II) have been added. Stir the mixture at room temperature while monitoring the reaction progress by TLC (CH_2Cl_2 : MeOH at 10 : 1, $R_f = 0.89$). After completion of the reaciton (4d) the enzyme is removed by filtration and washed with 2 × 20 ml *t*-BuOMe. The solvent is removed and the product isolated without further purification. Obtained was 1.7 g (96%) of the title compound, m.p., 113–114°, $[\alpha]_D^{20} = 98.7°$, $c = 1$ $CHCl_3$ stab. with 1% EtOH.

5.5.6.2. REGIOSELECTIVE ESTERIFICATION OF 1,4 : 3,6-DIANHYDRO-D-GLUCITOL (ANHYDROSORBITE).[58,71] For the selective preparation of isoglucitol (isosorbide) mononitrates, the selective protection of one of the two secondary hydroxy groups was required. This could be achieved either by the regioselective hydrolysis/alcoholysis of the corresponding diesters or the selective esterification of isosorbide itself, both methods being complementary regarding the stereochemical outcome (Fig. 30).

The reaction products are attractive starting materials for the physiologically active mononitrates. The preparative scale esterification of isosorbide,

[71] R. Seemayer, Ph.D. thesis, University Wuppertal, 1992.
[72] L. Panza, M. Luisetti, E. Criati, and S. Riva, *J. Carbohydr. Chem.* **12**, 125 (1993).

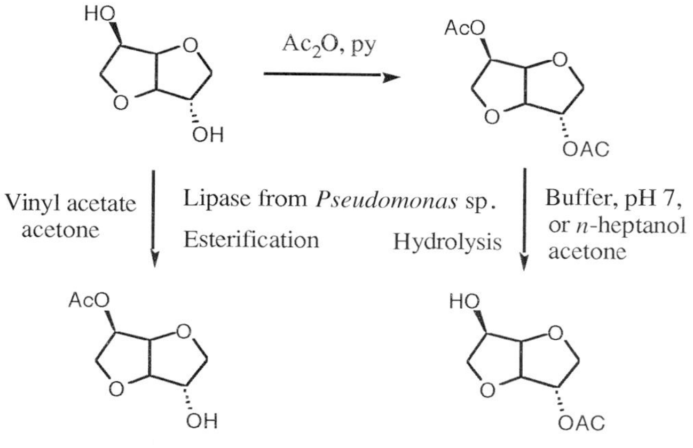

FIG. 30. Regioselective preparation of isosorbide esters.

leading to the 5-acetate, is described below. Note that to achieve sufficient solubility, acetone was used as a solvent. It is well established that lipases work in a wide variety of different solvents. Examples for this can be found in Chapter [17] in this volume. Vinyl acetate served as acyl donor, the reaction being another example for the method of irreversible acyl transfer.

5.5.6.3. SYNTHESIS OF ISOSORBIDE-5-ACETATE: EXPERIMENT. Dissolve 14.61 g (100 mmol) of isosorbide in 200 ml acetone containing 26.07 g (300 mmol) of vinyl acetate. Add 500 mg lipase from *Pseudomonas* sp. (SAM-II) and stir the mixture at room temperature while the reaction progress is monitored by GC. After total conversion of isosorbide (5 days), the enzyme is filtered off, the solvent removed *in vacuo,* and the product isolated isomerically pure (>99% de) and in excellent yield, 16.94 g (90%).

5.6. Regioselective Transformations: Nucleosides

The regioselective introduction of ester groups in nucleosides can also be achieved using lipase-catalyzed esterifications. Depending on the lipase employed, selective functionalization of either the primary or secondary hydroxy group can be achieved.

It is well established that the lipases from *Pseudomonas* sp. (Amano lipases PS, SAM-II, etc.) are selective toward secondary hydroxy functions, whereas lipases from porcine pancreas or *C. antarctica* preferentially transfer the ester functions onto the primary hydroxy groups.

For the regioselective esterification of 2'-deoxynucleosides in the 3'-position the lipase from *P. cepacia* (Amano lipase PS) was chosen. Again,

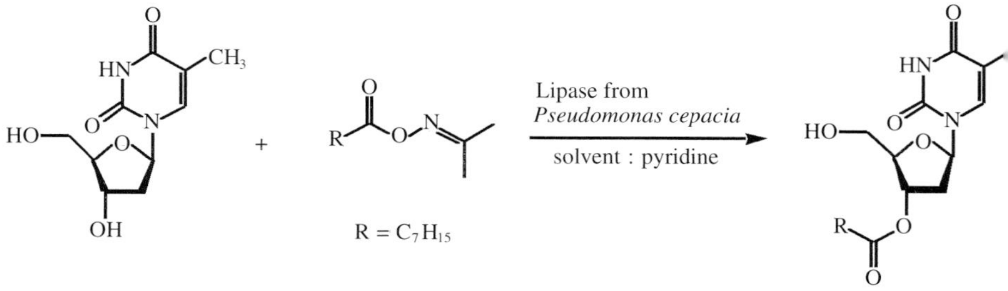

FIG. 31. Selective esterification of 2'-deoxynucleosides.

as in the case of underivatized carbohydrates, the problem of solubility had to be addressed and again pyridine proved to be the best choice. Under conditions of irreversible acyl transfer using oxime esters, completely regioselective esterifications were achieved[73] (Fig. 31). The oxime esters were prepared as described above.[9] The experimental procedure given below was transcribed from the original publication and is unchecked.

5.6.1. Synthesis of 3'-O-Caprinoyl-2'-Deoxythymidine: Experiment. Dissolve 4 mmol (1.58 g) of thymidine in 15 ml pyridine to which 5 mmol (1 g) of caprinoyl acetoxime and 1.5 g of the lipase have been added. Stir the mixture under nitrogen at 60° for 48 hr. The enzyme is filtered off and washed two times with 10 ml of MeOH. The solvents are removed under reduced pressure and the residue purified by flash chromatography with AcOEt:MeOH at 100:1. The reported yield was 1.76 g (82%).

Clearly there are nearly unlimited possibilities for exploring regioselective esterifications of a wide variety of natural products following the same general principle. Thus the regioselective esterifications of numerous steroids,[74] flavonoid mono- and disaccharides, shikimic acid derivatives, and many other systems have been reported in the recent literature.[75]

6. Conclusion

Lipases are probably at present the most widely used biocatalysts in organic synthesis. Many of them are commercially available, they are fre-

---

[73] V. Gotor and F. Moris, *Synthesis* 626 (1992).
[74] D. L. Delinck and A. L. Margolin, *Tetrahedron Lett.* **31,** 3093 (1990); S. Riva, R. Bovara, G. Ottolina, F. Secundo, G. Carrea, *J. Org. Chem.* **54,** 3161 (1989); A. Bertinotti, G. Carrea, G. Ottolina, and S. Riva, *Tetrahedron* **50,** 13165 (1994); G. F. Ottolina, G. Carrea, and S. Riva, *Biocatalysis* **5,** 131 (1991).
[75] B. Danieli, P. De Bellis, G. Carrea, and S. Riva, *Helv. Chim. Acta* **73,** 1837 (1990); B. Danieli, P. De Bellis, L. Barzaghi, G. Carrea, G. Ottolina, and S. Riva, *Helv. Chim. Acta* **75,** 1297 (1992).

quently stable toward elevated temperatures and organic solvents, and they are easy to handle. Due to their high enantio- and regioselectivities, broad substrate tolerance and versatility in catalyzing both hydrolyses and esterifications, these enzymes have developed in the last decade into high efficient tools in the hands of the organic chemist.

With the main focus on lipase-catalyzed esterifications via acyl transfer (transesterification), we have described numerous examples for such reactions as applied to a wide variety of structurally different hydroxy compounds. The examples range from the preparation of enantiomerically pure molecules via kinetic resolutions and enantiotope differentiations, respectively, all the way to regioselective functionalizations of natural products such as carbohydrates and nucleosides.

With some exceptions, the examples provided were selected from experiments carried out in our laboratory. The included experimental details should facilitate the use of the described methods by interested members of the scientific community.

[20] Cross-Linked Enzyme Crystals of Lipases as Catalysts for Kinetic Resolution of Acids and Alcohols

By JIM J. LALONDE, MANUEL A. NAVIA, and ALEXEY L. MARGOLIN

Introduction

The ability of lipases to catalyze the enantioselective hydrolysis or synthesis of esters makes these enzymes powerful tools in the resolution of chiral carboxylic acids and alcohols.[1,2] The desired biological activity of chiral pharmaceuticals, food, and agriculture chemicals most often lies in only one enantiomer, the other enantiomer being inactive ballast at best and, in some cases, harmful. Nonetheless, efficient methods for the preparation of optically pure chiral molecules have lagged behind the demonstrated need. Crude lipase preparations have proved to be simple and effective

---

[1] C.-H. Wong and G. M. Whitesides, "Enzymes in Synthetic Organic Chemistry," Tetrahedron Organic Chemistry Series, Vol. 12. Pergamon, New York, 1994; K. Drauz and H. Waldmann, eds., "Enzyme Catalysts in Organic Synthesis—A Comprehensive Handbook." Weinheim, Germany, 1995.

[2] A. M. Klibanov and B. Cambou, *Methods Enzymol.* **136,** 117 (1987).

catalysts for the kinetic resolution of an enormous range of carboxylate esters, allowing the preparation of alcohols and acids in optically pure form. In some specific situations, however, the crude lipase form is not suitable. The low purity of these preparations can lead to low or variable enantioselectivity, the solubility of these proteins makes product recovery difficult, and the low activity (especially in organic solvents) can result in poor efficiency. In these cases, the use of a highly purified and heterogeneous form of the lipase catalyst is preferred over the impure and soluble protein catalyst.

A survey of the most commonly used commercial lipases reveals that most preparations contain less than 25% protein, which in turn may contain only a fraction of the lipase of interest contaminated with other hydrolases.[3] For example, analysis of the commercial preparations of two of the most synthetically useful lipases—from *Candida rugosa* lipase (CRL) and *Pseudomonas cepacia* lipase (PCL)—reveals that they contain less than 5–6 and 1% of the major lipase, respectively.[4] Often these inexpensive yeast and bacterial fermentation extracts are of sufficient activity and selectivity to be useful catalysts for chiral resolution. However, the presence of other hydrolases with opposing selectivity can lower the selectivity of these catalysts and complicate their use. Moreover, the low activity of crude lipase preparations makes their use in commercial-scale kinetic resolutions difficult due to the low volumetric productivity and complex downstream processing. Although highly purified forms of lipases have been shown to have maximal selectivity in some kinetic resolutions, they are rarely used because of their expense and instability. Purification of the lipase often improves the resultant enantioselectivity, but exacerbates the problem of low catalyst efficiency. The situation in organic solvents is worse. Organic solvent-based ester *synthesis* is attractive because it allows direct access to the opposite enantiomer to that derived from ester hydrolysis, but the activity of lipases in organic solvents can be orders of magnitude lower than that in water. In 1964, Quiocho and Richards found that glutaraldehyde cross-linking of enzyme crystals can stabilize these proteins toward mechanical stress and high concentrations of salts and urea.[5] More recently, we have shown that it is possible to have both a highly purified lipase and a highly active and stable catalyst in an easily recovered form by first crystallizing a lipase to homogeneity, followed by chemically cross-linking the crystals, such cross-

---

[3] H. K. Weber, H. Stecher, and K. Faber, *in* "Preparative Biotransformations" (S. M. Roberts, ed.), Section 5.2.1. Wiley, London, 1995.

[4] J. J. Lalonde, C. Govardhan, N. K. Khalaf, O. G. Martinez, K. Visuri, and A. M. Margolin, *J. Am. Chem. Soc.* **117,** 6845 (1995).

[5] F. A. Quiocho and F. M. Richards, *Proc. Natl. Acad. Sci. U.S.A.* **53,** 833 (1964).

linked enzyme crystals[6] (CLEC®) are active and stable in water[4] and in organic solvents.[7,8]

Preparation of CRL and PCL in the CLEC form involves crystallization to give platelike crystals about 30 μm in length, followed by glutaraldehyde treatment of the protein crystals while suspended in the crystallization medium. After cross-linking, the lipase CLECs are insoluble in aqueous buffer and organic solvents, so they can be used in a variety of reaction media and then recovered and reused. Lipase CLECs, similar to other protein crystals,[9] are macroporous, containing about 50% solvent by volume. Channels traverse the body of the crystal, permitting diffusion of solvents, substrates, and products. The large pore diameter of CLECs has been shown to allow rapid diffusion of relatively large molecules into the protein crystal.[10] These heterogeneous lipase catalysts are active as a suspension in water, in aqueous organic solvents mixtures, or in substantially water-free organic solvent. On completion of an ester synthesis or hydrolysis reaction, the heterogeneous catalyst can be recovered by filtration, to be stored or reused.

As one might expect, the crystal form of these lipases can influence both the selectivity and activity of the lipase CLEC. X-ray structure analysis[11,12] of lipases reveals an α/β hydrolase fold structure in which the active site is shielded by a helical polypeptide chain termed a "flap" or "lid." This flap is believed to undergo a conformational change when the enzyme is positioned at an oil/water interface, and this change is responsible for the interfacial activation observed in lipase catalysis. Crystallization of CRL in the presence of polyethylene glycol (PEG) 8000[13] results in a conformer in which the active site is shielded by the flap, while crystallization in the presence of 2-methyl-2,4-pentanediol[14] (MPD) results in an "open" conformation in which the flap extends into solvent (see Fig. 1). We have previously reported that CLECs prepared using PEG-derived lipase crystals show somewhat lower activity and enantioselectivity (0.48 μmol/min mg;

---

[6] N. L. St. Clair and M. A. Navia, *J. Am. Chem. Soc.* **114**, 7314 (1992).
[7] R. A. Persichetti, J. J. Lalonde, C. Govardhan, and N. Khalaf, **37**, 6507 (1996).
[8] N. Khalaf, C. G. Govardhan, J. J. Lalonde, R. A. Persichetti, Y.-F. Wang and A. L. Margolin, *J. Am. Chem. Soc.* **118**, 5494 (1996).
[9] B. W. Matthews, *J. Mol. Biol.* **33**, 491 (1968).
[10] R. A. Persichetti, N. L. St. Clair, J. P. Griffith, M. A. Navia, and A. L. Margolin, *J. Am. Chem. Soc.* **117**, 2732 (1995).
[11] B. Rubin, *Struct. Biol.* **1**, 568 (1994).
[12] J. D. Schrag and M. Cygler, *Methods Enzymol.* **284**, [4] (1997).
[13] P. Grochulski, Y. Li, J. D. Schrag, and M. Cygler, *Protein Sci.* **3**, 82 (1994).
[14] P. Grochulski, Y. Li, J. D. Schrag, F. Bouthillier, P. Smith, D. Harrison, B. Rubin, and M. Cygler, *J. Biol. Chem.* **268**, 12843 (1993).

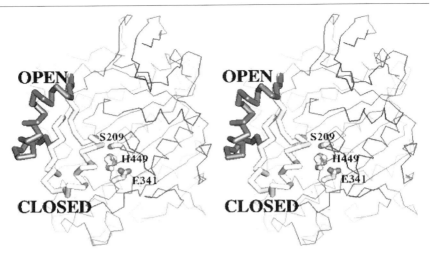

FIG. 1. Stereo view of the Cα tracing of CRL.[15] The serine, histidine, and glutamic acid residues that constitute the catalytic triad for CRL are labeled as S209, H449, and E341 in the figure. The helical lid is depicted in thick lines, with the closed conformer in white lines and the open one in gray.

$E = 25$) than those derived from MPD crystallization (1.4 μmol/min mg; $E = 64$) in the hydrolysis of chloroethyl ketoprofen.[4] It is possible that the "open" or activated conformation is maintained or favored in the MPD-derived crystals, thus accounting for the higher activity and selectivity (see, for example, the hydrolysis of methyl mandelate, Fig. 3B). However, the fact that the "closed" form can efficiently catalyze this reaction suggests that the lid movement may not be critical in the resolution of small organic substrates. All examples reported here employ CLECs derived from the "open" conformer crystals grown in MPD solutions.

Utility of CLEC Lipases

The utility of the CLEC form of the PC and CR lipases compared to that of the pure soluble or crude lipase forms is exemplified by the protocols given in this work. In the arylpropionate, aryloxypropionate and menthol resolutions, the high purity of the CR CLEC lipase results in a 4 to 50×

---

[15] Reproduced from data supplied from Refs. 13 and 14 and the Brookhaven Protein Data Bank: P. Grochulski and M. Cygler http://www.pdb.bnl.gov/, files "1crl" (open form) and "1trh" (closed form).

increase in enantioselectivity over the crude lipase preparation. The regioselective deprotection (not a kinetic resolution) of a highly insoluble steroid diacetate can be performed in 75% aqueous methanol by the PC CLEC, while hydrolysis by the soluble PC lipase is negligible, presumably due to denaturation of the soluble protein by the high concentration of organic cosolvent. Highly purified lipases, while free from competing hydrolases that can lower enantioselectivity, are often much less active than crude lipase preparations. High rates of catalysis and maximal enantioselectivity in organic solvent media are possible with the CLEC-stabilized lipase (CR-catalyzed menthol transesterification with vinyl butyrate). The principle of enantiocomplimentarity, the ability to produce either enantiomer through esterification or hydrolysis, is illustrated by comparison of ibuprofen, menthol, and phenethyl alcohol ester syntheses and hydrolyses. The potential for high catalyst productivity is shown by the large substrate to catalyst ratio (PC-catalyzed transesterification of phenethyl alcohol) and by the ability to recover and reuse the catalyst (multicyle ketoprofen resolution).

Characterization of CR and PC CLEC Lipases

Materials

Candida rugosa lipase (crude CRL) was obtained from Meito Sangyo (Tokyo, Japan). *Pseudomonas cepacia* lipase (PCL) was obtained from Amano (Osaka, Japan). Cross-linked enzyme crystals of CRL (ChiroCLEC-CR) and PCL (ChiroCLEC-PC) are commercial products of Altus Biologics (Cambridge, MA).

Preparation of Cross-Linked Enzyme Crystals

The major lipase in commercial CRL (designated CRL_1) was separated from the mixture by ion-exchange chromatography and then crystallized using 2-methyl-2,4-pentanediol.[11,16] Ion-exchange purification using Q-Sepharose (Pharmacia, Piscataway, NJ) was used to separate the two lipase activities. The mixture of proteins remaining after purification (fraction II), contained at least three major protein bands in addition to unrecovered

---

[16] B. Rubin, P. Jamison, and D. Harrison, *in* "Lipases: Structure, Mechanism and Genetic Engineering" (L. Alberghina, R. D. Schmid, and R. Verger, eds.) p. 63. VCH, New York, 1991.

 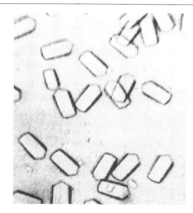

FIG. 2. Photomicrograph of cross-linked 30-μm CRL crystals (left) and 35-μm cross-linked PCL crystals (right).

CRL$_1$. The purified lipase was crystallized using 2-methyl-2,4-pentanediol as reported previously[11] with a minor modification: the extraction of protein with 50% isopropanol was omitted. Crystals of major lipase component CRL$_1$ are small uniform plates of 30 μm in length and no more than 2 μm thick.[17]

PCL crystals were produced by first dissolution of the soluble fraction of PS 30 (Amano), followed by purification[18] and crystallization to give 35-μm elongated plates as reported by Kim et al.[19] The crystals were then treated with glutaraldehyde in the presence of the crystallization medium.[6] Leaching of protein from the resultant cross-linked CR or PC lipase crystals into solution is less than 0.1 wt % after 4 days agitation at 40° in 10 mM Tris, 10 mM CaCl$_2$, pH 7.0, buffer. The CLEC lipase suspensions were stored in 10 mM Tris, 10 mM CaCl$_2$, 0.02% azide, pH 7.0 (see Fig. 2).

Manipulation of CLEC Suspensions. For hydrolytic reactions, CLEC lipases stored in aqueous buffer are used as a slurry. For organic solvent-based transformations, a dried form of the cross-linked lipase is used. To

---

[17] The size and shape of CRL crystals do not change after cross-linking.
[18] U. Bornscheuer, O.-W. Reif, R. Laush, R. Fretag, T. Scheper, F. N. Kolisis, and U. Menge, *Biochim. Biophys. Acta* **1201,** 55 (1994).
[19] K. K. Kim, K. Y. Hwang, H. S. Jeon, S. Kim, R. M. Sweet, C. H. Yang, and S. W. Suh, *J. Mol. Biol.* **227,** 1258 (1992).

transfer a representative sample of the lipase slurry, the suspension is agitated immediately before removing a sample. Samples can be transferred by automatic pipette (Eppendorf) fitted with large orifice pipette tips (Fisher Scientific). Reactions are shaken using a mechanical shaker/incubator (New Brunswick), stirred using a magnetic stirrer, or in the case of large reactions, by an overhead mechanical stirrer. Lipase CLECs can be recovered by filtration or low-speed centrifugation (Eppendorf microcentrifuge, 2000 rpm for 5 min) to afford complete recovery of the catalyst. Filtration is used to recover the catalyst using either gravity or vacuum filtration with a solid-phase extraction funnel fitted with a Teflon 10-μm filter such as the Poly-Prep 10-ml columns and the Bio-Spin 2-ml columns (Bio-Rad, Richmond, CA). For recovering larger quantities of the CLEC lipase, paper filters (Whatman No. 4) or glass fiber filter paper (MSI, Westborough, MA) can be used. Precipitated organic solids may be removed by washing the CLEC lipase with a nondeactivating solvent such as water-saturated t-amyl alcohol or aqueous 75% polyethylene glycol (average MW = 1000) and then stored in 10 mM Tris, 10 mM CaCl$_2$, 0.02% azide, pH 7.0.

Triacetin Hydrolysis. Determination of lipase CLEC activity is performed titrimetrically by following the release of acetic acid from the hydrolysis of triacetin. A suspension (6–25%) or solution (1–6%) of triacetin in 1 mM sodium phosphate, 1 mM calcium chloride, pH 7, buffer is incubated at 25° for 5 min. The CLEC lipase suspension is added (1 ml of 1.5 mg/ml) and the pH is maintained at 7 by the automatic addition of 0.05 N NaOH (Radiometer VIT90 autotitrator, Baegsvard, Denmark) over a period of 5 min.

Methyl Mandelate (5a) Hydrolysis. The initial rates of hydrolysis of racemic methyl mandelate 5a by the soluble lipase and CR CLECs is followed titrimetrically in 1 mm phosphate, pH 6. The temperature is maintained at 43° to prevent crystallization of the mandelate ester or acid.

Comparison of the triacetin esterolytic activity of the soluble CRL and the cross-linked CR crystals (Fig. 3A) indicates that there is no diffusional limit on the hydrolysis of triacetin; the specific activity of the soluble and CLEC forms of the lipase is equivalent. Only when the concentration of triacetin exceeds the solubility limit is there a difference between the two protein preparations, possibly due to a slight interfacial activation of the soluble enzyme. The rate of CLEC lipase-catalyzed hydrolysis of methyl mandelate (Fig. 3B) under mono- and biphasic conditions is three- to fourfold that of the soluble protein, indicating some activation or stabilization of the CLEC form.

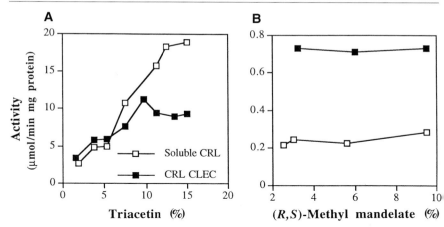

FIG. 3. (A) Effect of substrate concentration on rate of hydrolysis of triacetin at 25°. (*Note:* At substrate concentrations greater than 6% the system is biphasic.) (B) The effect of substrate concentration on the rate of hydrolysis of a synthetic substrate (**5b**) at 43°. The reaction mixture is in a biphasic condition at substrate concentrations greater than 6%. (Reprinted with permission from Ref. 4. Copyright 1995 American Chemical Society.)

Preparative Resolution of Racemic Acids and Alcohols by CLEC Lipase-Catalyzed Ester Hydrolysis

Materials

Methyl mandelate, (*R,S*)-ketoprofen, (*S*)-ketoprofen, (*R,S*)-ibuprofen, (*S*)-ibuprofen, (*R,S*)-flurbiprofen, and (*S*)-naproxen were purchased from Sigma Chemical Co. (St. Louis, MO). Trimethylsilyldiazomethane was purchased from Fluka (Ronkonkoma, NY). All other reagents were obtained from Aldrich Chemical Co. (Milwaukee, WI) and were of reagent grade. Methyl and chloroethyl esters of the 2-arylpropionic acids (**1b** through **4b**) were prepared by treatment of the carboxylic acid with thionyl chloride in dichloromethane, followed by addition of the appropriate anhydrous alcohol. Racemic **4b** was prepared by treating (*S*)-**4b** with 1% sodium methoxide in methanol.

General Procedure for Ester Hydrolyses

An emulsion of the substrate (unless specified otherwise, 100 mg) in the appropriate buffer (1 ml) and the CLEC lipase (2 mg) or the crude CRL (25 mg) in a 1.5-ml microcentrifuge tube is shaken in a G24 incubator (New Brunswick Scientific, Edison, NJ) at 200 rpm and 40°. Alternatively, the suspensions can be stirred with a magnetic stirrer with equivalent results.

The extent of hydrolysis is determined by periodically analyzing a 10-μl aliquot of the evenly suspended reaction mixture by high-performance liquid chromatography (HPLC) for compounds **1** through **5** [Microsorb MV 5 cm C_{18} column (Rainin, Woburn, MA) 60/40/0.1 acetonitrile, water, trifluoroacetic acid, flow = 1 ml/min, UV detection at 254 nm] or by capillary gas chromatography (GC) for compounds **6** through **8** [DB 1701 15-m × 0.25-mm GC column, 25-μm film thickness (J&W Scientific, Folsom, CA), split ratio 1:100, helium flow at 25 cm/sec]. Once the desired conversion is reached, the catalyst is isolated by filtration and stored for future use. The acid products (**1a–5a**) and the unreacted ester can be separated by solvent extraction. The pH of the suspension is adjusted to 2 with 1 N HCl and the acid and ester are extracted into diethyl ether. The acid is then extracted from the organic solution by shaking with pH 10 sodium carbonate solution, the pH of the aqueous layer adjusted to 2, and the acid extracted back into ether or ethyl acetate. Products from the hydrolysis of compounds **6–8** are purified by flash chromatography [9:1 hexane/ethyl acetate, 200–425 mesh Davisil 633 silica gel (Fisher Scientific, Fairlawn, NJ)].

The enantioselectivity of a first-order, irreversible kinetic resolution is best characterized by the E value developed by Sih and collaborators,[20] which numerically represents the ratio of the specificity constants (k_{cat}/K_M) of the catalyst for the faster converted enantiomer to that for the slower one. The enantiomeric excess of the substrate (ee_s) and that of the product may be used to determine the extent of conversion (c) using Eq. (1). To calculate E, the extent of conversion and the optical purity of either the remaining starting material and the product are needed.

$$E = \frac{\ln[(1-c)(1-ee_s)]}{\ln[(1-c)(1+ee_s)]} = \frac{\ln[1-c(1+ee_p)]}{\ln[1-c(1-ee_p)]} = \frac{k_{cat}/K_M \text{ (fast)}}{k_{cat}/K_M \text{ (slow)}} \quad (1)$$

Large-Scale Preparation of (S)-Ibuprofen (**1a**)

(*R,S*)-Ibuprofen methyl ester (**1b**) (100 g) and CR CLEC (1.5 g) are suspended in 1.5 liters of distilled water. The suspension is heated in a waterbath to 40° and stirred vigorously with an overhead stirrer. The pH will drop to 4.5 during the reaction, so no pH control is necessary. After 20 hr, the extent of hydrolysis is 38% (by HPLC). The reaction mixture is allowed to cool to room temperature and the CLECs recovered by filtration. The catalyst is washed with 100 ml of *t*-amyl alcohol and then stored for reuse in pH 7 Tris, 2 mM calcium acetate. The pH of the reaction mixture is adjusted to between 2 and 3, and then extracted with diethyl ether (2 ×

[20] C. S. Chen, C. J. Sih, Y. Fujimoto, and G. Girdaukus, *J. Am. Chem. Soc.* **104**, 7294 (1982).

800 ml). The combined ether extracts are extracted with saturated pH 9.5 sodium carbonate (5 × 300 ml), and then the combined aqueous layers are back extracted with diethyl ether (3 × 300 ml). The combined ether layers are washed with saturated sodium chloride and dried over anhydrous sodium sulfate and the solvent is evaporated under reduced pressure to give (*R*-**1b**) as a colorless oil (59.1 g, 95.3% of theoretical, 55.8% ee by chiral HPLC). The acid (*S*-**1a**) is precipitated from the combined sodium carbonate extracts by adjusting the pH of the aqueous layer to 2 with 6 *N* HCl, saturating with sodium chloride, then extracting with diethyl ether (2 × 500 ml). The ether extracts are extracted with saturated sodium chloride (100 ml) and then dried over anhydrous sodium sulfate and the ether is evaporated under reduced pressure to give (*S*-**1a**) as a crystalline white solid (30.6 g, 87% of the theoretical yield, 93% ee by chiral HPLC analysis). (*S*)-Ibuprofen of optical purity >90% can then be crystallized to an optical purity of >99% by simple *in situ* conversion to the sodium salt.[21]

Determination of Optical Purity

Assignments of absolute configuration are made from original citations and by comparison with authentic samples.

α-Methyl-4-[2-methylpropyl]benzeneacetic acid (ibuprofen, **1a**) and methyl ester (**1b**). *Chiral HPLC conditions for* **1b**: Chiracel OJ 25-cm column (Chiral Technologies, Exton, PA), mobile phase = 99.1% hexane, 0.9% isopropanol, flow rate = 1 ml/min, UV detection at 235 nm. *Retention times:* 8.13 min (*S*-**1b**) and 9.57 min (*R*-**1b**). Chiral HPLC conditions for **1a**: (*R,R*) Whelk-O1, 5-μm, 100-Å, 25-cm column (Regis Technologies, Morton Grove, IL). Mobile phase = 99% hexane, 1% isopropanol, and 0.5% acetic acid flow rate = 1 ml/min, UV detection at 254 nm. *Retention times:* 8.7 min (*S*-**1a**), 10.2 min (*R*-**1a**).

2-[3-Benzoylphenyl]propionic acid (ketoprofen, **2a**) and chloroethyl ester (**2b**).[22] *Chiral HPLC conditions:* Chiracel OJ 25-cm column mobile phase = 90% hexane (0.05% acetic acid) and 10% isopropanol, flow rate = 1 ml/min, UV detection at 235 nm. *Retention times:* 16.6 min (*R*-**2a**), 19.9 min (*S*-**2a**), 31.4 min (*S*-**2b**), and 34.5 min (*R*-**2b**).

2-Fluoro-α-methyl-4-biphenyl-acetic acid (flurbiprofen, **3a**) and chloroethyl ester (**3b**).[22] *Chiral HPLC conditions:* Chiracel OJ 25-cm

---

[21] T. Manimaran and G. P. Stahly, *Tetrahedron: Asymmetry* **4**, 1949 (1993).
[22] S.-H. Wu, Z.-W. Guo, and C. J. Sih, *J. Am. Chem. Soc.* **112**, 1990 (1990).

TABLE I
CR CLEC Lipase-Catalyzed Resolution of Arylpropionic Acid Esters

(R,S)-α-Arylpropionate ester $\xrightarrow{\text{CR CLEC}}$ (S)-Arylpropionic acid + (R)-Arylpropionate ester

Ar		% Enantiomeric excess (% conv.)		E	
		CR CLEC	Crude	CR CLEC	Crude
(4-isobutylphenyl)	**1a:** R = H **ibuprofen** **b:** R = Me	94.6 S-**1a** (22)[a]	81.9 S-**1a** (39.3)[a]	47	17
(3-benzoylphenyl)	**2a:** R = H **ketoprofen** **b:** R = CH$_2$CH$_2$Cl	91.1 S-**2a** (49.3)[b]	64.5 R-**2b** (66)[b]	66	5
(2-fluoro-4-biphenylyl)	**3a:** R = H **flurbiprofen** **b:** R = CH$_2$CH$_2$Cl	94.3 S-**3a** (34.4)[c]	61.1 S-**3a** (34)[c]	55	6
(6-methoxy-2-naphthyl)	**4a:** R = H **naproxen** **b:** R = Me	97.3 S-**4a** (39)[d]	76.2 S-**4a** (46.8)[d]	>100	12

[a] Reaction buffer 0.1 M pH 6 sodium acetate.
[b] Reaction buffer 0.1 M pH 5 sodium acetate.
[c] Reaction buffer 0.1 M pH 7 sodium phosphate.
[d] Reaction buffer 50% PEG 1000/50% pH 5 ammonium acetate.

column mobile phase = 99% hexane (0.05% acetic acid) and 1% isopropanol, flow rate = 1 ml/min, UV detection at 235 nm. *Retention times:* 45 min (*R*-**3a**), 54 min (*S*-**3a**), 79 min (*S*-**3b**), and 95 min (*R*-**3b**).

6-Methoxy-α-methyl-4-naphthaleneacetic acid (naproxen, **4a**) and methyl ester (**4b**).[22] *Chiral HPLC conditions:* (*R,R*) Whelk-O1, 5-μm, 100-Å, 25-cm column. Mobile phase = 90% hexane, 9.5% ethanol, and 0.5% acetic acid, flow rate = 1 ml/min, UV detection at 254 nm. *Retention times:* 16.8 min (*S*-**4b**), 19.2 min (*S*-**4a**), 21.8 min (*R*-**4b**), and 34.3 min (*R*-**4a**).

The results for the hydrolysis of compounds **1a**–**4a** are summarized in Table I.

α-Hydroxy-phenylacetic acid (mandelic acid, **5a**) and methyl ester (**5b**).[23]   **5a** and **5b** are resolved following the general protocol using 0.1 M pH sodium acetate as buffer. *Chiral HPLC conditions:* Chiracel OJ 25-cm column, mobile phase = 92% hexane (0.1% trifluoroacetic acid) and 8% isopropanol, flow rate = 0.8 ml/min, UV detection at 222 nm. *Retention times:* 18.6 min (*R*-**5b**), 21.8 min (*S*-**5b**), 27.2 min (*R*-**5a**), and 42.2 min (*S*-**5a**).

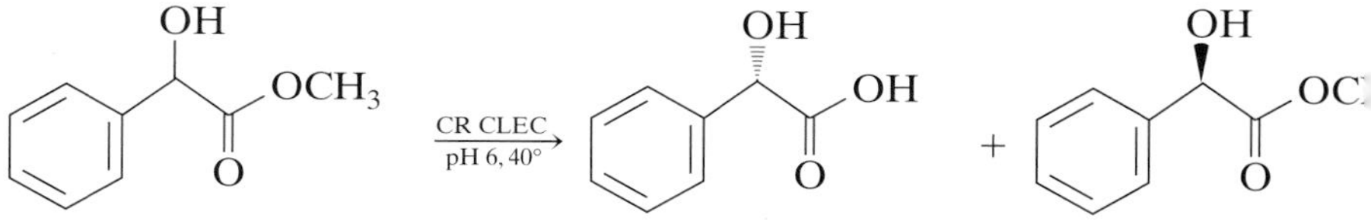

5b
(*R,S*)-Methyl mandelate

CR CLEC
pH 6, 40°

5a
(*S*)-Mandelic acid
96% ee, $c = 24\%$
$E = 64$ ($E_{crude} = 26$)

5b
(*R*)-Methyl mandel
30% ee

Menthol (**6a**) and menthol acetate (**6b**).[24]   Hydrolysis of **6b** is performed using the general protocol for ester hydrolysis with 100 mg of **6b** in 1 ml of 0.1 N, pH 7, phosphate buffer and CR CLEC (20 mg). *Chiral GC conditions:* Cyclodex B capillary GC 25-m column, 25-mm i.d.; He flow at 1 ml/min. *Temperature program:* Initial = 90° for 5 min, gradient rate = 1°/min, final = 115° for 10 min. *Retention times:* (+)-**6a**-menthol, 24.90 min; (−)-**6a**-menthol, 25.40 min.

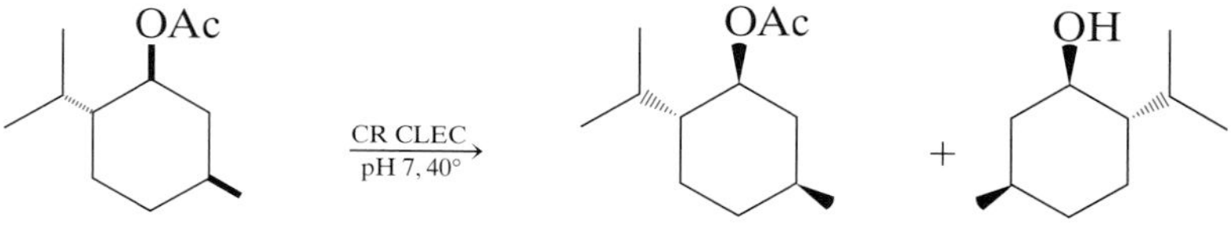

6b
(+/−) Menthol acetate

CR CLEC
pH 7, 40°

6b
(+) Menthol acetate
99.5% ee, $c = 51.3\%$
$E \cong 200$ ($E_{crude} \cong 200$)

6a
(−) Menthol
95.2% ee

Regioselective Hydrolysis of 5α-Androstane-3β,17β-Dihydroxy Diacetate (**7d**). To a solution of 37.6 mg 5α-androstane-3β,17β-dihydroxy diacetate (**7d**) in 2.5 ml methanol is added 0.84 ml of 0.1 M phosphate buffer,

---

[23] S. M. Ahmed, R. J. Kazlauskas, A. H. Morinville, P. Gorchulski, J. D. Schrag, and M. Cygler, *Biocatalysis* **9**, 1 (1994).
[24] G. Langrand, J. Baratti, G. Buono, and C. Triantophylides, *Tetrahedron Lett.* **27**, 29 (1986); G. Caron, W.-M. Tseng, and R. J. Kazlauskas, *Tetrahedron: Asymmetry* **1**, 83 (1994).

pH 6.5. To the resulting suspension is added 20 µL (1 mg) of PC-CLEC and the reaction mixture is agitated on a rotary shaker at 40°. Another 20 µL of catalyst slurry is added after 20 hr. The reaction is complete after a further 10 hr of mixing at the same temperature. After filtering off the catalyst, the product is extracted with ethyl acetate. The combined ethyl acetate extracts were washed with saturated sodium chloride and dried over sodium sulfate. The product is analyzed by GC, indicating complete conversion of **7d** and the selectivity of 3-hydroxy product (**7b**) is 100% by GC. None of the 17-hydroxy isomer (**7c**) was observed.[25] GC conditions: DB 1701 15-m × 0.25-mm capillary GC column, 25-mm film thickness, helium flow at 25 cm/sec. *Temperature program:* initial = 220° for 1 min, gradient rate = 15° per min, final 250° for 22 min. *Retention times:* 3,17-dihydroxy compound, 11.67 min; 3-hydroxy compound, 13.99 min; 17-hydroxy compound, 14.53 min; 3,17-diacetate, 17.66 min.

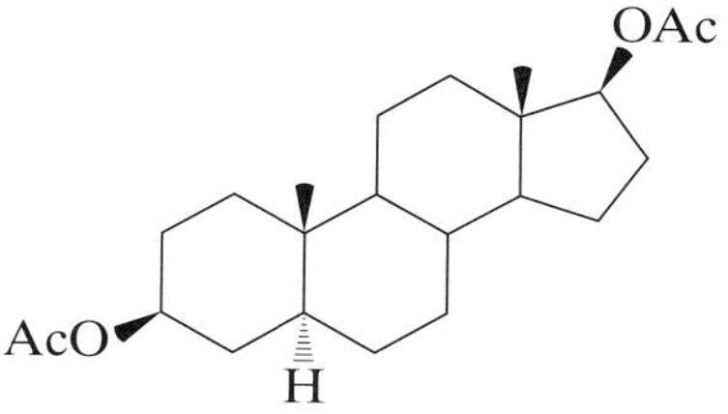

7d
5α-Androstane-3β,17β-
dihydroxy diacetate

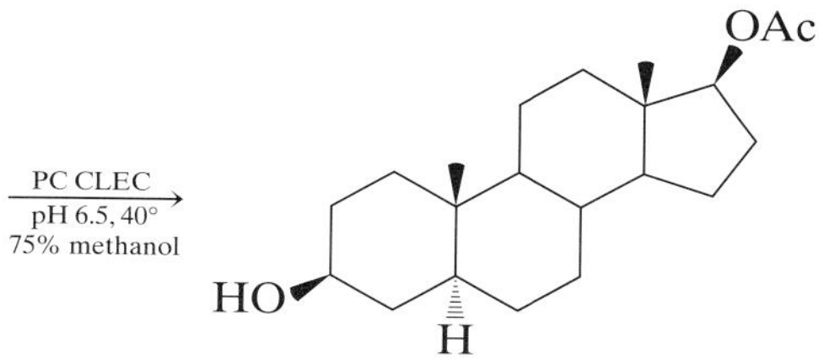

PC CLEC
pH 6.5, 40°
75% methanol

7b
5α-Androstane-3β,17β-
dihydroxy (17-O) monoacetate
>98%

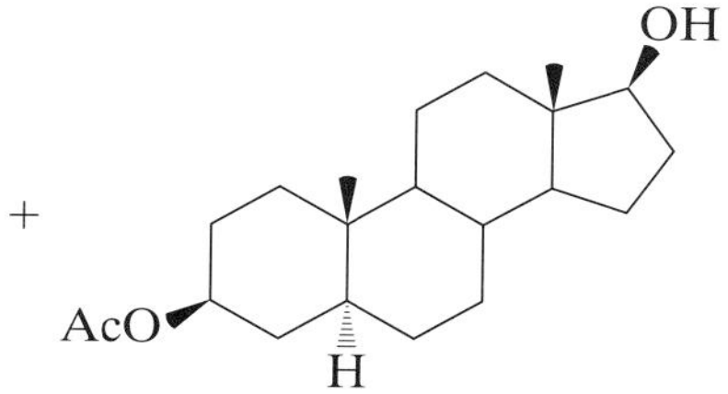

+

7c
5α-Androstane-3β,17β-
dihydroxy (3-O) monoacetate
0%

---

[25] Regioisomer was identified by comparison with selectivity of *Bacillus licheniformis* protease. S. Riva and A. M. Klibanov, *J. Am. Chem. Soc.* **110**, 3291 (1988).

Kinetic Resolution of Phenethyl Acetate (R,S)-8b. To a 10% solution of phenethyl acetate ((*R,S*)-**8b**) in 0.1 *M* phosphate buffer, pH 7.5, 8 mg of PC CLEC is added and the reaction mixture is stirred overnight at room temperature. The catalyst is then removed by centrifugation and the product, phenethyl alcohol **8a** and the unreacted ester **8b**, is extracted into ethyl acetate. Chiral GC is used to determine conversion and optical purity. At a conversion of 47.8%, the optical purity of product **8a** and substrate **8b** is 95.9 and 87.79%, respectively ($E = 140$). The ester and alcohol are conveniently separated by flash chromatography using 9:1 hexane/ethyl acetate as the elution solvent. *Chiral GC conditions:* Cyclodex capillary GC 25-m column, 25-mm i.d., N_2 1 ml/min. *Temperature program:* initial = 100° for 4 min, gradient rate = 5° per min to 135° and 2 min at 135°, 2° per min to 144° and 5 min at 144°, 5° per min to 150° and 2 min at 150°. *Retention times:* 12.08 and 12.47 min for (*S*) and (*R*) ester, respectively; 12.25 and 12.55 min for (*R*) and (*S*) alcohol, respectively.

8b
(*R,S*)-Phenethyl acetate

PC CLEC
pH 7.5, 23°

8b
(*S*)-Phenethyl acetate
87.8% ee, $c = 47.8\%$
$E \cong 140$

8a
(*R*)-Phenethyl alcohol
ee = 95.9%

Multiple-Cycle Ketoprofen (2a) Resolution. Ketoprofen chloroethyl ester (*R,S*-**2b**) (500 mg) and CRL CLECs (25 mg) are suspended in 5 ml of 50/50 0.5 *M* ammonium acetate/PEG (average molecular weight = 1000) in a capped 10-ml Poly-Prep solid-phase extraction funnel (Bio-Rad). The tube is shaken in a 40° incubator at 200 rpm, or the reaction is stirred by magnetic stirrer. Periodically, 10-μl aliquots of the evenly suspended reaction mixture are removed and analyzed by HPLC to determine the extent of conversion. Once approximately 50% conversion had been reached, the reaction mixture is removed by gentle suction filtration and the catalyst is washed once with 5 ml of water-saturated *t*-amyl alcohol, once with 5 ml of 20 m*M*, pH 7, Tris, 2 m*M* $CaCl_2$, and then with 5 ml of 50/50 0.5 *M* ammonium acetate/PEG 1000. The catalyst can be stored overnight in the reaction buffer, and a new hydrolysis cycle is initiated the following day by the addition of racemic **2b** (Fig. 4).

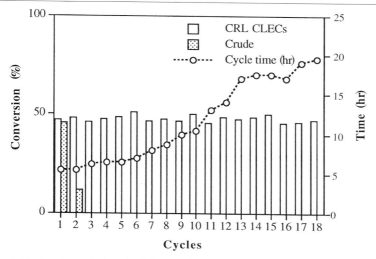

Fig. 4. Multicycle resolution of (S)-ketoprofen. Ten percent chloroethyl ketoprofen in 50/50 PEG 1000/pH 5 ammonium acetate buffer, 40°, 5 mg/ml CRL CLEC. (Reprinted with permission from Ref. 4. Copyright 1995 American Chemical Society.)

Resolution of Racemic Acids and Alcohols via CLEC-Catalyzed Esterification and Transesterification in Near Anhydrous Organic Solvents

Lipases have long been known to catalyze enantiospecific ester synthesis in substantially water-free organic solvents, although usually at rates much less than the corresponding hydrolytic reaction.[2] The absence of a bulk water phase allows reversal of equilibria, so that enantiospecific ester synthesis or transesterification is possible. Because the selectivity of the lipase is for the same absolute configuration in both the hydrolysis and synthesis reactions, the faster hydrolyzed ester is also the one preferentially formed in esterification. This *enantiocomplimentarity* allows direct access to either enantiomer (e.g., PC-catalyzed phenethyl alcohol ester hydrolysis and synthesis). Despite the attractiveness of performing ester synthesis in organic media, the rates are often orders of magnitude lower than the corresponding hydrolysis reaction. A highly purified lipase, while often more enantioselective, can be orders of magnitude slower still (see transesterification of menthol with vinyl butyrate). We have found that, by stabilizing the lipase in the CLEC form, rates of transesterification can equal and even exceed those of the corresponding hydrolysis reaction. Apparently, a bulk water phase is not a requirement for efficient protein catalysis, as long as there is sufficient hydration of the lipase and the catalyst is stabilized toward the solvent and reactants.

CLEC Drying Procedure[8]

PC CLECs or CR CLECs are suspended in 10 mM Tris, 10 mM CaCl$_2$, pH 7.0, and transferred to a sintered glass funnel (porosity ~5 μm). The supernatant is decanted or removed by suction. An equal volume (to the wet CLEC filter-cake) of 2-butanone with 30 g of the detergent N,N',N'-*polyoxyethylene(10)*-N-tallow-1,3-diaminopropane (EDT-20, Sigma) is added. The solvent and surfactant are removed by gentle suction. The mixture is transferred to a filter funnel after breaking up any lumps and dried in a stream of nitrogen to a water content of about 2–3% for the PC CLEC and about 10–13% for CR CLEC as determined by Karl Fisher titration. (*Note:* The drying process was stopped periodically and aggregates were broken apart.)

NOTE: Lipases, while active and stable in organic solvents, require a finite amount of water to maintain activity. High rates of catalysis require that an optimal amount of water, specific for the catalyst form and the organic solvent media, must be maintained during the reaction. Halling *et al.*[26,27] have used the concept of water activity to explain this phenomenon. We have found that the addition of a surfactant also improves the activity and stability of the heterogeneous CLEC catalyst. Thus the CLEC lipases in this work were dried to an optimal water level and could be used as such in hydrophobic solvents such as isooctane or toluene. For use in more polar solvents, additional water is required. For organic solvent reactions that consume or produce water, continuous control of water activity is suggested.[27]

Enantiospecific Esterification

Esterification of Ibuprofen **1a** *with n-Amyl Alcohol.*[28] (*R,S*)-ibuprofen **1a** (100 mg) is dissolved in 5 ml isooctane. To this solution is added 250

---

[26] G. A. Hutcheon, P. J. Halling, and B. D. Moore, *Methods Enzymol.* **286**, [21], 1997 (this volume).

[27] L. S. Kvittingen, B. Sjursnes, T. Anthonsen, and P. J. Halling, *Tetrahedron* **48**, 279 (1992); S. A. Khan, P. J. Halling, and G. Bell, *Enzyme Microb. Technol.* **12**, 453 (1990); P. J. Halling, and A. Macarae, European Patent Application 64,855 (1982). H. L. Goderis, *Biotechnol. Bioeng.* **30**, 258 (1987); (P. J. Halling and R. H. Valivetty, *in* "Biocatalysis in Non-Conventional Media" (J. Tramper, ed.), p. 13. Elsevier, New York, 1992).

[28] A. Mustranta, *Appl. Microbiol. Biotechnol.* **38**, 61 (1992).

μl *n*-amyl alcohol and 5 mg CR CLEC. The suspension is stirred at room temperature and the production of *n*-amyl ibuprofen (**1c**) is followed by chiral HPLC. After 24 hr the conversion will reach 28% and the optical purity of the product **1c** is >99.5. The acid and ester can conveniently be separated by simple extraction as described for the large-scale ibuprofen resolution. *Chiral HPLC conditions:* (*R,R*) Whelk-O1, 5-mm, 100-Å, 25-cm column. Mobile phase A = hexane 0.5% acetic acid, B = hexane. *Gradient elution:* at time zero mobile phase = 100% A, at 30 min mobile phase = 100% B. Flow rate = 1 ml/min, UV detection at 254 nm. *Retention times:* 16.7 and 19.2 min for the ester (**1c**), 21.8 and 28.1 min for the acid (**1a**).

NOTE: The reaction solvent and the acid/alcohol ratio is important in this reaction. Nonpolar solvents such as isooctane give the highest enantioselectivity, whereas solvents such as toluene or *t*-amyl alcohol, give lower enantioselection and rates of esterification. When higher proportions of alcohol are used, or when the alcohol is used as the reaction solvent, much slower conversions are observed.

1a
(*R,S*)-Ibuprofen

CR CLEC
n-Amyl alcohol
in isooctane

1c
(*S*)-Amyl ibuprofen
ee > 99.5%, $c = 28\%$
$E = 300$

+

1a
(*R*)-Ibuprofen
ee = 37.9%

Resolution of (R,S)-(4-Chloro)-2-Phenoxypropionic Acid by Esterification (R,S)-9a. (*R,S*)-(4-chloro)-2-phenoxypropionic acid (*R,S*)-**9a** (20 mg, 0.1 mmol) is dissolved in 1 ml of *n*-heptane with *n*-butanol (36.7 μl, 0.4 mmol). The resulting mixture is stirred at 25° while CR CLEC (10 mg) is added. Stirring is continued for 1.5 hr at which time the conversion is 47.5%. Conversion and optical purity are determined by Chiral HPLC. *Chiral HPLC conditions:* Chiracel OJ 25-cm column, mobile phase = 96.0% hexane (with 0.5% acetic acid), 4.0% isopropanol, flow rate = 0.7 ml/min, UV detection at 235 nm. *Retention times:* 18.84 min (*S*-**9a**) and 22.41 min (*R*-**9a**); 8.80 min (*S*-**9b**) and 9.79 min (*R*-**9b**).

$$\text{Cl-C}_6\text{H}_4\text{-O-CH(CH}_3\text{)-CO}_2\text{H} \xrightarrow[\text{butanol}]{\text{CR CLEC, 25°, Heptane}} \text{Cl-C}_6\text{H}_4\text{-O-CH(CH}_3\text{)-CO}_2\text{Bu}$$

9a
(*R,S*)-(4-Chloro)-
2-phenoxypropionic acid

9b
(*S*)-Butyl-(4-chloro)-
2-phenoxypropionate
99.5% ee, c = 47.5%
$E > 1000$

+ Cl-C$_6$H$_4$-O-CH(CH$_3$)-CO$_2$H

9a
(*R*)-(4-Chloro)-
2-phenoxypropionic acid
89.4% ee

Enantiospecific Synthesis of (−) *Menthol Butyrate* (−)-**6c** *from* (+/−) *Menthol* (+/−)-**6a**. A solution of (+/−) menthol **6a** (156.3 mg, 1 mmol) in 5 ml of toluene is stirred with 50 mg of CR CLEC. Vinyl butyrate (135 μl, 1 mmol) is added and the resulting suspension is stirred at 25°. After 30 min, the degree of conversion of **6a** is 47%. *GC conditions:* DB 1701 15-m × 0.25-mm GC column, 25-mm film thickness; helium flow at 25 cm/sec. *Temperature program:* initial = 119° for 1 min, gradient$_1$ rate = 5° per min to 130° for 0.3 min, gradient$_2$ rate = 70° per min to 175° for 1.86 min. *Retention times:* 2.85 min (**6a**), 4.77 min (**6c**). The reaction is stopped by filtration of the catalyst. Optical purity of the product (−)-**6a** is determined by chiral GC. *Chiral GC conditions:* Cyclodex B capillary GC 25-m, column, 25-mm i.d. (J&W Scientific). N$_2$ flow at 1 ml/min. *Temperature program:* initial = 90° for 5 min, gradient rate = 1° per min, final = 115° for 10 min. *Retention times:* 24.90 min ((+)-**6a**), 25.40 min ((−)-**6a**), 35.97 min ((−)-**6c**), 36.11 min ((+)-**6c**).

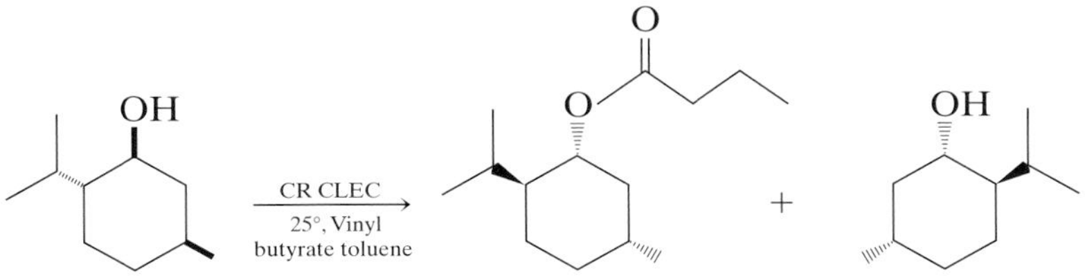

6a
(+/−) Menthol

→ CR CLEC, 25°, Vinyl butyrate toluene →

6c
(−) Menthol butyrate
>99.5% ee, $c = 47\%$
$E \cong 1000$ ($E_{crude} \cong 17$)

+

6a
(+) Menthol

Enantiospecific Synthesis of (−) Menthol Acetate (−)-6b from (+/−) Menthol (+/−)-6a. (−) Menthol acetate (−)-**6b** is prepared in a manner identical to that of menthol butyrate (−)-**6c** with the following exceptions: 15.6 mg of (+/−) menthol **6a** (0.1 mmol) is used in 1 ml isooctane, along with 25 mg CR CLEC and vinyl acetate (18.4 μl, 0.2 mmol). After 30 min, the degree of conversion of **6a** is 29% and the optical purity of **6b** is 99.9%. *Chiral GC conditions:* Cyclodex B capillary GC 25-m column, 25-mm i.d., He flow at 1 ml/min. *Temperature program:* initial = 90° for 5 min, gradient rate = 1° per min, final = 115° for 10 min. *Retention times:* (+)-**6a**-menthol, 24.90 min; (−)-**6a**-menthol, 25.40 min; (+)-**6b**-menthol acetate, 25.83 min; (−)-menthol acetate, 24.65 min.

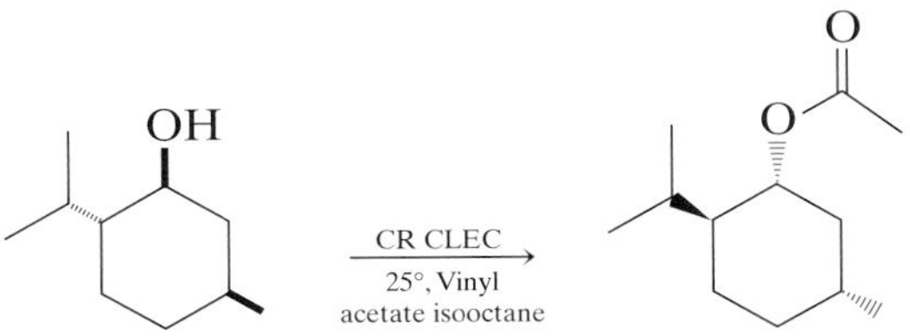

6a
(+/−) Menthol

→ CR CLEC, 25°, Vinyl acetate isooctane →

6b
(−) Menthol acetate
99.9% ee, $c = 29\%$
$E \cong 1000$ ($E_{crude} \cong 16$)

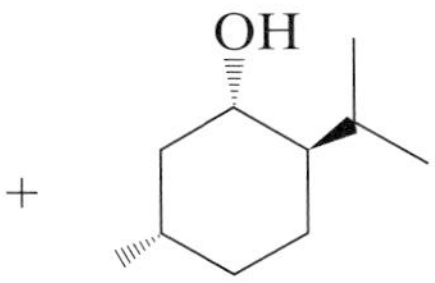

+

6a
(+) Menthol

TABLE II
COMPARISON OF RATE AND ENANTIOSELECTIVITY OF CR CLEC, PURE, AND CRUDE CRL IN THE TRANSESTERIFICATION OF MENTHOL[a]

Lipase form	Vinyl ester	% ee of (−) ester (% conversion)	Initial rate (μmol/mg solid/min)	E
CLEC	Butyrate	>99.5 (47)	4.7	~1000
Crude	Butyrate	86.6 (20)	0.008	17
Pure	Butyrate	96.8 (11.9)	0.00045	74[29]
CLEC	Acetate	99.9 (29)	1.2	~1000
Crude	Acetate	88.0 (8)	0.004	16

[a] Data from Pergichetti et al.[7]

From Table II we can see that all three lipase forms maintain selectivity for acylation of the same menthol stereoisomer, but differ widely in activity and enantioselectivity. The crude preparation is highly active, yet unselective while the pure lipase is selective, yet inactive. In contrast, the CLEC form of the lipase is the most enantioselective, while at the same time the most active lipase form. As in CR CLEC-catalyzed hydrolytic reactions,[4] the dramatic increase in enantioselectivity can be explained by the purity of the CR CLEC catalyst. Note that the enantioselectivity of the crude and the purified enzymes is equivalent in the hydrolysis, yet purified forms were more selective than the transesterification. This implies that only one of the hydrolases is active in organic solvents, while in the organic solvent transesterification, two or more of the hydrolases in the crude preparation may be active.

Resolution of Phenethyl Alcohol ((R,S) 8a) by Transesterification. To 1 ml of 200 mM phenethyl alcohol ((R,S)-**8a**) and 200 mM vinyl acetate in toluene is added to 1.2 mg of PC CLEC and stirred at room temperature for 30 min. The catalyst is removed by centrifugation to stop the reaction. Conversion and optical purity are determined by chiral GC. *Chiral GC conditions:* Cyclodex capillary GC 25-m column, 25-mm i.d., N_2 1 ml/min. *Temperature program:* initial = 100° for 4 min, gradient rate 5° per min to 135° and 2 min at 135°, 2° per min to 144° and 5 min at 144°, 5° per min to 150° and 2 min at 150°. *Retention times:* 12.08 and 12.47 min for (S) and (R) ester, respectively; 12.25 and 12.55 min for (R) and (S) alcohol, respectively. Conversion was found to be 41.7% and enantiomeric excess of product and substrate at this conversion was found to be 98.6 and 70.4%, respectively, with an E value of 297.

---

[29] Reaction conditions: Menthol (0.2 mmol) and vinyl butyrate (0.2 mmol) in 1 ml toluene, 1 μl water, and 40 mg pure CRL (49.5% protein).

OH   OH   OAc

8a
R,S)-Phenethyl alcohol   **8a**   **8b**
(S)-Phenethyl alcohol   (R)-Phenethyl acetate
70.4% ee, $c = 41.7$%   98.6% ee
$E \cong 300$

Preparative Scale Resolution of Phenethyl Alcohol (Productivity Ratio of 1 : >4500). A high substrate to catalyst ratio can be used to maximize the productivity of the catalyst. To a solution of 100 ml of a mixture containing 500 mM phenethyl alcohol **8a** and 500 mM vinyl acetate in toluene is added 1.3 mg of PC CLEC. The reaction mixture is stirred at room temperature until the conversion reaches about 50% (16 hr) based on GC analysis of the reaction mixture. The yield of phenethyl acetate (R)-**8b** per mg of catalyst under these conditions is found to be about 4.5 g.

*Regioselective Esterification of 5α-Androstane-3β,17β-Dihydroxy(**7a**).* To a suspension of 21.9 mg 5α-androstane-3β,17β-diol and 56 μl of vinyl acetate in 5 ml toluene is added 2 mg of PC CLEC. The resulting mixture is stirred at room temperature until **7a** is consumed (about 40 hr), after which time the reaction is halted by filtration of the catalyst. The product can be directly analyzed by GC. The yield of the 3-acetate product **7c** is quantitative; none of the 17-acetate isomer can be detected. *GC conditions:* DB 1701 15-m × 0.25-mm capillary GC column, 25-mm film thickness, He flow at 25 cm/sec. *Temperature program:* initial = 220° for 1 min, gradient rate = 15° per min, final 250° for 22 min. *Retention time:* 3,17-dihydroxy compound, 11.67 min; 3-hydroxy compound, 13.99 min; 17-hydroxy compound, 14.53 min; 3,17-diacetate, 17.66 min.

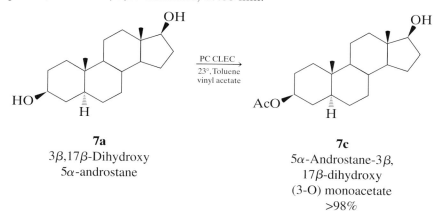

7a
3β,17β-Dihydroxy
5α-androstane

7c
5α-Androstane-3β,
17β-dihydroxy
(3-O) monoacetate
>98%

7b
5α-Androstane-3β,
17β-dihydroxy
(17-O) monoacetate
0%

Conclusion

The cross-linked crystalline form of a lipase is superior to the soluble lipase preparation under some specific circumstances. Situations where a CLEC lipase is preferred include (1) the highly pure catalyst gives higher enantioselectivity, (2) a recoverable form of the catalyst is needed, (3) high catalyst productivity is needed, and (4) organic-solvent-based reactions where high activity is needed. The highly pure, insoluble form of the lipase allows it to be used in either aqueous or organic media to resolve alcohols and esters, and then to be recovered by simple filtration and reused.

Acknowledgments

The research outlined in this chaper represents the dedicated efforts of Chandrika Govardhan, Nazer Khalaf, Aldo Martinez, Rose Persichetti, Nancy St. Clair, Kalevi Visuri, Yi Fong Wang, Kirill Yakovlevsky, and Bailing Zhang.

[21] Measurement and Control of Hydration in Nonaqueous Biocatalysis

By GILLIAN A. HUTCHEON, PETER J. HALLING, and BARRY D. MOORE

Introduction

The study of biocatalysis in organic rather than aqueous media has not only increased the range of possible synthetic reactions but has also promoted a greater understanding of the effect of water on protein structure and function. In these systems, enzymes are generally suspended rather than dissolved in the organic media and a bulk water phase is no longer present. Therefore, for example, the amount of water bound to the enzyme can be varied and the effects of this observed.

Biocatalytic activity is often very sensitive to the hydration state of the enzyme. Some lipases require very little bound water for activity; however, for many other enzymes, increased numbers of bound water molecules lead to increased activity. The hydration state of an enzyme varies with the water content of the reaction system and depends on the solubility of water in the various components of the reaction mixture, for example, solvent, support, and substrates. It is therefore important to control the amount of water present in each phase very carefully.

For some applications, addition of a known quantity of water to the reaction system will suffice; however, when studying enzyme activity in a nonaqueous environment, control of the water present is crucial for reproducible results and comparison of different reaction systems.

The thermodynamic water activity, a_w, is the most useful indicator of the extent of hydration of nonaqueous biocatalytic systems. By definition, at equilibrium, the thermodynamic water activity is equal in all the phases present in a reaction vial. This occurs because water will tend to move from a phase of high activity to phases of lower activity until they are all at equilibrium. This means that the vapor headspace above the reaction mixture is also at that same water activity. Hence the ratio of the vapor pressure of water above the reaction mixture to the vapor pressure of pure water, at the same temperature, can be used to define the water activity, a_w, which has a value between 0 and 1. (Theoretical details concerning water activity are detailed elsewhere.[1]) This means that water activity is a

---

[1] P. J. Halling, *Enzyme Microb. Technol.* **16**, 178 (1994).

useful predictor of optimum biocatalyst hydration conditions even if other reaction conditions are different.

In the course of a reaction, water may be produced or consumed. In these cases, it is often important to maintain the reaction system at the original water activity for the reaction to proceed efficiently. In this chapter, we describe methods for controlling the hydration of both the solid biocatalyst and the nonaqueous reaction media. Preequilibration methods and means of controlling the water activity during the course of a reaction are described and methods for measuring the water content in the different phases and the reaction system as a whole are also outlined.

Methods for Controlling Hydration

Preequilibration with Saturated Salt Solutions

A saturated solution of an inorganic salt has a well-defined water vapor pressure at a controlled temperature.[2] Provided excess solid salt is present, and the system is at equilibrium, the vapor pressure will stay the same regardless of any gain or loss of water. It is most convenient if the saturated solution is prepared as a solid-rich slush rather than as a solution containing a small amount of residual solid.

These saturated salt solutions are useful for preequilibrating the enzyme and/or reaction mixture to a defined water activity through the vapor phase. Any material in contact with the vapor will equilibrate to the defined water activity of the saturated salt solution. A variety of different salts can be employed to span hydration levels from water activities of 0.0657 to 1[2] (Table I). Below 0.0657, drying agents such as molecular sieves must be used; however, the water activities obtained are more uncertain.[3]

Equilibration of the Various Components of a Reaction Mixture. If the final reaction mixture is to be at a clearly defined water activity, preequilibration of all the components must be considered carefully. It is usually preferable to preequilibrate all of the components because addition of significant quantities of some anhydrous phases can significantly reduce the water activity of the whole system, especially at low hydration levels. Obviously the reaction system cannot be equilibrated as a whole because one component must be added to start the reaction. Therefore, one or more phases must be equilibrated separately. Two options commonly applied are (1) equilibration of the solid biocatalyst and the organic solvent containing the other components, and (2) equilibration of the biocatalyst in the organic

[2] L. Greenspan, *J. Res. Natl. Bur. Stand. A* **81A,** 89 (1977).
[3] R. H. Valivety, P. J. Halling, and A. R. Macrae, *FEBS Lett.* **301,** 258 (1992).

TABLE I
INORGANIC SALTS FOR THE PREPARATION OF
SATURATED SALT SOLUTIONS AT A CONTROLLED
WATER ACTIVITY

Inorganic salt	Water activity (25°)
LiBr	0.07
LiCl	0.11
Potassium acetate	0.23
K_2CO_3	0.43
NaBr	0.58
KI	0.69
NaCl	0.75
KCl	0.84
K_2SO_4	0.97

reaction mixture and one other reactant separately (usually choose the reactant with the lowest concentration). Equilibration of enzymes in the presence of nonpolar organic solvents can sometimes produce greater catalytic activity because addition of solvent postequilibration can displace water molecules bound to the biocatalyst, increasing the water activity and causing the enzyme to aggregate.[4]

Rates of equilibration for the different materials will vary depending on the quantity of water to be transferred to the medium. It is therefore important to check that the system is in equilibrium before starting the reaction. For the solid biocatalyst, this can be measured gravimetrically and can take several days. The solubility of water in an organic solvent determines the equilibration time for the reaction media. Hence equilibration of nonpolar solvents will be completed in a matter of hours, but polar solvents require much longer because more water has to be transferred and in this case it can be beneficial to add roughly the required amount of water to the solvent prior to equilibration. When equilibrating more polar solvents using saturated salt solutions, the water activity obtained can be significantly different from that expected from the saturated salt because dissolution of solvent in the solution can alter the water activity. Therefore salt hydrates are often a better option for reactions in polar media.

Equilibration Method

MATERIALS
Inorganic salts, analar grade (Table I)
Distilled water

---

[4] M. C. Parker, B. D. Moore, and A. J. Blacker, *Biotechnol. Bioeng.* **46,** 452 (1995).

15–100 cm³ sealed equilibration jars
Reaction components to be hydrated
1–10 ml reaction vials

PROCEDURE. The inorganic salts are mixed with the distilled water to produce a solution of a slushy consistency. For example 5 ml of water added to approximately 10 g of K_2CO_3 will produce a saturated slush covered by a thin film of water, which is ideal. The salts for the lower water activities will require the least water due to their higher solubilities. The slush is placed in the bottom of a jar and the samples to be equilibrated are positioned, in a reaction vial, on top (preferably jacketed in a second vial to aid clean weighing) (Fig. 1). Typically, the jar will have a diameter of 10 cm and the reaction vial volume will be 3–6 ml, which is usually sufficient to hold either the quantity required for one reaction or 30–50 mg of lyophilized enzyme.

The jars need to be sealed carefully using PTFE tape or an O-ring to prevent evaporation of solvents. Brief exposure to the laboratory atmosphere during transfer does not generally affect the equilibrated material except when the total water content is low, for example, when some nonpolar solvents are used. This can be solved by performing sample trans-

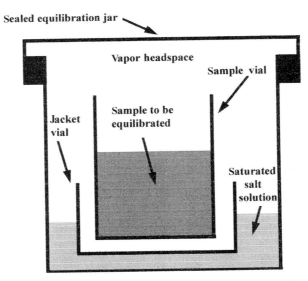

FIG. 1. Diagram of the apparatus used to hydrate either solid or liquid preparations by equilibration through the vapor phase using saturated salt solutions.

fer and mixing in a glove box preequilibrated to the desired water activity.

Throughout the equilibration period, the temperature has to be controlled to prevent cold surfaces from condensing liquid water, which will change the water activity of the system. It is preferable to use ambient temperatures whenever possible. Where heating is required, a water-jacketed vessel or an incubator can be used to control the temperature.

Prevention of Water Loss/Gain in a Preequilibrated Reaction Mixture. A reaction mixture will tend to exchange and equilibrate with the water vapor above it. In systems where the total water content is low, exposure to the atmosphere during the reaction or while sampling can cause significant changes in the water activity of the reaction mixture. Reaction vessels must be tightly sealed to prevent unnecessary exposure to the atmosphere, especially if long reaction times and a waterbath are employed. Materials used to seal the vessels must be resistant to the reaction components, for example, PTFE sleeves, silicone sealants, and O-rings.

Salt Hydrates for Water Activity Control during a Reaction That Generates or Consumes Water. Although reaction mixtures can be successfully preequilibrated to a defined water activity, if water is produced or consumed during the reaction, the system will continuously reequilibrate to a different water activity. The solubility of water in the various phases can also be altered by the disappearance of reactants or formation of new products leading to changes in the water activity.

Addition of pairs of solid salt hydrates to the reaction mixture allow the defined water activity to be maintained throughout the course of the reaction.[5-8] These pairs act as water "buffers," removing or adding water as required. Salt hydrate pairs (e.g., $NaH_2PO_4 \cdot 2H_2O$ and $NaH_2PO_4 \cdot 7H_2O$), should have some of each pair present at all times for effective buffering of the water content. Salt hydrate pairs should be added to the mixture prior to reaction and, provided that both hydrates are formed, the water activity of the system will reach equilibrium (Table II). Throughout the reaction, the lower hydrate will adsorb any water produced and the higher hydrate will release water if the water activity falls. Most salt pairs have not been tested for compatibility with the many different biocatalysts and reactants employed; therefore, care should be taken when first using them.

[5] P. Kuhl, U. Eichhorn, and H. D. Jakubke, in "Biocatalysis in Nonconventional Media" (J. Tramper, M. H. Vermue, H. H. Beeftink, and U. von Stockar, eds.), p. 513. Elsevier, Amsterdam, 1992.
[6] D. A. Robb, Z. Yang, and P. J. Halling, *Biocatalysis* **9,** 227 (1994).
[7] P. Berglund, C. Vorde, and H. E. Hogberg, *Biocatalysis* **9,** 123 (1994).
[8] P. J. Halling, *Biotechnol. Techniques* **6,** 271 (1992).

TABLE II
RECOMMENDED SALT HYDRATE PAIRS SUITABLE
FOR WATER BUFFERS

Salt hydrate pair	a_w (25°)
NaI 2/0	0.12
Na$_2$HPO$_4$ 2/0	0.16
NaAc 3/0	0.28
NaBr 2/0	0.35
Na$_2$HPO$_4$ 7/2	0.61
Na$_2$HPO$_4$ 12/7	0.80
Na$_2$SO$_4$ 10/0	0.80

Measurement of Water in Essentially Nonaqueous Systems

It may be necessary to measure the water content of the different components of a system prior to reaction or to monitor the water content of samples periodically removed from the reaction mixture. Care must be taken to avoid changes in the water content during the measurement process.

Measurement of Water in the Liquid Phase

Water Concentration by Coulometric Karl Fischer Titration. Coulometric Karl Fischer titration is the best method for measuring low water concentrations in organic solvents or reaction mixtures. The Metrohm 684KF and Fischer Instruments Accumet KF are examples of instruments commonly used. Samples are introduced into a titration vessel by a syringe through a rubber septum and should preferably be added directly from the reaction mixture. The syringe must be flushed several times with the sample to avoid changes in water concentration from contact with the syringe walls and analysis must be repeated several times until a constant value is obtained. The advantage of this method is that low water concentrations can be measured, although care has to be taken because instrument drift can produce large errors due to adsorption of atmospheric moisture into the titration vessel. Reproducibility is best when the sample contains between 100 and 300 μg of water. Therefore, with nonpolar solvents larger sample volumes may have to be analyzed. This, however, introduces a large volume of solvent into the titration vessel, which will eventually form a separate phase whereupon the reagents must be changed.

Indirect Measurement of Water Activity Using Sensors. Humidity sensors can be used to measure the relative humidity or water partial pressure above a reaction mixture. When all of the components in the reaction vessel are in equilibrium, these measurements can be used to determine the water activity of the reaction mixture. The sensors can be positioned inside the sealed reaction vessel in the gas headspace above the liquid phase. The sensor can be usefully sealed in a quick-fit cone section, which can then be easily attached to the reaction vessel. It usually takes an hour or more for the gas headspace to reach equilibrium with the liquid phase below; therefore, immediate readings obtained will represent the initial water vapor pressure of the headspace and not necessarily the water activity of the solution below. This method is therefore only suitable for measuring very slow changes in water activity.

With organic media volatile solvent present in the gas headspace may interfere with many of the available sensors but a few can still be used. The effects of organic solvents can be tested by comparing the readings for a saturated salt solution before and after the addition of a small amount of solvent.

Calibration of the sensor can be performed with either saturated salt solutions at a constant temperature or pure water over a range of temperatures. Water vapor pressure is highly dependent on temperature effects and immersion of the reaction vessel in a waterbath or incubator is advisable to prevent cold spots from developing. Examples of sensors that can be used are the Weiss LiCl humidity sensor[9] and Sina equihygroscope instrument.

Measurement of the Amount of Water Bound to the Biocatalyst

In many instances, the hydration state of the biocatalyst must be known. The amount of water tightly bound to a specific enzyme is a function of the water activity and temperature and is to a first approximation independent of the reaction media. However, the quantity of bound water will differ between different biocatalysts. Gravimetric measurements can be used to quantify bound water but require large amounts of enzyme and fail in the presence of organic media. A more sensitive ^{2}H solution state NMR technique can be employed to measure the amount of deuterated water bound to a protein that has been equilibrated to a specific water activity either in the presence of air or suspended in a nonpolar solvent.[4]

Materials

Enzyme
Deuterated water (99.9 at.%)
Deuterated sodium hydroxide (NaOD)
Phosphorus pentoxide

---

[9] Weiss Umwelttechnik GmbH, Postfach 1163, D-35445 Reiskirchen, Germany.

Inorganic salts
Anhydrous organic solvents
Anhydrous 1-propanol
3-ml Vials
Sealed jars

Procedure. The enzyme is dissolved in pure D_2O and the p^2H corrected using NaOD before lyophilization. To maximize exchange of the labile protons, it is advisable to then hydrate the lyophilized enzyme to a water activity of 0.86 using KCl/D_2O-saturated salt solutions for approximately 3 days. Then dry the enzyme over P_2O_5 for a further 3 days. Ten-milligram samples of the enzyme are weighed into 3-ml vials and solvent added if applicable. The samples are then placed in sealed jars containing D_2O-saturated salt solutions of the desired water activities for 4–5 days. A weighed amount of 1-propanol is added to the sample vials, shaken, centrifuged, and the solvent analyzed by ^2H nuclear magnetic resonance spectroscopy. The enriched propanol hydroxyl signal is measured relative to the natural abundance of the deuterium methylene signal and hence the amount of deuterated water present in the systems can be determined. For the samples containing solvent, this must then be corrected for the amount of D_2O present in the solvent. Correction may also be necessary for exchange of enzyme deuterons with the propanol protons, which will enhance the OD signal.

Conclusion

It is recognized that for biocatalysis in nonaqueous media, the small amount of water present is of great importance for catalytic activity. We have highlighted several methods for the control and measurement of hydration for biocatalysts in nonaqueous media.

Acknowledgments

We thank members of the working group on enzymes in nonaqueous media in the Departments of Bioscience and Biotechnology and Pure and Applied Chemistry at the University of Strathclyde, Glasgow (1996).

[22] Solvent Effect in Lipase-Catalyzed Racemate Resolution

By THORLEIF ANTHONSEN and JAAP A. JONGEJAN

Introduction

The preparation of enantiomerically pure compounds presents a major challenge to organic synthetic chemists. Different strategies have been explored by which enantiomerically pure building blocks, for example, for use as starting materials for the synthesis of pharmaceuticals and agrochemicals, may be obtained. A variety of organic compounds such as carbohydrates, terpenoids, alkaloids, and amino acids are available directly as pure enantiomers from natural sources, the so-called "chiral pool." Asymmetric synthesis allows prochiral starting materials to be converted into enantiomerically pure products. In racemate resolutions, the two enantiomers that comprise a racemic mixture have to be separated.

The ability of enzymes to catalyze enantioselective and enantiospecific reactions is evident from the large variety of enantiomerically pure natural products. Their potential as catalysts for asymmetric synthesis and racemate resolution in organic synthetic applications has obtained much attention.[1]

The majority of enzymes currently used for this purpose belong to the class of hydrolases, comprising lipases, esterases, and proteases. Following the pioneering work of Zaks and Klibanov[2,3] and others on the use of enzymes in organic solvents, it is now well established that the stereochemical features of hydrolases can be utilized both in hydrolytic and in condensation reactions.[4] Replacing bulk water with organic solvents has been reported to affect enzymes in various ways. Changes of stability, activity, and selectivity have been observed. Solvent effects of practical importance for the application of lipases in racemate resolutions are presented here.

Terminology

In this field, the selective properties of enzymes have been collectively addressed by the term enantio*selectivity*. However, concise definitions of

---

[1] K. Faber, "Biotransformations in Organic Chemistry." 3rd ed. Springer-Verlag, Berlin, 1997.
[2] A. Zaks and A. M. Klibanov, *Proc. Natl. Acad. Sci. U.S.A.* **82,** 3192 (1985).
[3] A. Zaks and A. M. Klibanov, *J. Biol. Chem.* **263,** 3194 (1988).
[4] A. M. P. Koskinen and A. M. Klibanov, "Enzymatic Reactions in Organic Media." Blackie Academic and Professional, Glasgow, UK, 1996.

the terms *stereoselective, stereospecific, enantioselective,* and *enantiospecific* have been proposed by Eliel and Wilen.[5] To avoid confusion, we support the use of enantio*specificity* for reactions where different enantiomers give rise to different products. Enantio*selectivity* will be reserved for reactions where a prochiral compound is converted into different amounts of enantiomers.

The chiral quality of the compounds involved in a resolution is routinely expressed as the enantiomeric excess. Of course, both the enantiomeric excess of the product formed (ee_p) and of the remaining substrate (ee_s) will be of interest. Enantiomeric excess values can be determined by several methods, such as polarimetry, nuclear magnetic resonance (NMR) spectroscopy using chiral shift reagents or solvents, or by chromatography, gas liquid chromatography (GLC) or high-performance liquid chromatography (HPLC), on chiral supports. Chromatography appears to be the most accurate and convenient method, in particular, because columns suitable for chiral separation of various compounds are commercially available. To establish absolute configurations comparison with compounds of known stereochemistry is required.

Although ee_p and ee_s are perfectly suited to characterize the products of a resolution, the enantiomeric ratio, E, has been generally accepted as the single most crucial process parameter of enzyme-catalyzed resolutions. Following the recommendations presented in an authoritative paper by Chen and co-workers,[6] E is defined as the ratio of specificity constants[7] for the enzymatic conversion of the two enantiomers:

$$E = \frac{k_{sp}^R}{k_{sp}^S} = \frac{(k_{cat}/K_M)^R}{(k_{cat}/K_M)^S}$$

By this definition the E value is a constant throughout an enzyme-catalyzed resolution. For reasons outlined later, we do not advocate the use of *apparent E* values to parametrize resolutions where other factors contribute to the final result.

Measurement of E Values

The enantiomeric ratio can be evaluated in several ways. If applicable, E can be calculated using the expressions derived by Chen *et al.*[6] for irreversible reactions following minimal Michaelis–Menten-type kinetics:

[5] E. L. Eliel and S. H. Wilen, "Stereochemistry of Organic Compounds." J. Wiley and Sons, New York, 1994.
[6] C.-S. Chen, Y. Fujimoto, G. Girdaukas, and C. J. Sih, *J. Am. Chem. Soc.* **104,** 7294 (1982).
[7] Nomenclature Committee of the International Union of Biochemistry, *Eur. J. Biochem.* **128,** 281 (1982).

$$E = \frac{\ln[(1 - \xi)(1 - ee_s)]}{\ln[(1 - \xi)(1 + ee_s)]} \quad (1a)$$

$$E = \frac{\ln[1 - \xi(1 + ee_p)]}{\ln[1 - \xi(1 - ee_p)]} \quad (1b)$$

From these relations, the importance of both the ee values as well as the extent of conversion, ξ ($0 < \xi < 1$), can be appreciated. When direct measurements of ξ are cumbersome or inaccurate, their value can be calculated using the enantiomer excesses (provided that side reactions can be excluded):

$$\xi = \frac{ee_s}{ee_s + ee_p} \quad (1c)$$

In general, determination of E from measurement of either ee_p or ee_s at a single value of ξ will lead to unacceptably low accuracies. Computer programs for the determination of E by nonlinear regression analysis of data obtained at several extents of conversion are available.[8,9] An alternative way of calculating the enantiomer ratio, again obviating the separate determination of ξ, uses the relation shown in Eq. (2)[10]:

$$E = \ln\left(\frac{1 - ee_s}{1 + \frac{ee_s}{ee_p}}\right) \bigg/ \ln\left(\frac{1 + ee_s}{1 + \frac{ee_s}{ee_p}}\right) \quad (2)$$

A graphic presentation of the relation between the enantiomeric excesses and the extent of conversion during an enzyme-catalyzed kinetic resolution of a racemic substrate (Fig. 1) serves to emphasize two important points. When the compound of interest is the *product* of the resolution process, the enantiomeric ratio of the enzyme places an upper bound on the maximum attainable enantiomeric purity. For the *remaining substrate*, however, there appears to be no such limitation. For high enantiomeric ratios (e.g., $E > 100$), the reaction virtually stops when half of the substrate has reacted and the enantiomer ratio of the remaining substrate is close to 100%. Using enzymes of medium selectivity (e.g., $E = 10$), ee_s will be only 70% at 50% conversion. However, high degrees of enantiomeric purity can still be obtained by simply extending the conversion beyond 50%, but the chemical

---

[8] H. W. Anthonsen, B. H. Hoff, and T. Anthonsen, *Tetrahedron: Asymmetry* **6**, 3015 (1995).
[9] H. W. Anthonsen, B. H. Hoff, and T. Anthonsen, **7**, 2633 (1996). "E&K calculator, version 2.03." http://Bendik.kje.ntnu.no.
[10] J. L. L. Rakels, A. J. J. Straathof, and J. J. Heijnen, *Enzyme Microb. Technol.* **15**, 1051 (1993).

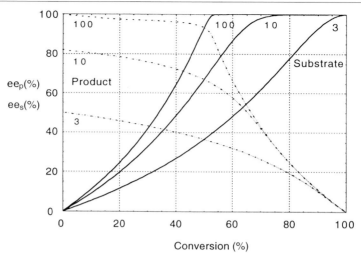

FIG. 1. Enantiomeric excess values, ee_s and ee_p, versus conversion, ξ, for an irreversible enzyme-catalyzed kinetic resolution of racemic substrate ($E = 3$, 10, and 100).

yield of the resolution will decrease. The trade-off between quality and quantity is a characteristic of resolutions as opposed to asymmetric synthesis. In asymmetric synthesis, the ee values remain constant throughout the reaction.

Note, however, that the assumptions underlying the derivation of Eqs. (1a) and (1b) limit their use to enzymes obeying minimal Michaelis–Menten-type kinetics. Lipase-catalyzed reactions do not fall into this class. In particular, the bi-bi ping-pong (two reactants and two products) kinetic scheme applicable to most lipases can give rise to deviations. Although the absolute magnitude of such deviations is of limited extent, they may enforce the breakdown of the trade-off principle.

This phenomenon has been clearly demonstrated for the PPL-catalyzed enantiospecific hydrolysis of fatty acid esters of the commercially attractive chiral synthon, 2,3-epoxy-1-propanol, glycidol.[11] The original publication by Ladner and Whitesides[12] constitutes an early tribute to what we have since recognized to be common features of the use of lipases in the kinetic resolution of racemates, notably, experimental simplicity, easy processing of laboratory-scale amounts, and good versatility. By addressing the enantiospecificity as the enantiomer excess value at 60% conversion, Ladner and Whitesides did not explicitly claim the preparation of enantiopure

[11] J. B. A. van Tol, J. A. Jongejan, and J. A. Duine, *Biocatl. Biotransform.* **12,** 99 (1995).
[12] W. E. Ladner and G. M. Whitesides, *J. Am. Chem. Soc.* **106,** 7250 (1986).

remaining ester. Considering the fact that by running the experiment at pH 7.8 equilibrium is displaced completely in favor of the hydrolyzed products, absolute enantiomeric purity is expected to result after a sufficient amount of ester becomes hydrolyzed (see Fig. 1).

Preliminary experiments in collaboration with Andeno BV, Venlo, The Netherlands, to investigate the feasibility of this process for large-scale production of glycidyl butyrate of industrially acceptable enantiomeric purity proved this expectation to be wrong. Preparative runs at the indicated composition of the mixture showed that the ee values of remaining glycidyl butyrate no longer increased once a plateau value close to 96% ee_s was reached (Fig. 2).

After considerable experimentation, it became clear that the description of lipase enantiospecificity requires a more demanding equation [Eq. (3)]:

$$\frac{r_A^R}{r_A^S} = \frac{c_A^R(E \cdot K_{eq} \cdot c_B + \alpha^R \cdot c_P^S) - c_P^R \cdot (E \cdot c_Q + \alpha^R \cdot c_A^S)}{c_A^S(K_{eq} \cdot c_B + \alpha^R \cdot c_P^R) - c_P^S \cdot (c_Q + \alpha^R \cdot c_A^R)} \quad (3)$$

with:

$$E = \frac{k_1^R k_2^R/(k_{-1}^R + k_2^R)}{k_1^S k_2^S/(k_{-1}^S + k_2^S)} = \frac{k_{sp}^R}{k_{sp}^S} \left(= \frac{\alpha^R}{\alpha^S} \right)$$

$$\alpha^R = \frac{k_1^R k_2^R/(k_{-1}^R + k_2^R)}{k_{-3} k_{-4}/(k_{-3} + k_4)} = \frac{V_1^R/K_{M,A}^R}{V_{-1}^R/K_{M,Q}^R}$$

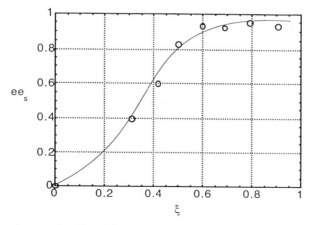

FIG. 2. Plot of ee_s as a function of the extent of conversion for the PPL-catalyzed hydrolysis of glycidyl butyrate at pH 7.8 (uni-uni plot).

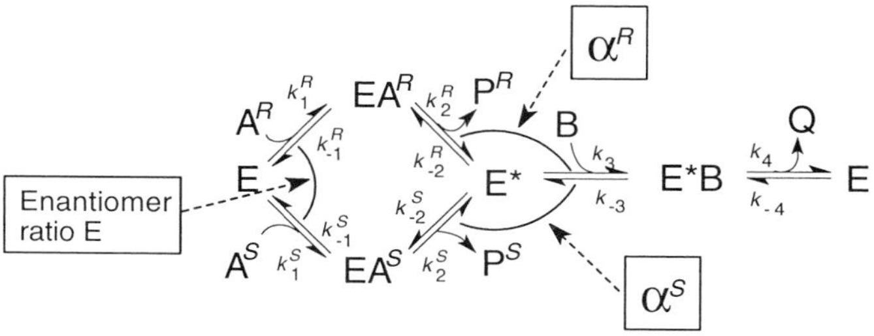

SCHEME 1. Bi-bi ping-pong mechanism for the hydrolysis of an ester of a chiral alcohol. Partitioning of the acyl enzyme complex (E*) is governed by E and α^R. For the kinetic resolution of (R,S)-glycidyl butyrate by PPL: A^R, A^S, (R,S)-glycidyl butyrate; P^R, (S)-glycidol; P^S, (R)-glycidol; B, water; Q, butyric acid; E, PPL; E*, butyryl–PPL; E*B, butyryl–PPL–water complex.

$$K_{eq} = \frac{k_1^R k_2^R k_3 k_4}{k_{-1}^R k_{-2}^R k_{-3} k_{-4}} = \frac{k_1^S k_2^S k_3 k_4}{k_{-1}^S k_{-2}^S k_{-3} k_{-4}}$$

In addition to the enantiomeric ratio E (defined as the ratio of specificity constants, as before) and the overall equilibrium constant K_{eq}, a third parameter, α^R, measuring the ratio of nucleophilic attack of the acylated enzyme by (R)-alcohol and water, respectively, is introduced. As indicated (Scheme 1), applying irreversible conditions by titrating the acid formed does neither affect α^R nor the value of K_{eq} as defined here.

Despite the overall thermodynamic drive toward complete hydrolysis, partitioning of the acylated enzyme with respect to the (three) nucleophilic species present during the reaction constitutes an important kinetic control mechanism. Clearly, this control will lead to (partial) equilibration of chiral ester at high concentrations of the released alcohol enantiomers, leveling off the observed ee value. Although this phenomenon appears to be general for bisubstrate-type enzymatic mechanisms,[13] lipases are particularly vulnerable to this effect due to the rate-controlling character of the hydrolysis of the acyl-enzyme intermediate. A second round of hydrolysis of 96%-enriched ester will normally solve the problem with moderate costs in terms of the overall yield.

The α-selectivity constant as well as the overall equilibrium constant will also play an important role when lipases are employed for esterification or transesterification reactions in organic solvents. In such cases, the reaction may well become reversible. If an enzyme prefers one enantiomer in the forward reaction, it will prefer the same enantiomer in the reverse reaction, thus the reversibility will eventually reduce ee_s when the reaction

[13] A. J. J. Staathof, J. L. L. Rakels, and J. J. Heinen, *Biocatalysis* **7**, 13 (1992).

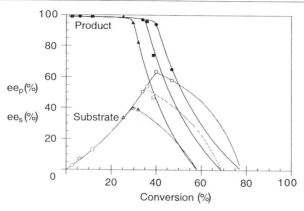

FIG. 3. Variation of relative concentrations of 2-chloroethyl butanoate in transesterification of 1-phenoxy-2-propanol. Circles, 5× excess; squares, 3× excess; triangles, 1.5× excess. Filled symbols represent product fraction; open symbols remaining substrate fraction.

is run to a high conversion. Equations (4a) and (4b) for biocatalytic resolutions will then involve the equilibrium constant also in cases where minimal Michaelis–Menten kinetics apply.[14]

$$E = \frac{\ln\{1 - (1 + 1/K_{eq})[\xi + ee_s(1 - \xi)]\}}{\ln\{1 - (1 + 1/K_{eq})[\xi - ee_s(1 - \xi)]\}} \quad (4a)$$

$$E = \frac{\ln[1 - (1 + 1/K_{eq})\xi(1 + ee_p)]}{\ln[1 - (1 + 1/K_{eq})\xi(1 - ee_p)]} \quad (4b)$$

A typical example of a reversible reaction is the transesterification of 2-chloroethyl butanoate with a secondary alcohol (1-phenoxy-2-propanol). The enantiomeric ratio was determined by regression analysis of ee_s and ee_p values at several extents of conversion. Because apparent values of the equilibrium constant, $K = K_{eq} \cdot c_{donor}$, were used, different K values resulted for different amounts of acyl donor. However, calculations based on a bi–bi mechanism gave $K_{eq} = 0.32$.[9]

As shown in Fig. 3, even with a very high E value (here $E > 100$), the enantiomeric excess of the remaining substrate fraction may be low. The production fraction is affected to a lesser extent.

Optimization of Resolutions

Because a biocatalytic reaction comprises substrates, products, possible modifiers, enzyme and medium, it is obvious that optimization has to take

[14] C.-S. Chen, S. Wu, G. Girdaukas, and C. J. Sih, *J. Am. Chem. Soc.* **109**, 2812 (1987).

TABLE I
Log P Values of Organic Solvents Used in Biocatalytic Processes

Dimethyl sulfoxide	−1.3	Chloroform	2.0
Dioxane	−1.1	Benzene	2.0
N,N-dimethylformamide	−1.0	Toluene	2.5
Methanol	−0.76	Dibutylether	2.9
Acetonitrile	−0.33	Styrene	3.0
Ethanol	−0.24	Tetrachloromethane	3.0
Acetone	−0.23	Pentane	3.0
Tetrahydrofuran	0.49	Xylene	3.1
Ethylacetate	0.68	Cyclohexane	3.2
Pyridine	0.71	Hexane	3.5
Diethylether	0.85	Heptane	4.0
Cyclohexanone	0.96	Decane	5.6
Cyclohexanol	1.5	Hexadecane	8.8

all participants into account. The organic chemist's approach might be to alter the substrate by introducing protective groups, which may induce better resolution and can be readily removed afterward. The biochemical way to optimize the process could be to change the enzyme. There are many options, discussion of which is beyond the scope of this chapter.[15,16] To change the medium of the reaction, however, is by far the simplest way to change the process. Considerable effort has been put into this approach.

Solvents for Biocatalytic Purposes

Classically, organic solvents have been characterized according to their polarity. Alternative parameters include the dielectric constant and the electron pair acceptance index, E_T^N.[17] For biocatalytic purposes, the log P value that quantifies the partitioning between water and 1-octanol has afforded reasonable correlations[18,19] (see Table I).

Solvents with log $P < 2$ appear to be less suitable for biocatalytic purposes. Most probably, water molecules essential for the structural integrity of the enzyme are affected provoking denaturation. Solvents with log $P > 4$ are considerably less harmful. Solvents with intermediate log P values show more unpredictable properties.

[15] K. Faber, G. Ottolina, and S. Riva, *Biocatalysis* **8,** 91 (1993).
[16] C. J. Sih and S.-H. Wu, *Topics Stereochem.* **19,** 63 (1989).
[17] C. Reichardt, "Solvents and Solvent Effects in Organic Chemistry." Verlag Chemie, Weinheim, 1990.
[18] C. Laane, *Biocatalysis* **1,** 17 (1987).
[19] C. Laane, S. Boeren, K. Vos, and C. Veeger, *Biotechnol. Bioeng.* **30,** 81 (1987).

TABLE II
ENANTIOMERIC RATIOS ON ADDITION OF 10% OF COSOLVENTS TO THE
CANDIDA ANTARCTICA LIPASE B CATALYZED HYDROLYSIS OF THE BUTANOATE OF
3-METHOXY-1-(PHENYLMETHOXY)-2-PROPANOL

Without cosolvent	20	Diethylether	15
Hexane	5	Acetonitrile	21
DMSO	6	Dimethylformamide	22
Methanol	10	Tetrahydrofuran	29
2-Methyl-2-butanol	14	Ethanol	58
1-Propanol	15	Acetone	70
1-Butanol	15	*tert*-Butanol	98

Hydrolysis Reactions and Significance of Cosolvents

Most of the earlier work in the field was carried out using proteases. One early example is by Bryan Jones[20] who found large differences in enantiospecificity of chymotrypsin-catalyzed hydrolysis of pure enantiomers of methyl 2-acetamido-2-phenylacetate when 0–30% organic cosolvent was used.

Appreciable solvent effects have been observed on addition of different cosolvents to the *Candida antarctica* lipase B-catalyzed hydrolysis of the butanoate of 3-methoxy-1-(phenylmethoxy)-2-propanol.[21] As shown in Table II, lower E values are observed for some solvents. However, addition of acetone and *tert*-butanol leads to values that are considerably higher. In subsequent experiments these two cosolvents were used for the hydrolysis reactions of the two target compounds.

The stereochemical preference of lipase B from *C. antarctica* was found to be in agreement with the predictive rules developed by Kazlauskas *et al.* (Fig. 4).[22] The amount of *tert*-butanol and acetone as cosolvents varied over the range 0–50% also for the hydrolysis of the butanoate of the more synthetically useful chloro derivative 3-chloro-1-(phenylmethoxy)-2-propanol. For *tert*-butanol, an optimal E value was observed around 20% v/v for both substrates. The enantiomeric ratio for hydrolysis of the butanoate of 3-methoxy-1-(phenylmethoxy)-2-propanol increased from 21 to 106, whereas the corresponding increase for the butanoate of 3-chloro-1-(phenylmethoxy)-2-propanol was from 7 to 64.

---

[20] J. B. Jones and M. M. Mehes, *Can. J. Chem.* **57,** 2245 (1979).
[21] T. V. Hansen, V. Waagen, V. Partali, H. W. Anthonsen, and T. Anthonsen, *Tetrahedron: Asymmetry* **6,** 499 (1995).
[22] R. J. Kazlauskas, A. N. E. Weissfloch, A. T. Rappaport, and L. A. Cuccia, *J. Org. Chem.* **56,** 2656 (1991).

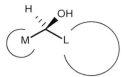

Fig. 4. Model for lipase stereospecificity.

With acetone as the cosolvent, slightly different results were obtained for the two compounds. The maximum E value for the butanoate of 3-methoxy-1-(phenylmethoxy)-2-propanol was obtained around 15% acetone in water while the maximum for the chloro compound was at 30%. For the latter compound, a tremendous increase of the enantiomeric ratio from 7 to ≈200, was observed. These very valuable synthons, (R)- and (S)-3-chloro-1-(phenylmethoxy)-2-propanol, can be easily prepared under these conditions in good yields and with very high enantiomeric excesses.

Esterifications and Transesterifications

Apart from hydrolysis the most obvious way to utilize the stereochemical properties of lipase-catalyzed reactions is synthesis of chiral esters by the condensation of alcohols and acids in organic solvents. To prevent the reaction from running into equilibrium, the water produced in the reaction has to be removed. To avoid this problem, transesterification is usually preferred. Fundamental aspects of nonaqueous enzymology[23] and reviews on solvent dependence have been published recently.[24,25]

Significance of Water Activity

It is well documented that the water content of the organic solvent in which an enzyme-catalyzed reaction takes place may be of great importance. It is important for enzyme stability and reactivity and also for enzyme specificity. Solvent effect on K_M has been explained on the basis of substrate solvation. Substrates with low affinity for solvent bind strongly to enzyme and K_M is low. Substrates with high affinity for solvent give the opposite result. The content of water in an enzyme, nonaqueous solvent, substrate system is best expressed as the thermodynamic activity of water, a_w.[26] By definition

---

[23] Z. Yang and A. J. Russel, *in* "Enzymatic Reactions in Organic Media" (A. M. P. Koskinen and A. M. Klibanov, eds.), p. 43. Blackie Academic and Professional, Glasgow, UK, 1996.
[24] C. R. Wescott and A. M. Klibanov, *Biochim. Biophys. Acta* **1206**, 1 (1994).
[25] G. Carrea, G. Ottolina, and S. Riva, *TIBTECH* **13**, 63 (1995).
[26] P. J. Halling, *Enzyme Microb. Technol.* **16**, 178 (1994).

at equilibrium the water activity a_w will be the same in the whole system, the hydrated enzyme, solvent, and headspace although concentrations of water will be different.

Different enzymes perform optimally at different water activities. Thus it is important to provide this information for enzymes that are going to be used for biocatalysis. A list of water activities of several inorganic salts at different temperatures has been published.[27]

The enantiomeric ratio of esterification of 2-methylalcanoic acids in cyclohexane catalyzed by *Candida rugosa* lipase (CRL) has been shown to increase when the water activity a_w is increased from 0 to 0.76.[28] Increasing water activity also increased E in the transesterification of 1-phenoxy-2-propanol with 2-chloroethyl butanoate.[8] Increasing water activity will, however, introduce water as a competing nucleophile and thus influence the total result in a negative way.

Use of Salt Hydrates to Buffer Water Activity

One practical way to keep the water activity at a constant value during a biocatalytic reaction is to add solid inorganic salts to the reaction medium. A mixture of more and less hydrated forms of the same salt will keep the a_w at a constant level. A mixture of $Na_2SO_4 \cdot 10H_2O$ and $Na_2SO_4 \cdot 0H_2O$ will give an a_w of 0.76. Various effects of addition of salt hydrates in CRL-catalyzed esterifications have been tested.[29,30] When water was needed for enzyme activity at the start of the reaction, the higher hydrated salt released water. Because water is produced during an esterification, the lower hydrated form takes up this water. Thus near-optimal conditions were maintained throughout the reaction.

Transesterification of Racemic Glycidol with Vinyl Acetate: A Case Study

The three-parameter model derived for the kinetic resolution of racemic esters in lipase-catalyzed hydrolytic conversions [Eq. (3)] appears to be equally valid for the description of transesterifications carried out in organic media. As an example we report the transesterification of racemic glycidol

---

[27] P. J. Halling, *Biotechnol. Tech.* **6**, 271 (1992).
[28] H.-E. Högberg, H. Edlund, P. Berglund, and E. Hedenström, *Tetrahedron: Asymmetry* **4**, 2123 (1993).
[29] L. Kvittingen, B. Sjursnes, T. Anthonsen, and P. J. Halling, *Tetrahedron* **48**, 2793 (1992).
[30] B. Sjursnes, L. Kvittingen, T. Anthonsen, and P. Halling, *in* "Biocatalysis in Non-Conventional Media" (J. Tramper, M. H. Vermüe, H. H. Beeftink, U. von Stockar, eds.), p. 451. Elsevier Science Publishers, Amsterdam, 1992.

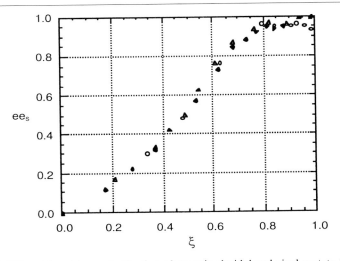

FIG. 5. PPL-catalyzed transesterification of racemic glycidol and vinyl acetate in organic solvent. Enantiomeric excess values of glycidol, ee_s, and the extent of conversion, ξ, for: ◆, diisopropyl ether; △, tetrachloromethane; and ○,2-butanone. See experimental section for detailed description.

with vinyl acetate.[31] When partitioning of free enzyme and acyl donor species is not taken into account, major effects of the solvent on the enantiomeric ratio would seem to apply (Fig. 5).

However, plateau levels of the enantiomer excess value for the remaining glycidol, similar to those observed for the hydrolytic case, do occur. Consequently, a more consistent picture is obtained when α-selectivity is taken into account. A minor modification of Eq. (3) is required. The selectivity factor α^S will now represent the ratio of specificity constants for the acylation of free enzyme with vinyl acetate and (S)-glycidyl acetate, respectively. It is important to notice that (R)-glycidol and (S)-glycidyl acetate possess identical configurations at the stereo center.

At present, it is not clear whether a failure to recognize this phenomenon represents a main cause for the numerous reports claiming a change of the intrinsic enantiospecificity when lipases are employed in organic media.[32]

For the example given, even better correlations result when solvent effects are introduced explicitly. Two types of contributions have been

[31] J. B. A. van Tol, J. A. Jongejan, and J. A. Duine, in "Biocatalysis in Non-Conventional Media" (J. Tramper, M. H. Vermüe, H. H. Beeftink, U. von Stockar, eds.), p. 237. Elsevier Science Publishers, Amsterdam, 1992.
[32] J. A. Jongejan, J. B. A. van Tol, and J. A. Duine, Chimica Oggi **12(7/8)** 15 (1994).

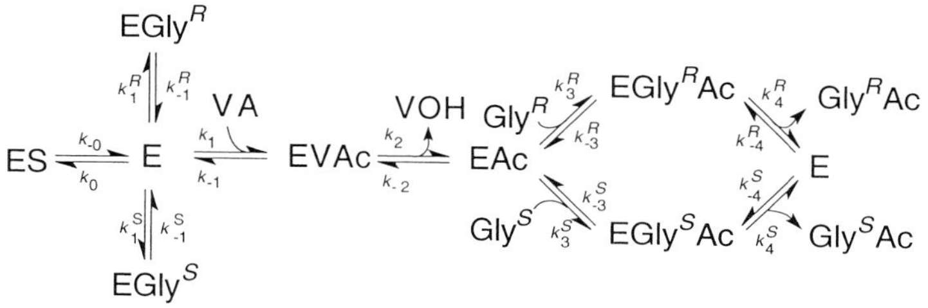

SCHEME 2. PPL-catalyzed transesterification of (R,S)-glycidol with vinyl acetate. Bi-bi ping-pong mechanism including competitive substrate and solvent inhibition. VA, vinyl acetate; VOH, vinyl alcohol/acetaldehyde; $Gly^{R,S}$, (R,S)-glycidol; $Gly^{R,S}$ Ac, (R,S)-glycidyl acetate; ES, PPL-solvent complex.

investigated so far. Kinetic inhibition by the solvent is apparent from initial rate measurements.[33] For the system considered here, this mechanistic effect appears to be well represented by invoking a competitive interaction between the solvent and the free enzyme species (Scheme 2).

Thermodynamic effects of the solvent become clear when *thermodynamic activities* are introduced as the mass action equivalents of the *concentrations* of all species involved in regular rate equations. Rationalization for this choice has been given elsewhere.[34-37] Also in this case, representation of rate equations in terms of thermodynamic activities narrows the range of α-selectivities observed in the solvents studied, emphasizing the intrinsic value of this parameter (Table III).

We must stress that the example presented here has been designed specifically to emphasize the *intrinsic* properties of E and α, as opposed to the various *ad hoc* effects of organic solvents. Unambiguous interpretation of resolution curves and accurate determination of E and α pose restrictions that should (and, in most cases, can) be avoided in practical resolutions. Thus, to suppress side reactions (e.g., hydrolysis of esters), the reaction required low water activity. Since various lipases, such as *Pseudomonas cepacia* lipase (Amano PS), show appreciable catalytic activity only at

---

[33] J. B. A. van Tol, D. E. Kraayveld, J. A. Jongejan, and J. A. Duine, *Biocatal. Biotransform.* **12**, 119 (1995).

[34] J. B. A. van Tol, J. B. Odenthal, J. A. Jongejan, and J. A. Duine, in "Biocatalysis in Non-Conventional Media" (J. Tramper, M. H. Vermüe, H. H. Beeftink, U. von Stockar, eds.), p. 229. Elsevier Science Publishers B.V., Amsterdam, 1992.

[35] A. J. J. Straathof, J. L. L. Rakels, and J. J. Heinen, in "Biocatalysis in Non-Conventional Media" (J. Tramper, M. H. Vermüe, H. H. Beeftink and U. von Stockar, eds.), p. 137. Elsevier Science Publishers, Amsterdam, 1992.

[36] A. E. M. Janssen and P. J. Halling, *J. Am. Chem. Soc.* **116**, 9827 (1994).

[37] W. P. Jencks, "Catalysis in Chemistry and Enzymology." McGraw-Hill, New York, 1969.

TABLE III
KINETIC PARAMETERS, E, AND α OBTAINED FROM NONLINEAR REGRESSION OF THE DATA PRESENTED IN FIG. 5[a]

Solvent	E	$\alpha(x)$	$\alpha(\gamma)$
Diisopropylether	5.4($\pm$0.2)	0.14($\pm$0.01)	0.22
Tetrachloromethane	5.5($\pm$0.3)	0.16($\pm$0.02)	0.19
2-Butanone	5.6($\pm$0.3)	0.32($\pm$0.04)	0.29

[a] The selectivity factor based on molar fractions, $\alpha(x)$; the selectivity factor based on activity coefficients, $\alpha(\gamma)$. Activity coefficients were obtained from references.[38,39]

$a_w > 0.2$, they were not suited for this demonstration. Similarly, E and α values were chosen to be (far) below 20 to enable observation of mechanism-inherent intricacies. Suitable screening protocols designed to obtain optimal E values for lipase-catalyzed resolutions of racemates should be considered for practical applications.

It is commonly accepted that the E values observed in the experimentally more accessible hydrolytic mode provide good guidance for the values to be expected for the reverse reaction.[14] Comparison of the examples discussed here leads to the conclusion that an appreciable reduction of the E value of PPL occurs when the resolution of racemic glycidol is conducted in organic solvents. There are compelling reasons to rationalize this observation in terms of the properties of the lid structure that has been detected by X-ray crystallography of several lipases. Elaboration of possible solvent effects on the status of the lid structure present in PPL is given later.

Enantiospecificity and Interfacial Activation

Interfacial activation has long been the hallmark of lipase catalysis. Until recently, the 5- to 10-fold stimulation of the hydrolytic activity of lipases acting at the aqueous/organic interface formed the subject of numerous speculations. Elucidation of the crystal structure of the lipase from *Rhizomucor miehei* in its "closed" and inhibitor-derived "open" form led to the recognition of the importance of the lid structure.[40] Similar changes have been proposed to take place on interfacial adsorption of related lipases.

[38] A. Fredenslund, J. Gmehling, and P. Rasmussen, "Vapor-Liquid Equilibria Using Unifac." Elsevier, Amsterdam, 1977.
[39] "Dechema Database-PC." DDBST, Dortmund, Germany, 1991.
[40] Z. S. Derewenda and A. M. Sharp, *Trends Biochem. Sci.* **18**, 20 (1993).

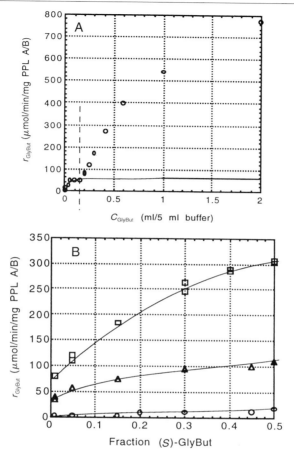

FIG. 6. (A) Interfacial activation of PPL-catalyzed hydrolysis of glycidyl butyrate. Solubility limit of (R,S)-glycidyl butyrate is indicated at 0.18 ml/5 ml buffer. (B) E values for monophasic and biphasic systems, as obtained by nonlinear regression according to the initial rate method,[41] $E = 8, 12$, and 16, for 0.024, 0.2, and 0.39 ml of (R,S)-glycidyl butyrate per 5 ml of buffer, respectively.

Because conformational changes have been commonly called on by enzymologists to explain otherwise elusive phenomena, the change of the intrinsic enantiospecificity observed for the PPL-catalyzed hydrolysis of glycidyl butyrate (Fig. 6) would not seem to require an alternative rationalization.

[41] J. A. Jongejan, J. B. A. van Tol, A. Geerlof, and J. A. Duine, *Recl. Trav. Chim. Pays-Bas* **110,** 247 (1991).

Yet, formation of enzyme-substrate Michaelis complexes requires "open" conformations for both enantiomers. Because there is little reason to expect enantiodifferentiation for the opening of the lid, and considering that the subsequent processing of either dissolved or interface-exposed substrate will probably trace the same steps, a change of E on interfacial activation is not expected.

An elegant explanation for this intriguing paradox can be deduced from inspection of the free energy profiles. Following the formalism advocated by Chen, and taking for granted the largely unchallenged "theory of absolute reaction rates" proposed by Eyring in 1935 and thereafter addressed as the (Eyring) TS theory, the lumped microscopic constants that make up the enantiomeric ratio can be presented as in Eq. (5). In this equation, E is a ratio of exponentials, $\beta = 1/RT$, preexponential factors have been cancelled on the assumption of equality, and ΔG values are as presented in Fig. 7A.

$$E = \frac{(k_{cat}/K_M)^R}{(k_{cat}/K_M)^S} = \frac{k_2^R k_1^R/(k_{-1}^R + k_2^R)}{k_2^S k_1^S/(k_{-1}^S + k_2^S)} = \frac{e^{-\beta \Delta G_3^R - \beta \Delta G_1^R}/(e^{-\beta \Delta G_2^R} + e^{-\beta \Delta G_3^R})}{e^{-\beta \Delta G_3^S - \beta \Delta G_1^S}/(e^{-\beta \Delta G_2^S} + e^{-\beta \Delta G_3^S})} \quad (5)$$

On minor algebraic manipulation, disregarding possible differences of the preexponential κ, one arrives at Eq. (6). Note the reciprocal occurrence of S- and R-related TS energies.

$$E = \frac{e^{\beta \Delta G_A^S} + e^{\beta \Delta G_B^S}}{e^{\beta \Delta G_A^R} + e^{\beta \Delta G_B^R}} \quad (6)$$

Proper identification of the free-energy differences involved leads us to equate the enantiomeric ratio as a ratio of summed Boltzmann-type contributions arising from the individual free energies of activation as they present themselves to the reacting (free) enzyme and substrate species (Fig. 7B).

On the assumption that the specificity constants [Eq. (1)] can be identified with a (virtual) kinetic barrier (Fig. 7C), a simple relation for E and a difference of TS energies has been proposed (see Ref. 42). It will be clear from the present analysis that the *pseudo*-character of the specificity constants does not by itself invalidate such a simplification; however, the exponents $-\beta \Delta G_C$, cannot be straightforwardly identified with a physically realized TS energy [Eq. (7)]. In Eq. (7) complex dependence of "virtual" TS energies related to specificity constants and "genuine" TS energies involved in M–M free-energy profiles are shown.

---

[42] R. S. Phillips, *Enzyme Microb. Technol.* **14**, 417 (1992).

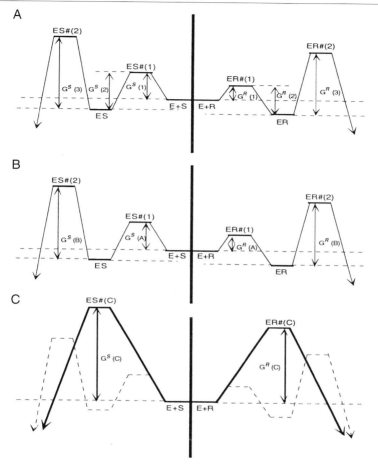

FIG. 7. Free-energy profiles associated with irreversible minimal Michaelis–Menten kinetics for the kinetic resolution of racemates.

$$E = \frac{k_{\text{sp}}^R}{k_{\text{sp}}^S} = \frac{e^{-\beta \Delta G_C^R}}{e^{-\beta \Delta G_C^S}} \left(= \frac{e^{\beta \Delta G_A^S} + e^{\beta \Delta G_B^S}}{e^{\beta \Delta G_A^R} + e^{\beta \Delta G_B^R}} \right) \quad (7)$$

In general, the numerical value of the numerator and denominator of Eq. (6) will be dominated by the contribution from the largest activation energy. In this respect, the picture supports earlier assumptions underlying attempts to calculate E values from force field energy differences of the diastereomeric tetrahedral intermediates in serine-type hydrolase catalysis.[43] However, the contribution of the activation energy involved in the formation of the Michaelis complex, normally taken to be equivalent to the diffusion-controlled rate of encounter, will certainly be raised when the substrate has to open the lid by itself. This in contrast to the situation where interfacial adsorption and concomitant lid opening precede substrate binding. It follows that the extra energy involved, the magnitude of which will most probably be independent of the chiral nature of the substrate, will lower the differentiation of $\Delta\Delta G$ for the overall process, resulting in a decreased E value for the monophasic system.

This argument may well be stretched to include the low E values observed for the transesterification of glycidyl esters in organic solvents (see earlier discussion). Apparently, the lid of PPL assumes its closed conformation under these conditions.[44] Whether advantage of this analysis can be taken to persuade a lipase to adopt its more highly enantioselective "open" conformation during nonaqueous kinetic resolutions is not clear. This phenomenon, however, may be responsible for the pretreatment-induced "memory effects" on the enantiospecific properties of lipases that have been claimed by several authors.[45]

Origin of Solvent Effects

A lot of speculation has surrounded the origin of solvent effects. Several factors must be taken into account and the significance of each contributing factor will be different from example to example. Solvent effects have been observed not only in lipase-catalyzed reactions, but also in other hydrolase-catalyzed reactions. The explanations offered have been varying, but so far no rationale of general validity has emerged. First of all, note that many of the examples of solvent effects reported are based of enzyme preparations consisting of more than one enzyme. Different enzymes may behave differently when the medium is changed. A few examples may serve to illustrate the complexity of the problem.

The effect of organic solvents on the initial rates of subtilisin-catalyzed transesterifications of chiral amines using 1,1,1-trifluoroethyl butyrate has been investigated.[46] Significant effects were observed in the case of chiral 1-(1-naphtyl)ethylamine for which v_S/v_R increased for 1.6 in octane to 22 in 3-methyl-3-pentanol. When the enantiomeric ratio of hydrolysis of chlo-

---

[43] M. Norin, A. Mattson, K. Hult, and T. Norin, *Biocatalysis* **7**, 131 (1993).
[44] A. Louwrir and A. M. Klibanov, *Biotechnol. Bioeng.* **50**, 1 (1996).
[45] A. M. Klibanov, *Nature* **374**, 596 (1995).
[46] H. Kitaguchi, P. A. Fitzpatrick, J. E. Huber, and A. M. Klibanov, *J. Am. Chem. Soc.* **111**, 3094 (1989).

roethyl esters of some N-acetylated amino acids catalyzed by proteinases was compared to transesterifications using the same esters and propanol in butyl ether, it was observed that E values were significantly higher in hydrolysis.[47] When different solvents were used for the same transesterification reaction, the E value was found to be inversely proportional to the log P value of the solvent in the range -0.8 to 3.1; that is, the enantiospecificity is higher in more polar solvents. This observation was rationalized on the following basis. When the substrate binds to the active site of the enzyme, adsorbed water molecules need to be expelled into the surrounding medium. The more apolar the medium, the more unfavorable the repulsion of water molecules. On the assumption that the faster reacting enantiomer requires removal of extra water molecules, apolar media would lead to a decrease of the E value.

In another study, transesterification of 1-phenylethanol with vinyl butyrate catalyzed by subtilisin Carlsberg showed the opposite dependence.[48] $E = 61$ was found in dioxane, while $E = 3$ was found in acetonitrile. The lower value of E resulted from increased reactivity of the slower reacting enantiomer in a solvent with higher dielectric constant, a polar solvent. It was concluded that a solvent with a high dielectric constant gives the enzyme more flexibility and consequently allows for greater reactivity of the slower reacting enantiomer.

To trace the origin of the cosolvent effect in the hydrolysis of butanoates of 3-methoxy-1-(phenylmethoxy)-2-propanol and 3-chloro-1-(phenylmethoxy)-2-propanol, preliminary NMR studies were undertaken. The ^{1}H NMR spectra of dissolved lipase B revealed no detectable changes when going from D_2O to acetone-d_6 : D_2O at 50:50. This strongly indicates that the conformation of the enzyme at the used acetone concentration does not change. Thus the observed solvent effect is not due to changes in the enzyme. When the phenyl protons were irradiated, no nuclear Overhauser effect was observed on the other resonances. When dissolved enzyme was added a nOe was observed indicating that the correlation time of the substrate had become longer due to interaction with the enzyme.

As stated earlier, the basis for enzyme-catalyzed resolution is the weighted difference in free energy of activation for reaction of the two enantiomers [Eq. (6) or analogous expressions for more elaborate kinetic models]. This difference, which can be related to the enantioselectivity E by invoking the difference of "virtual" TS energies [Eq. (7)]–$\Delta\Delta G = -RT \ln E$—may well reflect a change in the interaction energy between the

---

[47] T. Sakurai, A. L. Margolin, A. J. Russel, and A. M. Klibanov, *J. Am. Chem. Soc.* **110**, 7236 (1988).
[48] P. A. Fitzpatrick and A. M. Klibanov, *J. Am. Chem. Soc.* **113**, 3166 (1991).

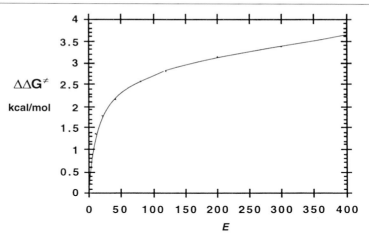

FIG. 8. Relationship between enantiomer ratio E and the difference in free energy of activation between two enantiomers in an enzyme-catalyzed reaction.

enzyme and the two enantiomers. Going from an E value of 7 to 220 corresponds to an increase in difference of free energy of activation of 2.0 kcal/mol (Fig. 8). It may be valid to relate this value to the binding energy of enzyme/substrate interactions measured by Fersht for tyrosyl tRNA synthetase.[49] The observed energy change may be in the range of changing one or two hydrogen bonds. Conversely, the observed effect may be due to the combined contribution of the rate of dissolution and diffusion of the substrates. In the absence of cosolvents, the reaction rates for both enantiomers may be dominated by the limiting availability in the dissolved state, thus suppressing the kinetic resolution.

The crystal structure of lipase B from *C. antarctica* has been solved.[50] Molecular modeling of enzyme with the two enantiomeric substrates confirms the observed stereochemical preferences.[51] A search for suitable hydrogen bonds that may be changed might shed light on the origin of the observed solvent effect in this case.

Note that although the kinetic resolution is generally independent of the absolute concentration of the racemic mixture, low rates of dissolution will lead to a steady-state situation in which the dissolved pool of the faster reacting enantiomer will be consistently depleted. On addition of cosolvent,

[49] A. R. Fersht, *Trends. Biochem. Sci.* **12**, 301 (1987).
[50] J. Uppenberg, M. T. Hansen, S. Patkar, and T. A. Jones, *Curr. Biol. Structure* **2**, 293 (1994).
[51] J. Uppenberg, N. Öhrner, M. Norin, K. Hult, G. Kleywegt, S. Patkar, V. Waagen, T. Anthonsen, and T. A. Jones, *Biochemistry* **34**, 16838 (1995).

concentrations (and possibly diffusion coefficients) of both dissolved substrates will be raised.

Solvent effects in the corresponding transesterification reaction have also been investigated. Reactions with 3-chloro-1-(phenylmethoxy)-2-propanol and vinylbutyrate were performed in hexane and in hexane:acetone at 70:30. The value of E increased from 7 to 15. Also in this case the (S)-alcohol was the faster reacting enantiomer, but the result is not suitable for practical purposes.

Explanations of the origin of solvent effects are provided in the previously mentioned reviews.[23–25]

Experimental

Hydrolysis and One- and Two-Phase Systems

Fresh solutions of lipase from porcine pancreas are prepared as the supernatant of crude PPL (100 mg, Sigma Type II) in potassium phosphate buffer (10 mM, pH 7.8, 30 ml) after centrifuging at 27,000g for 10 min. Conversion experiments using racemic glycidyl butyrate are conducted in a pH-stat (Metrohm Instruments, Herisau, Switzerland) using enzyme solution (30 ml) and ester (5–30 ml). A crude estimate of the extent of conversion is calculated from the amount of titrant (2 M sodium hydroxide) consumed. Accurate values are determined by HPLC (HP-1, Hewlett Packard, Nederland, B.V., Amstelveen) after calibration with chemically pure ester and alcohol. Enantiomer excess values are measured by GC analysis of suitably small samples extracted with dichloromethane and dried over anhydrous magnesium sulphate. Baseline separation of (R)- and (S)-glycidol and glycidyl butyrate is obtained on γ-TA (Astec, Whippany, NJ). Both compounds can be resolved in a single run by applying a temperature program (85° for 8 min; 10°/min; 90° for 10 min).

Initial rate measurements are conducted in triplicate for crude PPL as well as for purified PPL A/B. Ten- to 100-μl aliquots of enzyme solution are added to 5 ml of buffer containing the indicated amount of glycidyl butyrate, obtained by mixing (R)- and *rac*-glycidyl butyrate in the appropriate ratio. Rates are calculated from the consumption of titrant (0.01 M sodium hydroxide).

Kinetic Resolution of Glycidol in Organic Solvents

Organic solvent (Baker, reagent grade, 150 ml) is mixed with vinyl acetate (Janssen Chimica, Belgium) and dried over 4A zeolite (Aldrich Chemical Co. Inc., Milwaukee, WI). 2-Octanone (50 μl) is added as an internal standard. Equilibration of PPL is conducted by shaking PPL (700 mg) with 2-butanone (0.04% v/v water, 20 ml). After overnight treatment, the suspension is centrifuged (27,000g for 15 min) and decanted. The sediment (<0.1% v/v of water, by Karl Fischer titration) is slurred in 3 ml of dry 2-butanone. Conversion experiments are conducted by applying 0.5 ml of the PPL preparation to 20 ml of the organic solvent/acyl donor mixture. After the reaction is started by addition of the appropriate amount of racemic glycidol, solvent equilibrated N_2 is flushed (5 ml/min) through the stirred suspension to remove acetaldehyde that formed. The enantiomeric composition of alcohol and ester is determined by GC analysis (γ-TA) of samples as before. Extents of conversion were calculated using Eq. 1c.

Hydrolysis and Addition of Cosolvents

Enzymes. Lipase B from *C. antarctica* Novo-Nordisk SP 435 immobilized on Lewatit had specific activity of 19,000 PLU/g. The lipase used for NMR experiments was 50% pure and had specific activity of 240 LU/mg. The 1 LU releases 1 mmol of fatty acid per minute from a tributyrin emulsion at pH 7.0 and 30°. The 1 PLU is a corresponding unit for immobilized lipase measured during transesterification.

General Experimental Procedure for Enzymatic Hydrolysis. Butanoates are suspended in 5 ml of 0.20 M phosphate buffer at pH 7.2 and the cosolvent is added in various amounts (10, 20, 30, 40, and 50%). Lipase (50 mg) is added and the reaction mixture stirred at room temperature. The hydrolysis is stopped by repeated extraction with Et_2O and the reaction mixture was analyzed directly by HPLC.

Use of Salt Hydrates to Buffer Optimal Water Activity

Determination of Water in the Organic Phase. The concentration of water in the standard organic phase of the reaction mixture (1.0 mmol BuOH, 0.50 mmol butanoic acid, 8 ml hexane) after equilibration at 20° with $Na_2SO_4 \cdot 10H_2O$ (a_w = 0.76) is determined by coulometric Karl Fischer titration using a Metrohm Coulometer 684 KF as 100 mg liter^{-1} (5.7 mmol-liter^{-1}).

General Experimental Procedure for Enzymatic Esterification with Salt Hydrates. A typical reaction mixture consists of lipase (5.0 mg, when not otherwise stated), *n*-hexane (8 ml), *n*-butanol (1.0 mmol), butanoic acid (0.50 mmol), and, as internal standard, *n*-decane (0.31 mmol). Salt hydrates in pairs (0.5–1.5 g) are added to the reaction mixtures, when not otherwise stated. The reactions are carried out at room temperature in closed vials (10 ml, 20-mm i.d.) with shaking 130 strokes/min. All experiments are conducted at least twice and the maximum deviations are less than 10%.

Analytical Methods. Enantiomeric excess (% ee) of alcohols and esters is determined by HPLC using a Varian 9000 system equipped with UV/VIS detector on a chiral column, Chiralcel OB, delivered by J. T. Baker, Deventer, Holland. Optical rotations are determined using an Opical Activity Ltd. (Ramsey, Huntingdon, England). AA-10 automatic polarimeter, concentrations (*c*) are given in g/ml. Enantiomeric excess of epoxides is determined by GLC using a chiral column, Chiraldex G-TA, delivered by Astec, Whippany, NJ. NMR spectroscopy was performed on Bruker AM-500 operating at 500 MHz for ^{1}H and 125 MHz for ^{13}C.

[23] Lipases in Supercritical Fluids

By ENRICO CERNIA and CLEOFE PALOCCI

Introduction

During the past few years, nonaqueous biocatalysis has established itself as a standard approach in the development of new strategies for organic synthesis.[1,2] For example, enzymes suspended in organic solvents have been used in the synthesis of many pharmaceutical compounds,[3] the resolution of racemic mixtures,[4] the synthesis of optically active polymers,[5] the development of biosensors,[6] and for peptide synthesis.[5,7] Basic studies on these reaction systems have led to an understanding of enzyme mechanism activity,[8] specificity,[9] stability,[10] and structure[11] in anhydrous organic solvents. Furthermore, there is currently a good deal of interest in using "solvent

---

[1] J. S. Dordick, "Biocatalysis for Industry." Plenum, New York, 1991.
[2] A. Zacks and A. M. Klibanov, *Science* **224**, 1280 (1984).
[3] A. L. Margolin, D. L. Delinck, and M. R. Whalon, *J. Am. Chem. Soc.* **109**, 3802 (1986).
[4] G. Kirchner, M. P. Scellar, and A. M. Klibanv, *J. Am. Chem. Soc.* **107**, 7072 (1985).
[5] A. L. Margolin, P. A. Fitzpatrick, P. L. Dubin, and A. M. Klibanov, *J. Am. Chem. Soc.* **113**, 4693 (1991).
[6] R. Z. Karandijand, J. S. Dordick, and A. M. Klibanov, *Biotechnol. Bioeng.* **107**, 5448 (1986).
[7] H. Ooshima, H. Mori, and Y. Harano, *Biotechnol. Lett.* **7**, 789 (1985).
[8] A. J. Russel and A. M. Klibanov, *J. Biol. Chem.* **263**, 11624 (1988).
[9] K. Ryu, D. R. Stafford, J. S. Dordick, J. R. Whitaker, and P. Sonnet, "Biocatalysis in Agricultural Biotechnology." American Chemical Society, Washington, DC, 1990.
[10] R. Tor, Y. Dror, and A. Freeman, *Enzyme Microb. Technol.* **12**, 299 (1990).
[11] P. A. Burke, S. O. Smith, W. W. Bachouchin, and A. M. Klibanov, *J. Am. Chem. Soc.* **111**, 8290 (1989).

engineering" to manipulate the activity and specificity of enzymes in anhydrous media.[12] In fact, in anhydrous organic solvents, the absence of a continuous aqueous phase surrounding the enzymes makes it possible for them to interact directly with the solvents, and very recently it has been found that the physicochemical properties of the solvents can dramatically influence enzyme activity and selectivity.[13]

Very recently, the use of microbial lipases in nonconventional solvents, such as supercritical fluids, has been proposed as a means of improving the activity and utility of such enzymes in anhydrous environments.

The potential use of supercritical fluids (SCFs) in chemical separation processes has been of considerable research interest since the mid-1980s. The fundamentals of SCF extraction technology and a number of potential applications have been described in several reviews.[14-16] One very interesting and, as yet, not fully tested offshoot of SCF extraction technology is the use of an SCF solvent as a reaction medium in which an SCF either actively participates in the reaction or functions only as the solvent medium for reactants or as catalysts and/or products.

By exploiting the unique solvent properties of SCFs (wide variations in density and viscosity are possible with small changes in pressure and/or temperature) it may be possible to enhance reaction rates while maintaining or improving selectivity. Also, separating products from reactants can be greatly facilitated by the ease with which the solvent power of the SCF can be adjusted.

This chapter presents a review of the field of reactions in SCFs and covers several examples of experimental studies on lipase-catalyzed reactions in SCFs.

Supercritical Fluids

Introduction

Supercritical fluids are those that exist at temperatures and pressures above their critical points.[17] Figure 1 shows the three classical states of

[12] J. S. Dordick, *Enzyme Microb. Technol.* **11,** 194 (1989).
[13] E. Cernia, C. Palocci, and E. Catoni, *J. Mol. Catal.* **105,** 79 (1996).
[14] M. A. McHugh and V. J. Krukonis, *in* "Supercritical Fluid Extraction: Principle and Practice." Butterworth, Stoneham, MA, 1985.
[15] D. F. Williams, *Chem. Eng. Sci.* **136,** 1769 (1981).
[16] M. A. McHugh, *in* "Recent Development in Separation Science" (N. N. Li and J. M. Calo, eds.), Vol. IX. CRC Press, Boca Raton, Florida, 1985.
[17] J. M. H. Levelt Sengers, *in* "Supercritical Fluids, Fundamental for Application" (E. Kiran, J. Levelt Sengers, eds.), p. 39, Nato ASI Series, Vol. 273, 1993.

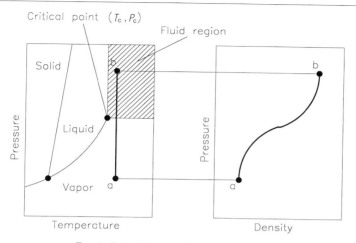

FIG. 1. State diagram of a pure compound.

matter: solid, liquid, and vapor. The critical point is defined as the point where the two phases, liquid and vapor, become indistinguishable. A fluid is supercritical above the critical temperature (T_c) and critical pressure (P_c) (Table I); its properties are between those of liquids and gases. It is almost as dense as a liquid, which allows high levels of solubility. It also has high molecular diffusivity and low viscosity, which make it an ideal medium for efficient mass transfer (Table II). Moreover, its unusually high compressibility gives large density variations with very small pressure changes, yielding extraordinary selectivity characteristics, which are very important in solubility-based separation processes.

TABLE I
CRITICAL CONSTANT OF SUPERCRITICAL FLUIDS

Fluid	P_c (atm)	T_c (°C)
He_2	226.0	−267.9
H_2	12.8	−239.9
N_2	33.5	−146.8
CH_4	45.8	−82.0
C_2H_6	50.5	9.9
CO_2	72.7	31.0
NH_3	111.0	132.3
H_2O	218.0	374.2

TABLE II
PHYSICAL PROPERTIES OF SUPERCRITICAL FLUIDS

	Gas	SG	Liquid
Density (g/cm^3)	10^{-3}	0.1–1	1
Diffusion (cm^2/sec)	10^{-1}	10^{-3}–10^{-4}	$<10^{-5}$
Viscosity (g/cm sec)	10^{-4}	10^{-3}–10^{-4}	10^{-2}

Gas-Like	Liquid-Like
* High diffusivity (mass transport in complex matrices) * Low viscosity (favorable flow characteristics)	* High solvating power (dependent on density)

The properties of these supercritical fluids are so particular that they have the potential to become the process solvents of the future. Indeed, SCFs have already been used commercially in the extraction of oils and aromas from plants, de-asphalting of petroleum, decaffeination of coffee beans, and extractions from hops. Many supercritical fluids, such as carbon dioxide ($T_c = 32°$ and $P_c = 72$ atm), are not only nontoxic and nonflammable, but can also be readily recycled. Releasing the pressure in an SCF-based reactor will cause a selective precipitation of solutes, which facilitates downstream recovery of valuable components and reduces the volume of liquid and vapor effluent in any process.

These special properties of SCFs make them particularly well suited for use in new chemical engineering processes. During the last 15 years a great deal of attention has been given to the use of supercritical fluids in various chemical processes including extraction,[18] atomization,[19] crystallization,[20] reactions,[21] waste destruction,[22] and chromatography.[23]

Biochemical Reaction in Supercritical Fluids

Many of the properties of SCF solvents that make them suitable for use in extraction processes also render them attractive as a medium for biocatalytic reactions:

[18] G. Hotier, F. X. Cormerais, and C. Margerin, in "Proceedings of the International Symposium on Supercritical Fluids" (M. Perrut, ed.), p. 851, 1989.
[19] D. W. Matson, K. A. Norton, and R. D. Smith, *Chemtech. Ang.* **480** (1989).
[20] S. J. Teichner, *Adv. Colloid Intface Sci.* **5**, 245 (1976).
[21] S. Bela and A. M. McHugh, *Ind. Eng. Chem. Process. Des. Dev.* **25**, 1 (1986).
[22] E. Stahl, K. W. Quirin, and G. Gerard, "Dense Gases for Extraction and Refinery." Springer Verlag, New York, 1986.
[23] B. O. Brown, J. Kisbaugh, and M. E. Paulatis, *Fluid Phase Equilibria* **36**, 247 (1987).

- High diffusivity and low surface tension lead to reduced internal mass transfer limitations for heterogeneous chemical or biochemical catalysts.
- The dramatic sensitivity of SCF solubility to changes in pressure and temperature can be used to manipulate the selectivity of a reaction and to recover a product from reactants.
- The use of SCF is not accompanied by the problem of solvent residues in the reaction product because SCF solvents are gases under atmospheric conditions. In particular, the low toxicity and reactivity of $SCCO_2$ make it attractive as a nonaqueous reaction medium for food and pharmaceutical products.
- The pressures associated with suitable SCFs such as supercritical CO_2 ($SCCO_2$) are not so high as to damage biopolymers, and yet the temperatures required are appropriately low for thermally labile biomolecules.

Figure 2 outlines the operating principle of a process that uses CO_2 as the reaction medium. The reactor is a packed-bed, column-type system with fluid recirculation. It can operate either batchwise or continuously. The contents of the reactor are fractionated by supercritical extraction. The desired product is finally crystallized from the $SCCO_2$ by letting the fluid expand under lower pressure.

In the literature, there are many examples of SCFs as media for chemical reactions especially with regard to hydrolysis reactions, thermal and photochemical reactions, isomerization, and heterogeneous catalytic reactions. In processes involving conventional solvents, successful correlation of the physical properties of the solvent with the behavior of the enzyme is hindered by inevitable differences in the chemical structure between one solvent and another. Indeed, one solvent may interact with an enzyme, causing inhibition, while another may not. Supercritical fluids offer the flexibility of changing the physical properties of the system without drastic changes in multiple properties (including structure). It is often possible to change these properties in a predictable, continuous, and controllable fashion simply by varying the pressure (or temperature) of the system.

However, SCF processes may have some limitations. This is because the solubility of molecules in supercritical fluids is, in some cases, relatively low, and very high pressure (5000–10,000 psi) may be needed to achieve solubility similar to liquids. The lack of solubility and thermodynamic property data for supercritical fluids makes process design difficult; more data are needed for even the most widely studied supercritical fluid, carbon dioxide. The need for costly high-pressure equipment could make some technically feasible processes "economically unattractive," but carrying out

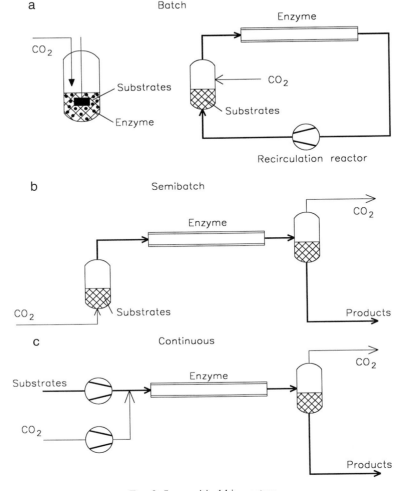

FIG. 2. Supercritical bioreactors.

a process with fluid recirculation can improve the economic situation substantially.

In the following section, results achieved to date with enzymatic reactions in SCF, with particular reference to lipases, are discussed drawing attention to some fundamental problems such as substrate solubility, pressure and moisture effects, mass transfer effects, and enzyme stability in supercritical media.

Mass Transfer Effects

One of the potential benefits of using supercritical fluids as reaction media is their better mass transfer properties compared to liquids.

External mass transfer occurs between the bulk of the reaction mixture and the surface of the enzyme support. The mass transfer resistance across this external film depends on the flow conditions. External diffusion in liquid media is rate limiting in enzyme-catalyzed reactions only when the liquid flow rate is approximately 1 cm/sec.

Internal diffusion occurs in the pores of the enzyme support and is independent of flow conditions. Internal diffusion is the reaction rate-limiting phenomenon when enzyme activity is high, substrate concentration is low, and the particles of the enzyme support are large. Internal diffusion usually controls enzymatic reaction rates more than external diffusion. In this way, if the reaction rates in $SCCO_2$ are limited by external mass transfer they can be speeded up by increasing flow rates in an enzyme column. No optimum flow rate has been found in any of the published data on this argument.

Solubility of Substrate and Water in SCF

The application of SCFs in biocatalyzed processes or extractions requires extensive information on the phase behavior of the components. Although thermodynamic modeling of phase behavior of SCFs has been carried out,[24] it is still difficult to predict the behavior of complex mixtures of biological molecules in SCFs in this way. This is partly due to the lack of data on the physical properties of the molecules. A simple example, in which the solubility of a single component is considered, is given below. This takes into account the effects of temperature and pressure, and it is calculated using the solubility equation proposed by Chrastil. In this equation, the molecules of the SCF and the solute are assumed to form a complex and the law of mass action is applied to the equilibrium between the complex and the unassociated molecules.

$$C = d^k \exp(a/T + b) \quad (1)$$

where C is the concentration of solute in SCF; d, the density of SCF; T, the absolute temperature; and $a = \Delta H/R$, $b = \ln(M_A + kM_B) + q - k \ln M_B$, where M_A and M_B are the molecular weight of solute and solvent, respectively, and k is the solvation number (number of solvent molecules associated with one solute molecule to form a complex).

[24] T. T. Randolph, *Trends Biotechnol.* **36,** 78 (1990).

Water is a reactant in hydrolytic reactions, and it also influences the catalytic activity of enzymes. It is therefore necessary to know the solubility of water, as well as that of other substrates and products, in the SCF solvents. The solubility of water in supercritical carbon dioxide calculated using the preceding equation was compared with the experimental data reported by Wiebe et al.[25] Agreement between data was found to be good, with the solubility largely dependent on the choice of temperature and pressure. This means that a supercritical CO_2/H_2O mixture can be used as a reaction medium for either hydrolytic or synthetic reactions catalyzed by hydrolases. The solubility of water in n-hexane is only 0.878 mM at 40° and normal pressure. This solubility is 100 times higher in $SCCO_2$ compared to n-hexane at 15 MPa. The high capacity of $SCCO_2$ for dissolving water may prove to be a favorable feature, in that it avoids the wetting of immobilized enzymes in supercritical carbon dioxide.

The solubility of substrates in SCF can also be improved by the addition of cosolvents, such as ethanol, although cosolvents should be used with caution because they may participate actively in the reaction, thus yielding by-products.

Stability of Enzymes in SCF

Solvent Effect

Chemical or biochemical reactions in SCF are carried out at high pressure. Transition-state analysis is introduced to explain the rate enhancement observed at high pressure in SCFs. As described by Eckert,[26] for a bimolecular reaction, a chemical equilibrium is assumed between the reactants A and B and the transition state M

$$A + B \Leftrightarrow M^{\neq} \Rightarrow \text{products}$$

The variation of the reaction rate constant k with pressure is given by

$$\delta \ln k / \delta P = -\Delta V^{\neq}/(RT)$$

where $\Delta V^{\neq}$, the activation volume, is the difference in the partial molar volumes of the activated complex and the reactants and is given by

$$\Delta V^{\neq} = V_M - V_A - V_B$$

If the volume of activation is positive, then the reaction will be hindered by pressure. However if $\Delta V^{\neq}$ is a negative quantity, then the reaction rate

---

[25] R. Wiebe and V. L. Gaddy, *J. Am. Chem. Soc.* **63**, 475 (1941).
[26] C. A. Eckert, *Annu. Rev. Phys. Chem.* **23**, 239 (1972).

will be enhanced by pressure. The unusual reaction behavior of ethylene polymerization in supercritical ethylene, for example, was ascribed to a very large negative partial molar volume of the activated complex near the critical point of the solvent.

The effect of pressure on an enzymatic reaction can also be determined by considering the activation volume $\Delta V^{\neq}$ as consisting of two parts: the volume change of the catalytic step, $\Delta V^{\neq}_c$ and the volume change of the binding step, $\Delta V^{\neq}_b$:

$$\Delta V^{\neq} = \Delta V_c^{\neq} + (K_M/(K_M + C)) \Delta V_b^{\neq}$$

where C and K_m are substrate concentration and the Michaelis–Menten constant, respectively. The temperature and pressure dependence of the activation volume $\Delta V^{\neq}$ can be expressed if the free-energy change of the reaction is considered:

$$\Delta V^{\neq} = \Delta V_0^{\neq} + \Delta \beta (P - P_0) + \Delta \alpha (T - T_0)$$

The volume changes $\Delta V_c^{\neq}$ and $\Delta V_b^{\neq}$ were measured at pressures of up 200 MPa for several proteases: if both values are negative, this indicates that the reaction rate is enhanced at high pressure.

Another very important factor that may cause enzyme activity loss in $SCCO_2$ is the depressurization step. Nakamura et al.[27] have reported that *Rhizopus delemar* lipase loses its activity with increasing pressure and contact time. In this case, the lipase contained more than five times its weight in water. The mechanism of denaturation in this case may have been the release of CO_2 from the bound water during pressurization. However, lipase of low water content is maintained more than 98% of its initial activity, even after 154 hr of exposure to $SCCO_2$ (50°, 29.4 MPa).

The reported results indicate that the stability of lipase in $SCCO_2$ is significantly influenced by its water content.

Moisture Effect

Enzymes require water to function as catalysts. However, even a small amount, perhaps even a "monolayer" on the enzyme molecules, can suffice.[28] To prevent dehydration of the enzyme, the fluid in contact with it must contain water. Moisture equilibrium between the fluid and the enzyme support is specific for each system. The most hydrophobic hydrocarbons, such as *n*-hexane, dissolve approximately 0.01% water, whereas $SCCO_2$ can

---

[27] K. Nakamura, in "Second Conference to Promote Japan/U.S. Joint Project" (A. Fichter, H. Okada, and R. D. Taylor, eds.), p. 257. Springer Verlag, Berlin, 1989.
[28] A. M. Klibanov, *Chemtech.* **16**, 354 (1986).

dissolve more than 0.3% water. The more hydrophilic the fluid, the more likely it is to dehydrate the enzyme.

Van Eijs et al.[29] determined the moisture equilibrium between $SCCO_2$ and a commercial immobilized lipase, Lipozyme (Novo). The commercial lipase contains 8% water. To maintain this content in the enzyme support the $SCCO_2$ should contain 0.3% water at 100 bar and 60°.

Some authors[29-31] have optimized the moisture of $SCCO_2$ to maximize the lipase-catalyzed alcoholysis and the esterification rate. The optimum moisture of 0.1 to 0.2 wt% in $SCCO_2$ was found for myristic acid esterification with ethanol and ethyl acetate alcoholysis with amyl alcohol. The equilibrium water content of the *Mucor miehei* lipase was 5%. However, the optimum moisture seemed to depend on the substrate. Alcoholysis between ethyl acetate and nonyl acetate with the same enzyme showed no moisture optimum.[29,30] The drier the enzyme support, the greater the initial reaction rate. The optimum water content of lipases seems to depend on both support materials and the substrates. Optimum water contents ranging from 4 to 110% have been reported.[32,33] The differences may be due to the different water adsorption isotherms of the enzyme carriers.

The optimum water content of enzyme supports in $SCCO_2$ is much higher than in the more hydrophobic organic liquids. This is generally attributed to differences in water partitioning between the solvent and the enzyme support. The more hydrophilic the solvent, the more evenly the water will be distributed between it and the enzyme.

For optimum enzyme activity, it seems that CO_2 should contain water from 0.1 wt% up to saturation (approximately 0.3%). Too much water in the enzyme will probably reduce the accessibility of substrate molecules to the active sites of the enzyme and eventually make the thermodynamic equilibrium less favorable.

Clearly, the humidity of the $SCCO_2$ must be controlled. Excess water produced, for example in batch esterification, must be removed from the reaction mixture. If the solubility of substrates and reaction products in $SCCO_2$ is low ($\leq 0.1\%$) and CO_2 is saturated with water, there will be a lot of water in the reaction mixture after CO_2 separation. If some of the

---

[29] A. M. M. Van Ejis, J. P. L. de Jong, H. J. Doddema, and D. R. Lindebeam, *in* "Proceedings of the International Symposium Super. Flu." (M. Perrut, ed.), p. 933, 1988.

[30] A. M. M. Van Ejis, J. P. L. de Jong, H. H. M. Ooostrom, H. J. Doddema, M. A. Visser, and R. Stoop, *in* "Proceedings of the Sec. Netherl. Biotech. Congr." Amsterdam, Oct. 20–21, p. 581, 1988.

[31] T. Dumont, D. Barth, and M. Perrut, "2nd Int. Symp. High Pressure Chem. Eng." Erlangen, Germany, Sept. 24–26, 1990.

[32] A. Marty, W. Chulalaksananukul, J. S. Condoret, R. M. Willemot, and G. Durant, *Biotechnol. Lett.* **12**, 11 (1990).

[33] Y. M. Chi, K. Nakamura, and T. Yano, *Agric. Biol. Chem.* **52**, 1541 (1986).

substances in the reaction mixture are water soluble, they will have been separated from an aqueous solution. This would require traditional downstream processing steps, and some of the technical benefits of using supercritical fluids as reaction media would be lost.

Lipase-Catalyzed Reactions in Supercritical Fluids

From the first demonstrations of Randolph et al.[34] and Nakamura et al.,[35] which showed that enzymes are active and stable in supercritical fluids, several studies have reported on enzyme-catalyzed reactions in supercritical fluids.[36-44] Owing to its numerous advantages, most of the studies employed carbon dioxide as the supercritical solvent. The lipase-catalyzed reaction for fats and steroid modification was first selected as a model reaction due to the lipophilic nature of carbon dioxide and the high stability of the enzyme.[34-42]

Douglas et al.[45] investigated the *Rhizopus arrhizus* lipase-catalyzed interesterification of triglycerides, a reaction that can be used to produce upgraded fats and oils. In the interesterification reaction (acidolysis), the mixture of triglycerides becomes enriched with the new fatty acids, thereby changing the physicochemical properties of triglycerides. Interesterification is well suited for study in $SCCO_2$ because the nonpolar reactants are soluble in $SCCO_2$ and no cofactors are required for enzymatic activity. The lipase from *Rhizopus arrhizus* is stable under operational conditions in $SCCO_2$ (80 hr at 1400 psi and 35°). The water content of $SCCO_2$ has little effect on enzyme activity and the interesterification rate increases much more rapidly than the overall rate, indicating that the selectivity of the reaction for interesterification over hydrolysis improves at higher pressure.

Cernia et al.[13] investigated on lipolytic microbial enzyme stability in supercritical carbon dioxide: *Pseudomonas cepacia* lipase incubated in a

---

[34] T. W. Randolph, H. W. Blanch, and J. M. Prausniz, *Biotechnol. Lett.* **7**, 325 (1985).
[35] K. Nakamura, Y. M. Chi, Y. Yamada, and T. Yano, *Chem. Eng. Commun.* **45**, 207 (1985).
[36] D. A. Muller, H. W. Blanch, and J. M. Prausniz, *Ind. Eng. Chem. Res.* **30**, 939 (1991).
[37] S. Kamat, J. Barrea, E. J. Beckman, and A. J. Russel, *Biotech. Bioeng.* **40**, 158 (1992).
[38] A. Marty, W. Chulalaksananukul, J. S. Condoret, R. M. Willomot, and G. Durant, *Biotechnol. Bioeng.* **39**, 273 (1992).
[39] T. Dumont, D. Barth, C. Corbier, G. Braulant, and M. Perrut, *Biotechnol. Bioeng.* **39**, 329 (1992).
[40] P. Pasta, G. Mazzola, and G. Carrea, *Biotechnol. Bioeng.* **9**, 643 (1989).
[41] D. A. Hammond, M. Karel, and A. M. Klibanov, *Appl. Biochem. Biotechnol.* **11**, 393 (1985).
[42] T. W. Randolph, D. S. Clark, H. W. Blanch, and J. M. Prausniz, *Science* **239**, 387 (1988).
[43] D. C. Steyler, P. S. Moulson, and J. Reynolds, *Enzyme Microb. Technol.* **13**, 221 (1991).
[44] K. Nakamura, *Tibtech.* **8**, 288 (1990).
[45] A. M. Douglas, H. W. Blanch, and J. M. Prausniz, *Ind. Eng. Chem. Res.* **30**, 939 (1993).

high-pressure reactor with $SCCO_2$ at a temperature of 40° and a pressure of 20 MPa shows, also for a long incubation time, higher stability than with any other organic solvents. Different esterification reactions, employing as model substrates 1-phenylethanol with different substitution moieties, were carried out and the results were compared to those achieved in supercritical CO_2. From the experimental results it appears that in organic solvents, the percentage of conversion and the enantioselectivity of the reaction dramatically depend on the physicochemical characteristics of solvent, showing that the conversion rate increases with increasing log P_{ow} values; for the reactions carried out in $SCCO_2$ the reaction rates and the enantioselectivity are significantly higher than in any conventional solvent tested.

An interesting application of enzymes in $SCCO_2$ could be the production of pure optical isomers either by resolution of a racemic mixture via stereospecific derivation or by chiral synthesis. Ikushima et al.[46] used a Candida cylindracea lipase in the transesterification of (±) citronellol with oleic acid in $SCCO_2$. They noticed that the reaction rate was enhanced by increasing the pressure and this is particularly pronounced near the critical region; the rate at 8.41 MPa is greater by a factor of 3 than that at 7.58 MPa. They also reported on the optical purity of the product calculated from the specific rotation. Optical purity was found to be sensitive to reaction conditions: it is nearly 100% at 8.41 MPa and 304.1 K, indicating that the (S) ester is stereoselectively formed, while it is much less at higher pressures and temperatures.

Rantakyla et al.[47] studied the esterification reaction of racemic ibuprofen, (R,S)-2-(isobutylphenyl)propionic, with n-propanol catalyzed by an immobilized lipase from M. miehei in supercritical carbon dioxide. In the conditions employed, the initial reaction rate increased with the initial concentration of ibuprofen and the enantiomeric excess decreased with increasing conversion; as the conversion reached 50%, the enantiomeric excess decreased sharply. The initial esterification rate of ibuprofen and the enantiomeric excess as a function of pressure at two ibuprofen concentrations showed that pressure has a great effect on the reaction rate in terms of substrate concentration, in particular, near critical pressure. To compare the esterification rates in supercritical carbon dioxide and in an organic solvent at atmospheric pressure, the reaction was also carried out in n-hexane in the same conditions. Reaction rates in $SCCO_2$ are comparable with the rates in n-hexane.

---

[46] Y. Ikushima, N. Saito, T. Yokoyama, K. Hatakeda, S. Ito, M. Arai, and H. W. Blanch, *Chem. Lett.* **109**, 341 (1993).

[47] M. Rantakyla and O. Aaltonen, *Biotechnol. Lett.* **110**, 825 (1994).

The control of organic solvent residues is crucial in the production of chiral intermediates for the pharmaceutical industry. Martins et al.[48] reported the resolution of racemic glycidol through esterification with butyric acid catalyzed by porcine pancreatic lipase, free or immobilized, and an immobilized lipase from *M. miehei*.

The solubility of glycidol, the least soluble reactant, was measured in CO_2 at 35° and pressure in the range 70–180 bar. For the porcine pancreatic lipase (free form), the maximum reaction rates were obtained for an enzyme water content of 10 ± 2% w/w. This is the same as in organic solvents, although the corresponding initial rate was much lower. Even though the selectivity observed is comparable to the values obtained in organic solvents, $SCCO_2$ offers the advantage that, due to its adjustable solvent power, a series of separators can be used to recover the product of interest in pure form.

The physical properties of supercritical fluids are greatly dependent on pressure. Lipases, however, are generally unaffected by small pressure increases. Clearly there is the possibility of performing biocatalysis in SCFs at different pressures. In such a system, the effect of pressure on the reaction rates should be the result of changes in the physical properties of the fluid rather than changes in the structure of the bulk solvent. Naturally, for this approach to be successful, the physical properties of the fluid must be tunable over a range that will affect the activity of an enzyme.

Kamat et al.[49] reported the effect of pressure on *C. cylindracea* lipase-catalyzed transesterification of methylmethacrylate and 2-ethyl-1-hexanol in near-critical propane, supercritical ethane, carbon dioxide, fluoroform, sulfur hexafluoride, and ethylene. The authors studied the dependence of the physical properties of some supercritical solvents on pressure in the range within which we wanted to test enzyme activity. The use of fluoroform as a solvent is very interesting because there is a marked effect of pressure on the solvent dielectric constant. Experimental data have shown that the dielectric constant of fluoroform is dependent on pressure increasing from 1 to 8 over a small pressure range (850–4000 psi). In this regard, we must remember that some authors have reported that the majority of significant alterations in the flexibility of a protein occur when the solvent dielectric increases from 1 to 10. Given that changes in the dielectric constant have also been related to the flexibility, hydration, intrinsic activity, specificity,

---

[48] F. J. Martins, I. Borges de Carvaho, T. Correa de Sampaio, and S. Barreiros, *Enzyme Microb. Technol.* **16**, 785 (1994).

[49] V. S. Kamat, S. Iwaskewycs, E. J. Beckman, and A. J. Russel, *Proc. Natl. Acad. Sci. U.S.A.* **90**, 2940 (1993).

and stability of enzymes in anhydrous media, fluoroform appears to be an ideal solvent in which to study such effects in detail.

Variations of either SCF temperature or pressure induce marked changes in the enzyme reaction rates. With regard to the effect of the Hildebrand solubility parameter, dielectric constant, and density on enzymatic activity in supercritical ethane, fluoroform, and sulfur hexafluoride, it is clear that the solvent effect on lipase-catalyzed transesterification in SCFs can be predicted only by focusing on the dielectric constant rather than the solubility parameter.

Conclusion

It should be clear from the several examples cited in this review that supercritical fluids can be advantageously used as reaction media. It may be possible to carry out enzymatic reactions reducing interphase mass transfer limitations, and labile reaction products could be more readily isolated from the reaction mixture by adjusting the pressure and/or the temperature to induce a phase split, thus avoiding unwanted side reactions. Moreover, reaction rates may be advantageously enhanced by running the reaction at conditions close to the critical point of the pure SCF.

Of the enzymatic reactions in SCFs investigated so far, the use of microbial lipases shows most commercial promise and many opportunities exist for applications in organic synthesis. Greater understanding of the characteristics and mechanism of lipase-catalyzed reactions in SCFs could be achieved by studying different reaction conditions, particularly near the critical point.

A change in temperature or pressure can influence the solvation state of the substrate, product, and even the enzyme, and improve results in terms of reaction velocities and selectivity. Integration of reaction and separation bioprocesses will further demonstrate the superior characteristics of SCFs and will lead to use of these fluids in many new application areas.

[24] Stabilization of Lipases against Deactivation by Acetaldehyde Formed in Acyl Transfer Reactions

By H. K. WEBER and K. FABER

Introduction

Lipases have been employed for several years to catalyze a variety of synthetically useful reactions performed in monophasic organic solvents at low water activity.[1–5] The most striking advantages of these reactions are (1) the possibility of enhancing the stability of the enzymes in nonaqueous media compared to that in water,[6] (2) the repression of undesired side reactions being caused by water,[7] and (3) the easy recovery of the (insoluble) enzymes from the organic solvent by simple filtration or centrifugation. Conventional ester synthesis from an alcohol and carboxylic acid via reversal of ester hydrolysis is impeded by the formation of one molar equivalent of water. Due to the low solubility of the water in the lipophilic organic solvent, it tends to be collected at the polar (hydrophilic) enzyme surface by forming a discrete aqueous phase. This results in restricted access of the (lipophilic) substrate, coagulation of the enzyme preparation, and the reversibility of the reaction. The latter fact is particularly undesirable in view of its depletion of the enantioselectivity of the reaction.[8] To overcome this problem, the following techniques can be employed: Continuous removal of water by evaporation,[9,10] by azeotropic distillation,[11] or via entrapment using molecular sieves.[12]

Recently, water extraction into a saturated salt solution through silicon tubing has been recommended.[13] A more elegant approach—acyl trans-

---

[1] G. Carrea, G. Ottolina, and S. Riva, *Trends Biotechnol.* **13,** 63 (1995).
[2] F. Theil, *Chem. Rev.* **95,** 2203 (1995).
[3] E. Santaniello, P. Ferraboschi, and P. Grisenti, *Enzyme Microb. Technol.* **15,** 367 (1993).
[4] S. Riva, G. Ottolina, and K. Faber, *Biocatalysis* **8,** 91 (1993).
[5] K. Faber and S. Riva, *Synthesis* 895 (1992).
[6] P. Adlercreutz and B. Mattiasson, *Biocatalysis* **1,** 99 (1987).
[7] Y.-F. Wang, S.-T. Chen, K. K.-C. Liu, and C.-H. Wong, *Tetrahedron Lett.* **30,** 1917 (1989).
[8] C. J. Sih and S.-H. Wu, *Topics Stereochem.* **19,** 63 (1989).
[9] M. Trani, A. Ducret, P. Pepin, and R. Lortie, *Biotechnol. Lett.* **17,** 1095 (1995).
[10] G. Fregapane, D. B. Sarney, and E. N. Vulfson, *Biocatalysis* **11,** 9 (1994).
[11] G. Lin, S.-H. Liu, S.-J. Chen, F.-C. Wu, and H.-L. Sun, *Tetrahedron Lett.* **34,** 6057 (1993).
[12] H. F. de Castro, W. A. Anderson, R. L. Legge, and M. Moo-Young, *Indian J. Chem., Sect. B* **31B,** 891 (1992).
[13] E. Wehtje, I. Svensson, P. Adlercreutz, and B. Mattiasson, *Biotechnol. Tech.* **7,** 873 (1993).

fer—avoids the undesirable formation of water at all. To provide a highly irreversible reaction,[14] so-called "activated esters,"[15] such as oxime,[16,17] 2-haloethyl,[18] cyanomethyl,[19] alkylthio,[20] and carbonate esters,[21] have been investigated as acyl donors. Alternatively, acid anhydrides can be used,[22] so long as the polarity of the solvent remains low to prevent spontaneous (nonspecific) acylation. Nevertheless, the majority of these acyl donors have several disadvantages: some of them develop toxic intermediates (e.g., formaldehyde-cyanohydrin) or low volatile by-products during the course of reaction that impede workup (trichloroethanol). Others are too expensive to be used on a large scale (trifluoroethyl esters). A particular class of acyl donors is enol esters, such as vinyl[23] and *iso*-propenyl esters.[24] Both have gained wide popularity for several reasons: the unstable enol liberated during the course of the reaction tautomerizes immediately to form the corresponding aldehyde or ketone, respectively, which shifts the reaction out of the equilibrium and the removal of the by-products is generally easy due to their high volatility. Although *iso*-propenyl esters would be advantageous over vinyl esters since they liberate acetone (which is more innocuous with respect to enzyme stability), the reaction rates are usually lower due to steric hindrance.[25]

As a consequence, inexpensive vinyl esters such as the acetate or butanoate have become the most widely used acyl donors. In contrast to earlier assumptions,[26] the generation of one molar equivalent of acetaldehyde may cause serious problems with respect to enzyme stability and selectivity. Because early observations on the possible reuse of lipase were rather

---

[14] J.-M. Fang and C.-H. Wong, *Synlett.* 393 (1994).
[15] G. Kirchner, M. P. Scollar, and A. M. Klibanov, *J. Am. Chem. Soc.* **107,** 7072 (1985).
[16] V. D. Athawale and S. R. Gaonkar, *Biotechnol. Lett.* **16,** 149 (1994).
[17] M. Mischitz, U. Pöschl, and K. Faber, *Biotechnol. Lett.* **13,** 653 (1991).
[18] A. L. Margolin, D. L. Delinck, and M. R. Whalon, *J. Am. Chem. Soc.* **112,** 2849 (1990).
[19] L. Blanco, G. Rousseau, J.-P. Barnier, and E. Guibe-Jampel, *Tetrahedron: Asymmetry* **4,** 783 (1993).
[20] C. Orrenius, N. Öhrner, D. Rotticci, A. Mattson, K. Hult, and T. Norin, *Tetrahedron: Asymmetry* **6,** 1217 (1995).
[21] E. Guibe-Jampel, Z. Chalecki, M. Bassir, and M. Gelo-Pujic, *Tetrahedron* **52,** 4397 (1996).
[22] D. Bianchi, P. Cesti, and E. Battistel, *J. Org. Chem.* **53,** 5531 (1988).
[23] M. Degueil-Castaing, B. De Jeso, S. Drouillard, and B. Maillard, *Tetrahedron Lett.* **28,** 953 (1987).
[24] Y.-F. Wang, J. J. Lalonde, M. Momongan, D. E. Bergbreiter, and C.-H. Wong, *J. Am. Chem. Soc.* **110,** 7200 (1988).
[25] Acyl transfer reactions employing vinyl esters are usually about four times faster than those of *iso*-propenyl esters. See Y.-F. Wang and C.-H. Wong, *J. Org. Chem.* **53,** 3127 (1988).
[26] S.-H. Hsu, S.-S. Wu, Y.-F. Wang, and C.-H. Wong, *Tetrahedron Lett.* **31,** 6403 (1990).

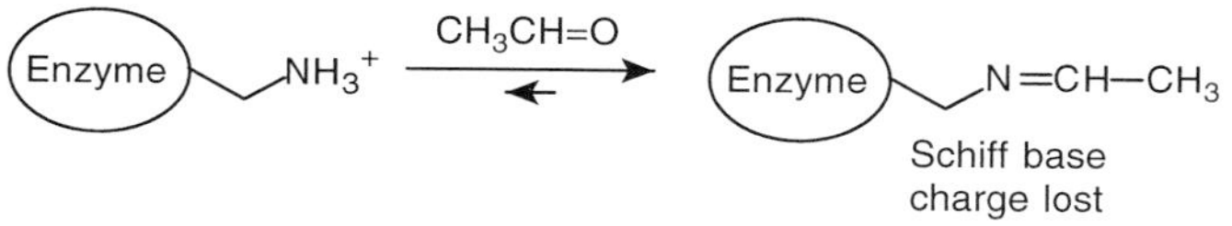

$$\text{Enzyme} - NH_3^+ \underset{}{\overset{CH_3CH=O}{\rightleftarrows}} \text{Enzyme} - N=CH-CH_3$$
Schiff base
charge lost

FIG. 1. Enzyme deactivation by acetaldehyde through Schiff base formation.

contradictory,[27-29] a detailed study revealed that the extent of lipase deactivation strongly depends on structural elements of the enzyme, which are in turn determined by the microbial source.[30]

Acetaldehyde is known to act as an alkylating agent on enzymes by forming Schiff bases in a Maillard-type reaction[31] in particular on N-terminal amino residues of lysine.[32] As depicted in Fig. 1, a positive charge is removed from the enzyme's surface during the course of this reaction, possibly leading to deactivation.

The extent of the deactivation is determined by the relative reactivity (i.e., nucleophilicity) of particular lysine residues within an enzyme. The latter parameter, in turn, is determined by specific structural properties of lysine residues, such as the pK_a value and its accessibility.[33] As a consequence, those lipases that possess highly exposed lysine residues having increased pK_a values are likely to be more readily deactivated by acetaldehyde than others, whose lysine groups are buried and exhibit only moderate basicity. The results from a detailed investigation of the acetaldehyde sensitivity of the lipases used most commonly for acyl transfer reactions are depicted in Fig. 2. It can be seen clearly that the amount of deactivation strongly depends on the microbial source of the enzyme. Whereas the majority of enzymes from *Aspergillus, Chromobacterium, Humicola, Mucor, Penicillium, Pseudomonas, Rhizopus,* and *Thermomyces* sp., as well as from *Candida antarctica,* are remarkably stable toward acetaldehyde, candidates from *Candida rugosa* and *Geotrichum candidum* are extremely sensitive. As a consequence, they lose up to ~80% of activity in a typical acyl transfer reaction when vinyl acetate is employed as acyl donor. Change in activity in Fig. 2 is given in percent, with the native enzyme corresponding

[27] E. W. Holla, *J. Carbohydr. Chem.* **9**, 113 (1990).
[28] U. Ader, D. Breitgoff, P. Klein, K. E. Laumen, and M. P. Schneider, *Tetrahedron Lett.* **30**, 1793 (1989).
[29] S. Mitsuda and S. Nabeshima, *Recl. Trav. Chim. Pays-Bas* **110**, 151 (1991).
[30] H. K. Weber, H. Stecher, and K. Faber, *Biotechnol. Lett.* **17**, 803 (1995).
[31] F. Ledl and E. Schleicher, *Angew. Chem. Int. Ed. Engl.* **29**, 565 (1990).
[32] T. M. Donohue, D. J. Tuma, and M. F. Sorrell, *Arch. Biochem. Biophys.* **220**, 239 (1983).
[33] H. K. Weber, J. Zuegg, K. Faber, and J. Pleiss, *J. Mol. Catal. B* in press (1997).

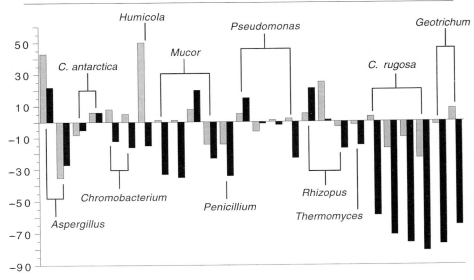

FIG. 2. Deactivation of crude industrial lipase preparations by acetaldehyde. to 100% as standard. Gray bars indicate the variation in activity by treatment with toluene alone (blank); black bars show the results after acetaldehyde treatment. For experimental conditions and enzyme source, see Refs. 30 and 34.

Because *C. rugosa* lipase has been shown to be an extremely useful enzyme particularly for the resolution of secondary alcohols having a sterically demanding structural framework, which are generally not accepted by *Mucor* or *Pseudomonas* sp. lipases,[35] the following techniques for the stabilization of this enzyme have been developed.

Stabilization by Covalent Immobilization

Enzyme immobilization is a very common procedure to enhance the stability of an enzyme toward physical and chemical influences.[36–40] In some

---

[34] Crude microbial lipase preparations were exposed to a toluene solution of acetaldehyde (0.1 M) for 17 hr at room temperature. These conditions are a mimic for acyl transfer reactions.

[35] K. Faber, "Biotransformations in Organic Chemistry," 3rd ed., p. 92. Springer, Heidelberg, 1997.

[36] P. Monsan and D. Combes, *Methods Enzymol.* **137**, 584 (1988).

[37] T. H. Maugh II, *Science* **223**, 474 (1984).

[38] A. M. Klibanov, *Science* **219**, 722 (1983).

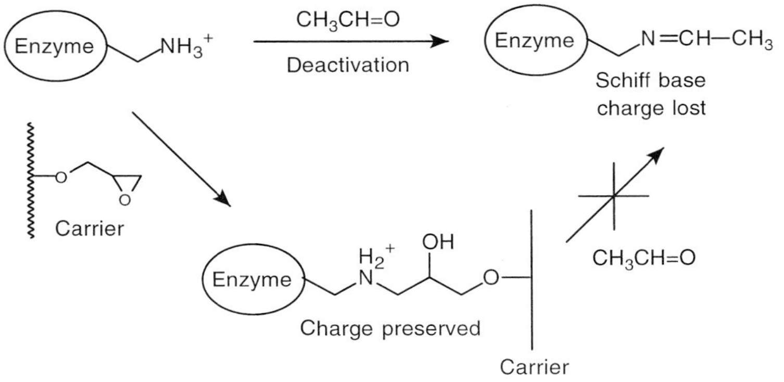

FIG. 3. Covalent immobilization of enzyme onto an epoxy-activated carrier.

cases, it was also shown that the selectivity can be modulated through immobilization.[41] The large number of different immobilization techniques can be classified into three categories: covalent linkage, entrapment in macromolecules, or adsorption onto a carrier.

Covalent linkage onto macroscopic carriers is achieved via several types of chemical bonds. Commercially available epoxy-activated carriers are most popular due to the ease of the immobilization procedure. During the immobilization, nucleophilic residues located at the enzyme's surface (most likely the ε-amino groups of lysine residues) attack the oxirane ring and irreversibly bind the enzyme onto the carrier (Fig. 3). During this reaction, the amino group is blocked via alkylation and, as a consequence, Schiff base formation by acetaldehyde is severely impeded.

For this reason, it was anticipated that *C. rugosa* lipase covalently immobilized through epoxy groups should exhibit an enhanced stability toward acetaldehyde. As can be deduced from Fig. 4, this assumption proved to be true to some extent. The resolution of ($\pm$)-*endo*-bicyclo[2.2.1]oct-5-en-2-ol (norbornenol)[42] via acyl transfer from vinyl acetate proceeded with low selectivity—expressed as the enantiomeric ratio[43]—when native lipase was used ($E \sim 8$). The reuse of the enzyme was impossible due to severe loss of activity (-90%) during the first run. In addition, the initial selectivity

---

[39] A. M. Klibanov, *Anal. Biochem.* **93**, 1 (1979).
[40] J. Lavayre and J. Barratti, *Biotechnol. Bioeng.* **24**, 1007 (1982).
[41] D. S. Clark, *Trends Biotechnol.* **12**, 439 (1994).
[42] T. Oberhauser, M. Bodenteich, K. Faber, G. Penn, and H. Griengl, *Tetrahedron* **43**, 3931 (1987).
[43] C.-S. Chen, Y. Fujimoto, G. Girdaukas, and C. J. Sih, *J. Am. Chem. Soc.* **104**, 7294 (1982).

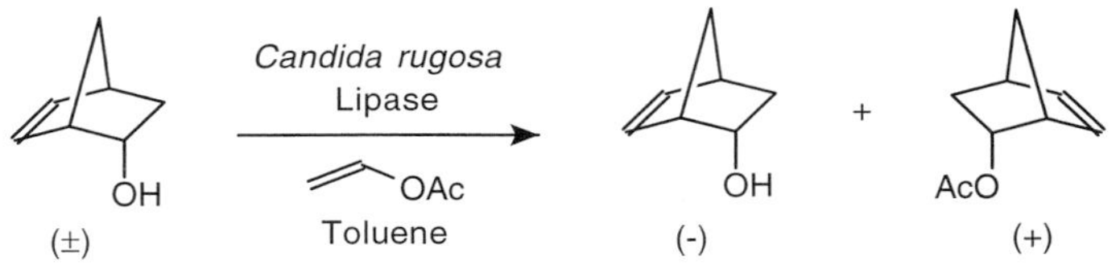

FIG. 4. Resolution of *endo*-norbornenol via acyl transfer using *C. rugosa* lipase.

dropped even further to almost nil. When the reaction was performed with *C. rugosa* lipase covalently immobilized onto the epoxy-activated carrier VA-Epoxy-Biosynth,[44] the selectivity was enhanced by a factor of 5 (Table I). In addition, the reuse of the enzyme now became feasible because the loss of selectivity was completely overcome and the activity was greatly stabilized compared with the performance of the native enzyme. The extent of the stabilization strongly depends on the nature of the epoxy-activated matrix. For instance, when a different commercial carrier was used,[45] the residual activity of the lipase was very low and the extent of solubilization negligible. A possible explanation for this discrepancy is the different density of reactive epoxy groups on the carrier. Whereas VA-Epoxy-Biosynth has an average epoxy content of ~300 μmol/g, Eupergit C is much more reactive (~600–800 μmol/g). The latter support binds via increased multipoint attachment, which makes the immobilized enzyme more rigid by impeding small conformational changes on formation of the enzyme–substrate complex. This is particularly important considering lipases, which are typically "induced-fit" enzymes. The actual effect of enzyme stability on multipoint attachment resides in the method being used, since diverse effects, leading to either enhanced[46] or reduced stability,[47] are known.

Stabilization by Adsorption

The covalent immobilization described earlier is not feasible on a large scale (in particular, for industrial applications) due to the high cost of the carrier material. For such applications, stabilization by adsorption is advantageous.

---

[44] Epoxy-activated polymer support VA-Epoxy-Biosynth from Riedel-de-Haen (Germany), product 329354.
[45] Eupergit C 250L from Röhm-Pharma (Darmstadt, Germany), batch 11778-14.
[46] J. M. Guisan, R. Fernandez-Lafuente, V. Rodriguez, A. Bastida, R. M. Blanco, and G. Alvaro, *in* "Stability and Stabilization of Enzymes" (W. J. J. van den Tweel, A. Harder, and R. M. Buitelaar, eds.), p. 55. Elsevier, Amsterdam, 1993.
[47] K. Burg, O. Mauz, S. Noetzel, and K. Sauber, *Angew. Makromol. Chemie* **157**, 105 (1988).

TABLE I
SELECTIVITIES OBTAINED WITH NATIVE AND COVALENTLY IMMOBILIZED
Candida rugosa LIPASE IN REPEATED BATCH REACTIONS

Carrier and run number	Relative rate (%)[a]	Selectivity (E)[b]
None		
1	100	8
2	11	4
3	10	1.6
VA-Epoxy-Biosynth		
1	124	42
2	76	42
3	40	42
Eupergit C 250L		
1	<10	10
2	<10	6.6
3	<10	4.9

[a] The initial rate using native enzyme was set as standard (100%).
[b] Expressed as enantiomeric ratio; see Ref. 42.

One of the most widely employed methods for enzyme immobilization to be used in organic media is adsorption onto a solid support. Because enzymes are not soluble in such solvents, there is no need for a covalent linkage between the support and the enzyme, because desorption into the medium is impossible. Among the various carriers, diatomaceous earth[48] ("diatomite," Celite, Filtercel) has been used most widely, in particular for industrial-scale applications due to its low cost and facile immobilization procedure.[49] The general structure of this material consists of an internal vacuole with a network of pores through their silica walls. Several studies have shown that this type of carrier serves mainly as a "disperser" of the enzyme by leading to enhanced reaction rates due to the better distribution of the catalyst in the heterogeneous system.[50] This is particularly important when the reaction is performed in column or packed-bed reactors.[51]

The results of the resolution of *endo*-norbornenol performed with *C. rugosa* lipase immobilized by adsorption onto Celite (see Fig. 4) are shown

---

[48] K. R. Engh, in "Kirk-Othmer's Encyclopedia of Chemical Technology," 4th Ed., Vol. 8, p. 108. Wiley, New York, 1993.
[49] R. A. Wisdom, P. Dunhill, M. D. Lilly, and A. Macrae, *Enzyme Microb. Technol.* **6**, 443 (1984).
[50] R. Bovara, G. Carrea, G. Ottolina, and S. Riva, *Biotechnol. Lett.* **15**, 937 (1993).
[51] R. Seemayer and M. P. Schneider, *J. Chem. Soc. Chem. Commun.* 49 (1991).

TABLE II
SELECTIVITIES OBTAINED WITH *Candida rugosa* LIPASE IMMOBILIZED BY ADSORPTION ONTO CELITE IN REPEATED BATCH REACTIONS (FOR THE REACTION SEE FIG. 4)

Carrier	Run number	Relative rate (%)[a]	Selectivity (E)[b]
Celite 545	1	80	31
Celite 545	2	55	29
Celite 545	3	30	29
Celite 545	4	29	29
Celite 545	5	31	28

[a] The initial rate using native enzyme was set as standard (100%).
[b] Expressed as enantiomeric ratio; see Ref. 42.

in Table II. Adsorption of *C. rugosa* lipase onto Celite was almost as effective as the covalent immobilization onto an epoxy resin. The selectivity was about fourfold enhanced as compared to the native enzyme, and the activity was largely stabilized. However, in contrast to the covalent immobilization method, the stabilizing effect by adsorption onto Celite is less clear on a molecular basis. Related studies on the effect of adsorption of lipase onto Celite on the selectivity show that it may either be unaffected[52] or enhanced.[53]

Material and Methods

Chemicals

Lipase AY-30 from *C. rugosa* was supplied by Amano (Japan, batch LAYN 12515, protein content 8.3%) and used as received without further purification. Celite 545 (Fluka, No. 22140) and VA-Epoxy Biosynth (Riedel-de-Haen, No. 39354) were commercially available. Vinyl acetate was used as received and the stabilizer was not removed to prevent polymerization. According to our experience, the lipase is not deactivated by the stabilizer. Toluene was saturated with water prior to use. Other chemicals used were of analytical grade.

Immobilization of Candida rugosa Lipase onto VA-Epoxy Biosynth

Crude *C. rugosa* lipase (90 mg, Amano AY-30) is dissolved in phosphate buffer (20 ml, 1 M, pH 7.2). VA-Epoxy Biosynth (3 g) is added and the

[52] G. Ottolina, G. Carrea, S. Riva, L. Sartore, and F. M. Veronese, *Biotechnol. Lett.* **14**, 947 (1992).

[53] H. Takahata, Y. Uchida, Y. Ohkawa, and T. Momose, *Tetrahedron: Asymmetry* **4**, 1041 (1993).

mixture gently agitated on an orbit shaker (60 rpm) for 72 hr at room temperature. Then sodium chloride solution (50 ml, 1 M) is added and the immobilized enzyme removed by filtration. The solids are washed twice with phosphate buffer (10 ml each, 0.05 M, pH 7.0). The immobilized enzyme is stored in phosphate buffer (0.05 M, pH 7.0).

The residual activity (as determined by the hydrolysis of tributyrin via the pH-stat method[54]) was in the range of 50–65% from several batches. When Eupergit C 250L was used as the carrier, the residual activity never exceeded ~10% regardless of the conditions employed during the immobilization procedure.

Adsorption of Candida rugosa Lipase onto Celite

Crude *C. rugosa* lipase (500 mg, Amano AY-30) is dissolved in phosphate buffer (10 ml, 0.1 M, pH 7.2). Celite 545 (2 g) is first washed with water, then with phosphate buffer (0.1 N). The washed Celite is added to the lipase solution and the mixture thoroughly stirred. The resulting paste is then dried at room temperature under vacuum (12–15 mm Torr) with occasional shaking until visibly dry (~12 hr).

General Procedure for the Lipase-Catalyzed Acyl Transfer Reaction

To a solution of racemic alcohol substrate (5 mmol) in toluene (water-saturated, 10 ml), vinyl acetate (5 ml) and the appropriate amount of immobilized lipase (~50% of weight versus substrate corresponding to the crude lipase used for immobilization) are added. The mixture is agitated on an orbit shaker (180 rpm) at room temperature. When the reaction has proceeded to about 50% conversion (as judged by TLC or GLC), filter the solids and remove the solvent *in vacuo*. Acylated product is separated from remaining substrate alcohol by column chromatography. The recovered lipase is washed with toluene and dried at reduced pressure (12–15 mm Torr) for reuse.

Conclusion

Acetaldehyde, which emerges as an unavoidable by-product in lipase-catalyzed acyl transfer reactions from vinyl esters (serving as acyl donors) can cause severe deactivation of enzyme activity and selectivity. The extent of the deactivation is dependent on the enzyme's structure, which in turn is governed by its microbial source. Whereas the majority of industrially

---

[54] H. K. Weber, H. Stecher, and K. Faber, *in* "Preparative Biotransformations" (S. M. Roberts, eds.), p. 5:2.1. Wiley, New York, 1995.

produced lipases are remarkably stable toward acetaldehyde, the *C. rugosa* and *G. candidum* enzyme is highly sensitive. Two methods for the stabilization of *C. rugosa* lipase have been employed. (1) Covalent immobilization onto an epoxy-activated carrier resin (VA-Epoxy-Biosynth) is suitable for small-scale reactions. This technique stabilizes the enzyme activity and increases its enantioselectivity about fivefold compared to the native enzyme. (2) Adsorption onto Celite 545 may serve as a more economic alternative for large-scale applications. The stabilizing effect on activity and enantioselectivity is only slightly less pronounced compared to the covalent immobilization.

Author Index

Numbers in parentheses are footnote reference numbers and indicate that an author's work is referred to although the name is not cited in the text.

A

Aaltonen, O., 506
Aalto-Setälä, K., 72
Ab, E., 156
Abe, R., 10
Abergel, C., 127
Abola, E., 392, 393(15), 394(15), 396(15)
Abouakil, N., 84, 98(8)
Abousalham, A., 129, 135(15), 136(15), 316, 323(36)
Abramow-Newerly, W., 74
Abrams, C. K., 14, 219
Abumrad, N. A., 17
Achiwa, K., 424
Achour, A., 394, 396(24)
Ackerman, E. J., 120
Adam, N. K., 161
Adams, M. R., 12
Adelhorst, K., 436
Adelman, S. J., 241
Ader, U., 351, 364, 375, 425, 511
Adlercreutz, P., 509
Ahle, S., 104
Ahmad, S., 218
Ahmed, S. M., 454
Ahrens, E. H., 79
Ahrens, E. H., Jr., 23
Ailhaud, G., 200, 329
Aires-Barros, M. R., 290, 291(75)
Akesson, B., 287
Alagona, G., 393, 397(22)
Alavi, M., 22
Albers, J. J., 28, 31

Alberts, J. J., 23
Albertsson, P. A., 15
Allietta, M., 185
Almeida, P. F. F., 174
Almog, S., 186
Almsick, A., 378
Alpers, D. H., 27, 31
Als-Nielsen, J., 289
Alström, P., 171
Alvarez, R., 59
Alvaro, G., 514
Ameis, D., 127, 192
Amherdt, M., 28
Amic, J., 127, 316
Amorico, M. G., 43
Amri, E. Z., 17
Anastasiadis, S. H., 307
Andalibi, A., 192
Andelman, D., 289
Andersch, P., 351, 373, 375, 406, 428
Anderson, R. A., 125, 127
Anderson, W. A., 509
Andreas, J. M., 307
Andrew, J., 36
Anel, A., 26
Angal, S., 135(35), 144, 191
Angel, A., 40
Angelin, B., 61
Anthonsen, H. W., 193, 475, 481, 483(8)
Anthonsen, T., 458, 475, 481, 483, 483(8), 492
Antonioli, A., 43
Aoubala, M., 126, 129, 131(14, 19), 133–134, 135(15, 26), 136(15), 141, 142(30), 143(30), 144(30), 148(19), 285–286,

287(45), 294, 295(12), 306, 316, 319(10), 323(36)
Aousalham, A., 134
Aoyagi, T., 253
API SYSTEM, 353
Apitz-Castro, R., 179, 180(32), 181(32), 182(32), 218, 279
Aponte, G. W., 10
Applebaum-Bowden, D., 23
Arai, M., 506
Arai, T., 227
Argraves, W. S., 104
Armand, K., 42
Armstrong, M. J., 25
Arner, P., 61, 62(63)
Arnett, E. M., 289
Arvidson, G., 185
Asahara, T., 288
Asano, M., 253
Ashton, R. W., 181
Astier, M., 316
Athawale, V. D., 510
Atkinson, D., 158, 159(27), 160(27)
Atrens, D. M., 8
Aubala, M., 144, 146(33), 148(33)
Avrameas, S., 143, 286

B

Baba, N., 289, 424
Baba, T., 117
Bachouchin, W. W., 495
Baekmark, T., 170
Bai, F. L., 8
Baier, L. J., 26
Baker, M. L., 181
Baker, N., 237, 241(50), 245(50)
Balasubramanian, K. A., 201
Balderas, V., 37
Balint, J. A., 6, 19, 21
Bangham, A. D., 276
Baptista, A., 193
Bar, N., 434
Baratti, J., 357, 359, 454
Barbara, L., 43
Barbato, G., 41
Barber, D. L., 25
Barberini, G., 43
Barbier, P., 253
Barenholz, Y., 102

Barnett, B. L., 235
Barnier, J.-P., 510
Barratti, J., 512(40), 513
Barrea, J., 505
Barreiros, S., 507
Barrowman, J. A., 31, 68, 126, 191, 225(4)
Barth, D., 504–505
Barzaghi, L., 442
Basdevant, A., 103
Baskin, D. G., 9
Bass, N. M., 26–27
Bassilian, S., 123
Bassindale, A., 395
Bassir, M., 510
Bastida, A., 514
Batenburg, A. M., 254
Bates, M. L., 27
Bates, T. E., 25
Battistel, E., 510
Bauer, R., 184
Baur, A. R., 43
Bayerl, T. M., 184
Beardshall, K., 10
Bearnot, H. R., 28, 29(211)
Beavo, J. A., 45
Beckman, E. J., 288, 505, 507
Beg, O. U., 104
Beglinger, C., 10, 190, 225
Beisiegel, U., 102–104, 104(8), 105(18), 108, 108(7), 113(8, 9)
Bekassy, A., 14
Bela, S., 498
Belfrage, P., 45–46, 46(2, 6, 10), 47(2), 48, 48(6, 9), 49(6, 9, 10, 23), 51, 53, 55(19), 56, 56(35), 57(2, 6), 59(3), 60(20–23), 61, 63, 63(44), 64(68,69), 65(9, 10, 44), 66(44)
Bell, G., 458
Bell, J. D., 179, 181
Bengtsson-Olivecrona, G., 102–103, 104(2), 106, 106(10), 107–108, 108(7), 113(9, 10), 114(2), 192
Bénicourt, C., 134, 135(26), 216
Benkouka, F., 193, 200(22), 202(22)
Bennet, W., 191
Bennett, B. D., 30
Bennett, F., 74
Bensadoun, A., 143
Bent, E. D., 181
Benton, W. D., 73
Berendsen, H. J. C., 171, 392, 394(18)

Berezovsky, V. M., 196
Berg, O. G., 169, 181(5), 184(5), 186(5), 272, 273(17), 274(17), 279(17), 282(17), 323
Bergbreiter, D. E., 424, 510
Berger, J. E., 45
Berger, M., 351, 358, 375, 381, 384, 406, 428–429, 429(53), 430(55), 431
Berger, S. L., 84
Bergholz, C. M., 3(8), 4
Berghy, 10, 13(341), 44
Berglund, P., 287, 390, 403, 405, 405(45), 469, 483
Bergö, M., 103
Beri, R. K., 50
Berk, P. D., 3(2), 4, 17
Berlowitz, M., 9
Bernard, A., 18
Bernard, G. D., 22, 23(145)
Bernbäck, S., 191–192, 231
Bernhardt, C. A., 12
Bernier, J. J., 14
Bernlohr, D. A., 27
Bernstein, F. C., 392, 393(14, 15), 394(14, 15), 396(14, 15)
Bertinotti, A., 442
Bertrand, J. M., 42
Besler, B. H., 396
Besnard, P., 18
Besohn, I., 22, 23(145)
Best, C. J., 27
Betteridge, D. J., 21
Bevington, P. R., 250
Bezzine, S., 126, 136
Bhat, S. G., 69, 231, 232(9), 285
Bianchi, D., 510
Biasco, G., 43
Bill, C. A., 113
Billington, C. J., 9
Biltonen, R. L., 168, 175, 179, 181(20, 34), 182, 184(34, 41), 185–186, 186(34, 41), 187(61)
Birchbauer, A., 65
Bird, D. A., 163(51), 166
Bisgaier, C. L., 28
Bivas, I., 184
Björgell, P., 45, 46(6), 48(6), 49(6), 57(6), 63, 64(68)
Björkling, F., 204, 209, 209(64), 210, 211(77), 409, 436
Björkling, R., 237

Bjørnholm, T., 285
Black, D., 32
Bläckberg, L., 191
Bläckburg, L., 231
Blacker, A. J., 467
Blackhall, N. W., 25
Blanch, H. W., 505–506
Blanchette-Mackie, E. J., 281
Blanco, L., 510
Blanco, R. M., 514
Blattner, W. E., 43
Bligh, E. G., 227
Bliss, C. M., 15
Block, E., 218
Blond, J.-P., 321, 323(40)
Bloom, M., 168, 170(3)
Bloom, S. R., 10, 25
Blow, D. M., 193, 235
Blum-Kaelin, D., 16, 224, 344
Bochenek, W. J., 23
Bode, C., 35
Bode, J. C., 35
Bodenteich, M., 513, 515(42), 516(42)
Bodmer, M. W., 135(35), 144, 191
Boel, E., 335
Boelens, R., 155, 345
Boeren, S., 480
Böhm, C., 289
Bohr, V. A., 113
Bois, A., 282, 283(33), 284(34), 307, 311(16), 312(16)
Bolscher, J., 28
Bolte, H. F., 12
Bonicel, J., 134, 135(26), 202–203
Bonifacino, J. S., 28
Boots, J. W., 204, 209(67), 252
Borch, K., 345
Bordia, A., 218
Borges de Carvaho, I., 507
Borgström, B., 14–15, 105, 123, 220, 223(92), 225(92), 252, 253(6), 256(6), 262(6), 263(6), 268, 287, 321
Borgström, H., 357
Borha, C. R., 232
Bornscheuer, U., 448
Borthwick, A. C., 51, 53(32)
Boruvka, L., 307
Bos, J. W., 254
Bosner, M. S., 69, 99
Bougis, P. E., 203

Bouloumié, A., 61
Bourgeois, D., 335
Bourne, Y., 195
Bousset-Risso, M., 202
Bouthillier, F., 200, 210, 237, 445, 446(14)
Bouvry, M., 14
Bovara, R., 442, 515
Bovier-Lapierre, C., 193
Bowyer, R. C., 15
Boyart, J. P., 35
Boyle, E., 164, 166(49)
Bradbury, M. W., 17
Bradley, D. C., 56
Brady, L., 8, 133, 200, 258, 341
Braiman, M., 175
Brasitus, T. A., 19, 22, 24, 24(113), 29, 40, 40(143)
Brass, O., 311, 315(32), 323, 326(32)
Braulant, G., 505
Bravo, T., 410
Bray, G. A., 9
Breitgoff, D., 365, 367, 419, 421, 421(29), 422(34), 424–425, 511
Bremer, K., 16, 224
Brentel, I., 185
Brescher, P., 167
Breslow, J. L., 72
Breuer, N., 43
Brewer, H. B., Jr., 103–104
Brice, M. D. J., 392, 393(14), 394(14), 396(14)
Bricklehurst, K. J., 50
Brieva, R., 409
Briggs, J. E., 9
Brillanti, S., 43
Brockerhoff, H., 231, 234(2), 287, 306
Brockman, H. L., 68–70, 73(16), 93, 95(14), 117, 231–232, 232(4, 9), 234, 234(4), 238(4), 284–285, 287, 292–295, 295(13), 298, 300, 301(14), 304, 304(13, 14, 17–20), 305(14, 15, 25), 306, 319, 323, 334, 357
Brodt-Eppley, J., 70, 98, 232
Brown, B. O., 498
Brown, M. S., 72
Brown, W. J., 28, 116
Bruce, R. D., 12
Bruce, W. R., 43
Brumm, T., 184
Brunzell, J. D., 102–104, 108
Bryant, S. H., 392, 393(15), 394(15), 396(15)
Brzozowski, A. M., 133, 200, 210, 211(77), 237, 252, 258, 341, 409

Buchnea, D., 287
Buchwald, P., 344
Buffa, R., 25
Bull, A. W., 43
Buono, G., 156, 190, 204–205, 205(65), 206(74), 207, 208(75), 282, 454
Burack, W. R., 175, 179, 181(20, 34), 184(34), 185, 186(34)
Burg, K., 514
Burgdorf, T., 346
Burks, N. K., 161
Burnette, W. N., 78
Burns, D. K., 72
Burton, D. J., 237, 238(46), 240
Burton, G. W., 235
Buset, M., 43
Bush, V. J., 287
Bustamante, S., 35–36, 36(268), 37(269)

C

Cabral, D. J., 163(51), 166
Cabral, J. M. S., 290, 291(75)
Cabrera, S., 218
Cacace, A. M., 25
Cagna, A., 306–308, 310(29), 311, 311(29), 315(32), 323, 324(47), 325, 326(32)
Cain, J., 23
Calam, J., 10
Calme, K. B., 117
Calogeropoulo, T., 237
Cambillau, C., 103, 107, 127, 133, 155, 192, 192(25, 27), 194–195, 200, 206(74), 207, 207(25, 27), 208(25, 75), 212, 216(25), 237, 292, 319, 335, 341, 343(18, 19)
Cambou, B., 443, 457(2)
Campbell, D. G., 45, 48(8), 49, 49(8), 50(25), 53(25)
Campbell, R. L., 44
Campieri, M., 43
Campion, J. M., 189
Camulli, E. D., 70, 117
Capecchi, M. R., 71
Capella, C., 25
Capuzzi, D., 30
Carafoli, E., 39
Cardelli, J. A., 32
Carey, F., 50
Carey, M. C., 3(4), 4, 15, 100, 226
Carling, D., 49–50, 50(25), 53(25, 27)

Carlquist, M., 7
Carmichael, H. A., 36, 37(272)
Caron, G., 454
Carpenter, B. K., 249
Carpenter, J. F., 422
Carpenter, Y., 37, 38(275)
Carraway, R. A., 25
Carraway, R. E., 25
Carrea, G., 401, 442, 482, 493(25), 505, 509, 515–516
Carrero, P., 113
Carrière, F., 68, 126–127, 190(2), 191, 192(116), 194, 195(2), 216, 225(4), 227, 321, 323(40), 327, 335, 344
Carrilo, 218
Carroll, K. K., 20, 23
Carter, C. P., 70, 74, 77(23), 78(23), 79(23), 100, 232
Carter, K. J., 22, 23(145)
Carter, P., 409
Carter, W. J., 40
Case, D. A., 393, 397, 397(22)
Cassidy, M. M., 41
Catalfamo, J. L., 218
Catoni, E., 496, 505(13)
Cerger, R., 216
Cernia, E., 287, 495–496, 505(13)
Cesti, P., 510
Cevc, G., 153
Cezard, J. P., 35
Chaffotte, A. F., 133
Chahin, N. J., 43
Chahinian, H., 195, 197(30), 218(30), 220(30), 221(30), 222(30), 225(30)
Chaillan, C., 134
Chajek-Shaul, T., 102
Chalecki, Z., 510
Chan, L., 192
Chandrasekhar, J., 397
Chang, M.-K., 46, 56(11), 57(11), 59, 65(11)
Chapman, D., 39
Chappell, D. A., 104
Chapus, C., 134, 193, 195, 200(46), 203, 293, 298(3), 316
Charles, M., 200, 316
Chat, S. G., 294, 295(13), 304(13)
Chaudhuri, S., 65
Cheah, E., 193, 235
Chebotareva, N. A., 196
Cheema, S., 4, 6(10), 33(10), 40(10)
Cheeseman, C. I., 18, 32, 32(110, 252–254), 33

Chen, C.-S., 287, 363, 365, 395, 417, 451, 474, 479, 486(14)
Chen, J. K., 307
Chen, R., 29, 30(214)
Chen, S.-C., 513
Chen, S.-J., 509
Chen, S.-T., 509
Chen, S. T., 422
Chen, X., 84
Cheng, P., 307
Cheng, T., 33, 103
Cheriathundam, E., 117
Cheung, M. C., 23, 31
Chevreuil, O., 103, 115(15), 116(15)
Chi, Y. M., 504–505
Chiang, Y., 84
Chin, A. S., 8
Chomczynski, P., 65
Chopineau, J., 434
Christen, M., 406
Christensen, N. J., 31
Christiansen, L., 133, 258, 341
Chronwall, B. M., 8
Chuat, J. C., 192
Chulalaksananukul, W., 504, 505
Chung, J., 224(103), 225
Church, G. M., 65, 73, 75(30)
Ciccotti, G., 392, 394(18)
Cinader, B., 40
Cioletti, I. A., 37
Clandinin, M. T., 3, 3(6, 9), 4, 6(10), 16(9), 18, 19(102, 112), 23–24, 32(102), 33(10, 101, 104), 39(104, 105), 40, 40(10, 19, 104, 105), 257
Clark, D. E., 241
Clark, D. S., 505, 513
Clark, S. B., 29, 30(214), 68, 99, 231, 232(10)
Clark, S. C., 74
Clarke-Lewis, I., 103
Clarkson, T. B., 12
Clausen, I. G., 287, 389–390
Clauser, E., 95
Cleasby, A., 200, 254, 344
Cohen, B. I., 43
Cohen, P., 45, 48(8), 49(8)
Colbran, R. J., 49, 50(25), 53(25)
Colditz, G. A., 42
Coleman, R. A., 118
Colman, A., 81, 92(3)
Colson, C., 193, 344
Combes, D., 512

Commons, T. J., 237, 241
Condoret, J. S., 504–505
Connor, W. E., 23
Conrad, D. H., 114
Conti, M., 45
Contreras, J. A., 47, 51, 51(12), 52(33), 53(12, 33), 57(33), 63(12), 64(12, 33)
Copeland, E. M., 37
Corbier, C., 505
Cordes, E. H., 186
Corey, D. R., 409
Corey, M., 41–42
Cormerais, F. X., 498
Correa de Sampaio, T., 507
Corton, J., 50
Corvera, E., 189
Costlow, N. A., 66
Costro, G. A., 37
Craib, K. J. P., 42
Craik, C. S., 409
Craven, P. A., 43
Criati, E., 440
Crijns, H. J. M. J., 224
Cronan, J. E., 39
Crout, D. H. G., 406
Cruz, M. R., 218
Cruz, T. F., 33, 40
Cruzeiro-Hansson, L., 189
Cuccia, L. A., 481
Cudrey, C., 195, 197(31), 199(31), 204, 205(65), 206, 207(71), 213, 215(82), 216(31), 221(31), 287, 335
Cuisiner-Gleizes, P., 35
Cullis, P. R., 184
Cummings, J. H., 43
Cunningham, B. A., 300, 304(20)
Cusin, I., 8–9, 9(20)
Cygler, M., 133, 193, 200, 210, 235, 237, 346, 445–446, 446(13, 14), 454
Czubayko, F., 224(104), 225

D

Dahim, M., 300
Dahl, A., 209
Dammann, B., 171
Daniel, C., 129, 129(14), 131(19), 133, 135(15), 136(15), 148(19), 316, 323(36)
Daniel, D. R., 8
Danieli, B., 442

Daniels, D. V., 59
Danielsson, B., 51, 52(33), 53(33), 57(33), 64(33)
d'Arcy, A., 133, 135(21), 191, 193(9), 200(9), 202(9), 237, 341
d'Auriol, L., 192
Dauzats, M., 61, 62(63)
Davidson, N. O., 19, 24(113), 31
Davis, R. A., 78
Davis, R. C., 51, 52(33), 53(33), 57(33), 64(33), 120
Davis, R. W., 73
Dawe, E. J., 43
Dawson, R. M. C., 276
Day, J., 346
De Bellis, P., 442
Debnam, E. S., 34
de Caro, A., 14, 126–127, 129, 131(14, 19), 133–134, 135(15, 26), 136(15), 141, 142(30), 143(30), 144, 144(30), 146(33), 148(19, 33), 285–286, 287(45), 294, 295(12), 306, 316, 319(10), 323(36)
de Caro, J., 193, 200(22), 202, 202(22), 315
de Castro, H. F., 509
Dechema Database-PC, 486
Deckelbaum, R. J., 102, 163(44), 164
Deeb, S., 61
Deems, R. A., 120, 179
Dees, J., 43
Degerman, E., 45, 48–49, 51, 52(33), 53(33), 55(19), 57(33), 60(20–23), 64(33)
de Geus, P., 133, 200, 292
De Gier, J., 168, 182(1)
Degrace, P., 18
Degueil-Castaing, M., 401, 412, 419(17), 510
de Haas, G. H., 105, 132, 178–179, 180(32), 181(32), 182(32), 195, 199, 204, 206, 207(72), 209(67), 212, 219(118), 227, 229(33), 232, 252, 265, 267, 267(3), 268, 268(9), 269(9), 270, 270(12), 271(12), 272(12), 275–276, 277(20), 279, 281(20), 282, 282(12, 16), 285, 285(3, 19), 306, 323, 323(5), 333–334, 337(11), 344–345, 387
de Jeso, B., 401, 510
De Jeso, D., 412, 419(17)
de Jong, J. P. L., 504
de la Fournière, L., 129, 135(15), 136(15), 316, 321, 323(36), 323(40)
Deleuze, H., 359
Delfini, M., 287

Delinck, D. L., 418, 442, 495, 510
Delorme, B., 306
De Maria, D., 43
de Meutter, J., 292
Deng, C., 71
Dennis, E. A., 120, 179, 182, 195
Denton, R. M., 54
De Pover, A., 3
Deprez, P., 14
Derewenda, U., 193, 210, 211(77), 237, 252, 409
Derewenda, Z. S., 133, 192–193, 210, 211(77), 237, 252, 258, 341, 409, 486
DeRubertis, F. R., 43
Dervichian, D. G., 153, 157, 158(24), 161, 199, 275, 306
Deschner, E., 43
Desnuelle, P., 153, 161, 193, 200, 202(45), 257, 281, 285, 294, 316, 318, 328–330, 330(2), 331(2), 332(6), 333(6), 335(2, 6), 337(6)
Deutchler, J. T., 39
Deveer, A. M. T. J., 206, 207(72), 268, 270, 282(16), 344
de Willigen, A. H. A., 329
Deykin, D., 118
de Zoete, M. C., 409
Dhillon, G. S., 48, 54, 59(37)
Diamond, J. M., 35
Dibble, A., 185
Dichek, H. L., 103
Dickie, N., 19
Di Donato, P., 43
Dierckman, T. A., 12
Dietschy, J. M., 22
Di Febo, G., 43
Dijkman, R., 204, 209(67), 252, 282, 344–345
Dijkstra, B. W., 193, 235, 344
Dimaggio, D. A., 8
DiMaggio, E. P., 42
Di Magno, E. P., 225
Ditschuneit, H., 21–22
Djavadi-Ohaniance, L., 133
Doctor, B. P., 235
Doddema, H. J., 504
Dodson, E., 133, 341
Dodson, G. G., 133, 200, 210, 211(77), 237, 252, 335, 341, 409
Doetschman, T. C., 74, 76
Domin, J., 10
Dominguez, A. A., 22

Donaldson, D. D., 74
Donohue, T. M., 511
Dordick, J. S., 288, 390, 495–496
Doring, K., 18
Douchet, I., 126, 129, 135(15), 136(15), 141, 142(30), 143(30), 144, 144(30), 146(33), 148(33), 205, 207, 282, 285–286, 287(45), 288, 289(68), 294, 295(12), 306, 316, 319(10), 323(36)
Douglas, A. M., 505
Dourassa, D., 20, 37(127)
Dowell, R. F., 14
Downs, D., 117
Drabløs, F., 127, 193
Draper, J. P., 24
Draser, B. S., 43
Drauz, K., 351, 406, 443
Drent, M. L., 16, 224, 224(100, 104, 107), 225
Drevon, C. A., 19
Drewe, J., 10
Drisko, J., 78
Dror, Y., 495
Drouillard, S., 401, 412, 419(17), 510
Drtina, G. J., 345, 346(32)
Dryden, S., 9
Duax, W. L., 236
Dube, M. G., 9
Dubin, P. L., 495
Ducret, A., 436, 509
Dudrick, S. J., 37
Dueland, S., 78
Dufour-Larue, O., 42
Dugi, K. A., 103
Duine, J. A., 476, 484–485, 487
Dumont, T., 504–505
Dunaway, C. M., 346
Dunhill, P., 515
Duplus, E., 62
Durant, G., 504–505
Durie, P. R., 41–42
du Toit, P. S., 22, 23(145)
Dutta, S. K., 42, 219
Dutton, P. J., 235
Dvolaitsky, M., 289
Dyer, W. J., 227

E

Eagan, J. J., 59
Eaton, B. R., 179

Eckerskorn, C., 127
Eckert, C. A., 502
Edery, M., 95
Edgren, G., 51, 52(33), 53(33), 57(33), 64(33)
Edholm, O., 391, 393
Edlund, H., 403, 483
Edmonds, R. H., 19
Egan, J. J., 46, 56(11), 57(11), 65(11)
Egawa, M., 9
Eggers, H., 224(106), 225
Egloff, M.-P., 156, 192(25), 194, 206(74), 207, 207(25, 75), 208(25), 216(25), 237, 292, 319, 341, 343(19)
Egmond, M. R., 200, 204, 209(67), 212, 252, 254, 344
Eibes, T., 237, 238(44), 240, 240(44)
Eichorn, U., 469
Eikenberry, E. F., 177
Eilers, G. A., 43
Eisenberg, S., 102
Eisenhardt, J. M., 35
Eistetter, H., 74
Ekholm, D., 48, 55(19)
Eliel, E. L., 474
Ellerfeldt, K., 61
Ellis, L., 95
Emerson, J. A., 307
Emi, M., 103
Emmerich, J., 104
Emmison, N., 51, 53(32)
Emson, P. C., 8
Eng, V. W. S., 43
Engelman, D., 161
Engh, K. R., 515
Entressangles, B., 153, 200, 202(45), 330, 332(6), 333(6), 335(6), 337(6)
Epand, R. M., 178
Epstein, M., 23
Erickson, K. L., 40
Eriksson, H., 48, 55(19)
Eriksson, M., 61
Erlanson, C., 15, 68, 123
Erlanson-Albertsson, C., 14, 127
Erman, M., 236
Escalante, J., 218
Esposito, G., 306–308, 310(29), 311(29)
Esposito, S., 285
Etherton, T. D., 65
Etienne, J., 192
Etzerodt, M., 102, 104(8), 113(8)

Evans, E., 168, 170(3)
Evans, E. A., 184
Evans, M. J., 71

F

Faas, F. H., 40
Faber, K., 399, 412, 444, 473, 509–512, 512(30), 513, 515(42), 516(42), 517
Fahey, D. A., 158, 159(28, 29), 160(28, 29), 161(28)
Fainaru, M., 28, 102
Fallat, R. W., 12
Fang, J.-M., 413, 510
Farrar, G. B., 25
Fearon, E. R., 42
Feaster, S. R., 231, 235, 237, 241(50), 245(50), 250
Fedorak, R. N., 3(7), 4
Feldman, E. B., 29, 30(213, 214)
Feltkamp, C., 28
Fernández, M., 108
Fernandez, P., 30
Fernandez-Lafuente, R., 514
Fernandez-Puente, L., 184
Ferraboschi, P., 418, 509
Ferrato, F., 127, 132, 194, 206, 207(71), 266, 287, 327, 335, 344
Ferris, C. F., 25
Fersht, A. R., 396, 492
Field, C. J., 4, 6(10), 24, 33(10), 40, 40(10, 168)
Fife, T. H., 251
Figarella, C., 14, 127, 316
Figlewica, D. P., 9
Fisher, E. A., 69–70, 80, 84, 89(1), 90(1), 91(2), 94, 96, 97(19), 98, 98(19), 99, 101(24)
Fitzpatrick, P. A., 490–491, 495
Fleming, M., 22, 23(145)
Fleury, P., 16
Flipsen, J., 325(48), 326
Florant, G., 61
Flyn, M. A., 23
Flynn, T. C., 23
Fogedby, H. C., 171
Foglizzo, E., 134, 203, 316
Fontaine, R. N., 70, 73(16), 74, 93, 95(14), 117
Fontecilla-Camps, J. C., 127
Foot, M., 33, 40
Forest, C., 62

Forrester, J. M., 21
Forstner, G. G., 41–42
Forsythe, I. J., 103
Forte, T., 29, 30(213, 214)
Foster, D. O., 235
Fourcans, B., 18
Fourneron, J.-D., 195, 197(29), 203(29)
Fowler, S. D., 116
Frace, S., 28
Fraken, S. M., 235
Frandsen, E. K., 59
Franken, S. M., 193
Fransson, J., 114
Fraumeni, J. F., Jr., 43
Frayn, K. N., 56, 63(44), 65(44), 66(44)
Fredenslund, A., 486
Fredrikson, G., 45, 48, 48(9), 49(9, 23), 53, 56(35), 60(23), 63, 64(69), 65(9)
Freed, L. M., 231
Freeman, A., 495
Fregapane, G., 509
Frenken, L. G., 200, 254, 344
Fretag, R., 448
Freudenberg, E., 231
Fried, M., 21, 225
Friedman, E., 43
Friedman, G., 102
Frigerio, B., 25
Friguet, B., 133
Frolow, F., 193, 235
Fronowitz, S., 287
Frost, G., 10
Frost, P., 36
Frykman, H., 401, 409
Fujimoto, K., 32
Fujimoto, W. Y., 108
Fujimoto, Y., 395, 417, 451, 474, 513
Fujimura, M., 25
Fujisaki, J., 32
Fujisaki, Y., 22
Fujita, D. A., 162(42), 163(42), 164
Fukugawa, K., 32
Furrer, R., 43

G

Gad, M. Z., 118, 120, 123, 123(27), 124(27), 125(27)
Gadacz, T. R., 42
Gaddy, V. L., 502
Gadient, A., 10

Gains, G., 157
Gains, N., 224(105), 225
Galfré, G. H. S. E., 128
Galibert, F., 192
Galitzky, J., 61
Gallagher, D., 41
Gallo, L. L., 68, 84, 89, 99, 117, 231, 232(10), 234
Gallon, L., 13
Galluzzo, D. R., 287
Galton, D. J., 21
Gancet, C., 195, 229(33), 268, 270, 270(12), 271(12), 272(12), 282(12, 16), 306
Ganguly, J., 237, 238(48)
Gantz, D., 162(46), 164, 166(49)
Ganz, M. B., 25
Gaonkar, S. R., 510
Garg, M. L., 3(6), 4, 6(10), 18, 24, 33, 33(10, 101), 40(10)
Gargouri, Y., 126–127, 129, 132, 135(15), 136(15), 144, 190, 190(2), 191, 195, 195(2), 196, 196(34), 197(30, 31), 199(31), 200(34), 201–202, 202(57), 203(57), 206, 207(72), 213, 215(82), 216(31), 217(58, 59), 218(30), 219(118), 220(30), 221(30, 31, 34), 222(30), 223, 223(34), 224(34), 225(30), 227, 229(34), 252, 265, 268, 282, 282(7, 15), 283(33), 284, 284(7), 286, 293–294, 295(11), 307, 311(16), 312(16), 315–316, 323(36), 334
Garland, P. B., 55
Garton, A. J., 45, 48(8), 49, 49(8), 50(25), 53(25, 26)
Gaset, A., 409
Gaskin, K. J., 41–42
Gaudry-Rolland, N., 307–308, 310(29), 311(29)
Gearing, D. P., 74
Gebhard, R. L., 22, 40(144)
Geerlof, A., 487
Gelb, M. H., 169, 181(5), 184(5), 186(5)
Gelmann, E. P., 39
Gelo-Pujic, M., 510
Gentry, M. K., 235
Gerard, G., 498
Getz, G. S., 20
Gheriani-Gruszka, N., 186
Ghio, C., 393, 397(22)
Ghisalba, O., 428
Ghosh, D., 236

Ghosh, S., 116
Gibbs, J., 10
Gilbert, W., 65, 73, 75(30)
Giller, T., 344
Gillespie, J. G., 50
Gilman, A. G., 59
Ginzinger, D., 103, 105(18)
Giovardhan, C., 444, 445(4), 450(4), 457(4), 462(4)
Girault, H. H. J., 307
Girdaukas, G., 395, 417, 451, 474, 479, 486(14), 513
Giroux, A., 436
Gizzi, G., 43
Glebova, G. D., 196
Glick, J. M., 96, 97(19), 98(19)
Glickman, R. M., 28–29, 29(211), 31
Gliemann, J., 102–104, 104(8), 113(8, 9)
Glock, J. M., 70
Glomset, J. A., 118
Glueck, C. J., 12–13
Gmehling, J., 486
Go, V. L. W., 42, 225
Goderis, H. L., 458
Godtfredsen, S. E., 409, 436
Goebell, H., 43
Goergens, U., 351, 365, 375, 378
Gold, P. W., 8
Goldberg, M. E., 133
Goldfine, I., 10
Goldman, A., 193, 235
Goldstein, J. L., 72
Gómez-Coronado, D., 113
Gomez-Fernendez, J. C., 39
Goni, F. M., 39
Gonvález, C., 409
Goodford, P. J., 394, 396(23)
Goodman, D. S., 118
Goormaghtigh, E., 173, 182(17)
Gopaul, D., 26
Gorchulski, P., 454
Gordon, J. I., 26–27
Gordon, T., 23
Gormsen, E., 324(47), 325
Gorter, E., 153
Gosnell, B. A., 8
Gossler, A., 76
Gotoda, T., 108
Gotor, V., 409, 413, 413(9), 435, 435(9), 442, 442(9)

Götz, F., 344
Gough, N. M., 74
Gould, R. G., 22
Govardhan, C., 445
Grace, M., 8–9
Graf, L., 78
Graille, J., 277, 278(24), 307, 311(14)
Grainger, D. W., 181, 184(38)
Grand, R. J., 35
Grandbois, M., 181, 184(38)
Granger, N. D., 31
Gray, M. E., 30
Gray, T. S., 8
Greeley, G. H., Jr., 25
Green, P. H. R., 28, 29(211)
Greenberg, A. S., 46, 56(11), 57(11), 65(11)
Greenberg, D., 10
Greenspan, L., 466
Gregory, P. C., 10
Gremer, J. F., 35
Grendel, F., 153
Greten, H., 127
Grey, V. L., 42
Griengl, H., 513, 515(42), 516(42)
Griffith, J. P., 445
Grimaldi, M., 307
Grimaldi, P. A., 17
Grisenti, P., 418, 509
Grober, J., 62
Grochan, B. M., 113
Grochulski, P., 200, 210, 237, 445–446, 446(13, 14)
Grogan, W. M., 116
Grundy, S. M., 23, 79
Gruner, S. M., 168, 177, 177(2), 178(2), 182(1), 188
Gubernator, K., 192
Guedeau-Boudeville, M.-A., 289
Guerciolini, R., 224(103), 225
Guesdon, J. L., 143, 286
Guibe-Jampel, E., 510
Guidoni, A., 193, 200(22), 202(22)
Guilhot, S., 192
Guisan, J. M., 514
Gundlach, A. L., 8
Gunther, R. D., 123
Guo, Z.-W., 288, 452, 453(22)
Gupta, A. K., 210
Gurwitz, D., 41
Gutman, A. L., 410

Guy, O., 117, 127, 233, 316
Guy-Grand, B., 103
Guzelhan, C., 16, 224, 224(102), 225
Gyr, K., 225

H

Haalck, L., 200, 204–205, 209(68), 252, 254–256
Haase, B., 351, 378
Haberich, F. J., 35
Hackeng, T. M., 252
Hadvàry, P., 15–16, 220, 225, 225(91), 252, 253(7), 259(7)
Haffen, K., 35
Hæffner, E. W., 39
Hæffner, F., 391, 394, 396, 396(24), 405(29)
Hall, J. C., 25
Halling, P. J., 390, 458, 465–466, 469, 482–483, 485
Halperin, G., 102
Hamada, M., 253
Hamilton, J. A., 162(42, 43, 46), 163, 163(39, 40, 42, 44, 50, 51), 164–165, 165(39, 40), 166, 166(5, 49), 167, 185
Hamilton, R. M. G., 20, 23
Hamilton, S. R., 42
Hammer, C. F., 162(42), 163(42), 164
Hammer, R. A., 25
Hammond, D. A., 505
Hamosh, M., 14, 15(75), 126, 190(3), 191, 219, 231
Hamosh, P., 219, 231
Hamsten, A., 109, 111(37), 112(37)
Han, J. H., 80, 91(2), 92(2), 94, 117
Hansen, J. D. L., 22, 23(145)
Hansen, L. D., 181
Hansen, M. T., 200, 344, 492
Hansen, S. H., 28
Hansen, T. V., 481
Harada, K., 108
Harano, Y., 495
Hardie, D. G., 49–50, 50(25, 27), 53(25)
Harel, M., 193, 235
Hargreaves, K., 24, 40(168)
Harisson, Y., 307, 311(24)
Harmon, J. T., 120
Harries, J. T., 219
Harris, L. J., 218
Harris, T.J.R., 135(35), 144, 191

Harrison, D., 200, 237, 445, 446(14), 447
Harrison, E. H., 70, 80, 84, 88(5), 89(1, 5), 90(1), 91(2), 92(2), 96(5), 116–118, 120, 123, 123(27), 124(27), 125(27)
Harrison, L., 42
Hartmann, D., 15–16, 224, 224(102), 225
Hartsuck, J. A., 117, 122(5)
Harvey, N. G., 289
Hata, A., 103
Hatakeda, K., 506
Hauptman, J. B., 15, 224, 224(103), 225
Hauser, E. A., 307
Hauser, H., 16, 39, 154
Hauser, S., 21, 22(131)
Hawkins, C. B., 29, 30(213)
Hawley, S. A., 50
Hayashi, H., 32
Hayden, M. R., 103, 105(18)
Hayek, T., 72
Hayes, E. B., 118
Hazato, T., 253
Hazzard, W. R., 23
Healey, P. A., 8
Heaslip, R. J., 45
Heath, J. K., 74
Hecht, H.-J., 200, 254, 346
Hedenström, E., 403, 405(45), 418, 483
Hedváry, P., 190
Heijnen, J. J., 475
Heinen, J. J., 478, 485
Heinzmann, C., 45, 48(9), 49(9), 65(9), 192
Helin, I., 14
Heller, R. A., 22
Hellers, G., 5
Hellström, L., 61, 62(63)
Hemler, R., 74
Hemming, D. J. B., 181
Henderson, B. E., 43
Henderson, H. E., 103
Henderson, P. J. F., 241
Hendrickson, H. S., 306
Herkenham, M., 8
Hermann, J., 351, 406, 426
Hermans, J., 392
Hermetter, A., 204, 209(68), 252
Hermon-Taylor, J., 15
Hernell, O., 3(4), 4, 100, 191, 226, 231
Herr, F. M., 27
Herrera, E., 61
Herslöf, B., 287

Herz, J., 72
Hide, W. A., 192
Hildebrand, H., 14
Hildebrand, P., 190
Hill, J. S., 120
Hill, M. J., 43
Hilton, D. J., 74
Hiratake, J., 422–423
Hirayama, T., 227
Hirn, J., 131(19), 133, 148(19)
Hirn, M., 126, 129, 131(14), 135(15), 136(15), 141, 142(30), 143(30), 144(30), 286, 316, 323(36)
Hirsch, A. H., 63, 65(67)
Hjorth, S., 335
Høbger, T., 168
Hochschuli, E., 252
Hochuli, E., 225
Hoff, B. H., 475, 483(8)
Hoffman, F., 16
Hofmann, A. F., 6
Hofmann, B., 200, 254
Hogan, S., 16
Högberg, H.-E., 403, 405, 405(45), 418, 469, 483
Holcombe, K. W., 23
Holla, E. W., 434, 511
Hollenbach, E. J., 12
Holm, C., 45–46, 46(10), 47, 48(9), 49(9, 10), 51, 51(12), 52(33), 53(12, 33), 56, 57(33), 58, 60–61, 63(12, 44), 64(12, 33), 65(9, 10, 44, 45), 66(44)
Holmberg, E., 288
Holmquist, M., 287, 386–387, 389, 389(4), 390, 393, 396, 403, 405, 405(29, 45)
Holst, J. J., 10
Holt, P. R., 22, 24
Holwerda, K., 329
Hønger, T., 182, 184–185, 189
Honig, B., 181
Honnor, R. C., 54, 59(37)
Honnor, R. S., 48
Hopkirk, T. J., 54
Horrevoets, A. J. G., 105, 252
Hosie, L., 237, 238(45), 240, 240(45)
Hosokawa, T., 40
Hostetler, A. M., 10
Hostetler, B., 24
Hotier, G., 498
Housset, S., 307

Houston, H., 43
Howard, K. P., 154
Howe, G. R., 42
Howles, P. N., 70, 77(23), 78(23), 79(23), 100, 232
Hsu, S.-H., 510
Hu, P. J., 43
Hua, Y., 104
Huang, J. J., 47, 51(13)
Huang, Y., 68–70, 232
Hubbard, V. S., 14, 219
Hubbell, T., 27
Huber, J. E., 490
Hübner, W., 154
Huff, M. W., 23
Huge-Jensen, B., 133, 200, 210, 211(77), 237, 258, 341, 409
Hughes, A., 265
Hughes, F. B., 30
Hughes, L. B., 78
Hughes, L. L., 120
Hughes, R. H., 20
Hui, D. Y., 67–70, 73(16), 74, 77(23), 78(23), 79(23), 93, 95(14), 98, 100, 117, 232, 237, 241(50), 245(50)
Hui, R. A. H. F., 288
Hult, K., 287–288, 386–387, 389, 389(4), 390–391, 391(9), 393–394, 394(2), 395–396, 396(24, 26), 399, 400(37), 401, 401(37), 402–403, 405, 405(9, 45), 406, 490, 492, 510
Hultin, M., 104–105, 113(27)
Hunziker, W., 133, 135(21), 191, 193(9), 200(9), 202(9), 237, 341, 344
Hussain, Y., 224(102), 225
Hutcheon, G. A., 458, 465
Hutchins, J. E., 251
Hwang, J. K., 397
Hwang, K. Y., 448
Hyan, J., 68
Hydén, S., 227

I

Ibdah, J. A., 294
Ide, T., 22
Iitaka, Y., 253
Ikeda, I., 437
Ikeda, M., 363, 417
Ikushima, Y., 506

AUTHOR INDEX

Im, W. B., 39
Imai, C., 21
Imamura, S., 227
Impey, R. W., 397
Inaba, T., 108
Inagaki, M., 422–423
Ingold, K. U., 235
Innis, S. M., 18, 33(104), 39(104, 105), 40(104, 105)
Inoue, I., 103
Ipsen, J. H., 171–172, 173(16), 184, 188–189
Ishi, S., 10
Ishibashi, S., 72
Isler, D., 16, 224(105), 225
Isobe, M., 357
Isola, L. M., 17
Israelachvili, J., 178, 185(24)
Isselbacher, K. J., 30
Ito, S., 25, 506
Ivanova, M., 144, 146(33), 148(33), 263, 285, 287(45), 294, 295(12), 306, 319(10), 323
Ivanova, M. G., 129, 131(14, 19), 133, 148(19), 277, 278(24), 279, 282, 283(33), 284(34), 290, 291(75), 294, 307, 311, 311(14), 315(32), 321, 323(40), 326(32)
Ivanova, T., 279, 307, 323
Iverius, P.-H., 103
Iwai, M., 357
Iwakiri, R., 32
Iwasa, J., 289
Iwase, J., 424
Iwaskewycs, S., 507
Izumi, F., 10

J

Jackson, K. W., 117
Jackson, M. J., 30
Jackson, R. L., 104
Jacobson, A., 89
Jacobson, P. W., 89, 117
Jaeger, K.-E., 193, 344
Jagocki, J. W., 276
Jahagerdar, D. V., 179
Jain, M. K., 18, 42, 169, 179, 180(32), 181, 181(5, 32), 182(32), 184(5), 186, 186(5), 218, 219(118), 227, 272, 273(17), 274(17), 279, 279(17), 282(17), 323
Jakubke, H. D., 469
James, D., 422

Jamison, P., 447
Jandacek, R. J., 3(8), 4, 12–13
Janiak, M. J., 162
Janne, P., 37, 38(275)
Jansen, J. B., 225
Jansen-Zuidema, J. J. N., 16, 224
Janssen, A. E. M., 485
Jarocka-Cyrta, E., 3, 23
Jeanrenaud, B., 8–9, 9(29)
Jefferson, J. R., 26–27
Jehanli, A. M. T., 15
Jencks, W. P., 485
Jenkins, D. J. A., 43
Jenkins, S., 70, 98, 232
Jensen, L. H., 156, 157(18)
Jensen, M., 403
Jensen, R. G., 231, 234(2), 287, 306
Jeon, H. S., 448
Jeppesen, C., 171
Jersild, R. A., 28
Jeunet, F. S., 15, 224
Jivray, T., 43
Johnson, J., 29, 30(214)
Johnson, L. N., 200, 254, 344
Johnson, L. R., 37
Johnson, R. A., 59
Johnson, W. J., 69–70, 98–99, 101(24)
Johnsson, B., 115
Johnston, P. V., 40
Joly, R., 16, 224(106), 225
Jones, B. J. M., 36
Jones, J. B., 288, 481
Jones, P. A., 43
Jones, S. T., 171
Jones, T. A., 200, 344, 492
Jongejan, J. A., 476, 484–485, 487
Jonkman, J.H.G., 224
Jøogensen, K., 171
Jørgensen, K., 168, 170–171, 171(6), 172, 173(16), 174–175, 182, 184, 188
Jorgensen, W. L., 397
Jörnvall, H., 192, 236
Jorquera, A., 218
Joven, J., 108
Junien, J.-L., 201, 216

K

Kaffenberger, R. M., 12
Kahn, S. E., 9

Kahrs, R., 23
Kai, Y., 10
Kaiser, R., 236
Kakis, G., 7
Kalara, S. P., 9
Kalk, K. H., 156
Kalogeris, T. J., 10
Kamat, S., 505
Kamat, S. V., 288
Kamat, V. S., 507
Kamp, F., 167, 185
Kaneko, T., 424
Kannel, W. B., 23
Kaplan, M., 19
Kaptein, R., 345
Karandijand, R. Z., 495
Karasov, W. H., 35
Karel, M., 505
Karpe, F., 109, 111(37), 112(37)
Kaslow, H. R., 56
Kather, H., 56
Katsuki, T., 363, 417
Katz, M., 74
Kauerz, M. T., 186
Kaufman, M. H., 71
Kawamoto, T., 29
Kawamura, M., 108
Kazlauskas, R. J., 210, 237, 375, 454, 481
Kearney, J. F., 128
Kedinger, M., 35
Keelan, M., 3(6, 9), 4, 16(9), 17–19, 19(102, 112), 20, 23–24, 24(106, 107), 30(95, 98), 32, 32(102, 106, 107, 124, 163, 250, 251), 33, 33(101), 34(115), 40(166), 41
Keller-Rupp, P., 16
Kemler, R., 76
Kempner, E. S., 107, 116, 119–120, 123, 123(27), 124(21, 27), 125(27)
Kennard, O., 392, 393(14), 394(14), 396(14)
Kensil, C. R., 182
Kerfelec, B., 134, 195, 203
Kern, P. A., 62
Kerscher, V., 424
Kershaw, M., 9
Kersting, S., 104
Kézdy, F. J., 276, 334
Khalaf, N. K., 444–445, 445(4), 450(4), 457(4), 462(4)
Khalafi, R., 41
Khalil, T., 25

Khan, S. A., 458
Khoo, J. C., 47, 51(13)
Kiang, C. L., 17
Kieboom, A. P. G., 434
Kihl, B., 25
Kim, H. K., 20
Kim, H. S., 77
Kim, K. K., 448
Kim, S., 448
Kim, Y. S., 19
Kimelberg, H. K., 38
Kimmel, A. R., 84
Kinberg, J., 224(103), 225
King, D. J., 135(35), 144, 191
King, M. M., 231
Kinnunen, P. K. J., 116, 169, 178(4), 182, 184(43)
Kinsman, R. I., 25
Kirchgessner, T. G., 45, 48(9), 49(9), 65(9), 102, 192
Kirchner, G., 412, 495, 510
Kirk, O., 409, 436
Kisbaugh, J., 498
Kissel, J. A., 70, 73(16), 93, 95(14), 117
Kitaguchi, H., 490
Kitahara, M., 253
Kitayama, T., 288
Kjaer, K., 285
Klausner, R. D., 28
Klein, M. L., 397
Klein, P., 425
Kleinfield, A. M., 26, 27(188)
Klemets, R., 125
Kleywegt, G., 492
Klibanov, A. M., 345, 346(32), 410, 412, 434–435, 437, 443, 455, 457(2), 473, 482, 490–491, 493(24), 495, 503, 505, 510, 512, 512(39), 513
Klinger, A., 175
Kneip, J., 8
Kneser, K., 104
Knezevich, A. L., 12
Knoops-Mouthuy, E., 212
Koberstein, J. T., 307
Koch, H. B., 344
Kock-van Dalen, A. C., 409
Koda, Y., 10
Kodali, D. R., 158, 159(27), 160(27)
Koetzle, T. F., 392, 393(14), 394(14), 396(14)
Kohler, G., 128

Kojima, F., 253
Koksis, A., 287
Koldovsky, O., 35–36, 36(268), 37(269)
Kolisis, F. N., 448
Kollman, P. A., 393, 396–397, 397(22)
Kollmer, M. E., 31
Komaromy, M., 61
Kopelman, H., 42
Koretz, R. L., 36
Koskinen, A. M. P., 473
Kossiakoff, A. A., 396
Kothari, H. V., 231, 234
Kounnas, M. Z., 104
Kovac, A., 205
Kowtzle, T. F., 392, 393(15), 394(15), 396(15)
Kozaki, K., 108
Kozubek, A., 181
Kraayveld, D. E., 485
Krabisch, L., 63, 64(68)
Kraemer, F. B., 61–62
Krapp, A., 103–104, 105(18)
Krause, G., 23
Kraut, J., 234, 397
Krebs, J. J. R., 39
Kreiser, W., 424
Kremer, J. M. H., 333
Krishna, G., 59
Kritchevsky, D., 20, 231, 234
Krone, W., 21
Krukonis, V. J., 496
Kuhl, P., 469
Kuksis, A., 7
Kupfer, E., 225, 252
Kurganov, B. I., 196
Kushwala, R. S., 23
Kvittingen, L. S., 458, 483
Kwartler, J., 22
Kwong, L. K., 103
Kyger, J. A., 117

L

Laane, C., 480
Labarca, C., 88
Labourdenne, S., 306, 308, 310(29), 311, 311(29), 315(32), 323, 326(32)
Lacey, J. H., 15
Ladam, A. J., 28
Ladeford, C., 324(47), 325
Ladner, W. E., 476
Laemmli, U. K., 78
Lafont, H., 84
Lafontan, M., 60–61
Lagone, J. J., 387
Lahav, M., 289
Lai, C.-Y., 237, 241(49)
Lalonde, J. J., 424, 443–445, 445(4), 450(4), 457(4), 462(4), 510
Lalouel, J.-M., 103, 192
Lam, L. K., 288
Lamers, C. B., 225
Lancaster, N., 31
Landau, E. M., 289
Lang, D., 200, 254, 346
Lang, S., 438
Lange, L. G., 69, 99, 117
Langhans, W., 10
Langin, D., 45–46, 46(10), 49(10), 56, 60–62, 62(63), 63(44), 65(10, 44), 66(44)
Langmuir, I., 295
Langrand, G., 357, 454
Lankin, V. Z., 196
Lantz, J. L., 237, 238(44), 240, 240(44)
Laper, V., 21, 22(131)
Laposata, M., 26, 163(51), 166
Larrson, K., 156
Larsen, N. B., 285
Larson, S. B., 346
Larsson, I., 16, 224(104), 225
Larsson, K., 156
LaRusso, N. F., 29
Lasater, L. S., 123
Lasunción, M. A., 113
Lathrop, B., 186, 187(61)
Laugier, R., 68, 84, 98(8), 126, 141, 142(30), 143(30), 144(30), 191, 216, 225(4), 286
Laumen, K., 351, 365, 367, 376, 406, 419–421, 421(29), 422(34), 424–426, 428–429, 430(55), 511
Laurell, H., 46, 46(10), 49(10), 65(10)
Laurent, S., 277, 278(24), 294, 307, 311(14)
Lausch, R., 448
Lauwereys, M., 133, 200, 292
Lauwers, A., 254, 255(13)
Lauzon, J., 40
Lavayre, J., 512(40), 513
Law, J. H., 276, 334
Lawson, D. M., 200, 210, 211(77), 237, 252, 335, 409
Leal, O., 218

Leathes, J. B., 153
Lecat, D., 201
Lechene de le Porte, P., 84
Ledezma, E., 218
Ledl, F., 511
Lee, A. G., 39, 172
Lee, D.-S., 437
Lee, E. Y. C., 47, 51(13)
Lee, F. D., 36, 37(272)
Lee, H. S., 69, 99, 101(24)
Lee, J. J., 66
Lee, K., 237, 241(50), 245(50)
Lee, R., 36
Lee, W. T., 114
Lee, Y., 25
Leeman, S. E., 25
Legge, R. L., 509
Lehtonen, J. Y. A., 182, 184(43)
Leibler, S., 185
Leibowitz, S. F., 8–9
Leiserowitz, L., 289
Lemmich, J., 189
Lengsfeld, H., 15–16, 190, 220, 225, 225(91), 252
Lennard-Jones, J. E., 43
Lepage, G., 42
Lequire, V. S., 30
Leroy, C., 42
Leroy, M.-J., 48, 60(20)
Letellier, S., 308, 310(29), 311(29)
Leung, P. M. B., 10
Leuveling Tjeenk, M., 344
Levanon, L., 289
Levashov, A. V., 196
Levelt Sengers, J. M. H., 496
Levin, R. J., 34
Levine, A. S., 8–9
Levine, G. M., 30
Levison, H., 41
Levy, E., 3(5), 4, 42
Lewis, K. A., 181, 184(38)
Li, D., 307
Li, F., 70, 232
Li, L., 70, 98
Li, N., 236
Li, W. H., 192
Li, Y., 133, 200, 210, 237, 346, 445, 446(13, 14)
Li, Z., 65
Lichtenberg, D., 179, 182, 184(41), 186, 186(41)

Lichtenberger, L. M., 37
Liddle, R., 10
Liesegang, B., 128
Lilljeqvist, C.-C., 19
Lilly, H. S., 163(51), 166
Lilly, M. D., 515
Lin, D. S., 23
Lin, G., 237, 241(49), 509
Lin, M., 307–308, 310(29), 311(29)
Lindberg, A., 103, 105(18)
Lindblom, A., 114
Lindblom, G., 185
Lindebeam, D. R., 504
Lindquist, G., 115
Ling, L., 413
Ling, S., 24
Ling, V., 20, 37(127)
Linke, M. J., 70, 117
Lipkin, M., 42–43
Lippincott-Schwartz, J., 28
Lithell, H., 61
Liu, K.-C., 422, 509
Liu, M.-S., 103, 105(18)
Liu, S.-H., 509
Lobell, M., 406, 410, 414
Lock, D. R., 31
Löfås, S., 115
Lombardo, D., 70, 84, 98, 98(8), 117, 195, 233
Londos, C., 46, 48, 54, 56(11), 57(11), 59, 59(37), 65(11)
Longhi, S., 155, 212
Lookene, A., 102–104, 104(8), 105, 105(22), 106(10), 107, 107(22), 111, 112(22), 113, 113(8–10, 22, 27), 115(15), 116(15), 192(116), 227
Loomis, C. R., 162
Lopez, E., 17
Lopez-Candales, A., 69, 99
Lopez Ortiz, F., 435
Lortie, R., 436, 509
Louveau, I., 65
Louwrier, A., 345, 346(32), 490
Lowe, J. B., 27
Lowe, M. E., 202
Lowe, P. A., 132, 135(35), 144, 191
Lubbers, A., 103
Lugli, R., 41
Luisetti, M., 440
Lundberg, B., 125
Lundmark, M., 154, 155(7), 165(7)

Lusis, A. J., 45, 48(9), 49(9), 65(9), 78, 192
Lüthi-Peng, Q., 194, 220, 220(24), 221(24)
Lutmer, R., 12
Lynch, H. T., 43
Lyons, A., 135(35), 144, 191

M

Ma, Y., 103
Maatman, R. G., 26
Mabis, A. J., 156, 157(18)
Macarae, A., 458
Machidori, H., 32
Machleder, D., 78
Machmüller, G., 414, 433(22)
MacLeod, J., 20, 32(124), 32(250), 33
Macrae, A., 515
Macrae, A. R., 466
Madden, T. D., 184
Madura, J. D., 397
Maeda, K., 253
Maeda, N., 72
Maerki, H. P., 220
Maghrabi, M. R., 17
Magnaniello, V., 48, 55(19)
Magun, A. M., 29
Mahley, R. W., 23, 30
Maillard, B., 401, 412, 419(17), 510
Makamura, K., 288
Malagelada, J. R., 225
Malagoli, G., 43
Malakhova, E. A., 196
Maloney, K. M., 181, 184(38)
Mamonneas, T. G., 22, 40(143)
Mandel, A. M., 181
Manenti, A., 43
Manganiello, V., 45, 48, 49(23), 60, 60(20–23)
Manimaran, T., 452
Mannesse, M. L., 204, 209(67), 212, 252
Manning, J. A., 27
Mansbach, C. M., 6, 266
Mansbach, C. M. II, 14, 28, 28(74)
Mantsch, H. H., 154
Manzocchi, A., 418
Mao, S. J., 29
Marcel, Y. L., 42
Marcon, N., 43
Marecek, J. F., 179
Margerin, C., 498
Margolin, A. J., 410
Margolin, A. L., 418, 442–443, 445, 491, 495, 510
Margolin, A. M., 444, 445(4), 450(4), 457(4), 462(4)
Marguet, F., 156, 190, 204–205, 205(65, 66), 206(66, 74), 207, 207(66), 208(75), 282
Marks, J. L., 9
Marley, A. E., 50
Marnett, L. J., 43
Marsh, J. B., 70
Martel, P., 193
Martin, V. S., 363, 417
Martindale, M. E., 231
Martinek, K., 196
Martinelle, M., 287, 387, 389, 389(4), 390, 391(9), 394(2), 399, 400(37), 401, 401(37), 405(9)
Martinez, C., 107, 133, 155–156, 192(25), 194–195, 200, 207(25), 208(25), 216(25), 237, 319, 341, 343(19)
Martinez, O. G., 444, 445(4), 450(4), 457(4), 462(4)
Martin-Hidalgo, A., 61
Martins, F. J., 507
Marty, A., 504–505
Maruhama, Y., 10
Mas, E., 70, 84, 98, 98(8)
Massari, V. J., 8
Mastella, G., 41
Matarese, V., 27
Mathieu, H., 35
Matori, M., 288
Matson, D. W., 498
Matsuo, M., 424
Matthews, B. W., 445
Matthyssens, G., 133, 200, 292
Mattiasson, B., 509
Mattson, A., 395, 396(26), 399, 400(37), 401, 401(37), 490, 510
Mattson, F. H., 11–12, 23(62)
Maugh, T. H. II, 512
Maurer, J. K., 12
Mauz, O., 514
Maylié, M. F., 200, 316
Mazzola, G., 505
McCarthy, D. M., 19
McCollum, J. P. K., 219
McFarlane, A., 25
McGregor, I. S., 8
McHugh, A. M., 498

McHugh, M. A., 496
McIntyre, Y., 20, 32(124, 250), 33
McKean, M. L., 241
McMahan, M. R., 12
McMahon, D. M., 251
McMonagle, S., 30
McNeely, S., 13
McPherson, A., 346
McQuillan, J. J., 26
Mcquire, V., 43
Mehes, M. M., 481
Meier, M. K., 16, 224, 224(105), 225
Meister, W., 16, 220, 252, 253(7), 259(7)
Mejdoub, H., 213, 215(82), 334
Méléard, P., 184
Melia, A. T., 224(103, 106), 225
Melo, E. P., 290, 291(75)
Melone, J., 10
Melville, D. M., 43
Menashe, M., 179, 182, 184(41), 186(41)
Mendel, V. E., 10
Mendelsohn, D., 22, 23(145)
Menéndez, E., 409, 413(9), 435(9), 442(9)
Menendez, J. A., 8
Menge, U., 133, 200, 258, 341, 448
Mengoli, V., 41
Merchant, J. L., 22
Merkel, M., 127
Merz, K. M., Jr., 396
Metcalf, D., 74
Metzger, H., 435
Mewshaw, R. E., 237
Meyer, E. F., 392, 393(14), 394(14), 396(14)
Meyer, J. H., 36
Meyer, N., 102, 104(8), 113(8)
Meyers-Payne, S. C., 232
Meyn, R. E., 113
Michel, R., 127, 316
Michelsen, P., 287
Middleton, C., 21
Mieras, M. C. E., 195, 267, 268(9), 269(9), 323
Miettinen, T. A., 68
Miglioli, M., 43
Miller, A. B., 42
Miller, C. G., 45, 48(9), 49(9), 65(9)
Miller, K. W., 162, 163(35, 37, 38), 164, 167(35–38)
Milne, R. W., 42
Milstein, C. B. S. W., 128

Milstein, G., 128
Mirajovsky, D., 289
Misaki, H., 227
Mischitz, M., 510
Mishkel, M. A., 23
Misset, O., 193, 344
Mitov, M. D., 184
Mitsuda, S., 511
Mizuno, N. K., 68, 117
Moeglen, C., 224(105), 225
Moestrup, S. K., 103, 113(9)
Mohandas, T., 45, 48(9), 49(9), 65(9)
Möhwald, M., 289
Molines, J., 161
Momongan, M., 424, 510
Momose, T., 516
Momsen, M. M., 295, 319
Momsen, W. E., 285, 292, 298, 301(14), 304, 304(18), 305(14, 15), 319, 323
Monck, M. A., 154
Monsan, P., 512
Montalto, G., 84, 98(8)
Moore, B. D., 458, 465, 467
Moos, M. C., Jr., 46, 56(11), 57(11), 65(11)
Moo-Young, M., 509
Moran, T. H., 10
Morarji, Y., 10
Moreau, H., 126–127, 144, 190, 190(2), 191, 195, 195(2), 196(34), 197(30), 200(34), 201–202, 202(57), 203(57), 216, 217(58, 59), 218(30), 219(118), 220(30), 221(30, 34), 222(30), 223, 223(34), 224(34), 225(30), 227, 229(34), 265, 277(6), 282(15), 286, 294, 295(11), 306, 323(8), 334, 337(12)
Moreau, J., 74
Morel, D. W., 70, 98
Moreno-Rea, J., 218
Morgan, D. O., 95
Mori, H., 495
Mori, K., 365
Mori, N., 108
Morin, C. L., 20, 37(127), 42
Morinville, A. H., 454
Moris, F., 442
Morley, J. E., 8
Morley, N., 287
Morlock-Fitzpatrick, K., 70, 94, 98
Morre, D. J., 30

Morris, A. P., 25
Morris, B. J., 8
Morrow, C. J., 410
Morson, L. A., 24, 40(168)
Mortensen, K., 184, 189
Mosbach, E. H., 21, 22(131)
Moss, G., 25
Moulin, A., 195, 197(29), 201, 202(57), 203(29, 57)
Mouloungui, Z., 409
Moulson, P. S., 505
Mouritsen, O. G., 168–170, 170(3), 171, 171(6), 172, 173(16), 174, 178(4), 182, 184, 188–189
Muderhwa, J. M., 284, 294, 300, 304(7, 19)
Mui, B. L.-S., 184
Muller, D. A., 505
Muller, D. P. R., 219
Mulvey, G., 17, 30(95, 98)
Murad, F., 60
Murata, M., 424
Murata, T., 48
Muroya, H., 22
Murthy, S. K., 237, 238(48)
Mutt, V., 7
Myers, R. L., 103
Myers, S., 68, 99, 231, 232(10)

N

Naeshima, S., 511
Naganawa, H., 253
Nagri, A. D., 287
Nagy, A., 74
Nagy, R., 74
Nakamura, H., 253
Nakamura, K., 503–505
Naray-Szabo, G., 397
Navia, M. A., 443, 445
Naylor, S., 295, 305(15)
Neale, K., 43
Neelands, P. J., 33, 40(257)
Nelson, C., 23
Nelson, L. M., 36, 37(272)
Nemecz, G., 26–27
Neumann, A. W., 307
Neumann, J. M., 154
Nevin, P., 28
Newbill, T., 68

Newmark, H. L., 43
Ngooi, T. K., 378
Nguyen, D. T., 397
Nicholl, C. G., 10
Nicholson, D., 120
Nicholson, J. A., 19
Nicola, N. A., 74
Nicolas, A., 155
Nicolson, G. L., 38
Nielsen, M., 102, 104, 104(8), 113(8)
Nielsen, T. B., 120
Nigro, N. D., 43–44
Nilsson, A., 118
Nilsson, N. Ö., 53, 56, 56(35), 63, 64(69)
Nilsson, O., 6
Nilsson, S., 45, 48(9), 49(9), 65(9)
Niot, I., 18
Nishioka, T., 422–423
Noble, M. E. M., 200, 254, 344
Noël, J.-P., 201, 202(57), 203(57)
Noetzel, S., 514
Nolen, G. A., 12, 23(62)
Nolph, G. V., 23
Noma, A., 6
Nordin, O., 418
Norin, M., 391, 393–395, 396(24, 26), 490, 492
Norin, T., 287, 394–396, 396(24, 26), 399, 400(37), 401, 401(37), 406, 490, 510
Norskov, L., 133, 200, 258, 341
Northfield, T. C., 6
Norton, K. A., 498
Norum, K., 19
Novak, L., 351, 406
Nurit, S., 200, 202(45)
Nury, S., 207, 288, 289(68), 307, 311(16), 312(16)
Nutting, D. F., 32
Nyholm, P. G., 154, 155(7), 165(7)
Nykær, A., 102–103, 104(8), 113(8, 9)

O

Oabauitiv, I., 307
Oberhauser, T., 513, 515(42), 516(42)
O'Brien, R. J., 6
Ockner, R. K., 27, 30
Oda, J., 422–423
Odenthal, J. B., 485
Odink, J., 16, 224, 224(102), 225

O'Doherty, B., 235
O'Doherty, P. J. A., 7
O'Donohue, T. L., 8
Ogata, R. T., 26, 27(188)
Ohkawa, Y., 516
Ohneda, A., 10
Ohno, A., 288
Öhrner, N., 396, 399, 400(37), 401, 401(37), 492, 510
Ohsawa, S., 410
Ohta, H., 410
Okamatsu, H., 22
Okamoto, Y., 365
Okkels, J. S., 345
Okumura, S., 357
Olbe, L., 25
Olchowvsky, D., 9
Olive, J., 153
Olivecrona, G., 102–104, 104(8), 105, 105(18, 22), 107, 107(22), 109, 111, 111(37), 112(22, 37), 113, 113(8, 27), 115(15), 116(15), 192(116), 227
Olivecrona, T., 102–104, 104(2), 106–109, 111(37), 112(37), 114(2), 192, 287
Oliver, J. D., 346
Olivieri, D., 41
Ollis, D. L., 193, 235
Olsen, O. H., 393
Olson, R. E., 23
Olsson, H., 45, 46(2), 47(2), 51, 57(2)
O'Neill, B., 25
Ong, D. E., 123
Ong, J. M., 62
Ooostrom, H. H. M., 504
Ooshima, H., 495
Op den Kamp, J. A., 39, 168, 182(1), 186
Orci, L., 25, 28
Orland, H., 289
Orrenius, C., 396, 401, 510
Osborne, J. C., 89, 107, 117, 120
O'Shea, R. D., 8
Østergaard, P., 103, 115(15), 116(15)
Østerlund, T., 51, 52(33), 53(33), 57(33), 64(33)
Ota, Y., 288
Ottohina, G., 442
Ottolina, G., 442, 480, 482, 493(25), 509, 515–516
Owen, W. G., 295, 305(15)
Ozaki, S., 413

P

Padykula, A., 28
Paganelli, G. M., 43
Paigen, K., 88
Pak, M. S., 44
Pallas, N. R., 307, 311, 311(24)
Palocci, C., 287, 495–496, 505(13)
Paltauf, F., 154, 204–206, 207(72), 209(68), 252, 255, 268
Panaiotov, I., 263, 279, 282, 283(33), 284, 284(34), 323
Pangborn, W., 236
Panza, L., 440
Parida, S., 288
Park, H.-O., 437
Park, J. E., 28
Parker, M. C., 467
Parmar, V. S., 438
Partali, V., 481
Partharsarathy, S., 14, 28(74)
Pascher, I., 154, 155(7, 8), 156(8), 165(7, 8)
Pasta, P., 505
Patel, I. H., 224(103), 225, 225(106)
Patkar, S., 200, 204, 209, 209(64), 210, 211(77), 237, 287, 344, 389–390, 409, 492
Patsch, W., 23
Patton, J. S., 15
Pattus, F., 265, 268, 277(5), 285, 294, 321
Paulatis, M. E., 498
Pease, S., 74
Pedersen, S., 170
Peeters, P. A. M., 224
Penn, G., 513, 515(42), 516(42)
Pepin, P., 509
Perin, N., 23
Perini, M., 43
Perrut, M., 504–505
Persichetti, R. A., 445
Persson, B., 192
Peters, G. H., 285, 393
Petersen, C. M., 103, 113(9)
Petersen, S. B., 127, 193
Peterson, J., 104, 108
Peterson, W. A., 178
Pethica, B. A., 311
Phelps, C. P., 9
Phillips, M. C., 164, 294
Phillips, M. L., 120
Phillips, R. S., 488

Philo, J. S., 120
Phipps, J., 235
Pickei, V. M., 8
Pickersgill, R. W., 171
Piedrahita, J. A., 72
Piéroni, G., 132, 135(35), 144, 191, 195–196, 197(29), 199(42, 117), 200, 202, 203(29), 217(58, 59), 223, 227, 265–266, 268, 280, 282(7), 284, 284(7), 286, 293–294, 295(11), 307, 311(16), 312(16), 321
Pieterson, W. A., 345
Pike, M. C., 43
Pilon, C., 192
Pinate, F. M., 218
Pioch, D., 277, 278(24), 294, 307, 311(14)
Pitts, V., 31
Plee-Gautier, E., 62
Pleiss, J., 511
Pletnev, V. Z., 236
Ploegh, H., 28
Plump, A. S., 72
Poirier, H., 18
Poksay, K., 31
Polak, M. J., 10
Poley, J. R., 22
Ponich, T., 3(7), 4
Ponz de, L., 43
Ponz de Leon, M., 43
Poon, P. H., 120
Poppe, L., 351, 406
Popper, D. A., 30
Porte, D., Jr., 9
Pöschl, U., 510
Postma, J. P. M., 392
Potier, M., 120
Potter, B. J., 3(2), 4
Potthoff, A., 256
Powell, J. F., 8
Pozo, M., 413
Pradelli, J. M., 41
Prasad, A. K., 438
Prausniz, J. M., 505
Preclik, G., 21
Prestegard, J., 154
Preston-Martin, S., 43
Previato, L., 104
Prigge, W. F., 22, 40(144)
Privett, O. S., 39
Profeta, S., Jr., 393, 397(22)
Prozialeck, D. H., 241

Prydz, K., 28
Pulido, R., 413, 435

Q

Quaade, F., 224(104), 225
Quinn, D. M., 104, 231, 233, 235, 237, 238(44–46), 239, 239(20), 240, 240(44, 45), 241(50, 51), 245(50), 249–250
Quiocho, F. A., 444, 458(5), 462(5)
Quirin, K. W., 498

R

Raber, K., 480
Rabon, E. C., 123
Rachmilewitz, D., 28
Radbruch, A., 128
Radic, Z., 234
Rahn, T., 48, 49(23), 60(20, 23)
Raicht, R. F., 43
Rajewki, K., 128
Rajotte, R. V., 20, 24(122), 32, 32(122, 255), 33, 33(122, 123), 40
Rakels, J. L. L., 475, 478, 485
Ramakrishnan, R., 102
Ramirez, F. L., 179
Randle, R. J., 55
Randolph, T. T., 501
Randolph, T. W., 505
Raneva, V., 323
Rangheard, M. S., 357
Ransac, S., 127, 157, 190, 190(2), 191, 193, 195, 195(2), 196(34), 197(30), 200(34), 206, 206(74), 207, 207(72), 218(30), 220(30), 221(30, 34), 222(30), 223(34), 224(34), 225(30), 229(33, 34), 263, 265, 268, 270, 270(12), 271(12), 272(12), 277(6), 282(12, 15, 16), 287, 288(56), 306, 323(8), 334, 337(12), 344
Rantakyla, M., 506
Raphel, V., 216
Rappaport, A. T., 481
Rasmussen, P., 486
Raucci, D. T., 25
Raul, F., 35
Ravazola, M., 25
Rayner, D. V., 10
Read, N. W., 10, 25

Rebolledo, F., 409
Rechsteiner, M., 93
Reck, G., 426, 427(49)
Reddick, R. L., 72
Reddy, B. S., 43–44, 44(337)
Redgrave, T. G., 158, 159(27), 160(27)
Reed, M., 30
Reichardt, C., 480
Reidelberger, R. D., 10
Reif, O.-W., 448
Reinbolt, J., 334
Remaury, A., 61
Remington, S. J., 193, 235
Ren, K., 103
Rety, S., 200
Reuban, M. A., 123
Reue, K., 65
Reynisdottir, S., 61, 62(63)
Reynolds, J., 505
Reynolds, L. J., 120
Richards, F. M., 444, 458(5), 462(5)
Richieri, G. V., 26, 27(188)
Richou, J., 307
Ridderstråle, M., 48, 49(23), 55(19), 60(23)
Riddle, M. C., 118
Ridinger, D. N., 103
Rietsch, J., 285, 294, 318
Rigo, G., 43
Rigtrup, K. M., 123
Risbo, J., 171
Ritchie, J. K., 43
Ritter, R. C., 10
Riva, S., 412, 434, 440, 442, 455, 480, 482, 493(25), 509, 515–516
Rivière, C., 132, 135(35), 144, 191, 195, 223, 229(33), 265, 268, 270, 270(12), 271(12), 272(12), 277(6), 282(12, 16), 284, 294, 306–308, 310(29), 311(16, 29), 312(16), 323(8), 334, 337(12)
Robb, D. A., 469
Robblee, N. M., 40
Roberts, M. F., 179, 181, 184(38)
Roberts, S., 421
Robinson, M. I., 19
Rodbell, M., 54
Roder, J. C., 74
Rodgers, J. B., 6, 19, 23
Rodgers, J. R., 392, 393(14), 394(14), 396(14)
Rodriguez, V., 514
Rogalska, E., 127, 157, 190(2), 191, 195(2), 206, 206(74), 207, 207(71, 72), 268, 287–288, 288(56), 289(68)
Rogers, D., 74
Rogers, J., 169, 179, 181(5), 184(5), 186(5), 279
Rogers, S., 93
Rohm & Haas Surfactants and Dispersants-Handbook of Physical Properties, 239
Rohner-Jeanrenaud, F., 8–9, 9(29)
Roigaard, H., 102, 104(8), 113(8)
Rojas, C. J., 117, 120, 123, 123(27), 124(27), 125(27)
Rokaeus, A., 25
Romero, G., 179, 182, 184(41), 186(41)
Roncucci, L., 43
Ronist, G., 6
Rönnstrand, L., 48, 60(20)
Rose, P. L., 289
Rosell, S., 25
Rosen, M., 10
Rosen, O. M., 63, 65(67)
Rosner, B. A., 42
Ross, A. C., 70, 80, 89(1), 90(1)
Ross, P. K., 43
Rossant, J., 74
Rosseneu, M. Y., 333
Rossi, F. A., 41
Rotenberg, Y., 307
Roth, R. A., 95
Rotticci, D., 396, 510
Rouard, M., 200, 202(45)
Roudani, S., 84, 98(8)
Rouimi, P., 202
Rousseau, G., 510
Roussel, A., 103, 192
Rovati, L. C., 10
Rovery, M., 193, 200(22), 202, 202(22), 293, 298(3)
Rowe, E. S., 189
Rowston, W. M., 15
Roy, C. C., 42
Rubin, B., 200, 210, 237, 352, 378(2), 409, 445, 446(14), 447, 447(11), 448(11)
Rubin, C. E., 31
Rubin, E. M., 72
Rudd, E. A., 68, 231, 232(4), 234, 234(4), 238(4)
Rudel, L. L., 23
Rugani, N., 192(25), 194, 207(25), 208(25), 216(25), 319, 341, 343(19)

Ruggiero, D., 8
Russel, A. J., 288, 482, 491, 493(23), 495, 505, 507
Russell, B. S., 29, 30(213, 214)
Russell, R. I., 36, 37(272)
Rutter, W. J., 80, 91(2), 92(2), 94–95, 117
Ruy, K., 390
Ruysschaert, J.-M., 173, 182(17)
Ruyssen, R., 254, 255(13)
Ryckaert, J.-P., 392, 394(18)
Rydel, T. J., 346
Ryu, K., 495

S

Sabesin, S. M., 28
Sabra, M., 171, 188
Sacchettini, J. C., 26
Sacchi, N., 65
Sachs, G., 123
Sackmann, E., 185
Sagiv, J., 289
Sahu, A., 9
St. Clair, N. L., 445
Saito, N., 506
Sakai, K., 375
Sakamoto, T., 25
Sakata, T., 32
Sakurai, T., 491
Salen, G., 79
Salesse, C., 181, 184(38)
Salmon, A., 162(43), 164
Salomon, Y., 59
Salter, M. G., 25
Salvioli, G., 41
Sanacora, G., 9
Sanderman, H., 39
Sando, G. N., 125, 127
Sandvig, K., 28
Sanson, A., 154
Santamarina-Fojo, S., 103–104
Santaniello, E., 418, 509
Saouaf, R., 167
Sarda, L., 129, 131(14), 132, 191, 192(27), 194, 196, 200, 207(27), 223, 237, 257, 265, 282(7), 284, 284(7), 292–294, 298(3), 327–329, 330(2), 331(2), 335(2), 341, 343(18)
Sari, H., 200, 202(45)
Sarney, D. B., 509
Sartore, L., 516
Satherley, J., 307
Sauber, K., 514
Saunders, D. R., 28
Saunders, K., 10
Saunière, J.-F., 132, 191
Savonen, R., 104–105, 113, 113(27)
Sbarra, V., 84, 98(8)
Scapin, G., 26
Scellar, M. P., 495
Schachter, D., 19, 22, 24, 24(113), 40, 40(143)
Schaefer, V. J., 295
Scharrer, E., 10
Schaumburg, K., 285
Scheper, T., 448
Scheurinck, A., 9
Schick, H., 426, 427(49)
Schiffrin, D. J., 307
Schleicher, E., 511
Schlotterbeck, A., 438
Schmid, P. C., 300, 304(19)
Schmid, R. D., 200, 254–255, 344, 346
Schmidt, W., 74
Schmit, G. D., 295, 305(15)
Schneeman, B. O., 41
Schneider, A., 21–22
Schneider, M. P., 253, 351, 358, 364–365, 367, 373, 375–376, 378, 381, 384, 386, 406, 410, 414, 419–421, 421(29), 422(34), 423–426, 428–429, 429(53), 430(55), 431, 432(23), 434, 511, 515
Schoeller, C., 3(9), 4, 16(9), 17, 30(95, 98)
Schomburg, D., 200, 254, 346
Schonfeld, G., 23, 31
Schønheyder, F., 329
Schotz, M. C., 45, 47, 48(9), 49(9), 51, 51(12), 52(33), 53(12, 33), 57(33), 61, 63(12), 64(12, 13), 65, 65(9), 192
Schrag, D., 200
Schrag, J. D., 133, 193, 200, 210, 235, 237, 346, 445, 446(13, 14), 454
Schreiner, B., 19
Schröder, F., 56
Schroeder, F., 26–27
Schwarz, M. W., 9
Schwarz, S. M., 24
Scilimati, A., 378
Scollar, M. P., 412, 510
Scott, D. L., 156, 178, 181
Scow, R. O., 6, 14, 281, 294

Sebastiao, M., 193
Secundo, F., 442
Seemayer, R., 351, 365, 367, 375, 406, 420–421, 422(31, 34), 423, 434, 440, 515
Seetharam, S., 27
Segel, I. R., 233, 239, 239(18)
Seiffer, E., 21
Selineuf, J. D., 445
Sémériva, M., 193, 200(46), 285
Sepple, C. P., 10
Serfling, E., 76
Serreqi, A. N., 210, 237
Sewell, R. B., 29
Shamir, R., 69–70, 80, 98–99, 101(24)
Sharp, A. M., 486
Sharpless, K. B., 363, 417
Shefer, S., 21, 22(131)
Sheldon, R. A., 409
Sherman, J. R., 31
Shiau, Y.-F., 30
Shibata, T., 365
Shields, H. M., 27
Shim, S. C., 437
Shimada, M., 108
Shimano, H., 108
Shimanouchi, T., 392, 393(14), 394(14), 396(14)
Shin, H.-C., 249
Shiotani, Y., 8
Shipley, G. G., 162
Shirai, K., 104
Shively, J. E., 45, 48(9), 49(9), 65(9)
Shugar, D., 50
Shuhua, J., 103
Shuhua, Z., 103
Shyamsunder, E., 177, 188
Sidler, W., 16, 220, 252, 253(7), 259(7)
Siegel, A. F., 307
Sigler, P. B., 156, 178, 181
Sih, C. J., 287–288, 363, 365, 378, 395, 417, 451–452, 453(22), 474, 479–480, 486(14), 509, 513
Silk, D. B., 36
Silman, I., 193, 235
Sim, A. T. R., 49, 50(27)
Simmonds, W. J., 17(220), 30, 69, 232
Simon, B., 56
Simon, P. M., 35
Singer, M. A., 189
Singer, S. J., 38

Singh, A., 19
Singh, B., 40
Singh, D. V., 44
Singh, U. C., 393, 397(22)
Sipols, A. J., 9
Sircar, B., 37
Sjostrom, L., 224(104), 225
Sjursnes, B., 458, 483
Skottova, N., 104–105, 105(22), 107(22), 112(22), 113(22, 27)
Slavin, B. G., 62
Slotboom, A. J., 204, 209(67), 252, 270, 285, 325(48), 326
Smaby, J. M., 284, 304
Small, D. M., 6, 15, 153, 156, 157(21), 158, 158(21), 159(27–29), 160(27–29), 161(21, 28), 162, 162(46), 163, 163(35, 37–40, 47, 51), 164, 165(21, 39, 40), 166, 166(49), 167(35–38), 263, 264(1), 327, 335(1)
Smart, T. A.-M., 235
Smerieri, O., 43
Smith, A. D., 8, 39, 134
Smith, A. G., 74
Smith, B. D. V., 307
Smith, C. J., 48, 55(19), 60(21)
Smith, G. P., 10
Smith, J. D., 72
Smith, L., 3(9), 4, 16(9)
Smith, L. C., 281
Smith, M. A., 8
Smith, P., 200, 237, 445, 446(14)
Smith, R. D., 498
Smithies, O., 77
Smuckler, E. A., 118
Snook, J. T., 19
Soares de Araujo, P., 333
Sodhi, H. S., 22
Sohl, J., 237, 238(46), 240
Sohn, J. E., 307
Solcia, E., 25
Soll, A. H., 10
Solomon, T. E., 225
Sonnet, P., 495
Sorensen, M. D., 438
Sorrell, M. F., 511
Sorrentino, D., 3(2), 4
Sparkes, R. S., 45, 48(9), 49(9), 65(9)
Spector, A. A., 39
Speizer, F. E., 42
Spencer, P. S., 237, 238(45), 240, 240(45)

AUTHOR INDEX

Spencer, S. A., 396
Spener, F., 200, 204–205, 209(68), 252, 254–256
Sperotto, M. M., 171–172, 173(16)
Spilburg, C. A., 69, 99
Spooner, P. J. R., 162(46), 163(47), 164
Spritz, N., 23
Stabile, B., 10
Stadler, J., 43
Stadler, P., 204–205, 209(68), 252, 255
Stafford, D. R., 495
Staggers, J. E., 100, 226
Stahl, E., 498
Stahl, M., 74
Stahly, G. P., 452
Stampfer, M. J., 42
Stanley, B. G., 8
Stanssens, P., 292
Stark, R. E., 295, 305(15)
Stauffer, E., 307
Stecher, H., 399, 444, 511, 512(30), 517
Stein, E. A., 22, 23(145)
Stein, O., 6, 102
Stein, Y., 6, 102
Steinberg, D., 45, 47, 51(13)
Stenson Holst, L., 46, 46(10), 49(10), 65(10)
Stern, H. S., 43
Stewart, C. L., 74
Steyler, D. C., 505
Stinson, S. C., 357
Stokeberry, H., 10
Stokes, D., 168, 186–187, 187(61)
Stokesberry, H., 10
Stoop, R., 504
Storch, J., 27
Story, J. A., 20
Stout, J. S., 233, 237, 238(45), 239, 239(20), 240, 240(45)
Straathof, A. J. J., 475, 478, 485
Strålfors, P., 45, 46(2, 6), 47(2), 48(5, 6), 49(6), 51(5), 52(5), 53, 56(35), 57(2, 6), 63, 64(69)
Strange, E. F., 21–22
Stratowa, C., 94, 117
Strauss, E. W., 28
Stremmel, W., 17, 30(95, 98)
Strickland, D. K., 104
Strickland, L. C., 346
Strike, D. P., 237
Strobel, W., 224(104), 225
Stroud, R. M., 235
Stubbs, C. D., 39
Stubbs, R., 10
Stuckmann, M., 255, 258(18), 261(18), 262(18)
Stump, D. D., 17
Sturman, J. A., 231
Sugai, T., 410
Sugano, M., 22
Sugerman, J., 167
Sugiara, M., 357
Sugihara, A., 284
Suh, S. W., 448
Sullivan, A. C., 16
Sullivan, J. E., 50
Sumida, M., 65
Summerskill, W. H. J., 42, 225
Sun, H.-L., 509
Sundel, S., 154, 155(7), 165(7)
Sussman, F., 397
Sussman, J. L., 193, 235
Sutherland, S. D., 103
Sutton, L. D., 237, 238(44–46), 240(44, 45)
Sutton, S. D., 233, 239, 239(20), 240
Svendsen, A., 287, 345, 389–390, 393
Svenson, K. L., 45, 48(9), 49(9), 65(9), 192
Svensson, I., 509
Swart, R. M., 295
Sweers, H. M., 419
Sweet, R. M., 448
Szatlryd, C., 61–62

T

Tahara, S., 289, 424
Taira, M., 48
Takahara, H., 22
Takahata, H., 516
Takao, K., 227
Takebe, Y., 288
Takeuchi, T., 253
Talley, G. D., 103
Tan, A., 28
Taneva, S., 284
Tanford, C., 154, 158(6)
Tang, J., 117, 231
Taplitz, R., 10
Tassi, L., 43
Tasumi, M., 392, 393(14), 394(14), 396(14)
Tate, M. W., 177
Tate, R. L., 89, 117

Tatemoto, K., 7
Tavernier, G., 61
Tavernini, M., 18, 24, 32(163, 251), 33
Taylor, I. L., 10
Taylor, P., 234
Teichner, S. J., 498
Tepper, S. A., 20
Terao, Y., 424
Terminassian Sraga, L., 284
Térnynck, T., 143, 286
Terpstra, O. T., 43
Tessari, M., 155, 345
Theil, F., 426, 427(49), 509
Therisod, M., 434–435
Thim, L., 133, 200, 210, 211(77), 237, 258, 335, 341, 344, 409
Thirstrup, K., 335, 344
Thistlethwaite, W., 30
Thomas, I. K., 40
Thomas, K. R., 71
Thompson, G. N., 42
Thompson, J. C., 25
Thompson, K., 179
Thompson, T. E., 174
Thomson, A. B. R., 3, 3(6–9), 4, 11(340), 16(9), 17–19, 19(102, 112), 20–21, 21(125), 23–24, 24(106, 107, 122), 30(95, 98), 32, 32(32, 102, 106, 107, 122–126, 161, 163, 167, 250–255), 33, 33(101, 122, 123, 125), 34(115), 36(242, 243), 37(258), 40, 40(166), 41, 44
Thunnissen, M. M. G. M., 156
Thurnhofer, H., 16
Tierney, W. J., 12
Tipper, C., 28
Tiruppathi, C., 201
Tofilon, P. J., 113
Tohyama, M., 8
Tolley, S., 133, 258, 341
Tomlinson, A., 295, 305(15)
Tompeter, R. S., 219
Tor, R., 495
Tornqvist, H., 45, 48, 49(23), 55(19), 59(3), 60(20, 21, 23), 63, 64(68)
Tornvall, P., 109, 111(37), 112(37)
Townsend, C. M., Jr., 25
Toxvaerd, S., 285
Trabucchi, C., 41
Trani, M., 436, 509
Traylor, J. D., 37

Treadwell, C. R., 68, 84, 117, 232
Tremblay, N. M., 120
Triantaphylides, C., 357
Triantophylides, C., 454
Triscari, J., 16
Tseng, W.-M., 454
Tsetlin, L. G., 196
Tso, P., 3(3), 4, 17(220), 30–32
Tsujisaka, Y., 357
Tsujita, T., 117, 295, 300, 301(14), 304, 304(14, 20), 305(14)
Tuba, J., 19
Tucker, W. B., 307
Tugrul, Y., 104
Tuma, D. J., 511
Turck, C. W., 70, 73(16), 93, 95(14), 117
Turkenburg, J. P., 133, 210, 211(77), 237, 252, 258, 341, 409
Turner, D. C., 177

U

Uchida, Y., 516
Uezono, Y., 10
Ulshen, M. H., 35
Umezawa, H., 253
Unstetter, C., 30
Uotani, K., 253
Uppenberg, J., 200, 344, 492

V

Vaananen, H., 68
Vagelos, P. R., 47, 51(13)
Vahouny, G. V., 68, 84, 99, 117, 231–232, 232(10)
Valet, P. P., 61
Valivetty, R. H., 458
Valivety, R. H., 466
Van Blankenstein, M., 43
Van Campenhoud, M., 173, 182(17)
van Deenen, L. L. M., 168, 182(1), 186
van den Berg, B., 155, 345
van der Beek, E. J., 16, 224(102), 225
van der Hijden, H. T., 204, 209(67), 252, 325(48), 326
Vanderveen, E. A., 224, 224(104), 225, 225(100, 107)
van Deurs, B., 28
van Eijk, J. H., 345

Van Ejis, A. M. M., 504
van Gunsteren, W. F., 392
Vanheuvel, M., 344
van Oort, M. G., 282, 344
van Rantwijk, F., 409
van Tilbeurgh, H., 103, 156, 192, 192(25, 27), 194–195, 197(31), 199(31), 206(74), 207, 207(25, 27), 208(25, 75), 216(25, 31), 221(31), 237, 292, 319, 341, 343(18, 19)
van Tol, J. B. A., 476, 484–485, 487
van Zoelen, E. J. J., 333
Vasta, V., 48, 60(21)
Vaughan, M., 45, 60
Vaz, L., 193
Vaz, W. L. C., 174
Veeger, C., 480
Veerkamp, J. H., 26
Venkatraman, J., 4, 6(10), 33(10), 40, 40(10)
Verbiar, R., 289
Verger, R., 47, 51(12), 53(12), 63(12), 64(12), 68, 126–127, 129, 131(14, 19), 132–134, 135(15, 26, 35), 136(15), 141, 142(30), 143(30), 144, 144(30), 146(33), 148(19, 33), 157, 190, 190(2), 191, 192(25, 27, 116), 194–195, 195(2), 196, 196(34), 197(29–31), 199, 199(31, 42, 117), 200, 200(34), 201–202, 202(57), 203(29, 57), 204–205, 205(65), 206–207, 207(25, 27, 71, 72), 208(25, 75), 213, 215(82), 216(25, 31), 217(58, 59), 218(30), 219(118), 220(30), 221(30, 31, 34), 222(30), 223, 223(34), 224(34), 225(4, 30), 227, 229(33, 34), 232, 237, 252, 263, 265–267, 267(3), 268, 268(9), 269(9), 270, 270(12), 271(12), 272(12), 275–277, 277(5, 620), 278(24), 279–281, 281(20), 282, 282(7, 12, 15, 16), 283(33), 284, 284(7, 34), 285, 285(3, 19), 286–287, 287(45), 288, 288(56), 289(68), 290, 291(75), 292–294, 295(11, 12), 298(3), 306–308, 310(29), 311, 311(14, 16, 29), 312(16), 315(32), 316, 318–319, 319(10), 321, 323, 323(5, 6, 8, 36, 40), 326(32), 327, 334–335, 337(11, 12), 341, 343(18, 19), 344, 387
Verheij, B., 325(48), 326
Verheij, H. M., 105, 204, 209(67), 212, 252, 344–345
Verkade, P. E., 329
Verma, C., 200
Veronese, F. M., 516

Verrips, C. T., 254
Verschueren, K. H. G., 193, 235
Verstuyft, J. G., 72
Vesselinovitch, D., 20
Vetter, W., 16, 220, 252, 253(7), 259(7)
Vettor, R., 8–9, 9(29)
Vidal, J. C., 178, 345
Vilaró, S., 108
Vilella, E., 108
Visser, C., 254
Visser, M. A., 504
Visuri, K., 444, 445(4), 450(4), 457(4), 462(4)
Vogelstein, B., 42
Voliverk, J. J., 178
Volpenhein, R. A., 11–12
Volqvartz, K., 329
Volwerk, J. J., 345
Vonbergmann, K., 224(104), 225
Vorde, C., 469
Vos, K., 480
Vulfson, E. N., 509
Vuoristo, M., 68
Vural, J., 163(44), 164

W

Waagen, V., 481, 492
Wada, A., 10
Wade, R. C., 393
Wagner, E. F., 74
Wagner, F., 438
Wahrenberg, H., 61
Waite, M., 184
Waldinger, C., 351, 386, 406, 414, 432(23)
Waldmann, H., 351, 406, 443
Walker, J. P., 25
Walker, K., 18, 19(102, 112), 24(106, 107), 32, 32(32, 106, 107, 252–254), 33
Walkins, J. B., 24
Wallace, J. S., 410
Wallinder, L., 108
Walsh, A., 72
Wang, C.-S., 117, 122(5), 231
Wang, Q., 9
Wang, Y.-F., 422–424, 445, 510
Wargovich, M. J., 43
Warshel, A., 397
Watanabe, Y., 108
Watkins, D. W., 41
Watt, S. M., 69, 232

Wawrzak, Z., 236
Weast, R. C., 311, 312(30)
Weber, A. M., 42
Weber, H. K., 399, 444, 509, 511, 512(30), 517
Weber, W., 102–103, 108, 108(7), 113(9)
Weekes, J., 50
Wehtje, E., 509
Weibel, E. K., 225
Weinberg, L., 28, 29(211)
Weinberg, R. B., 31, 294
Weiner, P., 393, 397(22)
Weiner, S. J., 393, 397, 397(22)
Weissfloch, A. N. E., 481
Weizman, Z., 42
Wek, S. A., 46, 56(11), 57(11), 65(11)
Welch, I., 10, 25
Well, J. A., 409
Wells, R., 93
Wen, J., 120
Weng, J., 392, 393(15), 394(15), 396(15)
Wengel, J., 438
Wenzig, E., 344
Wernstedt, C., 48, 60(20)
Wescott, C. R., 482, 493(24)
Westerfield, P. W., 89
Westergaard, H., 22
Whalon, M. R., 495, 510
Whitaker, J. R., 495
White, J. D., 9
White, P., 70, 98, 232
Whitehouse, I., 225
Whitesell, J. K., 422
Whitesides, G. M., 351, 443, 476
Wiebe, R., 502
Wiebel, E. K., 252
Wiegand, R. C., 117
Wieloch, T., 268, 321
Wiemer, D. F., 237
Wierzbicki, E., 18
Wiesenfeld, P. W., 117
Wiggins, H. S., 43
Wijkander, J., 49
Wild, G. E., 19, 34(115)
Wilen, S. H., 474
Willemot, R. M., 504
Willems, G., 37, 38(275)
Willett, W. C., 42
Williamolsson, T., 224(104), 225
Williams, D. F., 496
Williams, G., 9
Williams, G. J. B., 392, 393(14), 394(14), 396(14)
Williams, J., 10
Williams, R. L., 74
Willomot, R. M., 505
Wilson, B. E., 61
Wilson, D. E., 103
Wilson, J. E., 134
Wilson, T. A., 74
Winawer, S., 43
Winkler, F. K., 133, 135(21), 191–192, 193(9), 194, 200(9), 202(9), 220(24), 221(24), 237, 341
Winkler, K. E., 70
Winkler, U. K., 255, 258(18), 261(18), 262(18)
Winter, G., 426, 427(49)
Wisdom, R. A., 515
Wissler, R. W., 20
Witek-Giannotti, J., 74
Withers-Martinez, C., 335
Wöldike, H., 335
Wolfer, H., 15–16, 220, 225(91), 252, 253(7), 259(7)
Wong, C.-H., 351, 413, 419, 422–424, 443, 509–510
Wong, G. G., 74
Wong, H., 104, 120
Wood, F. E., 12
Wood, S. A., 28
Woods, S. C., 9
Woods, S. L., 51, 53(32)
Woodward, S. S., 363, 417
Wray, V., 438
Wu, F.-C., 509
Wu, G., 295, 305(15)
Wu, S., 200, 237, 486(14)
Wu, S.-H., 288, 452, 453(22), 480, 509
Wu, S.-S., 510
Wynder, E. L., 43

Y

Yamada, H., 410
Yamada, K., 35–36, 36(268), 37(269)
Yamada, N., 108
Yamada, Y., 363, 417, 505
Yamagata, S., 10
Yamamoto, K., 32
Yamano, M., 8

Yanagihara, N., 10
Yanbe, A., 10
Yang, C. H., 448
Yang, D., 120
Yang, Y.-F., 509
Yang, Z., 469, 482, 493(23)
Yano, T., 504–505
Yarranton, G. T., 135(35), 144, 191
Yazaki, Y., 108
Yeaman, S. J., 45, 48(8), 49, 49(8), 50(25), 51, 53(25, 26, 32)
Yeh, K.-Y., 24
Yeung, K. S., 43
Yokoyama, T., 506
Yoneda, K., 289, 424
York, C. M., 231
Yoshimatsu, H., 9
Young, E. A., 37
Yox, D. P., 10
Yu, B.-Z., 181
Yuan, L. C., 28
Yuan, Q., 175, 181(20)
Yunker, R., 13

Z

Zacks, A., 495
Zahalka, H. A., 235
Zaks, A., 473
Zandonella, G., 204, 209(68), 252
Zarjevski, N., 8–9, 9(29)
Zhang, H. F., 103, 105(18)
Zhang, S. H., 72
Zhi, J. G., 224(103, 106), 225
Zhou, S. L., 17
Zilversmit, D. B., 29, 78
Ziomek, E., 210, 237
Zografi, G., 199, 275, 285(19), 306
Zolfaghari, R., 69–70, 80, 84, 89(1), 90(1), 91(2), 92(2), 96, 97(19), 98, 98(19), 99, 101(24)
Zsigmond, E., 24, 40(168)
Zuckermann, M. J., 171–172, 173(16), 188–189
Zuegg, J., 511
Zundel, M., 209
Zuobi, K., 410

Subject Index

A

Acetaldehyde, lipase sensitivity and stabilization, 511–518
(1S,4R)-1-Acetoxy-4-hydroxycyclopent-2-ene, synthesis with lipases, 426–427
(1S,2R)-1-Acetoxymethyl-2-hydroxymethylcyclobutane, synthesis with lipases, 425–426
Acid cholesteryl ester hydrolase, see Hepatic lipase
Activation volume, pressure effects on enzymatic reactions, 503
Acyl donor, screening of lipases for synthetic applications, 374–375
Adenylate cyclase, assay, 59
β-Adrenergic receptor blockers, precursor synthesis with lipases, 364
Ajoene
  gastric lipase inhibition, 218–220
  isolation from garlic, 218
  structure, 197
Alkaline phosphatase, dietary fat effects on activity, 19, 36
5α-Androstane-3β,17β-dihydroxy diacetate, regioselective hydrolysis with cross-linked lipase crystals, 454–455
Anesthetics, bilayer and phospholipase A_2 activity effects, 187–189
Anhydrosorbite, regioselective esterification with lipases, 440–441
Antibody
  immunoaffinity purification of human gastric lipase, 140–141
  immunoinactivation of lipases, 131–133
  monoclonal antibody production against human lipases
    antigen preparation, 127
    enzyme-linked immunosorbent assay for screening, 128–129
    hybridoma generation, 128
    isotype identification, 129
    purification, 130
  polyclonal antibody production against human lipases
    immunization, 130
    purification, 130–131
  structural domain mapping of pancreatic lipase, 136
API ZYM, screening of lipases for synthetic applications, 353–354
Apo A-I, see Apolipoprotein A-I
Apo A-IV, see Apolipoprotein A-IV
Apo B, see Apolipoprotein B
Apolipoprotein A-I, metabolism, 29, 31
Apolipoprotein A-IV, metabolism, 31–32
Apolipoprotein B, metabolism, 29, 31

B

Benzylacetate, synthesis with lipases, 418
trans-2-Benzylcyclohexanol, resolution with lipases, 422
2-O-Benzylglycerol, enantioselective esterification by lipases, 424
BFA, see Brefeldin A
Bilayer, see Lipid bilayer
Bile acid
  cystic fibrosis, metabolism in patients, 41–42
  gallstone treatment with dietary supplementation, 41
  lipase inhibition, 14–15
  transport, 41
  tumor promotion, 43–44

Bile salt, bilayer and phospholipase A_2 activity effects, 186
Bile salt stimulated lipase, *see* Carboxyl ester lipase
Brefeldin A, effects on lipid transport, 28
Brush border membrane, intestine
  aging effects, 24
  diet-induced changes, 3, 33–38
  lipid absorption, 17–19, 25
  membrane structure and function, 38–40
2-Butyloxy-1-chloro-3-phenoxypropan, lipase enantioselectivity in hydrolysis, 365, 367
2-O-Butyryl-4,6-O-Benzylidene-α-methyl-D-glucopyranoside, synthesis using lipases, 440

C

cAMP, *see* Cyclic AMP
cAMP-PK, *see* Cyclic AMP-dependent protein kinase
Candida antarctica lipase
  inhibition by alkylphosphonates, 209
  surface pressure effects on stereoselectivity, 288–289
Candida cylindracea lipase, reactions in supercritical fluid, 506–507
Candida rugosa lipase
  acetaldehyde
    protection by lipase immobilization
      adsorption onto carriers, 514–518
      covalent linkage, 513–514, 516–518
      effects on enantioselectivity, 515–516, 518
      sensitivity, 511–512, 518
  inhibition by alkylphosphonates, 210–212
  substrate specificity, 357–360
3′-O-Caprinoyl-2′-deoxythymidine, synthesis using lipases, 442
Carboxyl ester lipase
  bile salt activation mechanism, 117
  cholesterol absorption role, 68–69, 78–80, 231–232
  immunoblotting, 78
  inhibitors
    assay, spectrophotometric, 238–239, 243
    boronic acid inhibitors, 240–241
    carbamate inhibitors
      assays, 243–244
      mechanism, 241–242
      decarbamylation characterization
        numerical integration, 247–251
        residual velocity, 245, 247
    haloketone inhibitors, 240–241
    inhibition constant calculation, 239–241
    structural classes, 237
  kinetic parameters, calculation for steady state, 233, 239–241
  knockout mice
    cholesterol absorption efficiency, 78–80
    production
      chimeric mouse generation, 76
      gene cloning, 73–74
      gene disruption in embryonic stem cells, 74–75
      germline transmission test, 76–77
      materials, 72–73
      principle, 70–72
      Southern blot analysis in screening, 75–76
      vector construction, 74
  mechanism
    chemical mechanism, 233–234
    interfacial catalysis, 232–233
  messenger RNA of rat liver
    detection using *Xenopus laevis* oocyte translation system
      bile salt activation, 89
      frog maintenance, 85–86
      inhibition by pancreatic enzyme antibody, 89
      message level correlation with activity, 92–93
      oocyte collection, 86–87
      oocyte injection, 87–88
      RNA isolation, 84
      solution preparation, 84–85
    Northern blot analysis, 92–93
  physiological role, 68–70
  radiation inactivation
    inactivation curves, 121
    substrate dependence, 122–123
    target size, 122
  substrate specificity, 117, 234–235
  three-dimensional structure modeling, 235–237
  tissue distribution, 70, 83–84, 90–91, 117, 231

transfection studies of rat enzyme
  calcium chloride transfection, 95–96
  effects of transfection
    hepatic cell cholesterol ester metabolism, 96–98
    intestinal cell cholesterol uptake, 99–101
  materials, 94
  plasmids, 94–95
CEL, *see* Carboxyl ester lipase
CF, *see* Cystic fibrosis
(R,S)-(4-Chloro)-2-phenoxyproprionic acid, resolution with cross-linked lipase crystals, 459–460
Cholecystokinin
  antagonists and satiety, 44
  dietary fat stimulation of release, 10
  receptors
    antagonists as satiety factors, 10–11
    types, 10
Cholesterol
  bilayer and phospholipase A_2 activity effects, 186, 189
  diet composition effects
    synthesis, 21–23
    uptake, 20–21, 23
  digestion and absorption
    absorption efficiency in carboxyl ester lipase knockout mice, 78–80
    carboxyl ester lipase transfection effects
      hepatic cell cholesterol ester metabolism, 96–98
      intestinal cell cholesterol uptake, 99–101
    overview, 7, 68–69
  effects
    glucose uptake, 23–24
    lipid bilayer fluidity, 164
  inhibitors of synthesis and uptake, 21
Cholesterol esterase, *see* Carboxyl ester lipase
Chromobacterium viscosum lipase
  alkylphosphonate inhibition, 210
  purification, 254–255
  tetrahydrolipstatin inhibition
    binding mode, 259–261
    emulsified substrate hydrolysis, 255–257, 263

hydrolytic activity of enzyme, 262–263
2-propanol effects, 257–259
Chylomicron, metabolism, 28–32
CLEC, *see* Cross-linked enzyme crystals
Colipase, *see* Pancreatic lipase
Colorectal cancer, dietary fat correlation, 42–44
Conformation, *see* Three-dimensional structure, lipases
C11-P, *see* O-Methyl-O-(p-nitrophenyl) n-undecylphosphonate
Cross-linked enzyme crystals, lipase
  assays
    methylmandelate hydrolysis, 449
    triacetin hydrolysis, 449
  crystallization conditions and lid domain conformation, 445–446
  indications for use in organic synthesis, 446–447, 464
  manipulation of suspensions, 448–449
  preparation, 445, 447–448
  preparative resolution of racemic acids and alcohols by ester hydrolysis, *see also specific compounds*
    enantiomeric ratio calculation, 451
    materials, 450
    reaction conditions, 450–451
  preparative resolution of racemic acids and alcohols by esterification and transesterification
    drying of crystals, 458
    enantiospecific esterifications, 458
    reaction rates using crystals, 457
    solvent selection and control of water, 458
  stability against solvents, 444
$C_{12:0}$-S-NbS, *see* Dodecyldithio-5-(2-nitrobenzoic acid)
Cutinase, *Fusarium solani pisi*
  acylglycerol synthesis using monomolecular film technique, 290, 292
  inhibition by alkylphosphonates, 209, 212
Cuvette adsorption assay, lipases, 388–389
Cyclic AMP, determination of levels in adipocytes, 59
Cyclic AMP-dependent protein kinase
  activity ratio determination, 59
  hormone-sensitive lipase, activation and assay, 45, 49–53
Cyclic GMP-inhibited phosphodiesterase

assays, 60
hormone-sensitive lipase, inhibition by dephosphorylation, 45, 48–49
immunoisolation from adipocytes, 59–60
β-Cyclodextrin, monolayer assay of lipases, 277, 279, 294, 306–307
Cystic fibrosis, lipid malabsorption, 41–42

D

2-Deoxy-5-O-propionyl-D-ribofuranose, synthesis using lipases, 438–439
2-Deoxy-D-ribose, regioselective esterification with lipases, 438
Diacylglycerol, interface partitioning in lipid bilayer, 165
1,3-sn-Dicaprylin, synthesis using lipases, 430–431
Diethyl p-nitrophenyl phosphate
 covalent inactivation of lipases
  approaches
   inhibition during lipolysis, 199
   lipase/inactivator preincubation, 199
   monolayer technique, 199–200
   poisoned interface, 199
  gastric lipase, 201–202
  interfacial binding of inactivated lipases, 203
  kinetic modeling, 195–196, 229–231
  pancreatic lipase, 200–203
  radiolabeling of lipases, 201–203
 hydrolysis by lipases, 202–203
 structure, 197
1,3-Dilaurin, synthesis using lipases, 429–430
5,5′-Dithiobis(2-nitrobenzoic acid)
 covalent inactivation of lipases
  approaches
   inhibition during lipolysis, 199
   lipase/inactivator preincubation, 199, 212–213, 216–217
   monolayer technique, 199–200
   poisoned interface, 199
  gastric lipase, 216–217
  kinetic modeling, 195–196, 229–231
  pancreatic lipase, 212–213
 structure, 197
4,4′-Dithiopyridine
 covalent inactivation of lipases
  approaches
   inhibition during lipolysis, 199
   lipase/inactivator preincubation, 199, 216–217
   monolayer technique, 199–200
   poisoned interface, 199
  gastric lipase, 216–217
 structure, 197
6-O-Dodecanoyl-D-glucopyranoside, synthesis using lipases, 436–437
Dodecyldithio-5-(2-nitrobenzoic acid)
 covalent inactivation of lipases
  approaches
   inhibition during lipolysis, 199
   lipase/inactivator preincubation, 199, 212–213, 217–218
   monolayer technique, 199–200
   poisoned interface, 199
  gastric lipase, 217–218
  interfacial binding effects, 213, 216
  kinetic modeling, 195–196, 229–231
  pancreatic lipase, 212–213, 216
 structure, 197

E

E_{600}, see Diethyl p-nitrophenyl phosphate
ELISA, see Enzyme-linked immunosorbent assay
Enantiospecificity, lipases
 cosolvent effects in hydrolysis reactions, 481–482, 491, 494
 covalent inhibitors
  gastric lipase, 206–207
  pancreatic lipase, 205–206
 enantiomeric ratio calculation, 395–396, 416–417, 451, 474–479
 enantioselective inhibition in batch hydrolysis, 405
 equilibrium constant calculation, 477–479
 Fisher representation of triacylglycerol, 264–265
 hydrolysis versus esterification, 366
 interfacial activation role, 486–490
 microbial lipases, 415–417
 optimization approaches, 480
 prediction, 375–376, 378
 screening for synthetic applications
  combined screening efforts, 371–375
  enantioselectivity values, 363

enantiotopic ester group differentiation, 371
qualitative testing, 354, 356–357, 364–365
solvent dependence and screening, 366–371, 373
time/conversion curves, 361
selectivity constant α, 477–478
substrate engineering in chemical synthesis, 378, 380–381
surface pressure effects in monolayer assays, 287–290
terminology for chiral selectivity, 473–474
water activity effects in esterification reactions, 483–486
Enzyme-linked immunosorbent assay
cross-species reactivity assay of human gastric lipase antibody, 136–137
enantiomeric excess determination, 474, 495
epitope mapping, 133–134
gastric lipase, 141–143, 286–287
interfacial binding assay of lipases
biotinylation of proteins, 143
protein adsorption quantification, 144, 146, 148
interfacial lipase quantification, 286–287
pancreatic lipase, 318–319
screening of anti-lipase monoclonal antibodies, 128–129
Ethyl-D-glucopyranoside, synthesis using lipases, 436
Ethyl octanoate, resolution of racemic alcohol by lipase esterification, 400–401
S-Ethyl thiooctanoate, resolution of racemic alcohol by lipase esterification, 400–401

F

Fatty acid
absorption in intestine, 16–18
dietary types and uptake, 18–19
fat intake and uptake, 21
induction of neurotensin-like immunoreactivity, 25
screening of lipases for synthetic applications, 357–359, 372
tumor promotion, 43–44
Fatty acid-binding protein

cytosolic forms, 26–28
distribution in intestinal epithelium, 27
fatty acid binding affinity, 26–27
mediation of fatty acid uptake, 17–18, 25
Fatty acid transporter, mediation of fatty acid uptake, 17–18
Film, see Monolayer assay, lipases
Flurbiprofen, enantioselective synthesis with cross-linked lipase crystals, 452–453
Fructose
diet composition and uptake, 34–35
esterification with lipases using phenylboronic acid, 437–438

G

Ga, see Gum arabic
Galactose, diet composition and uptake, 20, 23–24, 33–34
Gas chromatography
batch hydrolysis reaction monitoring, 404
screening of lipases for synthetic applications, 352, 356, 474
Gastric lipase
covalent inactivation
ajoene, 218–220
diethyl p-nitrophenyl phosphate, 201–202
5,5'-dithiobis(2-nitrobenzoic acid), 216–217
4,4'-dithiopyridine, 216–217
dodecyldithio-5-(2-nitrobenzoic acid), 217–218
tetrahydrolipstatin, 222–223
immunoinactivation, 131–133
interfacial binding assay
biotinylation of proteins, 143
protein adsorption quantification by enzyme-linked immunosorbent assay, 144, 146, 148
monoclonal antibody
cross-species reactivity
enzyme-linked immunosorbent assay, 136–137
indirect competition assay, 136, 139
Western blotting, 136, 139
epitope mapping
enzyme-linked immunosorbent assay, 133–134

peptide mapping, 134–136
immunoaffinity purification of enzyme, 140–141
production
  antigen preparation, 127
  enzyme-linked immunosorbent assay for screening, 128–129
  hybridoma generation, 128
  isotype identification, 129
  purification, 130
  quantitative enzyme-linked immunosorbent assay, 141–143, 286–287
pH optimum, 191
polyclonal antibody production
  immunization, 130
  purification, 130–131
stereospecificity for covalent inactivators, 206–207
surface pressure effects on stereoselectivity, 288–289
tetrahydrolipstatin inhibition, 15–16
GC, *see* Gas chromatography
Geotrichum candidum lipase
  acetaldehyde sensitivity, 511–512, 518
  substrate specificity, 357–360
Glucose
  diet composition and uptake, 20, 23–24, 33–35
  esterification with lipases using phenylboronic acid, 437
Glycerol, *see* Diacylglycerol; Monoacylglycerol; Triacylglycerol
Glycidol
  enantiospecific lipase hydrolysis of esters, 476–477, 493
  kinetic resolution in organic solvents, 493–494
  transesterification with vinyl acetate, 483–486
GRID, lipase substrate-binding site modeling, 394–395
GROMOS, modeling of solvent effects on lipase structure, 392
Gum arabic, effects on interfacial activation, 330, 337, 339, 341

H

HDL, *see* High density lipoprotein
HDS, *see* Hexadecylsulfonylfluoride

Heparin, lipoprotein lipase interactions
  binding site, 103
  surface plasmon resonance analysis
    kinetics, 116
    principle, 113–114
    reagents, 114
    sensor chip coating, 114–115
Hepatic lipase, *see also* Carboxyl ester lipase
  radiation inactivation
    acid cholesteryl ester hydrolase, 123–126
    carboxyl ester lipase, 121–123
    enzyme assays, 120–121
    homogenate preparation, 120
    irradiation, 120
    neutral cholesteryl ester hydrolase, 123–126
    principle, 118–119
    types, 116–118
Hexadecylsulfonylfluoride, lipoprotein lipase inhibition, 105
High density lipoprotein, metabolism, 28–29
High-performance liquid chromatography
  enantiomeric excess determination, 474, 495
  screening of lipases for synthetic applications, 352, 356, 365
HMG-CoA reductase, *see* 3-Hydroxy-3-methylglutaryl CoA reductase
Hormone-sensitive lipase
  adipocyte preparation for activation assays, 54
  assay in crude extracts, 62–64
  dephosphorylation and inhibition assay, 53
    basal site phosphatases, 51
    cyclic GMP-inhibited phosphodiesterase, 45, 48–49
    insulin, 45, 48
  fatty acid release assay by pH stat titration, 56
  glycerol release assay, 55–56
  immunoprecipitation, 57–58
  levels in disease states, 61–62
  messenger RNA
    insulin effects, 62
    Northern blot analysis, 65
    regulation of levels, 61–62

ribonuclease protection assay, 66–67
phosphorylation
  β-adrenergic agonists, mechanism of activation, 47–48
  AMP-activated protein kinase, 49–50, 53
  assays, 51–53, 56–57
  basal site kinases, 49–50
  cyclic AMP-dependent protein kinase activation, 45, 49–53
  denaturing polyacrylamide gel electrophoresis, phosphoprotein analysis, 55, 57
  sites, 45, 49, 57
  substrate specificity effects, 46–47
  translocation induction, 46
  physiological role, 45
  substrate specificity, 63–64
  Western blot analysis, 58, 63–64
HPLC, see High-performance liquid chromatography
HSL, see Hormone-sensitive lipase
Humicola lanuginosa lipase, substrate specificity, 357–360
3-Hydroxy-3-methylglutaryl CoA reductase
  diet effects on activity, 21–23
  inhibitors of synthesis and cholesterol uptake, 21

I

(S)-Ibuprofen
  enantiospecific esterification with cross-linked lipase crystals, 458–459
  preparation with cross-linked lipase crystals, 451–452
Industrial applications, lipases
  acetaldehyde sensitivity and stabilization of lipases, 511–518
  acid resolution, 401–402
  batch reaction conditions, 402–405, 407–408
  cosolvents in hydrolysis reactions, 481–482, 491, 494
  cross-linked enzyme crystals as catalysts, see Cross-linked enzyme crystals, lipase
  enantioselective inhibition, 405
  hydration control in nonaqueous biocatalysis, see Organic solvent, nonaqueous lipase biocatalysis
  immobilization of enzymes, 407–408, 512–518
  lipase sources and purity requirements, 407, 443–444
  mechanism of catalysis
    acyl transfer
      irreversible, 412–413
      reversible, 412
    catalytic residues, 409–410
    esterification, 398–400, 411–412
    hydrolysis, 398–399, 411–412
  monoesters of 1,3-diols, enantiotope differentiation, 423–428
  organic solvent
    effects on enantiospecificity, 483–486
    origin of solvent effects, 490–493
    selection for esterification, 415, 480
  organic synthesis overview, 406–407
  racemic alcohol resolution
    esterification, 400–401, 417–419
    hydrolysis, 400, 419
    primary alcohols, 417–418
    secondary alcohols, 419–423
  regioselective esterifications
    carbohydrates, 433–441
    glycerol, 428–433
    nucleosides, 441–442
    solvent selection, 434–435
  vinyl ester synthesis, 413–414
Inositol derivatives, enantiotope differentiation with lipases, 427–428
Insulin, hormone-sensitive lipase
  effects on gene expression, 62
  inhibition, 45, 48
Interfacial activation, lipases
  conformational change in enzymes, 341, 343–346
  controls in analysis
    inhibited lipase control, 333
    interfacial quality and artifacts, 333–334
    linearity of assay, 332–333
  definition, 329–330, 332
  discovery, 153, 178, 328–329
  exceptions and utility in defining lipases, 344, 346
  gum arabic effects, 330, 337, 339, 341
  interfacial binding assay

biotinylation of proteins, 143
protein adsorption quantification by enzyme-linked immunosorbent assay, 144, 146, 148
ionic strength effects, 330, 339
kinetic models for interfacial lipolysis
competitive inhibition analysis, 269–270, 272, 281–282
covalent inhibition of lipases, 195–196, 229–231, 282
hopping mode of catalysis, 272, 274
scooting mode of catalysis
definition, 272
high enzyme/vesicles ratio, 274
low enzyme/vesicles ratio, 273–274
steady-state parameter calculation, 233, 239–241, 266–269, 387–388
mechanism, 335, 337
nonaqueous media studies, 345–346
phospholipase A_2, 344–345
role in enantiospecificity, 486–490
short-chain triacylglycerols, aqueous solubility and suitability for activation, 335, 339
substrate interfacial area, variation in assay, 389
Isosorbide-5-acetate, synthesis using lipases, 441

K

Ketoprofen, enantioselective synthesis with cross-linked lipase crystals, 452, 456
Knockout mice
applications in research, 72
carboxyl ester lipase mouse
cholesterol absorption efficiency, 78–80
production
chimeric mouse generation, 76
gene cloning, 73–74
gene disruption in embryonic stem cells, 74–75
germline transmission test, 76–77
materials, 72–73
principle, 70–72
Southern blot analysis in screening, 75–76
vector construction, 74
gene targeting by homologous recombination in embryonic stem cells, 70–72

L

Leptin
gene, 7
satiety factor, 7, 44
Lid domain, see Three-dimensional structure, lipases
Lindane, bilayer and phospholipase A_2 activity effects, 188
Lingual lipase
inhibition by bile acids, 15
lipid digestion role, 14
Lipid, classification by behavior in water, 263–264, 327–328
Lipid bilayer
composition and dynamic heterogeneity, 175–177
computer simulation of nanoscale dynamics, 170–171
curvature stress field, 177–178
instability to non-bilayer-forming lipids, 177–178
interface partitioning of lipids in bilayers
cholesterol ester solubility, 162–164
diacylglycerols, 165
flip rate in *trans* bilayer movement, 166–167
long-chain triglycerides, 161–162, 164
monoacylglycerols, 165–166
nuclear magnetic resonance analysis, 162–164
triacylglycerol solubility, 163–164
lateral organization, 170–177
main phase transition, 170
nonequilbrium phenomena, 174–175
phase diagrams, 172–175
phospholipase A_2, substrate physical properties and activity
bilayer organization
composition effects on bilayer and activity, 180–181
curvature and morphology effects, 184–186, 188–190
lipid-packing defects, 182
microheterogeneity effects, 182–184
modulators and activity effects
alcohols, 186–189
anesthetics, 187–189
cholate, 186
cholesterol, 186, 189
lindane, 188

Lipid monolayer, *see* Monolayer assay, lipases
Lipoprotein lipase
  concentration assays, 107–108
  heparin interactions
    binding site, 103
    surface plasmon resonance analysis
      kinetics, 116
      principle, 113–114
      reagents, 114
      sensor chip coating, 114–115
  inhibitors
    assay, 106
    hexadecylsulfonylfluoride, 105
    incubation conditions for inactivation, 107
    purification of inactivated enzyme, 111–112
    serine protease inhibitors, 104
    tetrahydrolipstatin, 104–105
  low-density receptor-related protein binding, 102
  monomers
    mediation of lipoprotein binding to cell surface, 103–104
    preparation by mild dissociation, 109–111
    purification, 111–112
    separation from dimers by density gradient ultracentrifugation, 112–113
  physiological role, 102
  radiolabeling with sodium-125, 107–109
  solubility, 105, 107
  surface pressure effects
    substrate recognition, 157–158
    stereoselectivity, 288–289
Liposome
  phospholipase A_2, comparative kinetics with monolayer substrates, 279–280
  transfection, 82
Low-density receptor-related protein, lipoprotein lipase binding, 102
LPL, *see* Lipoprotein lipase
LRP, *see* Low-density receptor-related protein

M

Mandelic acid, enantioselective synthesis with cross-linked lipase crystals, 454

Membrane, *see* Lipid bilayer
Menthol acetate, enantioselective synthesis with cross-linked lipase crystals, 461–462
Menthol butyrate, enantioselective synthesis with cross-linked lipase crystals, 460, 462
Menthol, enantioselective synthesis with cross-linked lipase crystals, 454
O-Methyl-O-(p-nitrophenyl) n-pentylphosphonate, inhibition of digestive lipases, 204–205
O-Methyl-O-(p-nitrophenyl) n-undecylphosphonate
  covalent inactivation of lipases
    approaches
      inhibition during lipolysis, 199
      lipase/inactivator preincubation, 199
      monolayer technique, 199–200
      poisoned interface, 199
    crystal structure of pancreatic lipase–inhibitor complex, 207, 209
    kinetic modeling, 195–196, 229–231
  structure, 197
MGL, *see* Monoacylglycerol lipase
Monoacylglycerol, interface partitioning in lipid bilayer, 165–166
Monoacylglycerol lipase
  assay, 59
  fractionation of adipocytes, 56
1(3)-*rac*-Monolaurin, synthesis using lipases, 431–432
Monolayer assay, lipases
  acylglycerol synthesis using monomolecular film technique and cutinase, 290, 292
  covalent inactivation of lipases, 199–200, 223
  enzyme-linked immunosorbent assay of interfacial lipases, 286–287
  film recovery of lipids and proteins
    applications, 304–305
    efficiency of recovery, 294–295, 301
    hydrophobic support method
      advantages, 295
      calibration curve, 301–302
      hydrophobic paper preparation, 296
      lipid recovery and analysis, 298, 300
      monolayer collection, 296, 298
      principle, 295–296

protein recovery and analysis, 300–301
surface excess of lipase, calculation, 302–304
overview of techniques, 286, 294–295, 304–305
kinetic models for interfacial lipolysis
competitive inhibition analysis, 269–270, 272, 281–282
covalent inhibition of lipases, 195–196, 229–231, 282
hopping mode of catalysis, 272, 274
scooting mode of catalysis
definition, 272
high enzyme/vesicles ratio, 274
low enzyme/vesicles ratio, 273–274
steady-state parameter calculation, 233, 239–241, 266–269, 387–388
long-chain substrates, 276–277, 279, 293–294, 306–307
medium-chain substrates, 276
mixed monolayer substrates, 280–281
oil-drop tensiometer assay comparison, 323–325
phospholipase A_2, comparative kinetics with liposome substrates, 279–280
protein inhibition assays, 282–284
rationale, 265–266, 293
surface pressure effects on stereoselectivity, 287–290
velocity-surface pressure profile optimum, 284–285
zero-order trough, 275–276, 280
1(3)-*rac*-Monoolein, synthesis using lipases, 431–433

N

Naproxen, enantioselective synthesis with cross-linked lipase crystals, 453
NbS$_2$, *see* 5,5'-Dithiobis(2-nitrobenzoic acid)
Neuropeptide Y
antagonists and satiety, 8, 44
feeding signal, 8–9
modulation in hypothalamus, 9
subtypes, 8
tissue distribution, 7–8
Neurotensin-like immunoreactivity, induction by unsaturated fatty acids, 25

Neutral cholesteryl ester hydrolase, *see* Hepatic lipase
p-Nitrophenyl acetate, spectrophotometric assay of lipases, 388
p-Nitrophenyl butyrate, spectrophotometric assay of carboxyl ester lipase, 238–239, 243
NMR, *see* Nuclear magnetic resonance
Northern blot
carboxyl ester lipase, rat liver, 92–93
hormone-sensitive lipase, 65
NPY, *see* Neuropeptide Y
NTL$_1$, *see* Neurotensin-like immunoreactivity
Nuclear magnetic resonance
biocatalyst water content measurement, 471–472
glycerol conformation analysis, 154
lipid behavior in bilayers, 162–164

O

Obesity
chemically defined diets and nutrient uptake in animals, 36–38
dietary modification, 4–5
risk factor for diseases, 5
tetrahydrolipstatin therapy, 16
treatment
approaches, 44
precautions, 3
Octanoic acid, resolution of racemic alcohol by lipase esterification, 400–401
Oil-drop tensiometer, lipase assay
accuracy, 311–312
apparatus, 308–310
continuous assay, 307–308
data collection and processing, 308–310
material preparation, 310–312
microbial enzyme assays, 325–326
pancreatic lipase
bile salt effects, 320–321, 323
biotinylated protein assays, 316–317
colipase effects, 317–319
comparison with monomolecular film technique, 323–325
linearity of assay, 313
recovery and quantification of enzyme, 318–319
phospholipase A_2 assay, 315

principle, 307, 311–315
reproducibility, 311
Olestra
  absorption, 12
  adverse effects, 13
  excretion, 12
  fat similarities, 12
  lipase inactivity, 11–12
  structure, 11
Organic solvent, nonaqueous lipase biocatalysis
  cosolvents in hydrolysis reactions, 481–482, 491, 494
  esterification reaction of lipases, see Cross-linked enzyme crystals, lipase; Industrial applications, lipases
  hydration control
    preequilibration with saturated salt solutions
      apparatus, 468
      component equilibration, 466–467, 494–495
      materials, 467–468
      principle, 466, 483
      rates of equilibration, 467
      salt selection, 466–467, 469
      temperature control, 469
      vapor equilibration prevention, 469
    water activity, 465–466
    water extraction approaches, 509–510
    water measurement
      biocatalyst content with nuclear magnetic resonance, 471–472
      coulometric Karl Fischer titration, 470, 494
      humidity sensors, 470–471
  lipase assay and kinetics, 390–391
  log P values, 480
  modeling of effects on lipase structure, 391–392
Orlistat, see Tetrahydrolipstatin
Ovalbumin, inhibition of lipases, 282–284

P

Pancreatic lipase
  classification, 344
  colipase activation, 14–15, 126–127, 317–318
  covalent inactivation

diethyl p-nitrophenyl phosphate, 200–203
5,5′-dithiobis(2-nitrobenzoic acid), 212–213
dodecyldithio-5-(2-nitrobenzoic acid), 212–213, 216
O-methyl-O-(p-nitrophenyl) n-undecylphosphonate, 207, 209
tetrahydrolipstatin
  activity on mixed films, 223
  duodenal enzyme inhibition, 226–227, 229
  effects on secretion, 224–227
  inhibition during lipolysis, 221
  mechanism, 220–221
  poisoned interface inhibition, 222
epitope mapping
  enzyme-linked immunosorbent assay, 133–134
  peptide mapping, 134–136
immunoinactivation, 131–133
inhibition by bile acids, 14–15
interfacial activation, 344
lipid digestion role, 13–14
monoclonal antibody production
  antigen preparation, 127
  enzyme-linked immunosorbent assay for screening, 128–129
  hybridoma generation, 128
  isotype identification, 129
  purification, 130
oil-drop tensiometer assay
  bile salt effects, 320–321, 323
  biotinylated protein assays, 316–317
  colipase effects, 317–319
  comparison with monomolecular film technique, 323–325
  linearity of assay, 313
  recovery and quantification of enzyme, 318–319
pH optimum, 191
polyclonal antibody production
  immunization, 130
  purification, 130–131
processing, 15
stereospecificity for covalent inactivators, 205–206
structural domain mapping with monoclonal antibodies, 136
substrate specificity, 357–360

tetrahydrolipstatin inhibition, 15–16
three-dimensional structure of human enzyme
  catalytic triad, 193
  colipase complexes, 194–195
  domains, 191
  α/β hydrolase fold, 192–193
  lid domain, 193–194
PC, see Phosphatidylcholine
PDE3, see Cyclic GMP-inhibited phosphodiesterase
4-PDS, see 4,4'-Dithiopyridine
Phenethyl acetate, resolution with cross-linked lipase crystals, 456
Phenethyl alcohol, resolution with cross-linked lipase crystals, 462–463
Phenyl-*n*-butylborinic acid, carboxyl ester lipase inhibition, 240–241
Phenylethanol, resolution with lipases, 420–421, 491
Phosphatidylcholine, digestion and absorption, 6–7
Phospholipase A_2
  interfacial activation, 344–345
  lag period in activity, 179–182
  oil-drop tensiometer assay, 315
  processing, 15
  substrate physical properties and activity
    applications to other lipases, 168–169
    bilayer organization
      composition effects on bilayer and activity, 180–181
      curvature and morphology effects, 184–186, 188–190
      lipid-packing defects, 182
      microheterogeneity effects, 182–184
    comparative kinetics with liposome and monolayer substrates, 279–280
    modulators and activity effects
      alcohols, 186–189
      anesthetics, 187–189
      cholate, 186
      cholesterol, 186, 189
      lindane, 188
    surface dilution, 179
    triacylglycerol active site conformations, 155–156
Phospholipid, see also Lipid bilayer; Phosphatidylcholine
  behavior in solvents, 153–154

digestion and absorption, 6–7
interface partitioning of lipids in bilayers
  cholesterol ester solubility, 162–164
  diacylglycerols, 165
  flip rate in *trans* bilayer movement, 166–167
  long-chain triglycerides, 161–162, 164
  monoacylglycerols, 165–166
  nuclear magnetic resonance analysis, 162–164
  triacylglycerol solubility, 163–164
PNPA, see *p*-Nitrophenyl acetate
PNPB, see *p*-Nitrophenyl butyrate
Portal vein, lipid transport, 14
Pseudomonas lipases
  Pseudomonas cepacia lipase, reactions in supercritical fluid, 505–506
  substrate specificity, 375–376, 378

R

Radiation inactivation
  hepatic lipases
    acid cholesteryl ester hydrolase, 123–126
    carboxyl ester lipase, 121–123
    enzyme assays, 120–121
    homogenate preparation, 120
    irradiation, 120
    neutral cholesteryl ester hydrolase, 123–126
  principle, 118–119
  target sizes of lipases, 119–120, 122
Regioselectivity, lipases
  classification of lipases, 383
  esterifications in organic synthesis
    carbohydrates, 433–441
    glycerol, 428–433
    nucleosides, 441–442
  microbial lipases, 417
  regioisomeric excess values, 382–384, 386
  screening for synthetic applications, 381–386
  solvent effects, 383–384
Rhizomucor miehei lipase
  conformational state modeling, 392–394
  inhibition by alkylphosphonates, 209–210
  interfacial activation in nonaqueous media, 345–346

surface pressure effects on stereoselectivity, 288–289
Rhizopus arrhizus lipase, reactions in supercritical fluid, 505
Rhizopus oryzae lipase
  alkylphosphonate inhibition, 210
  purification, 255
  tetrahydrolipstatin inhibition
    binding mode, 259–260
    emulsified substrate hydrolysis, 255–257, 263
    hydrolytic activity of enzyme, 262–263
D-Ribose, regioselective esterification with lipases, 438

S

SCF, *see* Supercritical fluid
Sodium/hydrogen exchanger, mediation of fatty acid uptake, 17–18
SPR, *see* Surface plasmon resonance
Staphylococcus hyicus lipase, inhibition by alkylphosphonates, 209
Stereoselectivity, *see* Enantiospecificity, lipases
Sucrose, feeding and intestinal sucrase induction, 35–36
Sucrose density gradient ultracentrifugation, lipoprotein lipase, 112–113
Supercritical fluid
  applications in industry, 498
  critical constants, 497
  definition, 496–497
  enzyme catalytic medium
    advantages, 496, 499, 508
    limitations, 499–500
    lipase reactions, 505–508
    mass transfer effects, 501
    moisture equilibrium, 503–505
    stability of enzymes, 502–503
  physical properties, 497–498
  solubility of substrate and water, 501–502
Surface plasmon resonance, lipoprotein lipase–heparin interactions
  kinetics, 116
  principle, 113–114
  reagents, 114
  sensor chip coating, 114–115
SYBYL
  enantiomeric ratio calculation, 396–397

lipase conformational state modeling, 393–394

T

T cell, functional effects of dietary fat, 40
Tetrahydrolipstatin
  inhibition of lipases
    approaches
      inhibition during lipolysis, 199, 221
      lipase/inactivator preincubation, 199
      monolayer technique, 199–200, 223
      poisoned interface, 199, 222
    Chromobacterium viscosum lipase
      binding mode, 259–261
      emulsified substrate hydrolysis, 255–257, 263
      hydrolytic activity of enzyme, 262–263
      2-propanol effects, 257–259
    gastric lipase, 222–223
    kinetic modeling, 195–196, 229–231
    lipoprotein lipase, 104–105
    pancreatic lipase
      activity on mixed films, 223
      duodenal enzyme inhibition, 226–227, 229
      effects on secretion, 224–227
      inhibition during lipolysis, 221
      mechanism, 220–221
      poisoned interface inhibition, 222
    Rhizopus oryzae lipase
      binding mode, 259–260
      emulsified substrate hydrolysis, 255–257, 263
      hydrolytic activity of enzyme, 262–263
  specificity, 15–16, 220, 252
  obesity treatment, 16
  solubility, 253
  structure, 197, 253
  synthesis, 252–253
  tolerance, 16
TG, *see* Triacylglycerol
Thin-layer chromatography, screening of lipases for synthetic applications, 352, 356
THL, *see* Tetrahydrolipstatin
Three-dimensional structure, lipases

crystallization conditions and lid domain conformation, 445–446
modeling
  carboxyl ester lipase structure, 235–237
  conformational states, 392–394
  solvent effects on lipase structure, 391–392
  substrate-binding site, 394–395
pancreatic lipase, human
  catalytic triad, 193
  colipase complexes, 194–195
  domains, 191
  α/β hydrolase fold, 192–193
  inhibitor complexes, 207, 209
  lid domain, 193–194
TLC, see Thin-layer chromatography
Transfection
  calcium salts, 82, 95–96
  carboxyl ester lipase complementary DNA from rat
    calcium chloride transfection, 95–96
    effects
      hepatic cell cholesterol ester metabolism, 96–98
      intestinal cell cholesterol uptake, 99–101
    materials, 94
    plasmids, 94–95
  electroporation, 82
  fate of transfected DNA, 82
  liposomes, 82
  selection with neomycin, 82–83
  vectors, 83
Triacylglycerol
  aqueous solubility of short-chain triacylglycerols, 335
  chain length and surface melting points, 158–161
  collapse pressure, 158–159, 161
  conformation
    acyl chains, 155
    air/water interface conformations, 156–161
    lipolytic enzyme active site conformations, 155–156, 375–376
    nuclear magnetic resonance analysis, 154
    terminology, 154–155
  content in Western diet, 5–6

digestion and absorption, 6–7, 126, 190
Fisher representation, 264–265
interface partitioning in bilayers
  flip rate in *trans* bilayer movement, 166–167
  long-chain triglycerides, 161–162, 164
  nuclear magnetic resonance analysis, 162–164
  solubility, 163–164
screening of lipases for synthetic applications, 357–359, 372

V

Very low density lipoprotein, metabolism, 28–31
Vinyl-12-hydroxydodecanoate, synthesis using lipases, 413–414
Vinyl myristate, synthesis using lipases, 413–414
Vinyl octanoate, resolution of racemic alcohol by lipase esterification, 400–401
Vinyl oleate, synthesis using lipases, 413–414
VLDL, see Very low density lipoprotein

W

Water activity
  effect on Michaelis constants, 482
  measurement
    biocatalyst content with nuclear magnetic resonance, 471–472
    coulometric Karl Fischer titration, 470, 494
    humidity sensors, 470–471
  prediction of hydration control
    nonaqueous biocatalysis, 465–466
    supercritical fluids, 503–505
  values for salts, 466–467, 483
Western blot
  cross-species reactivity assay of human gastric lipase antibody, 136, 139
  hormone-sensitive lipase, 58, 63–64

X

X-ray crystallography, see Three-dimensional structure, lipases

Xenopus laevis oocyte, *in vitro* translation system
  carboxyl ester lipase messenger RNA of rat liver, detection
    bile salt activation, 89
    frog maintenance, 85–86
    inhibition by pancreatic enzyme antibody, 89
    message level correlation with activity, 92–93
    oocyte collection, 86–87
    oocyte injection, 87–88
    RNA isolation, 84
    solution preparation, 84–85
  posttranslational modification of proteins, 81

ISBN 0-12-182187-0